CHEMISTRY RESEARCH AND APPLICATIONS

URANIUM

CHARACTERISTICS, OCCURRENCE AND HUMAN EXPOSURE

CHEMISTRY RESEARCH AND APPLICATIONS

Additional books in this series can be found on Nova's website
under the Series tab.

Additional e-books in this series can be found on Nova's website
under the e-book tab.

NUCLEAR MATERIALS AND DISASTER RESEARCH

Additional books in this series can be found on Nova's website
under the Series tab.

Additional e-books in this series can be found on Nova's website
under the e-book tab.

CHEMISTRY RESEARCH AND APPLICATIONS

URANIUM

CHARACTERISTICS, OCCURRENCE AND HUMAN EXPOSURE

ALIK YA. VASILIEV
AND
MIKHAIL SIDOROV
EDITORS

nova publishers

New York

For permission to use material from this book please contact us:
Telephone 631-231-7269; Fax 631-231-8175
Web Site: http://www.novapublishers.com

Additional color graphics may be available in the e-book version of this book.

Library of Congress Cataloging-in-Publication Data

Uranium : characteristics, occurrence, and human exposure / [edited by] Alik Ya. Vasiliev and Mikhail Sidorov.
p. cm.
Includes bibliographical references and index.
ISBN 978-1-62081-207-5 (hardcover)
1. Uranium. 2. Groundwater--Pollution. 3. Uranium--Environmental asepcts. 4. Uranium--Health aspects. I. Vasiliev, Alik Ya. II. Sidorov, Mikhail.
TD427.R3U73 2011
669'.2931--dc23
2012006516

Published by Nova Science Publishers, Inc. † New York

CONTENTS

Preface		**vii**
Chapter 1	Naturally Occurring Uranium in Groundwater of the North Western Sahara Aquifer System (Northern Africa) *Najiba Chkir, Friha Hadj Ammar and Kamel Zouari*	**1**
Chapter 2	Anthropogenic Sources of Uranium in Serbia - Risk Assessment on Environment and Human Health *Mirjana D. Stojanović, Jelena V. Milojković, Zorica R. Lopičić, Marija L. Mihajlović, Miloš B. Rajković and Gordana S. Vitorović*	**49**
Chapter 3	Uranium Mining in Portugal: From Unmined Uranium Ores to Mining Waste Deposits *Maria João Batista, Cátia Prazeres, Maria Ondina Figueiredo, Jaime Leote, Daniel P. S. de Oliveira, Augusto Filipe and Teresa Pereira da Silva*	**87**
Chapter 4	Laser-Induced Nuclear Decays in Uranium Isotopes *G. A. Shafeev*	**117**
Chapter 5	Uranium Occurrence and Behavior in Groundwater (The Example of North-west Russia) *A. I. Malov*	**155**
Chapter 6	Uranium: Natural Occurrence, Physical and Chemical Characteristics - Environmental Contamination and Human Exposure *Marcos Moisés Gonçalves, Marli Gerenutti and Marta Maria Duarte Carvalho Vila*	**185**
Chapter 7	Uranium Hydrothermal Deposits *Miloš René*	**211**
Chapter 8	The Legacy of Uranium Mining Activity in Cunha Baixa: A Portuguese Study Case *M. O. Neves, M. M. Abreu and M. J. Matias*	**245**

Chapter 9 Geophysical Studies to Evaluate Geological and Hydrogeological Features in Old Uranium Mines **269**
E. C. Ramalho, J. Carvalho, S. Barbosa and F. A. Monteiro Santos

Chapter 10 Uranium: From Occurrence, Mining, Human Exposure to Remediation **295**
O. Saïdou

Chapter 11 Removal of Uranium from Aqueous Solution, Uranium Effluent and Seawater Using Various Microbes Isolated from the Uranium Deposits. **305**
Takehiko Tsuruta

Chapter 12 One-Group Fission Cross Sections for Plutonium and Minor Actinides Inserted in Calculated Neutron Spectra of Fast Reactor Cooled with Lead-208 and Lead-Bismuth Eutectic **321**
G. L. Khorasanov and A. I. Blokhin

Index **329**

PREFACE

Uranium has both radioactive and chemical hazards and risk assessments must address both types. Its environmental distribution depends on natural occurrence and has been influenced by human activities such as energy production, coal, and phosphate fertilizer exploration, which may impact freshwater ecosystems. In this book, the authors present current research in the study of the characteristics, occurrence and human exposure to uranium. Topics include the anthropogenic sources of uranium in Serbia; uranium mining in Portugal; laser-induced nuclear decays in uranium isotopes; uranium hydrothermal deposits; and geophysical studies to evaluate geological and hydro-geological features in old uranium mines.

Chapter 1 - Uranium concentrations in water are controlled by a number of factors, among the most important being pH, Eh and concentrations of coexisting dissolved ions. Uranium as a redox-sensitive heavy metal occurs in water principally under oxic conditions in its hexavalent (U(VI)) form. It is usually complexed in solution, especially with carbonate ligands, but also less significantly with phosphate, fluoride or sulphate depending on their respective dissolved concentrations and ambient pH. Under anoxic conditions, U is reduced to tetravalent U(IV) state and its concentration in water is usually low due to the insolubility of this form. On the other hand, despite their long radioactive half-lives that should induce a secular equilibrium between ^{238}U and ^{234}U isotopes, large range of natural variation of $^{234}U/^{238}U$ ratios have been found in water (0.5-40). The mechanism of such fractionation is preferential leaching of ^{234}U compared with ^{238}U from solid phase, caused by radiation damage of crystal lattice upon alpha decay of ^{238}U, oxidation of insoluble tetravalent ^{234}U to soluble hexavalent ^{234}U during decay, and alpha recoil of ^{234}Th (and its daughter^{234}U) into solution phase. In the last few decades, the behaviour of uranium isotopes in groundwater has contributed to an understanding of the mechanisms of geochemical reactions and transport processes in rock water systems. Uranium concentrations and isotopic fractionation in groundwater has been related to lithology, geomorphology, hydrochemistry and geological conditions of the aquifer.

This chapter describes the concentrations and distributions of uranium isotopes (^{234}U and ^{238}U) in groundwater from the North-Western Sahara Aquifer System (NWSAS) of Northern Africa and discusses the most likely sources and controls on U mobility. The ranges of uranium are compared to concentrations observed in groundwater worldwide in order be placed within a wider context. The NWSAS, one of the greatest system in the world, covers an area of over 1 million Km^2 shared by Algeria, Libya and Tunisia and lodges huge water reserves which are, however, mostly non-renewable and not fully exploitable. The system is mainly constituted by two deep aquifers: the *Continental Intercalaire* and the *Complexe Terminal*, to which is connected a third less important one in term of reserves and extension, the *Djeffara* basin. Due to increasing exploitation, the resource is currently confronting many risks such as water salinity, declining artesianism, natural discharge depletion or piezometric level drawdown. The objective is to investigate uranium distribution as a natural radioactive element and natural variation of uranium isotopic activity ratio in such a context of a large water-rock system with low to no-renewable groundwater.

Chapter 2 - The continual process of increased industrialization, militarization, and urbanization are among the most powerful sources of environmental degradation all over the world. Environmental contamination caused by radionuclides, in particular, by uranium and its decay products, is a serious problem worldwide and has become the common subject for further scientist investigations.

It is estimated that the technologically enhanced naturally occurring radioactive materials (TENORM) caused by nuclear and non-nuclear technologies, significantly increased natural radioactivity in the last 30 years. The development of nuclear science and technology, uranium mining, production and application of phosphoric fertilizer, the coal industry (mining and combustion), the oil and gas production, metal mining and smelting, mineral processing and building industry are generating radioactive waste that increased natural radioactive level on a regional scale. Economic sanctions for Serbia in the last decade of the 20th century, as well as the transition of its entire economy (which is still ongoing process), the NATO conflict in 1999, and absence of legislation and institutions, are all additional reasons for radioactive contamination of the regional ecosystem of Serbia. Also, one of the main contributing factor of environmental contamination by uranium in Serbia involved the cheap, imported phosphate fertilizers that were often of quality poor and radiological unsafe.

This outcome was a direct consequence of the reduced capacity in Serbian national production. On the other hand, according to the available data, many states were documented with bombing sites; 112 sites in Kosovo and Metohija, 12 locations in southern Serbia with depleted uranium (DU) ammunition, during the NATO aggression of Yugoslavia. On this occasion around 10 tons of DU was introduced into environment. The degree of contamination ranges from the bottom limit of 200 Bq/kg to 235,000 Bq/kg in the soil samples of mainly agricultural land, or 1,000 times above the tolerable level. Within the international mission, UNEP and FOCUS and by the engagement of national experts, the decontamination of endangered sites with depleted uranium has only been partially carried out. Today, unfortunately, the authors encounter an "invisible threat" of depleted uranium ammunition use, with highly radioactive and chemo toxic effect on human health, causing in the last 12 years an enormous increase in cancer rates and a number of newborns with genetic changes. This paper reviews some relevant aspects of environmental contamination with uranium, and gives an overview of the different remediation processes available.

This study reviews some relevant aspects of environmental contamination with anthropogenic uranium, primarily from the application of phosphate fertilizers and the use of ammunition with depleted uranium. The authors' study will include a special accent on Serbia's area concerning the uranium uptake and how it behaves in the food chain and will explore how these indicators impact human health.

Chapter 3 - In Portugal there are three uranium provinces - Beiras, Trás-os-Montes and Alto Alentejo - from which uranium mining only took place in the first one. Extraction in the

Beiras province was undertaken in about 60 mines. The first license dates from 1907 to mine radium, and the interest in uranium only started after the Second World War when the market demand for defense purposes started to grow. The market decreased later due to international treaties of non-dissemination of weapons of mass destruction and the demand turned to energy purposes. Uranium production was never intended for the internal market but rather for export purposes. Known uranium deposits in Portugal mainly occur as veins in granitic rocks and in the metasediments of the contact. There is only one known sedimentary (alluvial) deposit located in the Urgeiriça mining area. Most uranium ore deposits are small (less than 500 t of U_3O_8) but there are some ore deposits of medium size. The generally accepted genesis model of the Portuguese vein-type ore deposits consider the primary ore hydrothermal deposits as resulting from magmatic differentiation and the secondary as leaching processes from the host rocks and from the primary deposits in a long period of enrichment. Ore concentration techniques varied according to the grade, the ore type and the mining process. Natural leaching was the first method used, after that and in related with the grade of the ores, in situ leaching and heap leaching was used. Until 1946 only radium was extracted and treated and about 15-20 kg/t of ore processed to radium was then recovered to uranium. Several experiments to improve the recovery of ore deposits in the poorest ores of Urgeiriça, Rosmaneira and Bica showed the diversity of its compositions and ways to improve the efficiency of ore treatment. The exploration of uranium stopped late in the eighties and mining finished with the closure of mines in the Beiras province in 2001. After this period a remediation program started to solve environmental problems. In this chapter a series of considerations based on the information archived at the Lab oratorio National de Energies e Geologic (LNEG, the Geological Survey of Portugal) will be made now that uranium exploration in Portugal is gaining the interest of mining companies.

Chapter 4 - Experimental results are reviewed on laser-induced accelerated nuclear decays in Uranium isotopes and nuclides of its branching under laser exposure of metallic nanoparticles in aqueous solutions of Uranium salts. Laser exposure of metallic targets (Au, Ag, Pd) in aqueous solutions of $^{238}UO_2Cl_2$ leads to generation of nanoparticles of corresponding materials in the bulk of the liquid and simultaneously shifts the secular equilibrium in the activity of nuclides of its branching, such as ^{234}Th and ^{234m}Pa. Real-time acquisitions of gamma-spectra of solutions indicate the presence of non-equilibrium gamma-activity in the range of photon energies between 50 and 60 keV. This activity is attributed to the laser-assisted generation of short-living isotopes ^{214}Pb and ^{214}Bi that stand at the end of ^{238}U branching. The experimental results are presented on laser-induced accelerated alpha-decay of ^{232}U under exposure of aqueous solution of its salt in presence of metallic nanoparticles. The results on the shift of secular equilibrium under laser action are compared with similar results on electro-explosion of metallic wires in aqueous solutions of Uranium salt. Possible mechanisms of laser-induced decays are discussed on the basis of amplification of the laser wave on metallic nanoparticles due to plasmon resonance.

Chapter 5 - In the northern-west part of Russia, the main aquifers are joined to the sedimentary basin of the Mezen syneclise, filled predominantly with sandy-argillaceous sediments. These deposits were formed at the expense of drifting of products of the Baltic Shield breaking. Transfer of uranium in suspended and dissolved states was carried out by surface water and groundwater. During of the continental periods within raised blocks of the syneclise there was redistribution of uranium with the partial removal beyond the uplifts. Elevated (up to 2.6 Bq/L) values of alpha-activity of U isotopes were found in groundwater of the waterbearing complex of sandstones of the Padun suite of the Vend and high (up to 20) $^{234}U/^{238}U$ activity ratios. Study of the distribution of uranium isotopes was made in the North Dvina basin. Also the groundwater residence time in aquifers was determined. To do this was used the generic parameter of *p*: probability of ^{234}U atom transition into water in the decay of ^{238}U, obtained on the basis of geological dating, hydrodynamic calculations and paleo-hydrogeological reconstructions. As a result, eight types of groundwater with ages ranging from 160 to 130,000 years were marked out. Analysis of the calculated dependences showed that high solubility of rocks with Clarke concentrations of uranium, even at the time of contact of water with the rock of 75 000 years it will exceed the maximum allowable level of its content. In almost all samples uranium content is largely determined by its chemical dissolution. The most intense it is exerted on the sides of the basin, where is the development of the youngest water. As ground water moves to the discharge zone and increases its salinity, water's aggressiveness reduces and radiation decay begins to dominate by the leaching of uranium. Approximate estimation of radiation safety of fresh and mineral waters of the North Dvina basin has shown that they can be used for drinking. At the same time requires the constant monitoring of their radionuclides content.

Chapter 6 - Uranium has both radioactive and chemical hazards and risk assessments must address both types. Its environmental distribution depends on natural occurrence and has been influenced by human activities such as energy production, coal and phosphate fertilizer exploration which may impact freshwater ecosystems. In terms of radiological exposition, focusing chemical toxicity, uranium that is ingested or inhaled in sufficient amounts can be harmful. Radiological hazards associated with uranium isotopes are usually less severe than those of plutonium. The industrial activities involved in the nuclear fuel cycle should be considered including the uranium ore mining, milling, refining, conversion, fuel production, nuclear power generation, and nuclear waste management that is subjected to licensing and control processes. According to the International Atomic Energy Agency, the waste generated by these activities require appropriate procedures for its treatment, before disposing them in the environment. The use of nuclear fuels based on uranium is closely related to energy production by nuclear power plants, design and production of atomic weapons. Uranium-238 and uranium-235 are the two most abundant isotopes with half-lives respectively with order of magnitude of couple of billion and hundreds of million years. Thus, the different uranium salts used in fuel cycle production represent a permanent risk of environmental exposure as nuclear accidents can spread them out of the nuclear facilities and, depending on the accident, these salts can be chemically dispersed as they dissolve in water sources. When the uranium chemical species are subjected to extremely high temperatures of fire they can merge to form a residue with only a part of it oxidizing providing airborne contamination risk. The fissile nuclide uranium-235 has been used as nuclear fuel due to the fact that nuclear fission reactions take place when slowed thermal neutrons hit ^{235}U nucleus releasing two lighter nuclides, three free neutrons and a tremendous amount of energy. However, in terms of weight, the naturally occurring uranium isotopes has respectively 99.284% of ^{238}U, 0.711 % of ^{235}U and traces of ^{234}U requiring physicochemical separation process that is able to increase the content of ^{235}U. Natural uranium is the term referred to the refined uranium with the same isotopic ratio found in nature and despite the fact that natural uranium can be used to fuel both low- and high-power reactors, but the enrichment processes based on gaseous diffusion and gas centrifuge using uranium hexafluoride UF_6 are very common. Unfortunately, these enrichment processes give rise to large amounts of uranium with a lower content of the fissile isotope ^{235}U named depleted uranium. About 95 % of the depleted uranium produced to date is stored as UF_6, in steel cylinders in open air yards close to enrichment plants causing a permanent environment risk contamination. The long-term storage of depleted uranium as UF_6 represents safety risks and potential danger to health and environment because of its chemical instability. When exposed to moist air, UF_6 reacts with the water in the air to produce uranyl fluoride UO_2F_2 and hydrogen fluoride both of which are highly soluble and toxic. Storage cylinders must be regularly inspected for signs of corrosion and leaks. The estimated lifetime of the steel cylinders is measured in decades. There have been several accidents involving uranium hexafluoride in the United States and the government has been converting depleted uranium UF_6 to solid uranium oxides for disposal. Such disposal of the entire depleted UF_6 inventory could cost from tens to hundreds of thousands of U.S. dollars.

Chapter 7 - Hydrothermal ore deposits represent a significant type of industrial important U mineralization occurring in highly variable rock environments. These deposits constitute 30% of global U reserves and about 30% of world U production. Hydrothermal U deposits can be grouped in the four main categories based on their geologic setting, namely in vein-like deposits, metasomatic deposits, breccia complex deposits and volcanic–associated deposits. The first three groups usually occur in metasedimentary and in granitic, partly structurally deformed, rocks series. The volcanic–associated deposits occur together with felsic volcanic rocks, mainly in continental extensional settings. The typical ore grades of vein-like and volcanic associated deposits are between 0.1 and 2% U, whereas the ore grade of metasomatic and breccia complex deposits is distinctly lower (0.03–0.2% U). The metasomatic and breccia complex deposits are restricted to the Proterozoic era (Ukraine, Australia), whereas the vein-like deposits are restricted from the Upper Paleozoic to Mesozoic era (Western- and Central Europe, China). The hydrothermal U deposits are commonly related to late phases of orogenic cycles. The ore-forming fluids can be of various types including juvenile, post-magmatic, connate, diagenetic and meteoritic waters. The deposition of uranium minerals may be caused by changes in Eh, pH, pressure, and temperature of the ore-forming fluids. A significant feature of hydrothermal alterations occurring in the area of hydrothermal U deposits is the occurrence of alkali-rich rocks described usually as episyenites. The episyenites often form irregular pipe- or lenses-shaped bodies and are formed by albite, chlorite and hematite in high variable concentrations. The hydrothermal U deposits are thought to have formed by the remobilization of uranium from a variety of pre-existing uranium-rich rocks as black shales and uranium–enriched granitic rocks. The main mineral phases of the hydrothermal U deposits are pitchblende, uraninite and coffinite. The majorities of hydrothermal U deposits are monometallic, but also occur complex-mineralized deposits with economically interested concentrations of Cu, Au, Ag, Pb and Zn.

Chapter 8 - In the 20th century Portugal had a significant history of radioactive mining industry. Unfortunately, some of the 60 mines that have been exploited and abandoned represent pollution sources, which are associated with some environmental changes that may place public health or safety at risk. The purpose of this chapter is to review the previous studies done at Cunha Baixa mine-site, one of the most important uranium mines exploited (1970-1993) in the centre of Portugal, in order to assess the environmental impact from the past mining activity on the region, which was nowadays classified by the authorities as a prior area requiring reclamation measures. The proximity of the population to the mine site (100 m

SW) called for the diagnosis of the level and extension of the environmental problems as well as an assessment of the related chemical and radiological risks. Among other chemical elements, the uranium and radioactive daughters concentration analyzed in rocks, mine-wastes, surface and groundwater, sediments, soils and plant samples (foodstuff) was the support to evaluate potential risks to the surrounding exposure population and to recommend site-specific strategies for rehabilitation actions.

Chapter 9 - Portugal is a country with a large tradition in uranium exploitation in the XXth century with 61 registered old uranium abandoned mines. Those uranium exploitations, mostly located in the central part of the country, created open pits that in most cases were simply abandoned, causing environmental problems that sometimes lasted for decades. More recently, following the establishment on 2001 of the "Concession Act for Environmental Remediation of Abandoned Mines", a public service addressed to "EDM", an owned Portuguese State company, a major old abandoned mines recovery plan for Mainland Portugal was developed comprising the possible use of old open pits from abandoned uranium mining sites as "Waste Containment Deposits", where mining waste from other adjacent mines would be stored. The old mining sites major plan and classification were carried out by "EDM" accordingly to its location, accessibility, geological and hydrogeological conditions and three mining areas were previously selected. Considering the purpose of the future use for the old open pits of uranium mining sites, detailed information about the development in depth of the shallow geological features was needed. Mining waste deposition in the open pits may however cause additional environmental problems related with geological and hydrogeological features that must be predicted and prevented before a particular site is chosen. The identification of faults and conductive zones that may promote groundwater circulation and the spread of contaminated waters are of great importance, since sometimes the surrounding areas are populated or used in agriculture. The ability of geophysical methods in identifying either faults or conductive zones was used to understand geological and hydrogeological critical zones near previously selected abandoned uranium mine sites in Portugal. The possible negative environmental impacts of the presence of such potential waste disposal sites were therefore assessed using geological, hydrogeological and geophysical methods.

Deeper studies where developed in two open pit mines, the abandoned Quinta do Bispo and Prado Velho uranium mines, initially considered favourable sites to be used in the near future as a "Waste Containment Deposit" due to their location and regional geological and hydrogeological characteristics. But the need for the local characterization to prevent any possible negative environmental impact required different baseline studies, such as the use of geophysical methods to complement geological and hydrogeological information. These methods comprised the joint use of electromagnetic, electrical and both refraction and reflection seismic. They allowed the development of 2D schematic models that highlight the most important characteristics that may jeopardize the adequate geotechnical solutions to adopt. These 2D schematic models have been constructed showing alteration and fault zones in depth that control groundwater circulation and therefore, future water circulation problems with negative environmental impact.

Mining waste deposition in open pits of old uranium mines may cause environmental problems that can be predicted using the joint use of geological, hydrogeological and geophysical methods, preventing future problems with adequate solutions. This chapter shows some of the environmental assessment works that have been conducted in Portugal with this purpose.

Chapter 10 - Energy is very important for the socio-economic development of the world requiring sufficient access to reliable energy resources. Renewable, water and fossil energy cannot totally cover the increasing needs in power supply. Nuclear energy is considered as a resort to overcome this limitation. Before the nuclear accident of Fukushima, many countries already considered nuclear power in their energetic grid. Operating nuclear power plants requires the availability of nuclear fuel.

Uranium exploration, re-assessment and mining are observed nowadays around the world to supply power plants in nuclear fuel. Moreover, for most human activities involving minerals and raw materials, the levels of radiological exposure are not significantly greater than normal background levels. Such exposures, while having been the subject of much research, are not of concern for radiation protection. However, certain activities can give rise to significantly enhanced exposures that may need to be controlled by regulation. It is for instance the case of the activities related to uranium mining.

The objective of this paper is to deal with the natural radioactivity, the uranium occurrence, mining and related radiological exposure. Finally, regulatory aspects of uranium mining and milling, the radiological protection of the public and workers, the site remediation will be taken into account. The case of the uranium deposit of Kitongo in Cameroon will be tackled.

Chapter 11 - Attempts were made to recover uranium (U) occurring in nuclear fuel effluents and mine tailings using microorganisms isolated from U deposits in Canada, the United States, Australia, and Japan. To establish which microorganisms absorb the most U, hundreds of strains of microorganisms were screened. Extremely high U absorbing ability was detected in some bacteria isolated from North American U deposits. Arthrobacter and Bacillus sp. absorbed approx. 2500 μmol U per gram dry wt. of microbial cells within one hour. Cells removed U from refining waste water with high efficiency. Cells also absorbed thorium with high efficiency. Lactobacillus cells isolated from Japanese U deposits removed more U from seawater than the other bacteria that had superior U removal capacity from non-saline U solutions. Cells immobilized with polyacrylamide gel had excellent handling characteristics and can be used repeatedly in U adsorption-desorption cycles. These bacteria from U deposits can be used as an adsorbing agent for the removal of the nuclear fuel elements, which may be present in nuclear effluents, mine tailings, seawater, and other waste sources.

Chapter 12 - The chapter is dedicated to one-group fission cross sections of Pu and MA in LFRs spectra with the aim to increase these values by choosing a coolant which hardens neutron spectra. It is shown that replacement of standard Pb-Bi coolant over proposed Pb-208 coolant in the fast reactor RBEC-M, designed at Kurchatov Institute, Russia, leads to increasing the core mean neutron energy. As concerns Pu isotopes used as fuel, their one-group fission cross sections become slightly changed, while in the core spectrum Am-241 one-group fission cross section is changed more essentially. Another situation occurs in lateral blanket contained small quantities of minor actinides. It is shown that as a result of lateral blanket neutron spectrum hardening the one-group fission cross sections of Np-237, Am-241 and Am-243 increases up to 8-11%. This result permits to reduce the time of minor actinides exposition in FRs for their burning.

In: Uranium: Characteristics, Occurrence and Human Exposure ISBN: 978-1-62081-207-5
Editors: Alik Ya. Vasiliev and Mikhail Sidorov

Chapter 1

NATURALLY OCCURRING URANIUM IN GROUNDWATER OF THE NORTH WESTERN SAHARA AQUIFER SYSTEM (NORTHERN AFRICA)

Najiba Chkir[1], Friha Hadj Ammar[2] and Kamel Zouari[3]
[1]Department of Geography, Fac. of Letters and Humanities, Sfax, Tunisia
Lab. of RadioAnalyses and Environment,
National School of Engineers, Sfax, Tunisia
[2]Lab. of Radio-Analyses and Environment, National School of Engineers, Sfax, Tunisia
[3]Department of Geology, Lab. of Radio-Analyses and Environment,
National School of Engineers, Sfax, Tunisia

ABSTRACT

Uranium concentrations in water are controlled by a number of factors, among the most important being pH, Eh and concentrations of coexisting dissolved ions. Uranium as a redox-sensitive heavy metal occurs in water principally under oxic conditions in its hexavalent (U(VI)) form. It is usually complexed in solution, especially with carbonate ligands, but also less significantly with phosphate, fluoride or sulphate depending on their respective dissolved concentrations and ambient pH. Under anoxic conditions, U is reduced to tetravalent U(IV) state and its concentration in water is usually low due to the insolubility of this form. On the other hand, despite their long radioactive half-lives that should induce a secular equilibrium between ^{238}U and ^{234}U isotopes, large range of natural variation of $^{234}U/^{238}U$ ratios have been found in water (0.5-40). The mechanism of such fractionation is preferential leaching of ^{234}U compared with ^{238}U from solid phase, caused by radiation damage of crystal lattice upon alpha decay of ^{238}U, oxidation of insoluble tetravalent ^{234}U to soluble hexavalent ^{234}U during decay, and alpha recoil of ^{234}Th (and its daughter ^{234}U) into solution phase. In the last few decades, the behaviour of uranium isotopes in groundwater has contributed to an understanding of the mechanisms of geochemical reactions and transport processes in rock water systems. Uranium concentrations and isotopic fractionation in groundwater has been related to lithology, geomorphology, hydrochemistry and geological conditions of the aquifer.

This chapter describes the concentrations and distributions of uranium isotopes (^{234}U and ^{238}U) in groundwater from the North-Western Sahara Aquifer System (NWSAS) of

Northern Africa and discusses the most likely sources and controls on U mobility. The ranges of uranium are compared to concentrations observed in groundwater worldwide in order be placed within a wider context. The NWSAS, one of the greatest system in the world, covers an area of over 1 million Km^2 shared by Algeria, Libya and Tunisia and lodges huge water reserves which are, however, mostly non-renewable and not fully exploitable. The system is mainly constituted by two deep aquifers: the *Continental Intercalaire* and the *Complexe Terminal*, to which is connected a third less important one in term of reserves and extension, the *Djeffara* basin. Due to increasing exploitation, the resource is currently confronting many risks such as water salinity, declining artesianism, natural discharge depletion or piezometric level drawdown. The objective is to investigate uranium distribution as a natural radioactive element and natural variation of uranium isotopic activity ratio in such a context of a large water-rock system with low to no-renewable groundwater.

Keywords: Uranium, $^{234}U/^{238}U$ activity ratios, groundwater, water-rock interaction

1. Introduction

Uranium is one of the actinides, the heaviest naturally occurring elements of the cosmosphere. Although thought as rare, uranium is more common in nature than many other trace elements such as cadmium or selenium. Uranium occurs naturally in variable concentrations in all soils, minerals, rocks and waters. It can also be derived from several anthropogenic sources. In natural system, actinides are always found in association with varying amounts of their radioactive decay products. Differences in chemical and physical properties of an actinide and its daughters often lead to fractionation, giving rise to a radioactive disequilibrium. Due to long radioactive half-lifes, ^{238}U and ^{234}U isotopes are in secular equilibrium in all minerals and rocks greater than one million years old and that are closed systems for U. The $^{234}U/^{238}U$ activity ratio (AR) is therefore unity in the bulk of such systems. In natural waters, radionuclides are carried via normal recharge (cosmogenic nuclides) and as a result of water–rock interaction (natural decay series nuclides).

In groundwater, the ratio of ^{234}U to ^{238}U is almost always found to differ significantly from its equilibrium value. This isotopic disequilibria results from geochemical sorting or differentiation processes that may be physical as well as chemical, and isotopic as well as elemental (Osmond and Cowart 1992, Osmond and Cowart 1976, Ivanovich and Harmon, 1982). Various mechanisms have been suggested to interpret the generation of high disequilirium in solution, the most important being the occurrence of enhanced chemical solution of ^{234}U due to radiation damage to crystal lattices or to oxidation from U^{4+} to U^{6+} resulting from the decay of the parent ^{238}U. Aquifers environments are systems where water and minerals are strongly mixed and thus where both chemical and physical differentiation processes are particularly enhanced according to rock-water ratio, surface area and residence time. These conditions are often observed on deep and fossil aquifers.

The application of uranium isotopic disequilibria to solve hydrogeological problems is not as well spread as stable isotope of the water molecule. However, in several studies (Osmond et al. 1971, Meyer 1989, Banner et al. 1991, Roback et al. 2001, Suski et al. 2001, Dabbous and Osmond 2001, Suski et al. 2006, Bonotto 2006), uranium series disequilibria successfully help in the description of aquifer processes. The main result of these investigations was to classify aquifers on steady-state, augmenting or decaying systems in terms of dissolved uranium content value and in term of $^{234}U/^{238}U$ activity ratio values. Specific data analyses also help to refine groundwater mixing processes (Suski et al., 2001, Suski et al., 2006). Although activity ratios of uranium isotopes could be used as radiochonological tools, the dating of groundwater by disequilibria techniques remains challenging. Radiometric age determination in any case is possible only if three conditions are present: the age of the system being dated is of the same magnitude as the lifetime of the radionuclide used, the initial state of disequilibria is known and the radionuclide content of the system has not changed other than by radioactive decay. In groundwater, the closed system requirement is very difficult to verify, and the most limiting problem remains the estimation of initial radionuclide contents.

This chapter describes the concentrations and distributions of uranium isotopes (^{234}U and ^{238}U) in groundwater from the North-Western Sahara Aquifer System (NWSAS) of Northern Africa and discusses the most likely sources and controls on U mobility.

Samples have been taken from operational boreholes supplying drinking water and agriculture. Data are presented for 30 samples of the *Continental Intercalaire* taken along the main flowpath across southern Algeria and Tunisia, for 20 samples of the *Complexe Terminal* located in the eastern compartment of the aquifer and for 20 samples of the *Djeffara* basin. Large variations in both U contents (up to 4 ppb) and $^{234}U/^{238}U$ activity ratios (0.4–15.4) are observed in relation with hydrogeological conditions of each aquifer.

We have adopted a two-step methodology. First, we attempt to characterize groundwater aquifer in terms of U-contents : oxidized aquifer on "normal" U-content strata (values between 1 and 10 ppb); oxidized aquifer on enhanced U-content strata (values higher than 10 ppb); reduced aquifer on low U-content strata (values lower than 1 ppb)) and in terms of $^{234}U/^{238}U$ activity ratios data : "normal" world-wide groundwater (values between 1 and 2), possibility of formation processes (values higher than 2) or remobilization (values lower than 1). Then, we attempt to delineate preferential flowlines across the aquifer based on a two dimensional diagram of reciprocal uranium *vs* uranium isotopes activity ratios.

The plot of a two-dimensional diagram ($^{234}U/^{238}U$ AR vs reciprocal of [U]) allows identifying different compartments submitted to different oxidising/reducing conditions and leads to conclude that uranium chemistry is mainly controlled by water-rock interaction enhanced by long residence time recognised for this system.

2. Uranium Geochemistry

In this first part of the chapter, the geochemistry of the uranium is described by examining possible sources, mobilization and retardation processes, speciation in solution and mechanisms of fractionation between its isotopes.

2.1. Chemical Properties

Uranium has a multivalent behaviour (from +3 to +6 oxidation state), oxidised hexavalent (+6) and reduced quadrivalent (+4) states being the most observed in minerals.

In its reduced form, uranium is almost chemically immobile in the near-surface environment at low temperatures (Langmuir 1978). Uranium, however, may be mobilized as hexavalent uranyl ion with much higher solubility in natural waters (Langmuir 1978, Ribera et al. 1996, Radgnarsdottir and Charlet 2000) even for low Eh conditions (Paquette and Lemire 1981, Kertes and Guillaumont 1985). In oxidized aqueous systems, uranyl complexes (UO_2^{2+}, +6 oxydation state) are much more soluble than uraneous species (UO_2, +4 oxidation state). Dominant uranium species in waters depend on Eh-pH conditions, temperature, and presence of other ions in the solution (Madic and Genet 2001, Bourdon et al. 2003). In pure water and in the absence of other complexing ions, UO_2^{2+}, and ($\left(UO_2\right)_3(OH)_5^+$) are the important uranyl species for low concentrations (≤100nm) and pH while hydroxyl forms (UO_2OH^+, $UO_2(OH)_2$, and $UO_2(OH)_4^{2-}$) appear when the pH increase. However, their significance diminishes in the presence of low concentrations of CO_2 and above pH = 5 (Figure 1a).

For higher concentrations, hydroxyl forms of uranium, such as schoepite, precipitate for pH ranging between 6,5 and 7,5 with an uranium concentration of 200 nM and for pH ranging between 4 and 11,5 with a concentration of 1 mM (Figure 1b).

Naturally occuring uranium is often found in association with other elements such as nitrogen, sulphur or oxygen in uranite and pechblende formed by uranous oxides (UO_2). On the other hand, uranyl ions (UO_2^{2+}) have high solubility in most acid waters such as nitric of fluorhydric acids producing uranyl complexes (Ivanovich and Harmon 1992). Water salinity is a controlling factor of uranium solubility in oxidised environments but not in reducing ones. Besides, uranium is easily adsorbed on particular phase such as organic matters and colloids (Ivanovich and Harmon 1992, Drever 1997).

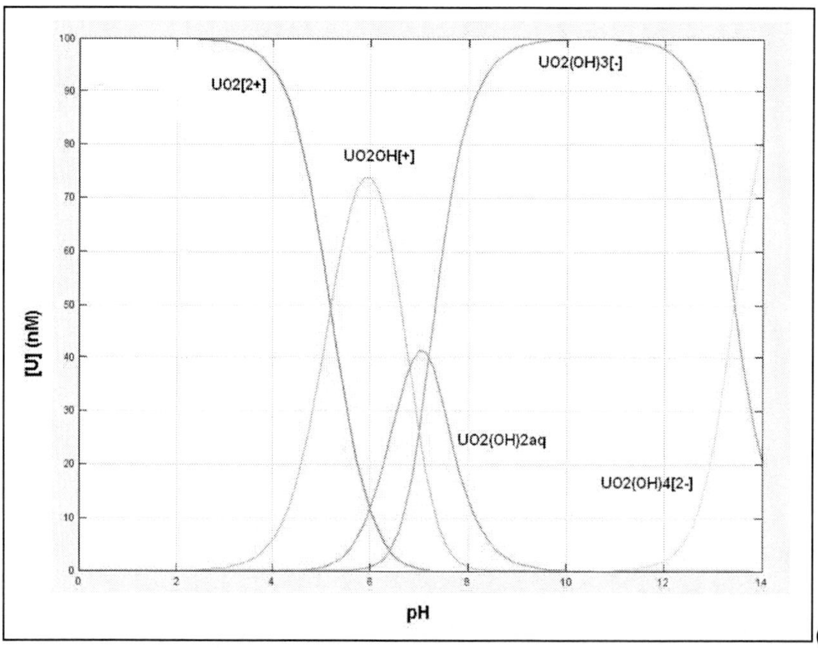

(a)

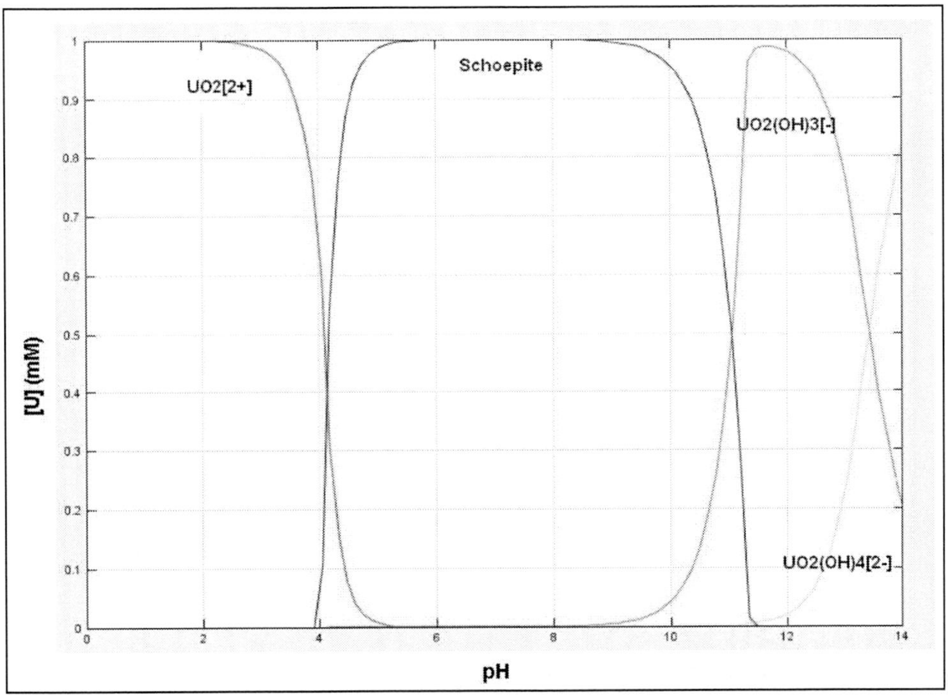

(b)

Figure 1. Theorical uranium species distribution for a total uranium concentration of 100 nM (a) and 1 mM (b) in pure water according to pH (in Untereiner 2008.

2.2. Mobilization and Transport of Uranium in Solution

Two types of geochemical aqueous environments distinguished according to their Eh could contain uranium (Shahandeh et al. 2001):

- Waters with an Eh related to the O_2/H_2O couple: This Eh is therefore dependant on pH and pCO_2. The environment is oxidised and uranium is mainly found in at +6 oxidation state in these waters.
- Anoxic waters with a Eh related to Fe(III)/Fe(II) and HSO_4^- / HS^- couples: This system is in reducing conditions and uranium is mainly found at +4 oxydation state. Uranium in this valence state is not soluble and therefore will precipitate.

Due to its complexation properties with oxydes, uranium could be associated with various dissolved or solid chemical species (Sheppard and Evenden 1988). These complexation mechanisms control uranium behaviour in natural systems and explain its heterogeous distribution. Soluble uranyl complexes that may exist in low concentrations groundwater (Figure 2) are formed with ions of carbonate (pH > 7.5), phosphate (pH 4.0 – 7.5), sulphate (pH < 4), fluoride (pH 3-4) and silicate (pH ~ 6) but are insignificant in natural waters. Sulphates are only formed in low concentrations below pH = 4. The most important ions are carbonate complexes that are stable over a wide range of pH (4-10), dependent on temperature and also on pCO_2. In high carbonate or high bi-carbonate aerobic groundwaters, the solubility of uranium may exceed 1 g per liter.

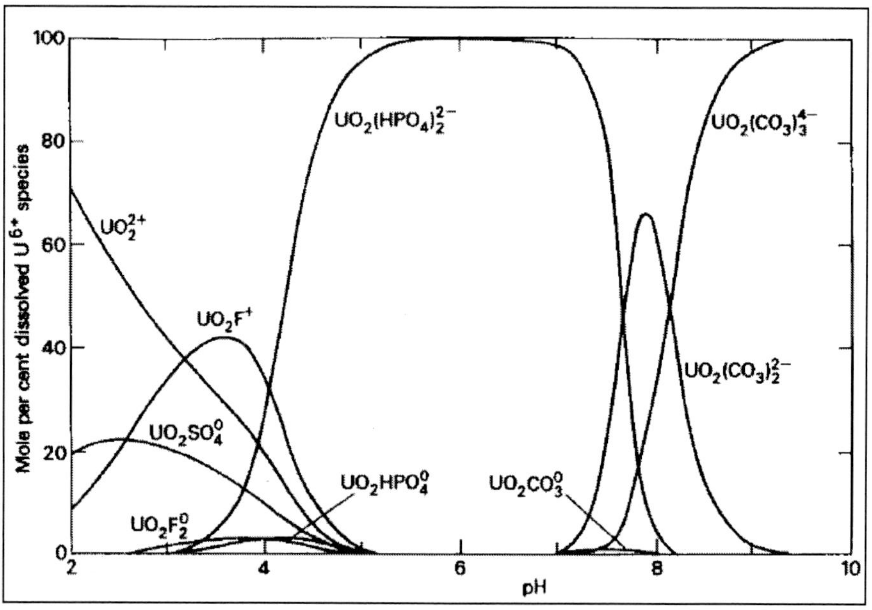

Figure 2. Distribution of uranyl complexes depending on pH for a $pCO_2 = 10^{-2}$ atm and a total uranium concentration of 10^{-8} M (2.38 ppb) at 25°C in presence of other ions: $\Sigma F = 0.3$ ppm, $\Sigma Cl = 10$ ppm, $\Sigma SO_4 = 100$ ppm, $\Sigma PO_4 = 0.1$ ppm, et $\Sigma SiO_2 = 30$ ppm (Gascoyne 1992).

However, natural waters containing uranium concentrations higher than 100 $\mu g.l^{-1}$ are quite rare, and have generally been found only in aquifers containing uranium mineralization (Singh et al., 1995).

Influence of organic complexes: Organic substances as the main vector of migration and redistribution are one of the most important components of the uranium complexation (Gueniot et al. 1988). In natural systems, organic substances constitute the reservoir of ligands that will associated with heavy metals and radionuclide such uranium, in particular for pH ranging between 5 and 7 (Jamet et al. 1993). Humic and fulvic acids are strongly bound with uranium. Uranyl humates are insoluble in the pH range 2.2-6 with maximum uranium adsorption by humic acids occurring at pH values of about 4-5. Uranyl fulvates are insoluble in the pH range 6-6.6. The possibility for solution, transport, and reprecipitation of uranium may easily occur for uranyl humates and fulvates in natural waters, due to small changes in pH (Schmeide et al. 1998, Lenhart et al. 2000).

Influence of organic acids: Carboxylic acids such as citrate, acetate or malate, are chemical specis found in most natural systems. As for humate and fulvate substances, uranium form strong and stable complexes with a wide variety of organic acids that are mobile and available for biological adsorption. Complexation degree depends on the organic acid properties (number of carboxylic and hydroxyl groups), of the concentration and of the pH (Jones 1998). The most common uranium organic complex is the uranyl citrate that form in a large range of pH (Rajan and Martell 1965, Markovits et al. 1972, Jones 1998, Ebbs et al. 1998, Keiling and Marx 1991, Allen et al. 1996, Dodge and Francis 1997, Huang et al. 1998,

Dodge and Francis 2003, Jiang et al. 2002, Shahandeh and Hossner 2002, Bailey et al. 2004, Bailey et al. 2005).

Influence of phosphates: Uranyl ions form different complexes when associated with phosphates depending on pH (Sandino and Bruno, 1992). Uranyl phosphate have low solubility and therefore will precipitate. The presence of phosphates in a U(VI)-system will induce precipitation for pH above 6 (Untereiner 2008) and will reduce its availability for biological adsorption (Arey et al. 1999). The uranyl phosphate ion is dominant for pH < 7.5. However, if the concentration of phosphate is not significant, urabyl carbonates become more important.

Influence of carbonates: Due to the abundance of CO_2 and carbonated minerals (calcite and dolomite), carbonates have a primordial role in environment. In a open system in contact with the atmosphere, the concentration of the three species (CO_3^{2-}, HCO_3^- and H_2CO_3) depend on pH and partial pressure of CO_2 that may be dissolved by the system.

For carbonate solution in the range of pH commonly observed in natural system (3 to 10), three ligands may associate with uranium (OH^-, and HCO_3^-). Thus, three complexes of uranyl carbonates $UO_2(CO_3)_{aq}$, and $UO_2(CO_3)_3^{4-}$ may be in the solution. In the absence of other ligands, uranium will preferentially be bound with carbonates in natural pH system (Ebbs et al. 1998, Echevarria et al. 2001).

The importance of the carbonate complexes and has been recognized for some time (Hostetler and Garrels 1962) with the bicarbonate species being the most stable over the pH range 4-10 depending on the pCO_2. In high-carbonate or -bicarbonate, oxidizing groundwater, the solubility of uranium may exceed 1 $g.l^{-1}$ (Allard et al. 1982, Puigdomenech and Bruno 1988)

However, groundwater containing greater uranium concentrations than 1 g l^{-1} are rare (Gascoyne 1989) probably due to the combined effects of low Eh and p_{CO2} (Gascoyne 1992). Higher concentrations are often observed in aquifers close to or lodged in ore deposits (Smedley et al. 2006).

Influence of water salinity: For reducing conditions (Eh < -200 mV and pH ≥ 7), uranium solubilities are quite constant while in mildly reducing (Eh > -100 mV) and in oxidising conditions, U^{6+} ions form complexes with Cl and SO_4 in saline groundwater (Lemire 1988). These charged species are more stable due to the higher ionic strength (Giblin and Appleyard 1987) and uranium solubilities become an order of magnitude higher than in dilute (fresh) groundwater.

2.3. Mobilization and Transport in the Colloidal Phase

The abundance of the colloidal phase, usually defined as particulate material that will pass through a 0.45 μm filter (Stumm and Morgan 1981), varies in groundwater and is commonly higher close to recharge areas and in karstic system. Colloids may consist of precipitates, bacteria, organic matter, and rock and mineral fragments from the fracture or access borehole. Colloid concentrations rarely exceed 100 $mg.l^{-1}$ except in extreme conditions (Degueldre et al. 1996) that means that their presence become significant only in relation with toxic species as radionuclide or in relation with organic substances (Van de Weerd and Leijnse 1997, Drever 1997). Uranium and daughter nuclides may sorb or precipitate onto these colloids reversibly or irreversibly. Radionuclides that are irreversibly sorbed may not be subsequently removed by changes in groundwater geochemistry, redox or solubility limitations and will migrate with the velocity of the groundwater (Ivanovich et al. 1987, Short et al. 1988, Vilks et al. 1993, Longworth et al. 1989, Payne et al. 1992). Adsorption on colloids could highly influence the isotope composition of the system (Grindrod 1993, Degueldre et al. 2000, Stagg et al. 1997, Zänker et al. 2007).

3. Uranium in Aquifer Systems

Uranium contents depend on the geochemistry of the environment. Concentrations of dissolved species are related to Eh, pH and coexisting ions. The concentration in aqueous phase of the U^{4+} ion never exceed the magnitude of $\mu g.l^{-1}$ while the U^{6+} ion concentration could reach the magnitude of some tens $mg.l^{-1}$ (Ragnarsdottir and Charlet, 2000).

Surface waters concentrations of uranium ranges between 0.01 and 5 $mg.l^{-1}$ (Gascoyne, 1992, Plater et al. 1992, Osmond and 1992, Riotte and Chabaux 1999). Groundwater in reducing environments are more enriched with 0.1 to 50 $mg.l^{-1}$ of uranium content, reaching 10 000 $mg.l^{-1}$ in highly mineralized systems (Osmond and Cowart 1992, Cochran 1992, Smedley et al. 2006). Some higher concentrations have been reported and associated with U-rich rock influences such as in Finland (Salonen and Huikuri, 2002), in Canada (Moss et al. 1983) or in Gabon (Salas and Ayora 2004). High levels of uranium in groundwater are often related to Na-HCO_3 waters (Dongarra 1984) where carbonate ions are available for efficient uranium leaching and for complexation of uranyl carbonates. However, high levels have also been recorded in sedimentary aquifers either consolidated (sandstone, or limestone) or not. In arid areas, high concentrations have been associated with high evaporation rates.

3.1. Isotopic Fractioning in Aquifer Systems

The hydrosphere presents the most extreme examples of uranium-series disequilibrium. This disequilibrium is due to differences in geochemical behavior of radionuclides in aqueous systems enhanced by isotopic fractionation processes related to the radioactive decay. Uranium isotopes commonly observed in natural waters are ^{238}U and ^{234}U, ^{235}U remains with low activities. The ratio of ^{235}U to ^{238}U is essentially constant in nature equal to 137.88±0.5, except in Oklo deposit and in some prehistoric volcanic materials (Fried et al. 1985). Whilst the ratio of ^{234}U to ^{238}U has a large variation range.

Radioactive disequilibrium results from geochemical differentiation that could occur at solid-liquid, solid-gas, and solid-solid phase boundaries. These differentiation processes may be physical as well as chemical, and isotopic as well as elemental. The mobility of each radionuclide depends on chemical properties such as solubility, adsorption, precipitation, and complexation and depends also on physical impulsion induced by radioactive decay in particular alpha particle emission (Fleischer 1988, Rama and Moore 1984). The most important sorting processes of geological interest are primarily at solid-liquid phase boundaries. In groundwater and aquifer environments where water and minerals are intimately mixed, rock-water ratio, surface area, and residence times have large values, and both chemical and physical differentiation processes have opportunity to operate (Hussain and Krishnaswami 1980, Banner et al. 1991, Osmond and Cowart 1992).

Range of U contents and $^{234}U/^{238}U$ activity ratios vary along the hydrological cycle (Figure 3).

The mean concentration of all oceans is about 3.33 ±0,8 $\mu g.l^{-1}$ (10^{-3} ppm) (Rogers and Adams 1969, Ku et al. 1974, Chen et al. 1986). This value is quite constant world wide except in coastal zones and in certain estuarine due to uranium supplied by surface water discharge (Brin 1979, Cochran 1992). Uranium flux supplied to oceans has been estimated to 0.85~0.88 10^{10} g/year assuming its conservative behaviour in estuarine zones (Sarin et al. 1990). Oceanic waters are characterised by a homogeneous activity ratio between uranium isotopes equal to $^{234}U/^{238}U=1.14 \pm 0.03$. In fact, the ocean could be considered as a huge reservoir reflecting all continental fractionation processes.

The isotopic disequilibrium of uranium is mainly explained par the one of continental waters. Surface waters of arid areas have a limited contribution to the ocean water budget but have a significant influence on the ocean uranium isotopes disequilibrium (Kronfeld and Vogel 1991, Kronfeld et al. 2004).

For fresh waters, uranium contents are highly variable depending on several factors such as duration of the contact with the uranium horizon uranium and the weathering rate (Moreira-Nordemann 1979, Murakami et al. 1997, Aguinaldo et al. 2003,), uranium concentration of this horizon, evaporation rate and the presence of complexing ions. Surface water commonly contain between 0.01 and 5 ppb of uranium (Osmond 1980, Palmer and Edmond 1993), the concentration is closely correlated with bicarbonate ions concentration and with the total of major cations (Na+K+Mg+Ca), with total dissolved charge (Borole et al. 1982, Barnes and Cochran 1993) and with the pH (Tieh-Chi and Jeng-Jong 2000). Groundwaters are more enriched in uranium with concentrations ranging between 0.1 and 50 ppb, up to 2000 ppb even in non mineralised areas (Betcher et al. 1988). In aquifer systems, large variation of concentration could be observed (Cowart and Osmond 1974) and explained by Eh-pH variation that cause precipitation/remobilisation of uranium along the flowpath of the aquifer.

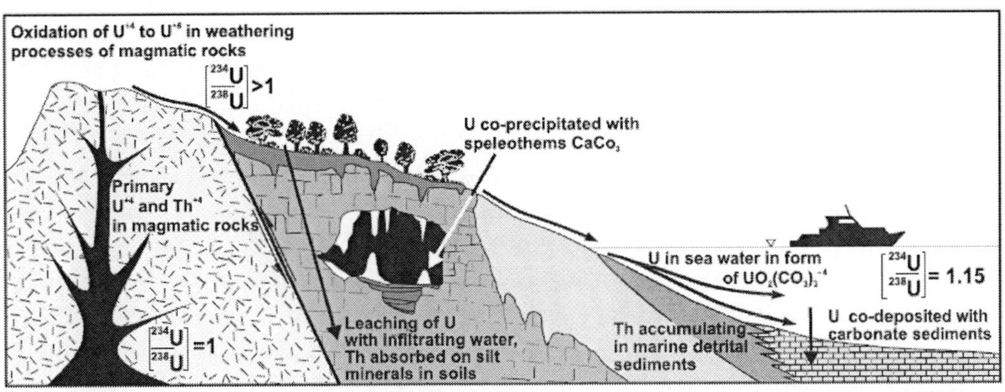

Figure 3. Simplified schema of uranium circulation and uranium isotopes activity ratios during weathering processes (from Hercman and Goslar 2002).

Various factors control the disequilibrium between uranium 234 and uranium 238. When a radionuclide is preferentially mobilised on the rock surface of the aquifer bearing-formation, the disequilibrium will be observed on both the solid and the liquid phases. Besides, precipitation could also be selective. All these processes depend on elemental geochemistry. The alpha-recoil fractionation mechanism is not element-specific. Each decay daughter of a radionuclide is more susceptible to be displaced than its immediate parent and therefore to be ejected across phase boundaries. Disequilibrium is the immediate result. In other cases, the displacement leads to unstable lattice sites where enhanced vulnerability to leaching results in deferred disequilibrium. The probability that a radionuclide is found in the liquid phase could be formulated by two terms: the first one is function of the alpha recoil geometry while the second one is function of geochemical conditions (solubility and sorption). The relative importance of these two terms depends on the half-life of the radionuclide. For short-live radionuclide, the alpha recoil geometry is tg factor while for lin-lived ones, the phenomenon is controlled by geochemical conditions (Osmond and Ivanovich 1992).

In fact, uranium isotopes fractioning in aqueous medium is a convergent effect of several mechanisms (Fleischer and Raabe 1978): i) Dissolution of the ^{234}U by oxidation to the hexavalent state; ii) direct ejection of ^{234}Th in the aqueous phase by alpha-recoil effect; iii) Recoil and diffusion interstitial medium, this effect is enhanced by the damage of the track fission. Therefore, the origin of the isotopic fractioning, in uranium open system, is essentially associated to physico-chemical processes taking place across the water-rock interface. Several recent studies are based on uranium content and isotopic disequilibrium as indicators of hydrodynamical and geochemical processes of aquifers (Sherman et al. 2007, Chalov et Tikhonov 1990, Meyer 1989, Hubert et al. 2006, Abdul-Hadi et al. 2001). Specificities of uranium that reacts to geochemical conditions variations by both content and activity ratio changes allow spatial differentiation of water compartments inside the aquifer. Characterization of each compartment may allow calculating mixing rates between different water masses and identifying origin of these waters. However, the application of this tracer remains constrained by the complexity of the uranium geochemistry, no typical model could be constructed and interpretations are local and specific to geographical conditions even if based on the same basic principles. It is recognised that the complexity of the uranium geochemistry is induced by the various variable factors controlling its solubility.

This fact led to the conclusion of Hollocher and Yuskaitis (1993), cited by several authors, that the presence of uranium in groundwater is not a function of a set of parameters but either a set of functions each one being related to a parameter set.

In natural waters, disequilibrium between ^{238}U and ^{234}U is commonly observed. $^{234}U/^{238}U$ activity ratios range between 0.5 and 40 (Gascoyne, 1992). ^{234}U excess is caused by radioactive decay in the crystalline matrix and by alpha recoil effect (Sun and Semkow 1998).

In fact, ^{238}U decay in ^{234}U induces the emission of one alpha particle and two beta particles. The crystalline matrix surrounding the ^{238}U radionuclide could be damaged by the ejection of these particles. Consequently, the ^{234}U daughter is more accessible to liquid phase leaching. On the other hand, ^{238}U decay could eject the ^{234}Th daughter in the surrounding environment such as water. Due to its short half-life (24.1 days), the radioactive decay of ^{234}Th enhances the $^{234}U/^{238}U$ isotopic ratio in the aquifer.

Concentrations and activity ratio span over a large range. In major natural waters, $^{234}U/^{238}U$ ratios vary between 0.7 and 2.0; however, it may reach 30 and more (Chkir et al.

2009). Besides, the activity ratio of seawater is considered constant equal to 1.14 with a uranium content equal to 3.3 $mg.l^{-1}$ (Ivanovich and Harmon, 1992).

3.2. Water-Rock Interaction in Groundwater

Different processes are grouped under this concept such as leaching, alpha recoil, precipitation et coprecipitation, etching, ionic exchanges and sorption. The qualitative model (Osmond 1980) used for the representation of chemical and physical transitions of the uranium-series at the water-rock interface has been quantified by Davidson and Dickson (1986) and Frohlich and Gellermann (1987). The fundamental concept of this model is implicitly used by all uranium mobility models (Hussain and Krishnaswami 1980, Krishnaswami et al. 1982, Cuttell et al. 1986, Ivanovich et al. 1991). This model assumes that uranium is on or close to the mineral surface.

Each disintegration produces a new element which position depends of the alpha recoil direction: inside the mineral far from its surface, toward the liquid phase or close to the interface. On the same time, decay daughter, whether it is on the interface or in the liquid phase, will be submitted to chemical forces that trend to solubility equilibration.

The chemistry effect of the surface layer itself is rarely considered when dealing with interface dynamics. Many models assume uniform uranium concentration from surface to inside of the mineral in contact with groundwater. This concentration is used to predict the flux of daughter radionuclides across the grain boundary via recoil processes. Such an hypothesis neglects the significance of the surface (chemistry and mineralogy) even it could often become a relatively high-concentration monolayer source of recoiled daughter. Thus, the surface layer is a different environment for radionuclides requiring distinct treatment in term of physical and chemical mobility modeling. This surface has a higher effective concentration than underlying layers of the solid or the overlying liquid phase. Any disruption of the near-surface layers will bring diffusion processes and disturb the solid-liquid equilibrium.

The ^{238}U is often concentrated on grain surfaces rather than within crystalline matrix. In most rock-water systems, even low concentrations of dissolved species represent a considerable total flux over geological time-scales. Uranium tends to be mobilized in the upper oxidized parts of an aquifer, and precipitated where reducing conditions prevail.

Much of this secondary accumulated uranium is found near the 'redox front' but some small fraction flow donstream and could accumulate on the rock-aquifer matrix. This process could increase the total accumulation of rock-water interface uranium without variation of the concentration of the bulk rock. A similar process often affects other redox-sensitive elements such as iron and manganese in deeper parts of the aquifer.

For this reason, a second solid phase must be considered when dealing with rock-water interaction chemistry-the surface coating phase (Lowson and Short 1986, Lowson et al.1988, Ivanovich et al. 1987, Romer and Rocholl 2004). In fact, most data available on the aquifer matrix reveal uranium-series equilibrium while related groundwaters have enhanced radioactive disequilibrium. However, detailed measurements on the coating layer of the mineral need high and cost analytical procedures (Bonotto et al. 2004, Dearlove et al. 1990, 1990).

The modelling of the removal rate process of uranium from the rock or mineral to the pore or fracture water depends conceptually of the process type. The effects of preferential removal of ^{234}U by etching or leaching and by alpha recoil can be subsumed using different rate-loss parameters for ^{234}U and ^{238}U. The parameters could be defined as *chemical decay constants* and act analogously to radioactive decay constants (Titayeva and Veksler 1977, Airey and Roman 1981, Andrews and Kay 1978, Andrews et al. 1982, Latham and Schwarcz 1989, Frohlich and Gellermann 1987, Alexander et al. 1990). Several of the models can be considered as extensions of the 'open-system' model proposed by Hille (1979) where the system is represented by a box exchanging radionuclide with its environment through parameters that have to be calibrated on experimental data. Andrews et al. (1982) succeed to explain the increase deep down of uranium concentrations and $^{234}U/^{238}U$ activity ratios of the Granit Stripa aquifer system (Sweeden) by using a model mainly based on alpha recoil.

This model allowed to highlight that this increase is related to the time residence of deep groundwater, as the alpha recoil effect due to the decay of ^{234}Th in ^{234}U as the preferential leaching of ^{234}U compared to ^{238}U is enhanced and induce $^{234}U/^{238}U$ activity ratio increase. Some exceptions of high activity ratios could also be observed in reducing conditions due to uranium precipitation.

3.3. Tracing of Aquifers

Tracing of mixing processes in aquifers is here based on uranium concentration and $^{234}U/^{238}U$ activity ratios variations. The basic hypothesis is that these parameters could be considered stable and constant for a long term and for a given area. This hypothesis is easily verified for most groundwater except for aquifers intensively exploited where long term perturbations have occurred. In applying uranium-series disequilibrium to hydrologeological systems, two kinds of system can be recognized: conservative and non-conservative. Conservative system means that the relative and absolute proportions of uranium in groundwater remain fixed as the water moves along the flowpath, thus any changes in such proportions could be interpreted as the result of mixing of different water masses. This definition applies to waters which achieved the equilibration in term radionuclides distribution between liquid and solid phase, and to waters which move rapidly compared to the hal-life of the dissolved radionuclides.

The use of dissolved radionuclides as tracers of mixing proportions assumes that each water masses has a identifiable uranium “concentration-activity ratio” fingerprints. The respective contribution of each water masses could therefore be calculated based on mass conservation equation by taking in account the ^{234}U excess (or deficit) (Ivanovich and Harmon 1992). This principle could be applied to uranium concentrations as well as activity ratios depending on the reliable characterisation of each end-members (Kronfeld et al. 1992, Daniel and Spalding, 1994).

Geochemical changes are not involved in this case although most groundwater masses encounter different geochemical conditions which cause changes in uranium concentrations enhanced by decay rates. These systems are considered “non conservative”. Such changes in uranium fingerprints decrease its interest as tracer of mixing processes but observed isotopic evolution could be used as a tracer of aquifer evolution and rock-water interactions.

3.3.1. Classification of Aquifers

Aquifers are classified according to uranium concentration or $^{234}U/^{238}U$ activity ratios (Cowart and Osmond 1980, Osmond and Cowart 1983, Toulhoat et al. 1988, Bonotto 1999, Bonotto and Andrews 2000, Roback et al. 2001). In term of U-contents, aquifers are classified according to three types: i) oxidized aquifer on ''normal'' U-content strata when uranium contents values vary between 1 and 10 ppb, ii) oxidized aquifer on enhanced U-content strata when uranium concentration values are higher than 10 ppb and, iii) reduced aquifer on low U-content strata when uranium concentrations values are lower than 1 ppb.

In term of $^{234}U/^{238}U$ activity ratios (AR), aquifers are classified on three types: i) "Normal" aquifers for AR ranging between 1 and 2, this is the case for most world-wide groundwater, ii) "Accumulation" aquifers with AR being greater than 2 and where precipitation could occur. This is the case of groundwater where leaching rate of ^{234}U are enhanced by both geochemical conditions (pH, Eh, coexisting ions) and/or the presence of uranium rich minerals with important alpha recoil effect, iii) "Remobilisation" aquifer with AR below 1 where uranium is highly dissolved.

Uranium concentrations activity ratios vary along the flowpath of the aquifer in response to radioactivity level of the aquifer-rock. This variation allows defining "steady-state", "decaying" and "decreasing" aquifers.

In steady-state system, the aquifer-rock has a uniform and low concentration of parent uranium, and the water moves very slowly through the system. This case is commonly observed for ancient water systems such as for a deep borehole in Ramallah (Palestine) (Kronfeld et al. 1975) for which a very high $^{234}U/^{238}U$ activity ratio had been explained by a long close interaction of water of low uranium solubility with rock of higher uranium concentrations.

On the other hand, other authors (Kraemer 1981, Kraemer and Kharaka, 1986) had explained high activity ratios observed for deep water in US Gulf Coast by high temperature that may cause rock-water isotopic re-equilibration.

In «decaying» system, groundwater percolates slowly through very rich uranium rocks such as granite. Uranium contents and $^{234}U/^{238}U$ activity ratios increase along the flow path. Activity ratios could also reach very high values bu uranium contents can remain quite low due to the retention of uranium on minerals or weathered materials (Wathen 1987, Kronfeld et al. 2004, Gascoyne 1989, Cherdyntsev 1971). "Decreasing" systems are the most common as deep groundwater percolating through low to moderate uranium rocks (Roback et al. 2001, Riotte and Chabaux 1999, Riotte et al. 2003).

3.3.2. Identification of Geochemical Processes in Aquifers

Geochemical processes taking place in the aquifer could be identified by a common method of analysis based on the plot of $^{234}U/^{238}U$ activity ratio against the reciprocal of uranium concentration (^{238}U-content). This diagram allows to the recognition mixing processes (conservative systems) and aquifer-water interactions (non-conservative systems) according to lines on which data are plotted (Figure 4) (Osmond et al. 1971, Osmond and Cowart 1992).

The first interpretative pattern is mixing of different uranium fingerprint water: the result of the dilution of an uranium-free water by an uranium-containing water is in the movement of the respective plotted point to the right; the reciprocal of the uranium concentration increases without affecting the $^{234}U/^{238}U$ activity ratio.

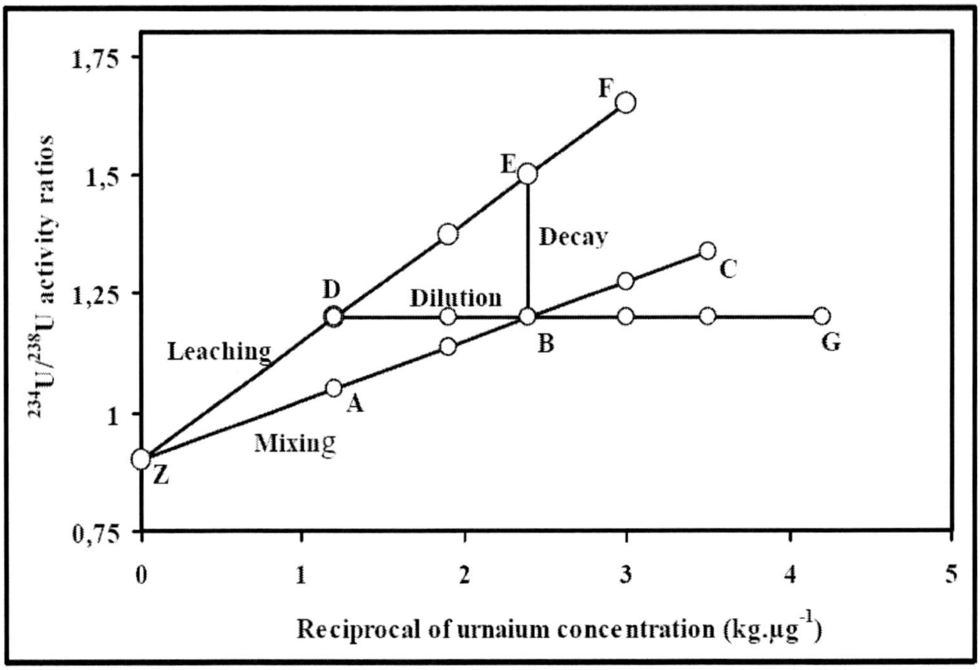

Figure 4. Diagram of $^{234}U/^{238}U$ activity ratio against the reciprocal of uranium concentration for the recognition of mixing processes (based on Osmond and Cowart, 1992).

The usual indication of the dilution process is a horizontal array of sample points, some of which are interpreted as having experienced more dilution than others.

Another commonly observed pattern is a linear line from lower left to upper right. This might be interpreted as a mixing line between two end-members, a water with high uranium concentration and low activity ratio mixed a water with low concentration and high uranium-isotope ratio.

However, this line could also be interpreted as an evolutionary pattern of rock-water interaction indicating a mixing of a water with a low uranium concentration and a high activity ratio with newly leached uranium of a lower $^{234}U/^{238}U$ value where the y-axis intercept is the leaching component of the aquifer, having an 'infinite' concentration and the $^{234}U/^{238}U$ activity ratio of the uranium existing at the rock surfaces. The slope of the line has can be interpreted as the ^{234}U-excess in the aquifer water (^{234}U per unit volume) which is mixing with the leached uranium. The dimension of this excess is in U-équivalent per ppb indicating concentration in µg per kg of an equilibrated quantity of ^{238}U (^{234}U activity per water weight (Dabbous and Osmond 2001).

In the figure 4 any straight-line as A-C can be interpreted as a mixing line. In this case sample B might be a mixture of samples A and C. A-C can also be mixture of C with variable amounts of leachate Z (water with 'infinite' uranium concentration). Water G may be a mixture of water B with water having no uranium at all (dilution). Sample B can have evolved in several ways: dilution of water like D, mixing of A and C, leaching of Z by C, or decay of excess ^{234}U in E. the entire trend of samples A-C might be the result of decay of an original array D-F in very old water in a deep aquifer system.

3.4. Redox Fronts

The mobility of radionuclides in aquifers is primarily depending on the host rock but is also controlled by geochemical conditions and by radioactive decay mechanisms in groundwater. The studies of variability of uranium isotopic composition between outcropping areas of aquifers and downstream confined zones showed that, in most cases, a redox front delimitate compartments inside the aquifer differing by geochemical conditions (Cowart 1980, Osmond and Cowart 1976, Osmond and Cowart 1983, Andrews and Kay 1983, Cuttell et al. 1986, Ivanovich et al. 1991).

Three zones could be identified (Figure 5), the underlying assumption is that the system is at a steady-state condition that is the front is stationary or moving only very slowly:

- Zone 1 corresponds to the recharge area characterized by oxidized conditions. Infiltrated water is charge by dissolved oxygen, thus relatively oxidised conditions enhance uranium dissolution. Uranium contents are elevated while $^{234}U/^{238}U$ remains moderate. The aquifer is commonly in an isotopic decay regime.
- Zone 2 is the redox front itself where changes in geochemical conditions occur. Oxygen is progressively consumed along the flow path toward the redox front, delineating the oxidised zone from the reducing deep one. In this zone, uranium precipitate and concentrate, thus, very low uranium contents are observed. Around the redox front, physico-chemical conditions are close to equilibrium, effects of decay and alpha recoil cause a preferential mobilization of ^{234}U and induce high $^{234}U/^{238}U$ activity ratios. In some large steady-state confined aquifers, activity ratios reaching 20 have been observed (Cowart and Osmond 1980, Kronfeld et al. 1994, Gilkerson et al. 1994, Andrews et al. 1994).
- Zone 3, located downstream the redox front, is characterized by reducing conditions. Uranium contents are low and quasi constant while $^{234}U/^{238}U$ activity ratios decrease along the flow path. Variation of uranium fingerprint in this downstream zone trend to an asymptotic adjustment toward rock-water equilibration values. The system is under an isotopic decreasing regime.

The redox front could migrate along the flow path with a velocity depending on that of the water, but also depending on Eh, pCo2 and physico-chemical conditions. This migration will induce cycles of uranium precipitation/remobilization with a isotopic fractioning: uranium precipite when water cross the redox front then uranium is progressively remobilised as oxidising conditions reappear when the front moves downstream. These cycles generate uranium deposit of « roll-front » type when lithological are uranifereous. However, in most area, these precipitation processes do not generate any accumulation of economical interest uranium contents remain to low of the naural background level.

3.5. Groundwater Datation

On of the most challenging problem of the uranium-series disequilibrium application is groundwater datation (Gholam *et al.* 1997). Given the long half-life of main radionuclides of this series, this datation could be a precious complement to tritium (12.32 years; Lucas and Unterweger, 2000) and/or to carbon 14 (5730±40 years; Godwin, 1962) that remains limited to relatively recent aquifers. However, the approach is quite different since the tritium is an intrinsic part of the water molecule and the carbon is incorporated in HCO_3 conservative ions while uranium is affected by all geochemical conditions of groundwater. Radiometric age determination rely on three conditions: the age of the system being dated is of the same magnitude as the lifetime of the radionuclide used, the initial state of disequilibria is known and the radionuclide content of the system has not changed other than by radioactive decay. For uranium, the closed system requirement is very difficult to verify, and the most limiting problem remains the estimation of initial radionuclide contents. Some authors (Osmond et Cowart 2000) suggest to determine the time since groundwater had crossed the redox front of the aquifer.

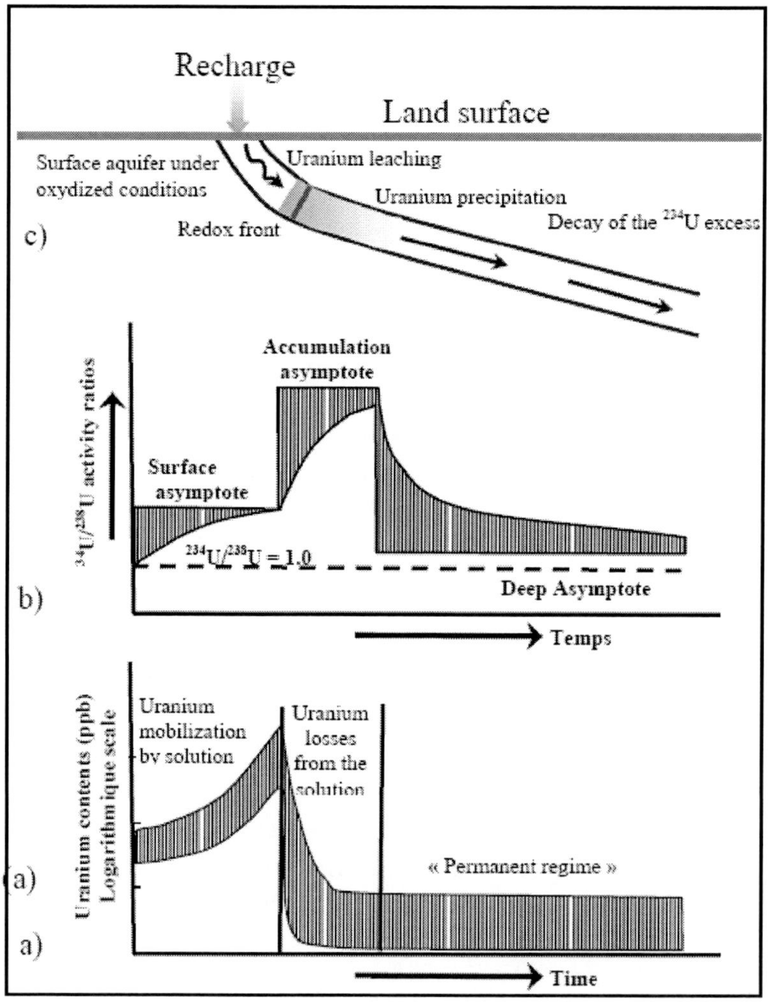

Figure 5. Schematic variation of uranium concentration (a) and activity ratios (b) along a flow path of a confined aquifer marked by a redox front (c) (from Ivanovich et al. 1991).

This time corresponds to the age of uranium in the groundwater and not to the water age. Indeed, due to adsoprtion/retardation processes at the rock-aquifer interface, uranium is in constant migration along the flow path with a velocity less that of water. Pearson *et al.* (1983) had shown by a numerical model that water could have a residence time 30 magnitude above that of the uranium migration.

On the other hand, the «closed system» condition is quite difficult to verify for uranium isotopes in groundwater. In fact, uranium could be considered as conservative in surface and near-surface environments since fractioning processes have long term dynamic and could be considered as non significant. This hypothesis is likely less plausible in deeper aquifers where uranium is suspected to be dissolved and reprecipitated in a cycle mode inducing a equilibration between the two phases. However, in very ancient groundwater assumed to be in chemical equilibrium with the rock-aquifer, dissolution/reprecipitation processes are neglectable. In permanent aquifers, an equilibrium state trend to be reached between water masses and rock-aquifer and water-rock interaction could also be neglected.

Several studies intended to apply the radioactive decay of dissolved ^{234}U for groundwater datation (Kronfeld and Adams 1974, Frohlich *et al.* 1984, Ivanovich *et al.* 1992, Cowart and Osmond 1974). In most cases, ages are widely above C-14 age (Rogojin et al. 1998). Ages could be corrected by taking into account the activity ratio of groundwater as well as of rock-aquifer (Osmond et Cowart 1992, Metcalfe et al. 1998). However, activity ratio of the rock is rarely available. Some models suggest to correct ages by including dispersion, radioactive decay and first order adsorption processes (Fröhlich and Gellermann 1987, Ivanovich et al. 1992, Andrews and Kay 1982, Andrews and Kay 1983, Gellermann et al, 1990). These models formulate the age by assuming an uniformity along the aquifer that could be weakly adapted to long periods covered by these processes as shown by Andrews et al. (1982) for groundwater percolating through Stripa granit in Sweden.

Due to the complexity of processes that have to be considered or measured, very few studies using uranium isotopes for groundwater datation could be reviewed. Most recent studies show that although processes have to be simplified and local measurements extrapolated for the aquifer, water ages could be reliable and contribute to refine the knowledge of the aquifer (Bonotto 2000, Bonotto 2006, Bonotto and Jiménez-Rueda, 2007).

4. ANALYTICAL METHODS

Several analytical chemistry techniques and methods are available for the determination of uranium activities depending on the matrices in which this element is found and on the level of its abundance. As seen, uranium concentrations in natural waters are usually to be found at the ppb to ppm levels. The majority of procedures used involve three main steps: a preconcentration or extraction; a purification scheme; and preparation of a source for radioactive counting. The preliminary extraction of the element from the host matrix can often be performed in a relatively simple procedure. However, the purification in order that specific radiometric counting may be performed, often leads to lengthy analytical procedures.

In the Laboratory of Radio-Analyses and Environment of the National School of Engineers of Sfax (Tunisia) where samples has been performed We used the protocol conceived and validated by Causse C. published in several works for different matrices

(Gaven 1975, Causse et Hillaire-Marcel 1989). This protocol has been modified in order to include new resins of uranium purification and new electrodeposition technology.

A know amount of spike is added to the sample in order to estimate the chemical yield. The spike is an artificial uranium isotope with well know activity. This spike is assumed to be conform to three conditions: i) to be different from natural radionuclides contained by the sample; ii) to emit alpha particles with an energy significantly different from those of the sample in order to be well differentiated in the counting spectra iii) to be stable during separation and purification protocol.

The spike used here is a Eckert and Ziegler Isotope Products standard solution of ^{232}U with an activity of 32.22 $Bq.g^{-1}$ at 01/07/2002). The alpha spectrum of the ^{232}U-spike (with a half-life of 69.85 years) is constituted of two alpha peaks with similar energy of around 5,32 MeV (68%) and around 5,27 (32%). The inconvenient of this spike is its ^{228}Th decay daughter which alpha energy emission (5340.54 keV) is too close to the ^{232}U one (5263.54-5320.30 keV) and could not be well distinguished according to the energy resolution of the PIPS detector used in alpha spectrometry. The alpha peak of ^{228}Th could be counted in the spectrum due to the presence of such element in the spike solution or due to the ^{232}U decay product in the sample. In this case, the activity if ^{228}Th could reach 3% of the activity of the spike one month after being added (Tome and Sanchez 1990). This contribution should be corrected from the uranium spectrum but need another spike, this correction should allow to gain 5% of precision in the calculation of the peak surface (Tomé and Sànchez 1990). We chose to proceed quickly to the source counting in order to avoid ^{228}Th product than to include another spike in the chemical protocol.

Sample volumes depend on uranium levels and can reach quite important values. Sample storage do not require particular precautions. However, waters are acidified immediately after sampling by nitric or hydrochloric acid to pH 2 in order to maintain trace elements in solution and to avoid biological proliferation. In fact, at pH=6-8 as for major natural waters, several metallic elements could form insoluble hydroxides that will precipitate due to the presence of particular matter. Waters samples should be filtered before acidification at 0.45 µm in order to eliminate particular matter and colloids (Grzymko et al. 2007, Reyes and Marques 2008). The amount filtered material will help decide whether it should be dissolved and added to the filtered water for total analysis or simply discarded. Given these simple precautions, water samples could be stored for several months after adding the spike (Smithson 1990).

In the case of low-uranium waters, large sample volumes need a preliminary treatment such as a preconcentration step after filtration. This preconcentration aims to reduce the sample volume to 500-600 ml without losses of existing elements (Singh and Wrenn 1988, Eichrom 1995). For Southern Tunisia waters, previous sampling campaigns shown that the optimal volume is about 20 liters given the uranium level in waters and the detection resolution of the alpha spectrometry.

The preliminary extraction of the uranium is made using a coprecipitation procedure what allow to reduce other coexisting element from the complex mixture. Iron hydroxide is commonly used for the isolation of uranium from acid aqueous solutions by the addition of carbonate-free ammonium hydroxide. It is essential that ammonia solution is free from carbonate as the uranium will form a carbonate complex which will remain in solution. In our protocol, the extraction consists on a coprecipitation with a $FeCl_3$ solution using ammoniac solution at pH 8, made on an acidified sample. The amount $FeCl_3$ added to the sample is about 1 mg but could be increases according to the sample quantity. However, this amount must be limited in order to avoid the degradation of the chemical yield of the following step (purification) by saturation. The chemical yield of this coprecipitation is higher in nitric acid (90~100%) than in hydrochloric acid (~50%) (Rodriguez-Alvarez and Sãnchez 1995). Various parameters are prone to reduce the chemical yield of the coprecipitation such as the presence of organic substances or carbonates (Hoshi and Hueno 1978, Radenkovic *et al.* 1996). Therefore, the sample is first boiled to remove dissolved carbonates and then burned in a mixture of nitric, hydrofluoric and perchloric acid to remove organic substances. The coprecipitation is a slow reaction and it is usual to allow several hours for its equilibration. The precipitate is recuperated by centrifugation and the extraction is repeated on the supernate in order to increase the chemical yield of this step what is important in low-uranium waters. The ferric hydroxide precipite, containing the uranium, is dissolved with 8 M HCl for the following step of the protocol.

The technique of liquid-liquid extraction has been widely applied to the separation and extraction of uranium from a wide range of accompanying elements by using organic acids such as ketones, ethers, esters, alcohols and organic derivatives of phosphoric acid. Nowadays, most chemical procedures make use of cation exchange resins for the separation of uranium from other elements including the actinides. Several studies had shown that ion exchange resins used for extraction/purification of uranium solution allowed to enhance chemical yield (Oliveira and Carvalho 2006, Tsai *et al.* 2008, Kyriakou and Pashalidis 2010). The purification step of our protocol is based on two ion exchange resins. The 8M HCl solution containing uranium obtained after the extraction step is first passed through an ion exchange resin column DOWEX AG-1X8 (200-400 mesh) of BIORAD. This resin had been washed by successive alternated passage of hydrochloric 8N and distilled water. According to distribution coefficient of this resin, the solution should cross the column with a flux of 10 ml.h^{-1} (Ghaleb 1988) in order to allow the UO_2Cl^- anionic complex to be retained The elution is obtained by a low normality hydrochloric solution (<0.1N). The resulting solution was evaporated and was taken back in a 3M HNO_3, then the solution was passed through a second ion exchange using UTEVA [Uranium and Tetravalent Actinide] resin (Eichrom, 1995). It is able to separate and concentrate uranium U(VI) from tetravalent actinides in aqueous solutions. After being washed with 3 M HNO3 and 0.01 M HCl, uranium is eluted from the column.

In order to ensure the best counting efficiency by alpha spectrometry, uranium has to be deposit onto a thin, flat, uniform deposit for a counting source. Ideally, the source would have a monatomic layer of the alpha emitter. Electrodeposition is well established as a reliable technique for the preparation of standard alpha source preparation, commonly made on stainless steel discs. The electrodeposition can be carried out from organic or aqueous solutions (Trautmann and Folger 1989, Zantuti et al. 1991). Many procedures are described in the literature (Lally and Glover 1984, Tome and Sanchez 1991, Sibbens and Altzitzoglou 2007, Aggarwal et al. 1991, Küppers 1998). Direct deposit is the easier technique and produces thin sources with limited energy degradation (Gaven 1975, Blanco Rodrígueza et al 1997). However, the deposit is irregular form and thickness what reduce the counting efficiency.

Nowadays, electrodéposition is most reliable technique. A cell is used in which the metal disc is the cathode with only one side exposed to the electrolyte. The anode is always platinum.

Most procedures for deposition use aqueous solutions in either slightly alkaline or acidic form. Uranium can be deposited as a hydrous oxide or as UF_4, depending on the electrolyte used. The electrodeposition yield, meant as the energy resolution allowed to be reached by the thin source (Sanchez et al. 1992), depends on several parameters such as courant voltage, distance between cathode and anode, duration, volume of solution (Hallstadius 1984, Tomé and Sanchez 1991). Electrodeposition had been carried out on a weakly acidic solution according to the methodology of Ghaleb (1990) with a current at 400 mA for a minimum duration of 2 hours. The distance between electrodes is about 5 mm. This parameter set alllows an electrodeposition yield of 60-85% without addition of a complexing agent. The electrodeposition is made on a SDEC-EDP 9000 equipment.

Alpha sources have been counted by and alpha-spectrometer calibrated with a known standard sources of natural radionuclides. The standard source (Eckert and Ziegler Analytics, SRC 7400) used is a mixture of alpha emitters: ^{238}U (99.4 dpm), ^{234}U (97.8 dpm), ^{239}Pu (93.1 dpm), ^{241}Am (83.6 dpm). This source is electrodeposit onto a stainless steel disc of 24.1 mm of diameter and 0.65 mm thickness.

Radioactive decay processes are random; they are described mathematically by an exponential decay function derived from probability theory (Laedermann, 2003). In alpha spectrometry, activity is detected through a number of disintegrations during a given time which frequency of occurrence could be approximated by a Poison distribution. The standard deviation for the Poisson distribution has a definite value in terms of the mean value, *n,* that is to say, the 'plus-or-minus' error incurred in measuring *n* disintegrations is simply equal to if *n* is large. Uranium sources had been counted at least for 2000 counts (disintegrations) for each isotope in order to ensure a relative measurement precision of about 2%.

5. Case Study: North-Wertern Sahara Aquifer System (North Africa)

5.1. Site Description

This chapter describes the concentrations and distributions of uranium isotopes (^{234}U and ^{238}U) in groundwater from the North-Western Sahara Aquifer System (NWSAS) of Northern Africa. This transboundary system, one of the greatest in the world, covers an area of over 1 million Km^2 shared by Algeria, Libya and Tunisia and lodges huge water reserves which are, however, mostly non-renewable and not fully exploitable (Figure 6).

The Saharan basin is a large multilayered deposit entity, the medium and long-term hydrodynamic behaviour of the system if highy influenced by hydraulic and chemical links and by exchanges among different aquifers. Previous geological and hydrogeological studies enabled to simplify the NWSAS onto two superposed aquifers that are the Continental Intercalaire (CI) and the Complexe Terminal (CT), separated by or linked through semiperpermeable layers. These two aquifers are linked through tectonical features to a third one, of less extension and reserves, the coastal Djeffarra basin.

Given the arid to saharian climate conditions, these formations are slightly recharged: the system recharge represent about 1 billion m3/year, mainly percolating in the Sahara Atlas piedmonts plains in Algeria, the Dahar in Tunisia and Djebel Nefoussa in Libya (Figure 7).

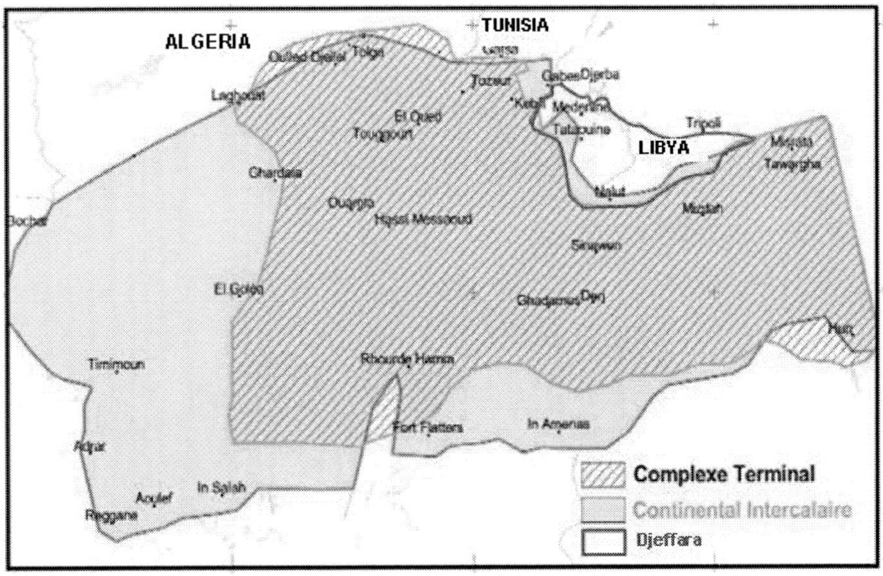

Figure 6. Extension of the NWSAS (from OSS, 2003).

However, the system extension and the layer thickness have facilitated the accumulation of important reserves over past centuries (OSS, 2003).

The Continental Intercalaire (CI) aquifer underlies all the Bas-Sahara in an arid region where rainfall amounts average around 100 mm/y.

This aquifer is lodged in lower Cretaceous continental formations (Neocomian, Barremian, Aptian and Albian) which thickness and lithology show significant lateral variation (ERESS, 1972; Gonfiantini et al., 1974; Aranyossy and Mamou, 1986; Mamou, 1990; Edmunds et al., 1997; Gries, 2000). The aquifer is however continuous over the whole basin from north to south from the Saharan Atlas to the Tassili mountains of the Hoggar, and west to east from western Algeria to the Libyan Desert. A major N–S structure, the M'zab dorsal (Cornet 1964, ERESS 1972, Bishop 1975) divides the basin into two subbasins (occidental and oriental).

The studied area is situated in the oriental basin (Great Oriental Erg), which crosses Algeria and Tunisia. It's limited on the north by the Saharian Atlas Mountains (900 to 1100 m), on the west by the M'zab dorsal, on the south by the Tinhert plateau, and on the east by the uplands of the Dahar Mountain (500–700 m). The aquifer roof becomes deeper from south to north. The minimum depth (0–500 m) is observed along the occidental border of the Dahar Mountain and near the Tademaït plateau while the maximum (1500–3500 m depth) is mainly located in the region of the endoreïc closed depressions of Chott Melrhir (Algeria) and Chott Djerid (Tunisia), giving rise to a high artesian pressure. Principal geological features in Algeria and Tunisia show important spatial variations. In the centre of the basin, the aquifer is hosted in purely continental formations; while first lacustrine and marine facies appear towards the borders, especially in the east. The aquifer thickness varies between 100 m in the recharge area in the western part of the basin region up to more than 1500 m in the Djérid region. A high degree of heterogeneity is observed in the heart of the CI aquifer due to the succession of different detrital sediments formations separated by clay-rich strata.

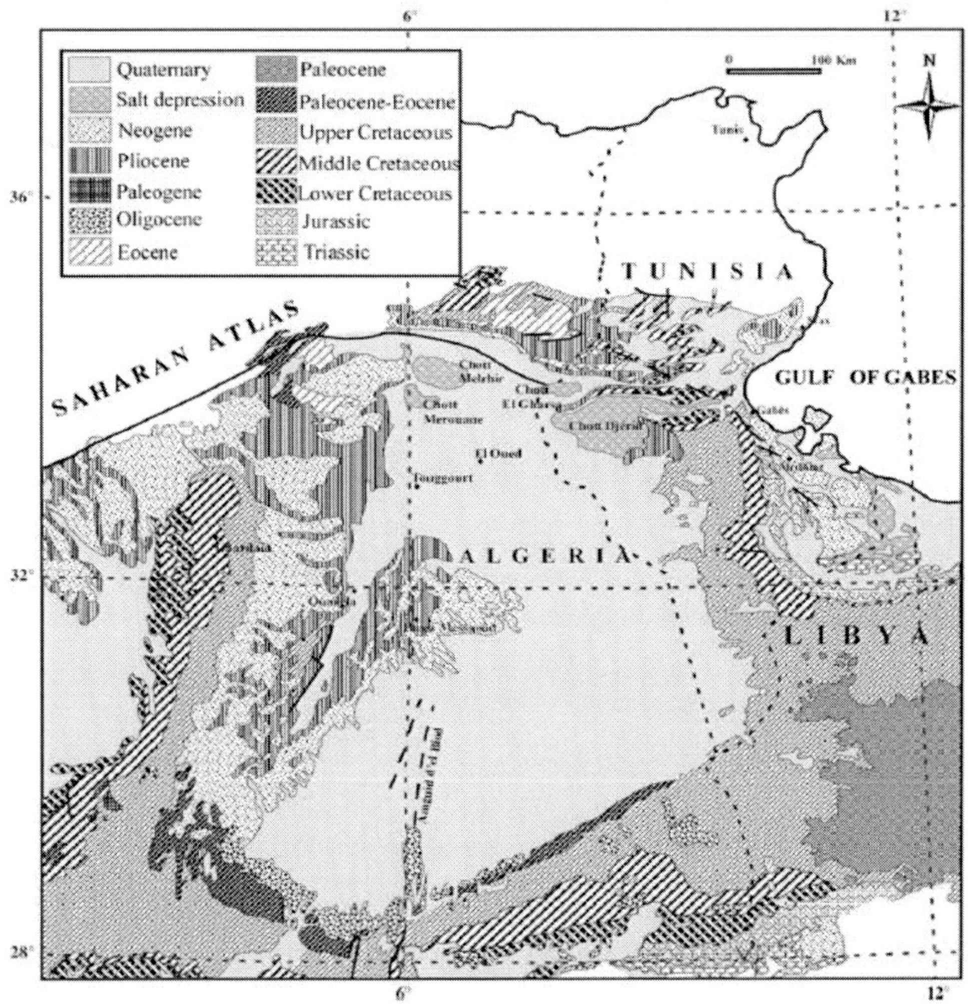

Figure 7. Location and geological features of the Saharan sedimentary basin in North Africa with NWSAS extensions (OSS, 2003).

Some thin but relatively permeable horizons may be very important in controlling groundwater flow and hydraulic continuity. This phenomenon is particularly marked where the Saharan platform passes to the lower-Sahara region in the south: thickening of the units and lithological changes give rise to changes in piezometry and chemical composition of water. Tectonic activities and thus faulting has played a major role in the compartmentalisation of the southern (El Amguid fault) and in the eastern (Elhamma fault) parts of the CI aquifer (ERESS, 1972; Edmunds et al., 1997).

Groundwater flows in the CI of the Great Oriental Erg converge towards a single discharge zone in Tunisia (Chott Djérid/Chott Fejej region and Gulf of Gabès). The main W-E flow path, comes from the Saharan Atlas eastward the Chott Djérid and the gulf of Gabès in Tunisia. Two secondary flow directions may be distinguished: SWNE from the M'zab dorsal southwestern Algeria eastward the chott Fejej/Gulf of Gabès in Tunisia and S–N from the Dahar region of Tunisia (Figure 3). Hence, two major regions have to be distinguished in the Tunisian basin of the CI aquifer: In the Djérid region, the aquifer is mainly lodged in continental formations extending from the Neocomian at the base to the Albian and the hydraulic continuity seems to be only ensured by the west-east flow component across the Algerian–Tunisian border (Mamou 1990, Edmunds et al. 1997). On the other hand, in the Nefzaou–Chott Fejej region, the aquifer is hosted in thick deltaic terrigenous sediments (Mamou, 1990) and groundwater flows along a north-north east line coming from a recharge (or paleorecharge) area located in the M'zab dorsal southwestern Algeria and in the plateau of Tinhert in southern Algeria and a second recharge area located in the Dahar uplands in southern Tunisia (Edmunds et al., 1997).

The Complexe Terminal aquifer overlies the Continental Intercalaire aquifer (Figure 7). The depth of the latter aquifer varies between 70 and 250 m in the Nefzaoua region and between 80 and 700 m in the Djerid region. The main outflow direction is south–southeast to north–northwest. The north–northwest discharge zone contains artesian is mainly represented springs that are mostsly dried nowadays. The CT aquifer is hosted in the Upper Cretaceous and Tertiary formations. The CT formations are relatively heterogeneous and are composed of three main aquifer horizons separated by semi-permeable to impermeable strata. In Tunisia, the main productive levels are located either in the carbonated levels of the Upper Cenomanian (Gafsa, west-northern part of the aquifer) and Upper Senonian (Kebili, east-south part of the aquifer) or in the Tertiary sandy formation (Tozeur, Western part of the aquifer). The Cenomanian-Turonian formation located in Tozeur is constituted by two distinct layers: at the base, dolomite and chalky dolomite, with a thickness varying between 120 and 150 m, are covered at the top by a layer of marl, clay, gypsum intercalations with thin levels of limestone. The thickness of the upper layer of this formation varies between 300 and 400 m. The Senonian formation also located in Tozeur region is constituted by the Upper Senonian, formed by flagwhite limestone and chalky marl levels reaching 100 to 200-m thickness and by the lower Senonian, formed by limestone and gypsum intercalations and reaching a thickness of 500 m in Hazoua region.

The Mio-Pliocene formation is constituted by coarse sands with thin levels of red clay and quartz. Its thickness ranges from 250 to 300 m. Transmissivity of the CT aquifer, obtained from pumping tests, ranges from $0.1\ 10^{-2}$ to $4.5\ 10^{-2}$ m^2/s. Most of the recharge of CT aquifer occurs at outcrops around the margins of the Grand Erg Oriental (south-western part of the aquifer in Algeria), some minor recharge is also ensured by direct infiltration through sand dunes of the Grand Erg Oriental. Discharge from the aquifer occurs by vertical percolation at the sabkha surfaces, upwelling at natural springs and spring mounds, and upwelling from water wells. Most of this discharge is concentrated in southern Tunisia around Chott Djerid, where natural upwelling occurs at a relatively constant rate throughout the year.

The Jeffara of Gabes is located in the southeastern Tunisia extending along the Mediterranean Sea. This zone is also characterized by an arid/semi-arid climate strongly influenced by the sea. Meteorological records indicate that interannual rainfall average is about 182 mm, and that mean monthly pan evaporation ranges from 40 to 124 mm. The mean annual temperature is of the order of 20°C varying between 6.7°C and 33.2°C.

The study area is characterized by outcrops ranging from Lower Cretaceous to quaternary units and by lithostratigraphic lateral variations. In the northern part of the zone, the geological outcrops consist in Mio-Plio-Quaternary formations overlying in discordance lower cretaceous formations constituted by continental deposits with intercalations of conglomeratic and sandy levels. While, in the Southern area, Cretaceous outcrops are dominating. Albo-aptian, Cenomanian, Turonian and Senonian formations outcrops at

Hallouga, Raggouba and Matmata Jebel and are characterized by big variations of lithologic facies. Albo-aptian units are intercalations of marno-chalks and sandstones. Cenomanian formations consist in an alternation of chalky dolomite, gypsum, chalky clay and sandstony to marly intercalations. Turonian formations are everywhere dolomites to chalky dolomites, whereas the Senonian is marked essentially by gypsy clay and carbonated formations. Moreover, the Jeffara basin is strongly affected by a complex system of NE/SW faults crossed by NW/SE ones that provoked a lateral compartmentalisation and connections between different aquifereous horizons. This is the case in the El Hamma Chenchou where a hydraulic continuity between the Continental Intercalaire and the Jeffara aquifers has been recognized. The Jeffara aquifer system is characterized by large changes in facies and thickness. In the north of Gabes, Jeffara waters are mainly lodged in the Miocene continental deposits with clays and sandy intercalation. The thickness of these sand deposits decreases from the north to the south of the regionPermeability of the Miocene horizon ranges from $2\ 10^{-5}$ to $3\ 10^{-4}$ m/s while transmissivities obtained from pumping tests, ranges from 3 to 33.8 106 3 m^2/s (Mekrazi, 1975). The principal aquifer of Gabes is hosted by the Senonian horizon represented by two stratigraphic units. The first one constitutes a limestone horizon with a thickness of approximately 300 m, and the second is formed by marly limestones characterized by variable thickness up to 500m. These water-bearing formations are strongly fissured with transmissivity ranging from $0.9\ 10^{-3}$ to $100\ 10^{-3}$ m^2/s. The Turonian aquifer, lodged in dolomites and fissured limestones which thickness does not exceed 50 m, constitutes the principal aquifer in Matmata regions. Transmissivities of the Turonian horizon vary according to the degrees of tectonic fracturing. Many studies had been concerned by hydraulic continuity between the Jeffara and Continental Intercalaire aquifers (CI). This latter is one of the largest confined aquifers in the world, considered as a typical case of large artesian basins. Groundwater reservoir of the CI is contained in the continental formations of Lower Cretaceous, formed by alternations of detrital levels with clayey silts horizons and gypsum intercalations. Thickness and lithology are highly laterally variable.

In the CI, Groundwater flow toward its discharge zone located in the Chott El Fejej region and toward the Jeffara system in El Hamma region trough El Hamma fault. According to piezometric levels, the Jeffara aquifer shows a general South-Western/North-Eastern flow direction, from El Hamma region and Matmata relief toward the Mediterranean Sea. This indicates that the natural discharge is located in the gulf of Gabes while the recharge area is located in the Tunisian Dahar in southeastern Tunisia. The region of El Hamma is characterized by hydraulic head around 40 m indicating the discharge of Continental Intercalaire groundwater through El Hamma faults.

5.2. Results and Discussion

For all samples, physical parameters (T, pH, conductivity) were measured in situ. For uranium analyses, a total of 20 liters have sampled and analysed according to the analytical protocol presented above. Data for the three aquifers are presented in the Table 1 with U-content, $^{234}U/^{238}U$ activity ratio, capture depth, salinity and physical data (pH, temperature) for all samples. Uranium fingerprints are summarized in Table 2.

Table 1. Chemical and uranium data of NWSAS groundwater

N°	Name	U ppb (2σ)	$^{234}U/^{238}U$ (2σ)	Depth (m)	RS ($mg.l^{-1}$)	T(°C)	pH
Continental Intercalaire Aquifer							
1	Berriane2	2.87± 0.01	3.19 ± 0.02	545	964	28.6	7.56
2	CedadaCI	0.025±0.003	11.02± 0.86	2371	2260	50	7.67
3	CFF3	2.354±0.238	1.65± 0.231	950	2770	63.5	6.67
4	CFF9	1.539±0.161	1.71± 0.245	886	3420	63.5	6.72
5	ChorfaCI4	0.056±0.005	7.84± 0.79	2190	2210	59.5	6.8
6	DebabchaCI4	0.054±0.005	8.84± 0.94	2295	2030	68.1	6.7
7	Djamaˆa	0.031±0.001	0.4± 0.22	1755	1719	46.5	7.82
8	ElHammaCI1Bis	0.12±0.001	5.7± 0.67	1550	2750	69.8	7.42
9	ElHammaCI4	0.015±0.005	8.37± 1.33	1475	2720	69.5	7.4
10	El-Oued (Berry)	0.52±0.03	1.496± 0.086	850	2449	35.9	7.8
11	Gassi-Touil	0.034±0.001	2.53± 0.09	955	4064	43.9	7.26
12	Ghardaia (Bouhra)	2.86±0.3	3.39± 0.169	390	1618	29.1	7.23
13	Ghardaia (Bend)	2.89±0.14	3.66± 0.02	467	1534	31.6	7.41
14	Hadjira	1.78±0.014	2.05± 0.02	1895	1767	50.8	7.6
15	Hassi Mess.	0.034±0.001	2.53± 0.09	1300	1687	52.8	7.64
16	HazouaCI	0.008±0.002	9.92± 1.37	2059	1800	59.1	7.24
17	KebiliCI6	0.483±0.018	3.97± 0.16	2774	2220	63.7	7.3
18	Laghouat (Assafia)	2.99±0.13	4.26±0.087	100	1181	20.6	7.32
19	LimagueusCI8	0.259±0.102	4.89± 0.2	1730	2420	67	7.0
20	MahacenCI1	0.089±0.001	5.55± 0.76	2557	2240	63.8	7.31
21	MansouraCI3	0.044±0.003	7.94± 0.58	2557	3000	62.9	7.3
22	NeftaCI2	0.006±0.001	15.38± 1.32	2584	2670	66.7	7.31
23	NeftaCI3	0.013±0.003	12.94± 1.27	2190	2820	57.9	7.68
24	Ouargla (Hadeb)	0.22±0.003	1.75± 0.029	1450	1745	48.5	7.58
25	Rhourde El Baguel	0.01±0.0005	3.17± 0.19	995	2713	40	7.42
26	SeftimiCI7	0.568±0.016	3.19± 0.1	1786	2440	67.6	6.9
27	TazraritCI	0.017±0.001	4.4± 0.38	2186	4120	52.4	7.38

Table 1. (Continued)

N°	Name	U ppb (2σ□)	$^{234}U/^{238}U$ (2σ)	Depth (m)	RS ($mg.l^{-1}$)	T(°C)	pH
28	TozeurCI4	0.009±0.002	7.9± 1.03	1981	4880	54.4	7.44
29	Zelfana2	3.39±0.019	2.71± 0.018	1000	1631	37.8	7.59
Complexe Terminal Aquifer							
1	HazouaCT1inter	2.10±0.25	2.53±0.38	528	2620	28.8	6.99
2	CedadaCT8	1.35±0.22	2.31±0.24	590	1600	33.6	7.32
3	TozeurCT2	1.56±0.21	2.29±0.19	622	1830	29.6	7.22
4	TelmineCT	2.40±0.19	1.93±0.018	201	2210	23	7.16
5	JemnaCT	1.60±0.16	1.67±0.12	118	1300	20	7.46
Djeffara Aquifer							
5	ICM1	3.44±0.38	1.63±0.25	149	3310	30.00	6.64
11	Sidi Boulbaba	2.95±0.32	1.73±0.26	250	2890	26.00	7.02
15	Kettana SONEDE	2.53±0.26	1.71±0.24	240	2580	22.50	7.28
17	Dahret Ettiour	2.54±0.27	1.79±0.27	130	2600	23.60	7.69
22	Matmata Nlle 2	2.68±0.27	1.63±0.23	258	2970	24.70	7.34
34	Zemlet el mharis	2.39±0.25	1.42±0.21	266	2600	24.10	7.34
24	Zouitinet	1.34±0.17	1.70±0.29	130	3100	16.80	7.77
14	Mnara 1	1.48±0.16	1.51±0.23	142	2.770	24.80	6.91
16	Oued Lihoudi	1.51±0.16	1.72±0.25	154	2770	23.90	6.82
18	Mareth	1.35±0.14	1.69±0.25	54	3900	22.40	7.37
21	Oued Taoujout	1.62±0.17	1.82±0.27	220	3070	26.00	7.35
25	Chenchou 14	2.43±0.24	1.71±0.23	60	3540	37.80	6.63
27	El Hamma SONEDE	1.68±0.16	1.74±0.23	152	3600	33.70	7.33
3	Ouedref SONEDE	1.75±0.18	1.79±0.25	66	3470	23.00	6.97
2	Oued El Maleh	1.89±0.19	1.62±0.23	122	4420	23.00	7.06
6	ICM2bis	2.38±0.21	1.66±0.21	90	3270	30.00	6.90
19	Béni Zelten	1.79±0.18	1.72±0.23	81	2130	22.40	7.43

Table 2. Variation ranges of uranium fingerprints of NWSNS groundwater

	U-content (ppb)			$^{234}U/^{238}U$ activity ratio		
	Range	Mean	Standard deviation	Range	Mean	Standard deviation
Continental Intercalaire aquifer	0.006-3.39	0.803	1.173	0.4-15.38	5.253	3.731
Complexe Terminal aquier	1.35-2.4	1.802	0.433	1.669-2.525	2.146	0.342
Djeffara aquifer	1.34-3.44	2.103	0.615	1.42-1.82	1.682	0.100

For the Continental Intercalaire aquifer, groundwaters were collected from boreholes along main flow paths over the Algerian and the Tunisian basins (Figure 7). The interval from which water is derived, are between 100 m in the recharge area to more than 2500 m in the deepest part of the aquifer in the area of the Tunisian chotts. Groundwater temperature increases from less than 20°C in the recharge area to above 70°C in the deep part of the basin.

The conductivity, as a measure of the total mineralization, increases from around 1500 to 6000 $mS.cm^{-1}$ across the aquifer but is influenced by different sources of salinity and different redox conditions in the aquifer (Edmunds et al., 1997; Guendouz, 1985). For the Complexe Terminal aquifer, groundwaters have been collected from five (5) boreholes along the Chott Djérid (hypersaline depression). Waters are sampled from depth ranging from 100 to 620 m with temperatures varying between 20°C and 34°C. Salinities are less than those of the Continental Intercalaire waters, ranging between 1.3 and 2.6 g/l bu could not be explained by lithofacies variations.

For the Djeffara aquifer, twenty seven samples were collected basin. Fourteen samples were collected from the Senonian aquifer horizon; three samples were collected from the Miopliocene aquifer horizon and only one sample from the Turonian aquifer horizon. The Jeffara waters are characterized by temperature lower than 30°C and therefore are qualified as hypothermal waters as well as for the Complexe Terminal aquifer (Ben Dhia, 1983). High temperatures observed at El Hamma and Chenchou areas (about 38°C) have been explained by the upwelling of the Continental Intercalaire hot water through the ElHamma fault that represent in this area the dominant origin of water recharge.

Uranium contents and uranium isotopes activity ratios cover large variation range, respectively 0.006-3.39 and 0.4-15.38, for the Continental Intercalaire in relation with the large extension of the concerned area. Forr the Complexe Terminal and the Djeffara, values have the same magnitude and are less scattered (Figure 8).

According to uranium concentrations and $^{234}U/^{238}U$ activity ratios (Cowart and Osmond 1980, Osmond and Cowart 1983, Toulhoat et al. 1988, Bonotto 1999) groundwater could be classified on different types: U-content values allow to classify aquifers in oxidized aquifer on ''normal'' U-content strata (values between 1 and 10 ppb), oxidized aquifer on enhanced U-content strata (values higher than10 ppb) and reduced aquifer on low U-content strata (values lower than 1 ppb) whilst $^{234}U/^{238}U$ activity ratios distinguish between ''normal'' world-wide groundwater (values between 1 and 2); possible U-accumulation (values higher than 2), or possible remobilization of a uranium (values lower than 1). High activity ratios (>2) can be due to a higher than normal ratio of leachable uranium in aquifer or amorphous uraninite, enhancing α-recoil. An activity ratio lower than 1 implies intense dissolution. Uranium concentration levels reflect both redox conditions and uranium concentrations in surrounding rocks (Cowart and Osmond 1980, Bonotto and Andrews 2000, Roback et al. 2001). In oxidizing conditions, uranium is very soluble as U(VI)-complexes, mainly carbonates. In reducing conditions, uranium is insoluble as U(IV)-oxides.

In order to highlight the different processes in the aquifers, data are interpreted using the diagram relating reciprocal U-content vs the $^{234}U/^{238}U$ AR (Osmond and Cowart 1983 or 1992). In this diagram, $^{234}U/^{238}U$ ARs plotted against reciprocal U concentrations on the mixing diagrams exhibit a linear trend extending from the lower left (low AR with high concentrations) to upper right (high AR with low concentrations). This can be interpreted as mixing lines (Dabous and Osmond, 2001; Chkir and Zouari, 2007); at one end is very dilute aquifer water, which has experienced little leaching of U (upper right), and at the other end member is the concentrated leachate of the host rock (lower left).

The dispersion of points between these two end members indicates water with various mixture ratios. The trend line is defined by two parameters: the y-intercept, which is the $^{234}U/^{238}U$ AR of the U being leached from the host rock, and the slope of the trend line, which is the ^{234}U-excess per kilogram (U-equivalent ppb) and is characteristic of the pre-leach water. ^{234}U-excess corresponds to the hypothetical concentration that the ^{234}U isotope has in excess to the ^{238}U concentration and has no real physical meaning as the actual mass of the ^{234}U isotope is 18 000 times less that the ^{238}U mass (Cizdziel et al., 2005). The unit for ^{234}U-excess is U-equivalent parts per billion, which is the concentration in micrograms per kilogram of an equilibrated amount of ^{238}U (^{234}U activity per mass of water) (Dabous and Osmond 2002). However, the interpretation of the data has to be supported by the hydrogeological and hydrochemical characteristics of aquifers.

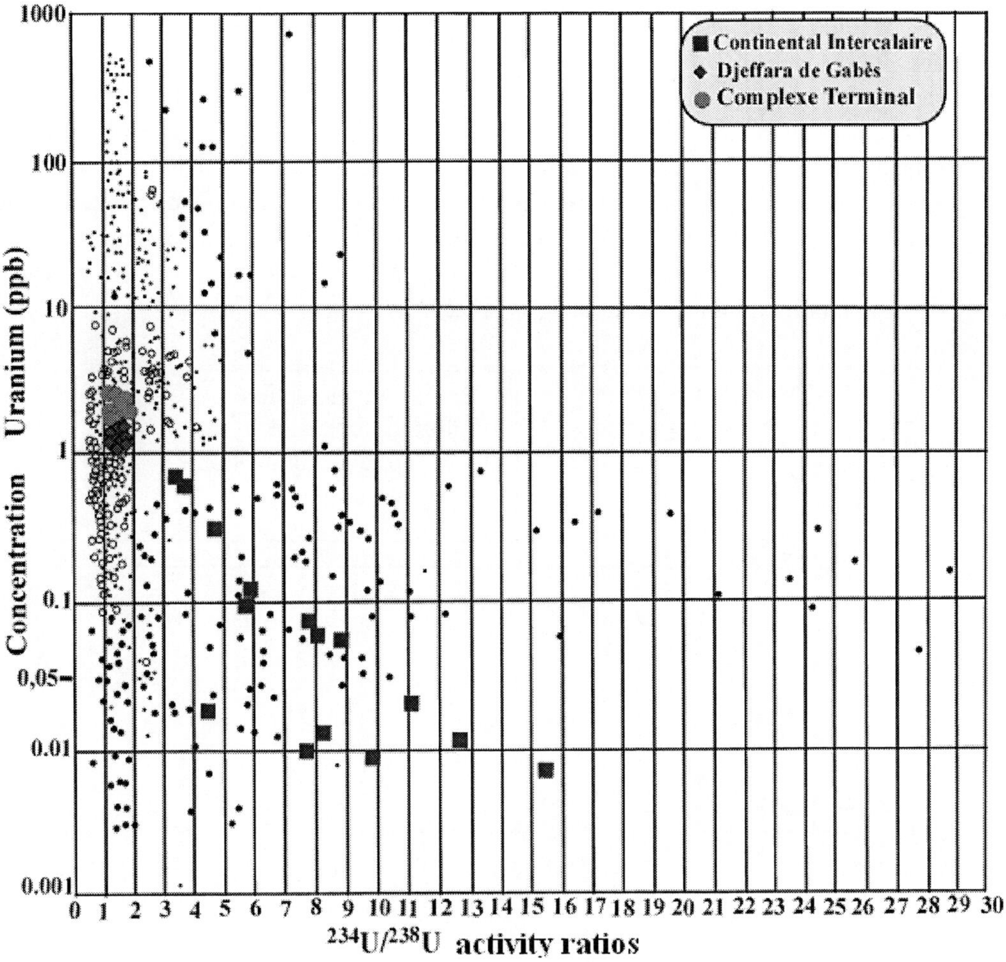

Figure 8. Variation range of uranium concentrations and $^{234}U/^{238}U$ activity ratios in worldwide groundwater (from Ivanovich and Harmon, 1992) on which Continental Intercalaire (Chkir et al. 2009), Complexe Terminal (Chkir and Zouari 2007) and Jeffara (Hadj Ammar et al. 2010) groundwater have been added.

The Continental Intercalaire Aquifer

According to the (U-content, $^{234}U/^{238}U$ AR) classification, the U content and AR measured in the CI aquifer are greatly variable, as shown in Figure 9 where samples are fitted in practically all fields.

However, along the main flow crossing the basin from west (Saharan Atlas in Algeria) to east (Djérid region in Tunisia), the CI aquifer can be shared in an upstream oxidized compartment and a downstream reduced compartment (Figure 9). The spatial distribution of groundwater data reveals that the oxidized portion, with relatively high uranium concentration (1.78–3.39 ppb) and moderate $^{234}U/^{238}U$ activity ratios (2.05–4.27) is restricted on the northwestern area of the survey zone. It is limited by a boundary located between 4°300E and 5°E of longitude and corresponding to the redox front previously identified by spatial relationships of several redox sensitive elements such as Fe, Mn or Cr (Guendouz, 1985; Edmunds et al., 1997). This compartment can be considered as a ''steady-state'' aquifer: oxidized aquifer on ''normal'' U-content strata (U-contents values ranging between 1 and 10 ppb) with possible U-accumulation processes (ARs >2) but not necessarily of economic size and grade. In this area, $^{234}U/^{238}U$ ARs are broadly decreasing along the main W-E flow line from recharge area situated on the southern side of Saharian Atlas toward the redox boundary. The only anomaly observed is for the ''Hadjira'' sample but it could be explained by a deeper capture depth (1900 m) compared with other samples of this upstream compartment for which sampling depth is increasing from the recharge area toward the redox boundary appearing at appreciatively 1000 m depth. The distribution of uranium content and uranium isotopic disequilibrium in the reduced part of the aquifer is less homogeneous and is highly influenced by lithostratigraphic conditions (Chkir and Zouari, 2007).

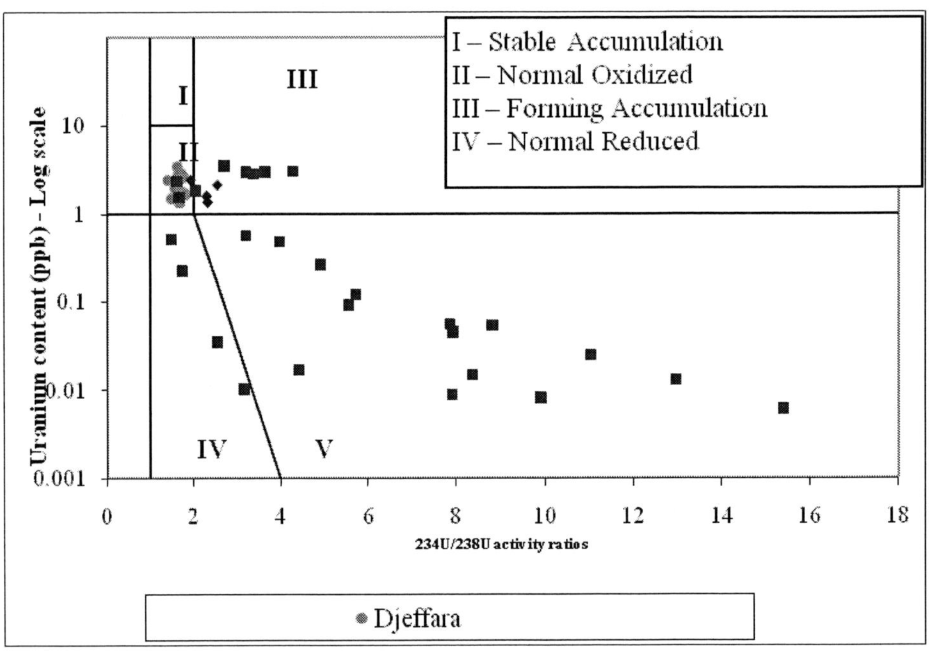

Figure 9. NWSAS groundwater plotted on the U concentration vs $^{234}U/^{238}U$ activity ratio diagram proposed by Cowart and Osmond (1980).

Uranium concentrations range between 0.006 and 0.52 ppb while $^{234}U/^{238}U$ ARs range between 1.49 and 15.38 in the Djerid region. Hence, this second part of the CI aquifer in the Djerid region can be defined as a reduced aquifer with possible remobilization of uranium or enhanced α-recoil processes.

Uranium concentrations remain very low along the main flow line while $^{234}U/^{238}U$ ARs reach very high values (2 or 3 magnitude of those of the upstream oxidized part of the aquifer). Such high $^{234}U/^{238}U$ ARs have already be found in other flow systems. In a slow-moving relatively stable system, high uranium concentrations are seldom found, this statement emphasizes that once precipitated at the barrier front of a permanent aquifer, uranium remains in place in the solid state for a long term (Osmond and Cowart 1992), the same is not true for ^{234}U.

If conditions at the front are near chemical equilibrium, the effects of decay and recoil may be enough to cause preferential mobilization of the daughter.

Therefore, high $^{234}U/^{238}U$ARs may be found in the groundwater near such stable front and also for some distance downflow. The size, climate, lithology, stratigraphy, hydrogeology, geochemical conditions, and extent of rock–water interactions are also factors responsible of the high variability of AR values in groundwater (Bonotto and Andrews 2000, Bonotto 1999). Moreover, alternative mechanisms have been suggested for the generation of enhanced $^{234}U/^{238}U$ AR ratios including preferential chemical dissolution of ^{234}U (Rosholt et al., 1963) and daughter recoil into solution by the alpha-recoil at the rock–water interface (Kigoshi, 1971). These mechanisms are favoured (Osmond and Cowart 1992, Ivanovich et al., 1991) by long period of interaction between water of low dissolved uranium contents and hosting formations of relatively higher uranium concentrations. Several studies have reported activity ratios reaching 12 in large confined aquifers with steady long term system such as 9.0 for deeper and older shallow groundwater of the Transvaal Dolomite Aquifer in South Africa (Kronfeld et al. 1994), 10.1 for groundwater sampled from a deep well in Ramallah (Kronfeld et al. 1975), 10.5 in confined aquifers in the Cambrian and Ordovician bedrock in the tri-state region of Missouri, Kansas and Oklahoma (Cowart and Osmond 1980), 11.8 at a redox front of the Milk River aquifer (Ivanovich et al. 1991) and 12.3 in a confined sandstone aquifer in Texas (Kronfeld 1974).

Activity ratios as high as 28 have also been observed in the brazilian part of the Guarani aquifer (Bonotto 2006) and have been explained by the large extension of the system enhancing the three main factors responsible of uranium isotopes disequilibria (rock–water interactions, preferential chemical dissolution and alpharecoil release of ^{234}U).

On the other hand, $^{234}U/^{238}U$ activity ratios for groundwater samples collected from the Cambrian and Ordovician (USA) range from 2.0 in the central part of the aquifers to 40.7 in the south. The lowest ratios occur in primary recharge zones in regions where the Cambrian and Ordovician are unconfined.

Activity ratios increase along flow path and with residence time of groundwater and ratios ranging from 20 to 25 reflect the location of a zone of groundwater discharge with a component coming from regions where the Cambrian and Ordovician is confined (Gilkerson et al. 1984). The highest activity ratios (values from 25 to 40) occur in the eastern part of the Cambrian and Ordovician aquifers where groundwater would have been recharged during the Wisconsinan. So high activity ratios are not unique since $^{234}U/^{238}U$ ARs reported for the continental intercalaire aquifer of the Irhazer Plain in Niger (Andrews et al. 1994) increase by preferential ^{234}U from 1.3–3.4 in the outcrop zone to 42 in the south-western confined zone close to U deposition zone or just downstream of it.

The two later cases are quite uncommon and reflect local and specific geological settings. Nowadays, 28 is a maximum value of uranium isotopes disequilibria that could be explained by hydrogeochemical settings of flow systems.

In the reduced part of the CI aquifer, along the main W-E flow line, the reducing conditions become dominant after redox boundary inducing an U-precipitation and groundwater with low dissolved uranium contents. However, even if uranium chemistry in this area remains quite stable, the increase of temperature (up to 70°C) are favourable to promote U dissolution at the greater depths of the CI aquifer. Since the α-recoil effect enhances ^{234}U dissolution, it may be thought that the long residence time of groundwater in this area leads to the highest uranium activity ratios observed downflow the aquifer.

The same data treatment has been applied along a secondary SW-NE flow line from the M'zab dorsal to the Nefzaou/chott Fejej area. The very low uranium concentrations are typically indicating reducing conditions in the confined aquifer beneath the Oriental Grand Erg in southern Tunisia and Algeria. Higher concentrations in the discharge area in the Nefzaoua/Chott Fejej region could be related to different redox conditions or complexation.

For the Continental Intercalaire aquifer, two mixing lines flowing from very dilute aquifer water with low dissolved U (upper right), to concentrated leachate of the host rock (lower left) appear to characterize the secondary SW-NE flow line groundwater (Figure 10). These two lines have different parameters but converge toward two samples from the Chott Fejej discharge area indicating that groundwater in this area have different origins. The first line extends from the M'zab dorsal recharge or paleorecharge area with a ^{234}U-excess of about 0.015 (U-equivalent ppb) and the AR of the host rock U is 1.82.

The second mixing line seems to be a more local flow line from the Nefzaoua area toward the same discharge area of the Chott Fejej. This line is also characteristic of a dissolution process with a ^{234}U-excess of about 0.27 (U-equivalent ppb) and the AR of the host rock U is 2.57. The first end member (upper right), which is expected to be very dilute aquifer water, corresponds to water sampled from the most confined and the deepest part of the aquifer as observed by Dabous and Osmond (2001) for Nubian aquifer (Egypt).

Low uranium contents could be explained by high reducing conditions enhancing uranium precipitation. The second end member (lower left) which is expected to be concentrated leachate of the host rock corresponds to water sampled in the discharge area of chott Fejej. The increase of dissolved uranium along these flow line indicates that the leaching process is predominant compared to alpha-recoil effect hence activity ratios decrease simultaneously. Thus, the water-exchange is the dominant mechanism that controls groundwater uranium chemistry as for other chemical elements (Edmunds et al. 1997, 2003). Uranium of groundwater sampled in the chott Djérid region indicated that uranium chemistry is mainly characterised by leaching process with preferential ^{234}U dissolution enhanced by a weak hydraulic gradient facilitating the water–rock interaction (Chkir and Zouari 2007).

For the Djérid area, ^{234}U-excess is about 0.045 (U-equivalent ppb) and the AR of the host rock U is 5.58. The two points (CF3 and CF9) have not been included in this leaching line because no hydrodynamic continuity can be proved. Moreover, this leaching line has not been back extended to the recharge zone because of data lack.

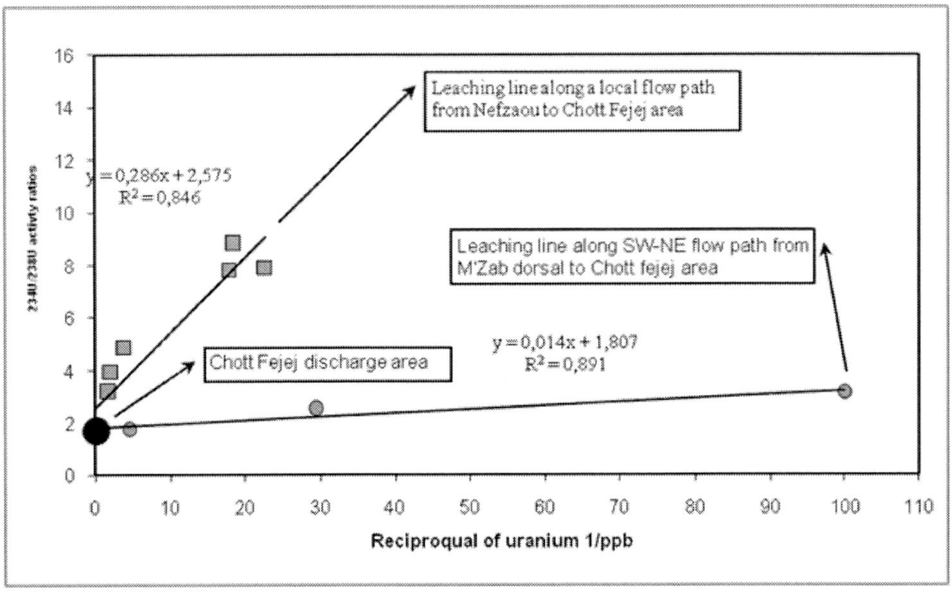

Figure 10. Reciprocal uranium content vs uranium isotopes activity ratios for the secondary SW-NE flow path.

In fact, this zone is an extremely remote area with difficult access and the sampling of groundwater has been possible mainly by using water from wells used for oil exploration and not all sampled water have been analysed for uranium investigation.

The Complexe Terminal Aquifer

Data from the Complexe Terminal aquifer are plotted in the area of oxidized aquifer on ''normal'' U-content strata with stable to forming U-accumulation (Figure 9). This indicates that uranium fingerprint of these groundwater are mainly controlled by geochemical conditions with probably a precipitation/remobilisation process that maintain uranium concentration quite constant with low variation of activity ratios induced by eaching/dilution processes.

On the reciprocal uranium content vs activity ratios diagram, a leaching trend does seem apparent even if no extrapolation could be based on only five points. However, when setted in the global context of the aquifer, some relationships between U isotopic variation and other geochemical data were referenced. U-contents are not significantly dependant on depth, but $^{234}U/^{238}U$ ARs show different trends according to host rock lithology. Samples collected from Pontien sands (northern shore of the Chott Djérid depression), uranium contents remain quite constant while ARs do decrease with increasing depth. Since samples have been collected along a flow line with similar geochemical conditions, ^{234}U precipitation along a flow path can be excluded as a possible explanation of the data (Chkir and Zouari 2007). However, a dilution process and subsequent mixing of groundwater, with some samples experiencing more dilution than others, may explain this trend. On the other hand, samples collected from Cenomanien limestone show that U-contents and ARs increase according to the local south–north flow of the aquifer in this zone. This trend is also observed for depth, temperature and mineralization. Since the aquifer lies in carbonates, a possible hypothesis is a preferential dissolution of ^{234}U from the matrix rock. However, two points do not provide sufficient information for a meaningful interpretation to be made.

The Djeffara Coastal Aquifer

Data from the Djeffara aquifer are plotted in the area of oxidized aquifer on ''normal'' U-content strata with forming U-accumulation (Figure 9). This indicates that uranium fingerprint of these groundwater are mainly controlled by geochemical conditions with probably a precipitation/remobilisation process that maintain uranium concentration quite constant with low variation of activity ratios induced by eaching/dilution processes.

A focused view on uranium contents highlighted that data can be discussed according to two groups related to lithofacies of hosting formations. The first group is represented by the Senonian carbonated horizon while the second one corresponds to water sampled from the Mio-pliocene clayey sandy horizon.

Even if aquifer classification mainly refers to 1 ppb and 10 ppb as U-content limits to distinguish aquifer types (Cowart and Osmond 1980, Roback et al. 2001, Bonotto and Andrews 2000, Chkir and Zouari 2007, Chkir et al. 2009), groundwater from the Senonian horizon are distinguished whether uranium contents are less or above 2 ppb (Hadj Ammar et al. 2010) according to relationships with other groundwater parameters (T, pH, Eh, depth, coexisting ions and isotopic data). These relationships suggest that the particular value of 2 ppb could be considered as a criterium for uranium content classification. Concentrations of uranium in water samples trend to increase with the decrease of the pH. Indeed, when the Jeffara waters become more acidic, U contents are slightly increased. Such relations between chemical properties of uranium and pH have been discussed in Tieh-Chi and Jeng-Jong (1997). In fact, they demonstrated that the concentrations of uranium in waters change with the pH. At pH 7, concentrations of uranium are extremely low while at pH <5, concentrations gradually increase. This indicates that, at pH 7, uranium cannot be leached much from the fractures or pores of the rock in the form of hydroxide. While, at pH <5, uranium is able to enter waters in the form of which are then suitable for leaching. Regarding their relationships with other parameters, we note that uranium contents increase proportionally to temperatures what indicates that the geochemistry of uranium is also related to this factor. For depth effect on uranium contents, a certain slight tendency indicates that the concentration of redox-sensitive elements increase with depth which prove the mobility of uranium into the aquifer and can be explained also by the increase of residence time and watererock reaction, what attests of a chemically active system (Chkir and Zouari 2007).

On the other hand, Chenchou 14 well is distinct from all other samples in demonstrating the variation of uranium content with temperature and pH. This well is located in an area where the contribution of the Continental Intercalaire groundwater to the recharge of the Djeffara aquifer by mixing processes has been proven to be significantly higher than 50% (Trabelsi et al. 2009), what could explain the high temperature and the low pH. This mixing effect is less obvious for other parameters (Eh and depth) but appears clearly on relationships between U-contents and chemical composition.

On the reciprocal uranium content vs activity ratios diagram, a dilution trend does seem apparent with points plotted on a horizontal line. This pattern could be explained by the mixing of uranium-rich and an uranium-free waters causing a variation of reciproqual of uranium but not of $^{234}U/^{238}U$ activity ratio. Points plotted along the horizontal line are interpreted as having experienced different mixing rates. The $^{234}U/^{238}U$ activity ratio of the originating water is about

1.682±0.1 corresponding to the mean of measured ARs. The spatial distribution of uranium concentrations and thus mixing rates could not be related to a particular direction of flowpath. However, the tectonic complexity of the region have structured the aquifer in several interconnected compartments what could explain so disturbed distribution.

CONCLUSION

Groundwaters occurring at the North-Western Sahara Aquifers System (North Africa) are very important resources, as they are extensively used for drinking purposes and agricultural activities, among others. This study was realized to investigate potential use of uranium isotopic variation for characterizing one of the greatest fossil groundwater in world, located over the Sahara, fromWestern Atlas Chain in western Algeria to Eastern North Africa along Mediterranean coast Tunisia and to southern to the Tassili of the Hoggar. We intend to use long-life radionuclide (^{234}U and ^{238}U) in order to identify different categories of groundwater in the two main aquifers of the North-Western Sahara Aquifer System, the Continental Intercalaire and the Complexe Terminal, and in the coastal interconnected Djeffara aquifer. The size, climate, lithology, stratigraphy, hydrogeology, geochemical conditions, and extent of rock–water interactions as well as alternative mechanisms related to radioactive phenomenon are among the factors responsible of the very variable uranium isotopes disequilibrium values in such hudge and complex aquifer. Our initial objective, which was to use the longlife radionuclides ^{234}U and ^{238}U for evaluating the residence times (age) of this groundwater, was achieved in part: the data obtained on the U isotopes in the groundwater from the two main aquifers of southern Tunisia are in good agreement with data from hydrogeology and with other geochemical and isotope studies. The U-isotopes data were favorable to apply the available models for the classification of groundwater and for the identification of two U-isotope controlling processes.

The Continental Intercalaire aquifer has a wide range of uraniumcontents (0.006–3.39 ppb) decreasing from the recharge area on the west (Algerian Atlas) to the discharge area of the east (Chott Djérid/Gulf of Gabès) and very scattered ^{234}U/^{238}U activity ratios (0.4–15.4) decreasing along this main W-E flow path. ^{234}U/^{238}U activity ratios have already been reported for deep groundwater and have been explained either by a long history of high surface area interaction between a water of low uranium solubility and an aquifer rock with relatively higher concentration of uranium (factor of 1000) (Kronfeld et al. 1975) or by an elevated temperatures thatmay cause rock–water isotopic re-equilibration to occur (Kramer and Kharaka 1986). High ^{234}U/^{238}U activity ratios are often observed in large confined aquifer with steady long-term flow systems such Continental Intercalaire aquifer.

The drop of uranium content values along the main W-E flow line is abrupt and allows to localize between 4°3E and 5°E of longitude the redox front that shares the CI aquifer on a oxidized compartment on its western part and a reduced compartment on its eastern part. Uranium contents and uranium isotopic disequilibrium values and variation in the oxidized compartment is characteristic of a ''steady-state'' aquifer lodged on enhanced uranium content strata and submitted to the decay of ^{234}U-excess along flow lines. Whilst the eastern part of the CI aquifer in the Djérid region has been found to be a reduced aquifer where uranium chemistry is controlled by possible remobilization or enhanced a-recoil effect.

On the other hand, different processes have been identified along the second flow line indicating different mechanisms. Indeed, more detailed investigations using reciprocal uranium vs activity ratios diagram have lead to distinguish two different origins for groundwater sampled in the Chott Fejej discharge area: the first preferential flow path is carrying groundwater from the recharge area situated in the M'zab dorsal in Southern Algeria while the second preferential flow path is thought to be a local flow line crossing the Nefzaoua region toward the Chott Fejej discharge one. The main mechanism controlling uranium chemistry along these two flow lines appears to be uranium leaching from the hosting formations enhanced by long residence time favourite by extended water–rock interactions.

The upper Complexe Terminal aquifer has a restricted range of U-contents (1.35–2.4 ppb) associated with relatively low $^{234}U/^{238}U$ ARs (1.7–2.5), both of which are characteristic of normal groundwater worldwide. The variations in U isotopes in terms of the hydrogeologic setting of the aquifer confirm that this aquifer is unconfined and suggest that the main process affecting U isotope patterns is groundwater mixing. This aquifer reveals to be lodged in "normal" U-contents stata with activity ratios mainly controlled by uranium leaching. Lower values or the activity ratios than those of the Continental Intercalaire indicate that water-rock interaction had taken place for lower residence time. However, the number of sampled points does not alow to extrapolate more.

Groundwater from the Djeffara aquifer lodged in Senonian limestones show that uranium range from 1.3 to 3.43 ppb with relatively low $^{234}U/^{238}U$ ARs (1.42-1.82). There are no observable geographic trends to the U concentration or to the $^{234}U/^{238}U$ activity ratios across the basin. Thus, the U-disequilibrium technique in this area as a hydrologic tool to trace water flow or to delineate areas where water bodies differ is not easily applied. On the other hand, due to the tectonic complexity that characterizes the Jeffara aquifer, the varying concentrations of uranium are influenced by several factors, including the geochemical properties of groundwaters. Nevertheless, the lack of significant correlations between uranium and both major chemical parameters and other chemical constituents in the groundwater suggests that water geochemistry has a very limited role in controlling uranium concentrations, and reflect the homogeneity of groundwater throughout the study area. The (1/U vs $^{234}U/^{238}U$) diagram reveals that the main factor that could explain the spatial distribution of uranium fingerprints in this aquifer is the dilution process between U-rich and U-free waters with different mixing rates.

The U-approach tested appears feasible and future application with more samples is being planned. Clearly, more experimental work is needed to firm up our conclusions, however, these observations have implications for the understanding uranium fingerprints of natural water systems and the interpretation of geological records.

REFERENCES

Abdul-Hadi A., Alhassanieh O., Ghafar M., 2001, Disequilibrium of uranium isotopes in some Syrian Groundwater, *Applied Radiation and Isotopes*, 55, 109–113.

Aggarwal S.K., Shah P.M., Duggal R.K., Jain H.C., 1991, Electrodeposition of milligram amounts of uranium on electropolished stainless steel disks, *J. Radioanal. Nucl. Chem. Letters,* 154 (2), 103-110.

Aguinaldo, N.M. Jr, AlGharib, I., Bernat, M., Fernex, F., 2003. Uranium and thorium isotopes in the rivers of the Amazonian basin: hydrology and weathering processes. *Hydrological processes*, 17, 17-31.

Airey P.L. and Roman D., 1981, Uranium series disequilibria in the sedimentary uranium deposit at Yeelirrie, Western Australia, *Journal of the Geological Society of Australia*, 28(3-4), 357-363.

Alexander, W.R., Mackenzie, A.B., Scott, R.D., McKinley, I.G., 1990. Natural analogue studies in crystalline rock: The influence of water-bearing fractures on radionuclide immobilisation in a granitic rock repository. *Nagra Technical Report* 87-08: 42 p.

Allard, B., Olofsson U., Torstenfelt B., Kipatsi H., Andersson K., 1982. Sorption of actinides in well-defined oxidation states on geologic media. *Materials Res. Soc. Symp. Proc.* 11:775-782.

Allen P.G., Shuh D.K., Bucher J.J., Edelstein N.M., Reich T., Denecke M.A., Nitsche H., 1996, EXAFS determinations of uranium structures : the uranyl ion complexed with tartaric, citic and malic acids. *Inorganic Chemistry*, 35, 784-787.

Andrews J.N. and Kay R.L.F., 1978. The evolution of enhanced 234U/238U activity ratios for dissolved uranium and groundwater dating. In Short Papers of the Fourth International Conference of Geochronology, Cosmochemistry, and Isotope Geology (ed. R. E. Zartman). *US Geological Survey Open File Report* 78-701, 11-13.

Andrews J.N. and Kay R.L.F., 1982. $^{234}U/^{238}U$ activity ratio of dissolved uranium in groundwaters from a Jurassic limestone aquifer in England. *Earth planetary Science Letter*, 57, 139-151.

Andrews J.N. and Kay R.L.F., 1983. The U contents and $^{234}U/^{238}U$ activity ratios of dissolved uranium in ground waters from some *Triassic sandstones in England. Isotope Geosciences*, 1, 101-117.

Andrews J.N., Giles I.S., Kay R.L.F., Lee D.J., Osmond J.K., Cowart J.B., Fritz P., Barker J.F., Gale J., 1982. Radioelement, radiogenic helium and age relationships for groundwaters from the granite at Stripa, Sweden, *Geochimica Cosmochimica Acta,* 46, 1533-1543.

Aranyossy, J.F., Mamou, A., 1986. Apport des Techniques Nucléaires à l'étude des Aquifères du Sud Tunisien, Projet RAF/8/007. *IAEA*, Vienne, p. 50.

Arey J.S., Seaman J.C., Bertsch P.M., 1999. Immobilization of uranium in contaminated sediments by hydroxyapatite addition. *Environmental Sciences Technology*, 33, 37-342.

Bailey E.H., Mosselmans J.F.W., Schofield P.F., 2004. Uranyl-acetate speciation in aqueous solutions-an XAS study between 25 and 250°C. *Geochimica et Cosmochimica Acta*. 68:1711-1722

Bailey E.H., Mosselmans J.F.W., Schofield P.F., 2005. Uranyl-citrate speciation in acidic aqueous solutions - an XAS study between 25 and 200°C. *Chemical Geology* 216:1-16.

Banner J.L., Wasserburg G.J., Chen J.H., Humphrey J.D., 1991. Uranium-series evidence on diagenesis and hydrology in Pleistocene carbonates of Barbados, West Indies. *Earth and Planetary Science Letters*, 107, 129-137.

Banner J.L., Wasserburg G.J., Chen J.H., Moore C.H., 1990. ^{234}U-^{238}U-^{232}Th systematics in saline groundwaters from Central Missouri. *Earth and Planetary Science Letters,* 101, 296-312.

Barnes C.E. and Cochran J.K., 1993, Uranium geochemistry in estuarine sediments : controls on removal and release processes. *Geochimica Cosmochimica Acta*, 57, 555-569.

Ben Dhia, H., 1983. Les provinces géothermiques en Tunisie méridionale. Thèse de doctorat Es e sciences naturelles, Univ. Bordeau I France, 196 pp.

Betcher R.N., Gascoyne M., Brown D., 1988. Uranium in groundwaters of southeastern Manitoba, Canada. *Can. J. Earth Sci*., 25, 2089-103.

Bishop, W.F., 1975. Geology of Tunisia and adjacent parts of Algeria and Libya. *Bulletin of American Association of Petroleum Geology* 59, 413–450.

Blanco Rodrígueza P., Martín Sánchez A., x Vera Toméa A., Ferrero Calabuig J. L., Roldán Garcíac C., Juanes Barberc D., Agere, F. J., da Silvad M. F. and Soares J. C., 1997, Characterization of alpha sources prepared by direct evaporation using Rutherford backscattering spectrometry, Nuclear Instruments and Methods in Physics Research Section B: *Beam Interactions with Materials and Atoms*, 132 (3), 501-506.

Bonotto D.M. and Andrews J.N., 2000. The transfer of uranium isotopes ^{234}U and ^{234}U to the waters interacting with carbonates from Mendip Hills area (England). *Applied Radiation and Isotopes*, 52, 965–983.

Bonotto D.M. and Jiménez-Rueda J.R., 2007, U-ages in soils and groundwater evidencing wet periods 400–600 kyr ago in southeast Brazil, *Applied Radiation and Isotopes*, 65, 776–783.

Bonotto D.M., 1999. Applicability of the uranium isotopic model as a prospecting technique in Guarany aquifer, South America. In: 9th Annual VM Goldschmidt Conf, abstr. 7021, *LPI contribution* no. 971. Lunar and Planetary Institute, Houston.

Bonotto D.M., 2000. Some constraints on the modelling using the uranium isotopes 238U and 234U for dating groundwaters from Guarany aquifer, South America, *Proceeding of the International conference on computer simulation in risk analysis and hazard mitigation* No2, Bologna , Italie (10/2000).

Bonotto D.M., Andrews J.N., Dardyshire D.P.F., 2004. A laboratory study of the transfer of ^{234}U and ^{238}U during water-rock interactions in the Carnmenellis granite (Cornwall, England) and implications for the interpretation of field data. *Applied Radiation and isotopes*, 54, 977-994.

Bonotto, D.M., 2006. Hydro(radio)chemical relationships in the giant Guarani aquifer, Brazil. *Journal of Hydrology* 323, 353–386.

Borole D.V., Krishnmaswami S., Somayajulu B.L.K., 1982, Uranium isotopes in rivers, estuaries and adjacent coastal sediments of western India : their weathering, transport and oceanic budget, *Geochimica Cosmochimica Acta*, 46, 125-137.

Bourdon B., Turner S., G.M. Henderson, C.C. Lundstrom, 2003. Introduction to U-series Geochemistry, *Reviews in Mineralogy and Geochemistry*; 52 (1); 1-21.

Brin A., 1979, Océan et énergie, *Ed. TECHNIP*, 120 p.

Chalov P.I. and Tikhinov A.I., 1990. Application of the natural fractionnation of ^{234}U and ^{238}U in studying underground-water dynamics under active water exchange conditions. N*uclear Geophysics,* 4(1), 1-15.

Chen J.H., Edwards R.L., Wasserburg G.J., 1986. U-238, U-234, and Th-232 in seawater. *Earth Planet Sciences Letters*, 80, 241-251.

Cherdyntsev V.V., 1971. Uranium-234. In: *Israël Program for Scientific Translations:* Jerusalem (Keter Press.). (Translated from Russian).

Chkir N. and Zouari K., 2007, Uranium isotopic disequilibrium for groundwater classification: first results on complexe terminal and continental intercalaire aquifers in Southern Tunisia, *Environ Geol* (2007) 53:677–685.

Chkir N., Guendouz A., Zouari K., Hadj Ammar F., Moulla A.S., 2009, Uranium isotopes in groundwater from the continental intercalaire aquifer in Algerian Tunisian Sahara (Northern Africa), *Journal of Environmental Radioactivity* 100 (2009) 649–656.

Cochran K., 1992, The Ocenic Chemistry of the Uranium- and Thorium-Series Nuclides. In: Uranium Series Disequilibrium: Applications to Earth, Marine, and Environmental Sciences. (M. Ivanovich and R. S. Harmon (eds.)) 2nd Ed. *Clarendon Press*, Oxford (1992).

Cornet, A., 1964. Introduction à l'hydrogéologie Saharienne. *Revue de Géographie Physique et de Géologie Dynamique* 61, 5–72.

Cowart J.B. and Osmond J.K., 1980. The relationship of uranium isotopes to oxidation/ reduction in the Edwards carbonate aquifer of Texas. *Earth Planetary Sciences Letters,* 48, 277–283.

Cowart J.B., 1980, The relationship of uranium isotopes to oxidation/reduction in the Edwards carbonate aquifer of Texas, *Earth and Planetary Science Letters*, 48 (2), 277-283.

Cowart, J. B., and Osmond, J. K., 1974. Uranium-234 and -238 in the Carrizo sandstone aquifer of south Texas. In Isotope Techniques in Groundwater Hydrology Proceedings Symposium, 2. *International Atomic Energy Agency*, Vienna, pp. 131-49.

Cuttell, J. c., Lloyd, J. W., and Ivanovich, M., 1986. A study of uranium and thorium series isotopes in chalk groundwater of Lincolnshire. *J. Hydrology*, 86, 343-65.

Dabbous A.A. and Osmond J.C., 2001. Uranium isotopic study of artesian and pluvial contributions to the Nubian aquifer, Western Desert, *Egypt. Journal of Hydrology*, 243, 242–253.

Dabous, A.A., Osmond, J.K., Dawood, Y.H., 2002. Uranium/Thorium isotope evidence for ground-water history in the Eastern Desert of Egypt. *Journal of Arid Environment*, 50, pp. 343-357.

Daniel D. S and Spalding R F., 1994, Uranium isotopes in the Platte River drainage basin of the North American High plains Region, *Applied Geochemistry*, 9 (3), 271-278.

Davidson M. R. and Dickson, B. L., 1986. A porous flow model for steady state transport of radium in groundwater. *Water Resourc. Res*., 22, 34-44.

Dearlove J.P., G. Longworth M. Ivanovich J.I., Kim B. Delakowitz P.Z., 1991, A Study of Groundwater-Colloids and their Geochemical Interactions with Natural Radionuclides in Gorleben Aquifer Systems. *In Radiochimica Acta*., 52/53, 83-89 (1991).

Dearlove J.P.L., Ivanovich M., Green, D.C., 1990. Partition coefficients (Rd) for the U series radionuclides in ion exchange sites and amorphous Fe phase in illite and granite samples.

In Proc. 6th Water/Rock Interaction Symp., Malvern, UK. Balkema, Rotterdam, pp. 191-5.

Degueldre C., Pfeiffer H.-R., Alexander W., Wernli B., Bruetsch R., 1996, Colloid properties in granitic groundwater systems. I : Sampling and characterisation, *Applied geochemistry*, 11 (5), 677-695.

Degueldre C., Triay I., Kim J.I., Vilks P., Laaksoharjuand M., Miekeley N., 2000. Groundwater colloid properties: a global approach, *Applied Geochemistry*, Vol. 15, Issue 7, 2000, pp 1043-1051.

Dodge C.J. and Francis A.J., 1997. Biotransformation of Binary and Ternary Citric Acid Complexes of Iron and Uranium. *Environ. Sci. Technol.* 31: 3062-3067

Dongarra, G., 1984. Geochemical behaviour of uranium in the supergene environment. In: Uranium geochemistry, mineralogy, geology, exploration and resources. pp. 18-22, *The Institution of Mining and Metallurgy*, London

Drever, J.I., 1997. The Geochemistry of Natural Waters. *Prentice Hall*, 306-309.

Ebbs S.D., Norvell W.A., Kochian L.V., 1998. Soil processes and chemical transport: The effect of acidification and chelating agents on the solubilization of uranium from contaminated soil. *J. Environ. Qual.* 27: 1486-1494.

Echevarria G., Sheppard M.I., Morel J.L., 2001. Effect of pH on the sorption of uranium in soils. *Journal of Environmental Radioactivity*. 53: 257-264.

Edmunds W.M., Guendouz A.H., Mamou A., Moulla A., Shand P., Zouari K., 2003, Groundwater evolution in the Continental Intercalaire aquifer of southern Algeria and Tunisia: trace element and isotopic indicators, *Applied Geochemistry,* 18, 805–822.

Edmunds W.M., Shand P, Guendouz AH, Moulla AS, Mamou A, Zouari K., 1997. Recharge characteristics and groundwater quality of the Grand Erg Oriental basin. In: Avicennes, technical report WD/97/46R, hydrogeology series. *British Geological Survey*, Keyworth, UK.

Eichrom, 1995. Eichrom Analytical Products Description. *Eichrom Industries Inc.*, Darien, USA.

ERESS, 1972. Etude des ressources en eau du Sahara Septentrional. *UNESCO*, Paris.

Fleischer, R.L., 1988, Nuclear track studies of alpha-recoil damage in nature: Relation to isotopic disequilibrium and leaching of radionuclides, International Journal of Radiation Applications and Instrumentation. Part D. *Nuclear Tracks and Radiation Measurements*, 14(4), 437–446.

Fleischer, R.L., Raabe, O.G., 1978, Recoiling alpha-emitting nuclei. Mechanisms for uranium-series disequilibrium, *Geochimica et Cosmochimica Acta*, 42(7), 973–978.

Fried A., Friedman M, Callis E., Screiner F., Hines J., Orllandini K., Nelson D. and Olsen E., 1985, Enrichment of 235U and the Concentration of 239Pu in Volcanic Samples. *Nature,* 313 (24), 301-303 (1985).

Fröhlich K. and Gellermann R., 1987. On the potential use of uranium isotopes for groundwater dating, Chemical Geology: Isotope *Geoscience section*, Vol. 65 (1), 67-77.

Frôhlich, K., Gellermann, R., Hébert, D., 1984. Uranium isotopes in a sandstone aquifer. In: Isotope Hydrology 1983,447-466. *IAEA*, Vienna.

Gascoyne M., 1989, High levels of uranium and radium in groundwaters at Canada's Underground Research Laboratory, Lac du Bonnet, Manitoba, Canada, *Applied Geochemistry*, 4 (6), 577-591.

Gascoyne M., 1992, Geochemistry of the actinides and their daughters. In Uranium series disequlibrium: Application to environmental sciences, Ivanovich M. and Harmon S., *Oxford Science Publications*, 2nd Ed., 34-61.

Gaven C., 1975, Déséquiibres dans les familles radioactives de l'uranium et du thorium au cours des proessus volcanogénétiques, Doctorat es Sciences Naturelles de l'Université Pierre et Maris Curie.

Gellermann R., Fröhlich K., Jordan H., 1990. Groundwater studies with uranium series disequilibrium, Groundwater Monitoring and Management/*IAHS Publ.* n°173, 113-120.

Ghaleb B., Hillaire-Marcel C., Causse C., Gariépy C., Vallières S., 1990. Fractionation and recycling of U and Th isotopes in a semi-arid endoreic depression of Central Syria, *Geochimica et Cosmochimica Acta*, 54, 1025-1035.

Giblin, A.M. and Appleyard, E. C., 1987. Uranium mobility in non-oxidizing brines: field and experimental evidence. *Appl. Geochem.*, 2, 285-95.

Gonfiantini, R., Conrad, G., Fontes, J.C., Sauzy, G., Payne, B.R., 1974. Etude Isotopique de la nappe du Continental Intercalaire et ses Relations avec les autres nappes du Sahara Septentrional, vol. I. IAEA-SM-182/25 in *Isotope Techniques in Groundwater Hydrology*, Vienne, pp. 227–241.

Gries, S., 2000. Etude isotopique et ge´ochimique des nappes profondes au Sahara-Sahel – Implications pour la gestion des Ressources en eau et les Reconstitutions Paléoclimatiques, Doct.es Sciences. Univ. Paris XI, *Faculté des Sciences* Orsay, p. 127.

Grindrod P., 1993. The impact of colloids on the migration and dispersal of radionuclides within fractured rock. *J. Contaminant Hydrol.*, 13: 167-181.

Grzymko T.J., Marcantonio F., McKee B.A., Stewart C.M., 2007, Temporal variability of uranium concentrations and $^{234}U/^{238}U$ activity ratios in the Mississippi river and its tributaries, *Chemical Geology* 243, 344–356.

Guendouz, A., 1985. Contribution à l'étude hydrochimique et isotopique des nappes profondes du Sahara septentrional, Algérie. Ph.D. Thesis, *Univ. Paris XI*, Orsay, France.

Gueniot B., Munier-Lamy C., 1988. Geochemical behavior of uranium in soils, part II. Distribution of uranium in hydromorphic soils and soil sequences. Applications for superficial prospecting. *J. Geochem. Explor.* 31: 39-55.

Hadj Ammar F., Chkir N., Zouari K., Azzouz-Berriche Z., 2010, Uranium isotopes in groundwater from the "Jeffara coastal aquifer" (southeastern Tunisia*), J. Environ.* Radioact., 101,

Hallstadius L., 1984. A method for the electrodeposition of the actinides. *Nuclear Instruments and Methods in Physics research*, 223, 266-267.

Hercman H. and Goslar T., 2002. Uranium-series and radiocarbon dating of speleothems – methods and limitations. *Acta Geologica Polonica*, 52 (1), 35-41.

Hille P., 1979. An open system model for uranium series dating, *Earth and Planetary Science Letters*, 42, 138-142.

Hollocher, K. and Yuskaitis, A., 1993. Chemical composition of surface and high uranium well water, Lake Sunapee Area, New Hampshire. Northeast. Geol. 15 (2), 159-169.

Hostetler P.B. and Garrels R.M., 1962. Transportation and precipitation of uranium and vanadium at low temperatures, with special reference to sandstone-type uranium deposits. *Econ. Geol.*, 57, 137-67.

Hubert A., Boudon B., Pili E., Meynadier L., 2006, Transport of radionuclides in an unconfined aquifer inferred from U-series disequilibria, *Geochimica et Cosmochimica Acta*, 20, 5437-5454.

Hussain N. and Krishnaswami S., 1980. U-238 series radioactive disequilibrium in groundwaters: implications to origin of U-234 excess and fate reactive pollutants. *Geochimica et Cosmochimica Acta*, 44, 1287-1291.

Ivanovich M., Frohlich K., Hendry M.J., 1991. Uranium-series radionuclides in fluids and solids from the Milk River aquifer, Alberta, Canada. *Applied Geochemistry*, 6, 405-418.

Ivanovich, M. and Harmon, R.S., 1992. Uranium-series Disequilibrium: Applications to Earth, Marine, and Environmental Sciences, *Oxford University Press New York*, 2nd Ed., USA.

Ivanovich, M., Longworth, G., Wilkins, M., Duerden, P., Payne, T., Nightingale, T., 1987. Natural analogue study of the distribution of uranium series radionuclides between the colloid and solute phases in the hydrogeological system of the Koongarra uranium deposit, N.T. Australia. In Proceedings of the Symposium Natural Analogues in Radioactive Waste Disposal (ed. B. Come and N.A. Chapman), CEC, Brussels, EUR 11037 EN, pp. 300-13. Graham and Trotman.

Jamet P, Hooker P.J., Schmitt J.M., Ledoux E., Escalier des Orres P., 1993. Hydrogeochemical modelling of an activee system of uranium fixation by organic soils and sediments (Needle's Eye, Scotland), *Mineralogica Deposita*, 28, 66-76.

Jiang J., Rao L., Di Bernardo P., Zanonato P. L., Bismondo A., 2002. Complexation of uranium(VI) with acetate at variable temperatures. *J. Chem. Soc. Dalton Trans*., 1832–1838.

Jones D.L., 1998. Organic acids in the rhizosphere - a critical review. *Plant and Soil,* 205, 25-44

Keiling C., Marx G., 1991. Investigations into the influence of organic complexing agents on the leaching behaviour and mobility of radionuclides solidified in ILW forms. *Radiochimica Acta*, 52/53: 287-290.

Kertes A.S. and Guillaumont R., 1985, Solubility of UO_2: A comparative review, *Nuclear and Chemical Waste Management*, 5(3), 215-219.

Kraemer T.F., 1981. 234U and 238U concentration in brine from geopressured aquifers of the northern Gulf of Mexico basin. *Earth planet. Sci. Left.*, 56, 210-16.

Kramer, T.F. and Kharaka, Y.K., 1986. Uranium geochemistry in U.S. Gulf coast geopressured-geothermal system. *Geochimica et Cosmochimica Acta* 50, 1233–1238.

Krishnaswami S., Graustein W.C., Turekian K.K., Dowd J.F., 1982. Radium, thorium, and radioactive lead isotopes in groundwaters: application to the in-situ determination of adsorption rate constants and retardation factors. *Water Resour. Res*., 18, 1663-75.

Kronfeld J. and Vogel J.C., 1991. Uranium isotopes in surface waters from southern Africa, *Earth and Planetary Science Letters*, 105, 191-195.

Kronfeld J., Godfrey-Smith D.I., Johannessen D., Zentilli M., 2004, Uranium series isotopes in the Avon Valley, Nova Scotia, *Journal of Environmental Radioactivity*, 73, 335–352.

Kronfeld J., Gradsztajn E., Müller H.W., Radin J., Yaniv A., Zach R., 1975. Excess 234U: an aging effect in confined water. *Earth Planet Sciences Letter* 27, 189–196.

Kronfeld J., Weinberger G., Yaniv A., Zafrir H., Vulcan U., Agami M., Rosenthal E., 1992, Uranium isotope disequilibria studies and the geohydrology of the Arava Rift Valley, Israel. *Nuclear Geophysics*, 6(4), 535-545.

Ku T.L., Knauss K.G., Mathieu G.G., 1974. Uranium in open ocean: concentration and isotopic composition. EOS, *Trans. Am. Geophys. Union*, 55,314.

Küppers G., 1998, Fast source preparation for alpha-spectrometry of uranium and transuranium isotopes, *Journal of Radioanalytical and Nuclear Chemistry*, Vol.230 (1-2), 167-174.

Laedermann J.P., 2003, Théorie bayesienne de la décision statistique et mesure de la radioactivité, *Thèse-Faculté des sciences de l'Université de Lausanne.*

Lally A.E. and Glover K.M., 1984, Source preparation in alpha spectrometry, *Nuclear Instruments and Methods in Physics Research*, 223 (2-3), 259-265

Langmuir, D., 1978. Uranium solution e mineral equilibria at low temperatures with applications to sedimentary ore deposits. *Geochim. Cosmochim. Acta* 42, 547-569.

Latham, A. G. and Schwarcz, H. P., 1989. Review of the modelling of radionuclide transport from U-series disequilibria and of its use in assessing the safe disposal of nuclear waste in crystalline rock. *Appl. Geochem.*, 4, 527-37.

Lemire, R.J., 1988. Effects of high ionic strength groundwaters on calculated equilibrium concentrations in the uranium-water system. Atomic Energy of Canada *AECL*-9549.

Lenhart J.J., Cabaniss S.E., MacCarthy P., Honeyman B.D., 2000. Uranium(VI) complexation with citric, humic and fulvic acids. *Radiochim. Acta* 88: 345-353.

Longworth G., Wilkins M.A., Ivanovich M., 1989, Characterisation of natural colloids in a shallow glacial aquifer and their radionuclide loading, *Materials Research Society Symposium Proceeding,* 127, 755-761.

Lowson, R. T. and Short, S. A., 1988. 238U decay series disequilibria in clay/quartz mineral phases. *Uranium,* 4, 275-8.

Lowson, R. T., Short, S.A., Davey, B.G., and Gray, D.J., 1986. $^{234}U/^{238}U$ and $^{230}Th/^{234}U$ activity ratios in mineral phases of a lateritic weathered zone. Geochim. Cosmochim. *Acta,* 50,1697-1702.

Madic C. and Genet M., 2001. Propriétés physiques et chimiques de l'uranium In: L'uranium, de l'environnement à l'Homme, Chapter 2, Metivier H., IPSN, Eds. *EDP Science.* pp 43-81.

Mamou, A., 1990. Caractéristiques, Evaluation, Gestion des Ressources en eau du Sud Tunisien. Thèse de Doct. Es Science. *Univ. Paris Sud*, p. 406.

Markovits G., Klotz P., Newman L., 1972. Formation Constants for the Mixed-Metal Complexes between Indium(III) and Uranium(VI) with Malic, Citric, and Tartaric Acids. *Inorganic Chemistry* 11: 2405.

Meyer F.W., 1989. Hydrogeology, ground-water movement, and subsurface storage in the Floridan aquifer system in southern Florida: *USGS Professional Paper* 1403-G, 59 p.

Moreira-Nordemann L.M., 1979. Use of $^{234}U/^{238}U$ disequilibrium in measuring chemical weathering rate of rocks, *Geochimica et Cosmochimica Acta*, 44, 103-108.

Moss, M.A., McCurdy, R.F., Dooley, K.C., Givner, M.L., Dymond, L.C., Slayter, J.M., Courneya, M.M., 1983. Uranium in drinking water – report on clinical studies in Nova Scotia. In: Chemical toxicology and clinical chemistry of metals. S.S. Brown and J. Savory (eds.), *Academia Press*, London, pp. 149-152.

Murakami T., Ohnuki T., Isobe H., Sato T., 1997, Mobility of uranium during weathering, *American mineralogist*, 82, 888-899.

Osmond J. K. and Cowart J. B., 1992, Ground Water, In Uranium Series Disequilibrium: Applications to Earth, Marine, and Environmental Sciences, M. Ivanovich and R. S. Harmon, 2nd Ed. *Clarendon Press*, Oxford.

Osmond J. K. and Ivanovich M., 1992, Uranium-Series Mobilization and Surface Hydrology. In: Uranium Series Disequilibrium: Applications to Earth, Marine, and Environmental Sciences. (M. Ivanovich and R. S. Harmon (eds.)) 2nd Ed. *Clarendon Press*, Oxford (1992).

Osmond J. K., Buie B. F., Rydell H. S., Kaufman M. I., Wallick E. I., 1971, Uranium and tritium as natural tracers in the floridan aquifer, *Florida Water Resources Research Center*, Publication n°14, 66p.

Osmond J.K., 1980. Uranium disequilibrium in hydrologic studies. In: Handbook of environmental Isotope Geochemistry (ed. P. Fritz and J.-Ch. Fontes), 259-82. Elsevier, Oxford.

Osmond J.K. and Cowart J.B., 2000. U-series nuclides as tracers in groundwater hydrology. In: Cook PG, Herczeg AL (eds) *Environmental tracers in subsurface hydrology*. Kluwer, Boston, 145–174.

Osmond, J.K. and Cowart, J.B., 1976. The theory and uses of natural uranium isotopic variations in hydrology. *Atomic Energy Rev.*, 14,621-79.

Osmond, J.K. and Cowart, J.B., 1983. Uranium disequilibrium in ground water as indicator of anomalies. *Journal of Applied Radiation and Isotopes*, 34, 283.

OSS, 2003. Système Aquifère du Sahara Septentrional. In: Hydrogéologie, vol. 2. OSS, *Tunis Int. Rep.* Project SASS.

Palmer, M. R. and Edmond J. M., 1993, Uranium in river water, *Geochimica et Cosmochimica Acta*, 57 (20), 4947-495.

Paquette, J. and Lemire, R.J., 1981. A description of the chemistry of aqueous solutions of uranium and plutonium to 200°C using potential-pH diagrams. *Nuclear Science and Engineering*, 79, 26-48.

Payne, T.E., Edis R., Seo T., 1992, Radionuclide transport by groundwater colloids at the koongarra uranium deposit, *Materials Research Society Symposium Proceeding*, 257, 481-488.

Plater, A.J., Ivanovich M. and Dugdale R.E., 1992, Uranium series disequilibrium in river sediments and waters: the significance of anomalous activity ratios, *Applied Geochemistry*, 7 (2), 101-110.

Puigdomenech I. and Bruno J., 1988. Modelling Uranium Solubilities in Aqueous Solutions: Validation of the Thermodynamic Database for EQ3/6 Geochemical Codes. *SKB Technical Report* 88–21.

Ragnarsdottir K.V. and Charlet L., 2000.. Uranium behaviour in natural environments. In "Environmental Mineralogy : Microbial interactions, anthropogenic influences. Contaminated land and waste management". *Mineralogical Society of Great Britain and Ireland*, 9: 333-377.

Rajan K.S., Martell A.E., 1965. Equilibrium studies of uranyl complexes. III. Interaction of uranyl ion with citric acid. *Inorg. Chem.*, 4: 462-469.

Rama and Moore W.S., 1984, Mechanism of transport of U-Th series radioisotopes from solids into groundwater, *Geochimica et Cosmochimica Acta*, 48, 395-399.

Reyes E. and Marques L.S., 2008, Uranium series disequilibria in ground waters from a fractured bedrock aquifer (Morungaba Granitoids—Southern Brazil): Implications to the hydrochemical behavior of dissolved U and Ra, *Applied Radiation and Isotopes* 66 (2008) 1531– 1542.

Ribera D., Labrot F., Tisnerat G., Narbonne J.F., 1996. Uranium in the environment: occurrence, transfer, and biological effects. *Revue of environmental contamination and toxicologie,* 146, 53-89.

Riotte J. and Chabaux F., 1999. (234U/238U) activity ratios in freshwaters as tracers of hydrological processes : The Strengbach watershed (Vosges, France). *Geochim. Cosmochimica Acta*, 63, 1263-1275.

Riotte J., Chabaux F., Benedetti M., Dia A., Gérard M., Boulègue J., Etamé J., 2003. U colloidal transport and origin of the 234U-238U fractionation in surface waters: New insights from Mount Cameroon. *Chem. Geol.* 202, 365-381.

Roback R.C., Johnson T.M., McLing T.L., Murrell M.T., Luo S., Ku T.L., 2001. Uranium isotopic evidence for groundwater chemical evolution and flow patterns in the eastern Snake River Plain aquifer, Idaho. *Geological Society of America Bulletin*, 113 (9), 1133-1141.

Rogers J.J. W. and Adams J.A.S., 1969, Uranium. In Handbook of geochemistry, Ed. K. H. Wedepohl, *Springer-Verlag*, Berlin.

Romer R. and Rocholl A., 2004, Activity disequilibrium of ^{230}Th, ^{234}U and ^{238}U in old stilbite: effect of young U mobility and a-recoil, *Geochimica Cosmochimica Acta,* 68(22), 4705-4719.

Rosholt J.N., Shields W.R., Garner E.L., 1963. Isotope fractionation of uranium in sandstone. *Science* 139, 224–226.

Salas J. and Ayora C. 2004. Groundwater chemistry of the Okelobondo uraninite deposit area (Oklo, Gabon): two-dimensional reactive transport modelling. *Journal of Contaminant Hydrology*, 69, 115-137.

SALAS, J, and AYORA, C. 2004. Groundwater chemistry of the Okelobondo uraninite deposit area (Oklo, Gabon): two-dimensional reactive transport modelling. *Journal of Contaminant Hydrology*, 69, 115-137.

Salonen L. and Huikuri P., 2002. Elevated Levels of Uranium Series Radionuclides in Private Water Supplies in Finland. Proceedings of the 5th *International Conference on High Levels of Natural Radiation and Radon Areas*, Munich, September 4-7, 2000, II, 87-91.

Sánchez A.M. and Tomé F.V., 1992. Optimal parameters for the electrodeposition of uranium, *Journal of RadioAnalytical Nuclear Chemistry*, 164(1), 23-28.

Sanchez A.M., Tomé F.V., Vargas M.J., 1992, GASP : A general-purpose program for environmental alpha spectra, *Nuclear Instuments and Methods in Physics Research*, A312, 211-216.

Sandino M.C.A. and Bruno J., 1992. The solubility of (UO2)3(PO4)2•4H2O(s) and the formation of U(VI) phosphate complexes: Their influence in uranium speciation in natural waters. *Geochimica et Cosmochimica Acta* 56, 4135–4145.

Sarin M.M., Krishnmaswami S., Somayajulu B.L.K., Moore W.S., 1990, Chemistry of uranium, thorium and radium isotopes in Ganga-Brahmaputra rivers systems: weathering processes and fluxes to the Bay of Bengal, *Geochimica Cosmochimica Acta*, 54, 1387-1396.

Schmeide K., Sachs S., Bubner M., Reich T., Heise K.H., Bernhard G., 1998. Interaction of uranium(VI) with various modified and unmodified natural and synthetic humic substances studied by EXAFS and FTIR spectroscopy. *Inorganica Chimica Acta*, 351: 133-140.

Shahandeh H. and Hossner L.R., 2002. Enhancement of uranium phytoaccumulation from contaminated soils. *Soil Science*, 167: 269-280.

Shahandeh H., Lee J.H., Hossner L.R., Loeppert R.H., 2001. Bioavailability of uranium and plutonium to plants in soil-water systems and the potential of phytoremediation. In "Trace elements in the rhizosphere", Gobran G.R., Wenzel W.W., Lombi E. Eds, Chap I.5, 93-124, *CRC Press,* Boca Raton London, New York, Washington D.C.

Sheppard S.C. and Evenden W.G., 1988. Critical compilation and review of plant/soil concentration ratios for uranium, thorium and lead. *J. Environ. Radioactivity*, 8: 255-285.

Sherman H.M., Gierke J.S., Anderson C.P., 2007. Controls on Spatial Variability of Uranium in Sandstone Aquifers, *Ground Water Monitoring and Remediation* 27, 2, 106–118.

Short S. A., R. T. Lowson, J. Ellis, 1988, $^{234}U/^{238}U$ and $^{230}Th/^{234}U$ activity ratios in the Colloidal Phases of Aquifers in Lateritic Weathered Zones, *Geochim. Cosmochim. Acta*. 52, 2555-2563.

Sibbens G. and Altzitzoglou T., 2007, Preparation of radioactive sources for radionuclide metrology, *Metrologia*, 44(4), S71.

Singh J., Singh L., Singh S., 1995. High U-contents observed in some drinking waters of Punjab, *India J. Environ. Radioact.*, 26 (3): 217-222.

Singh N.P. and Wrenn M.E., 1988, Determinations of actinides in biological and environmental samples. *Sciences Total Environnement*, 70, 187-203.

Smedley P L, Smith B., Abesser C. and Lapworth D., 2006. Uranium occurrence and behaviour in British groundwater, Keyworth, Nottingham *British Geological Survey* 2006.

Smithson G., 1990. Sampling and selection of analytical methods for radium, In: IAEA (International Atomic Energy Agency) (Ed.), *The environmental behaviour of Radium*. Darien, USA.

Stagg K.A., Kleinert U., Tellam J.H., Lloyd J.W, 1997. Colloidal populations in urban and rural groundwaters, UK. In: P.J. Chilton (Editor), Groundwater in the Urban Environment Vol. I, Problems, *Processes and Management Balkema*, Rotterdam: 187-192.

Stumm, W. and Morgan, J.J., 1981, Aquatic Chemistry: an introduction emphasizing chemical equilibria in natural waters, *John Wiley and Sons, Inc*., New York, N.Y.

Suksi J., Rasilainen K., Casanova J., Ruskeenieni T., Blomqvist R., Smellie J.A., 2001. U-seies disequilibria in a groundwater flow route as an indicator of uranium migration processes, *Journal of hydrology*, 47(2-4), 187-196.

Suksi J., Rasilainen K., Pitkänen P., 2006, Variations in $^{234}U/^{238}U$ activity ratios in groundwater - A key to flow system characterisation?, *Physics and Chemistry of the Earth*, Parts A/B/C, 31(10–14), 556-571.

Sun, H. and Semkow, T.M., 1998, Mobilization of thorium, radium and radon radionuclides in ground water by successive alpha-recoils, *Journal of Hydrology*, 205(1–2), 126–136.

Thurber D.L., 1962. Anomalous U^{234}/U^{238} in nature. *J. of Geophysical Resources*, 67, 4518-4523.

Tieh-Chi C. and Jeng-Jong W., 1997. Radioactive disequilibrium of uranium and thorium nuclide series in river waters from the Ta-Tun volcanic group area in Taiwan. *J. Appl. Radiat. Isotopes* 48 (8), 1149-1155.

Tieh-Chi C. and Jeng-Jong W., 2000. Radioactive disequilibrium of uranium and thorium nuclide series in Hot Spring and river waters from Peitou Hot Spring Basin in Taipei, *J. of Nuclear and Radiochemical sciences*, 1 (1), 5-10.

Titayeva, N.A. and Veksler, T.I., 1977. The state of radioactive equilibrium in the uranium and thorium series as an indicator of migration of radioactive elements and active interaction between phases under natural conditions: *Geokhimiya*, 1111-1120.

Toulhoat, P., Holliger, P., Menes, J., 1988. Analysis of lead isotopes and U-series disequilibrium in groundwater and possible sources rocks in the west Morvan area. *Uranium* 4, 307–325.

Trabelsi, R., Kacem, A., Zouari, K., Rozanski, K., 2009. Quantifying regional groundwater flow between Continental Intercalaire and Djeffara aquifers in southern Tunisia using isotope methods. *J. Environ. Geol.* 58, 171-183.

Trautmann N. and Folger H., 1989, Preparation of actinide targets by electrodeposition, Nuclear Instruments and Methods in Physics Research Section A: Accelerators, Spectrometers, *Detectors and Associated Equipment,* 282 (1), 102-106.

Untereiner G., 2008, Etude de l'impact de bactéries environnementales sur la spéciation de l'uranium en vue de processus de bioremédiation, Thèse - Institut des Sciences et Industries du Vivant et de l'Environnement (Agro Paris Tech), France.

Van de Weerd H. and Leijnse A., 1997. Assessment of the effect of kinetics on colloid facilitated radionuclide transport in porous media, *Journal of Contaminant Hydrology*; 26 (1-4), 245-256.

Vilks P., Cramer J.J. , Bachinski D.B. , Doern D.C. and Miller H.G. , 1993, Studies of colloids and suspended particles, Cigar Lake uranium deposit, Saskatchewan, Canada, *Applied Geochemistry*, 8 (6), 605-616.

Wathen, J.B., 1987, The effect of uranium siting in two-mica granites on uranium concentrations and radon activity in ground water, NWWA Conference on Radon, Radium and Other Radioactivity in Ground Water, Somerset, N.J. (USA), 7-9 Apr 1987.

Zänker H., Ulrich K.U., Opel K, Brendler V., 2007. The role of colloids in uranium transport: a comparison of nuclear waste repositories and abandoned uranium mines, IMWA Symposium 2007: *Water in Mining Environments*, R. Cidu and F. Frau (Eds), 27th-31st May 2007, Cagliari, Italy.

Zantuti F., Al-Medehem B., StLin V.I., Peretrukhin V.F., 1991, Electrodeposition of actinide traces from aqueous alkaline solutions and tributyl phosphate, *Journal of Radioanalytical and Nuclear Chemistry*, 147 (1), 51-58.

In: Uranium: Characteristics, Occurrence and Human Exposure ISBN: 978-1-62081-207-5
Editors: Alik Ya. Vasiliev and Mikhail Sidorov

Chapter 2

ANTHROPOGENIC SOURCES OF URANIUM IN SERBIA - RISK ASSESSMENT ON ENVIRONMENT AND HUMAN HEALTH

Mirjana D. Stojanović[1,*], Jelena V. Milojković[1], Zorica R. Lopičić[1], Marija L. Mihajlović[1], Miloš B. Rajković[2] and Gordana S. Vitorović[3]

[1]Institute for Technology of Nuclear and Other Mineral Raw Materials, Belgrade, Serbia
[2]Faculty of Agriculture, University of Belgrade, Belgrade-Zemun, Serbia
[3]Faculty of Veterinary Medicine, University of Belgrade, Belgrade, Serbia

ABSTRACT

The continual process of increased industrialization, militarization, and urbanization are among the most powerful sources of environmental degradation all over the world. Environmental contamination caused by radionuclides, in particular, by uranium and its decay products, is a serious problem worldwide and has become the common subject for further scientist investigations.

It is estimated that the technologically enhanced naturally occurring radioactive materials (TENORM) caused by nuclear and non-nuclear technologies, significantly increased natural radioactivity in the last 30 years. The development of nuclear science and technology, uranium mining, production and application of phosphoric fertilizer, the coal industry (mining and combustion), the oil and gas production, metal mining and smelting, mineral processing and building industry are generating radioactive waste that increased natural radioactive level on a regional scale. Economic sanctions for Serbia in the last decade of the 20th century, as well as the transition of its entire economy (which is still ongoing process), the NATO conflict in 1999, and absence of legislation and institutions, are all additional reasons for radioactive contamination of the regional ecosystem of Serbia. Also, one of the main contributing factor of environmental contamination by uranium in Serbia involved the cheap, imported phosphate fertilizers that were often of quality poor and radiological unsafe.

* Corresponding author: Mirjana D. Stojanović; e-mail address: m.stojanovic@itnms.ac.rs.

This outcome was a direct consequence of the reduced capacity in Serbian national production. On the other hand, according to the available data, many states were documented with bombing sites; 112 sites in Kosovo and Metohija, 12 locations in southern Serbia with depleted uranium (DU) ammunition, during the NATO aggression of Yugoslavia. On this occasion around 10 tons of DU was introduced into environment. The degree of contamination ranges from the bottom limit of 200 Bq/kg to 235,000 Bq/kg in the soil samples of mainly agricultural land, or 1,000 times above the tolerable level. Within the international mission, UNEP and FOCUS and by the engagement of national experts, the decontamination of endangered sites with depleted uranium has only been partially carried out. Today, unfortunately, we encounter an "invisible threat" of depleted uranium ammunition use, with highly radioactive and chemo toxic effect on human health, causing in the last 12 years an enormous increase in cancer rates and a number of newborns with genetic changes. This paper reviews some relevant aspects of environmental contamination with uranium, and gives an overview of the different remediation processes available.

This study reviews some relevant aspects of environmental contamination with anthropogenic uranium, primarily from the application of phosphate fertilizers and the use of ammunition with depleted uranium. Our study will include a special accent on Serbia's area concerning the uranium uptake and how it behaves in the food chain and will explore how these indicators impact human health.

Keywords: Natural uranium; depleted uranium; phosphate industry; NORM/TENORM; Serbia; health impact

1. Introduction - Natural Occurrence of Uranium

Uranium is a naturally occurring metal that consists of three radioactive isotopes: U^{238}, U^{235} and U^{234}. The concentration by weight is approximately 98.3% U^{238}, 0.72% U^{235} and 0.006% U^{234}, with half-life 4.5 billion yrs., 700 million yrs., 240,000 yrs. Uranium is ubiquitous in the environment and found in low levels within all rock, soil, and water. This is the highest-numbered radioactive element to be found naturally in significant quantities on earth. A typical concentration of U^{238}, as the main isotope, in the Earth's crust is between 0.5 to 10 g/t, with activity from 5 to 125 Bq/kg. For U^{235}, the activity concentration is 0.2 to 5 Bq/kg (UNEP, 1999).

According to the United Nations Scientific Committee on the Effects of Atomic Radiation (1993) the normal concentration of uranium in soil has a 300 μg/kg to 11.7 mg/kg range. The concentration in seawater contain about 3 μgU/kg; as uranium (VI) forms soluble carbonate complexes, with an estimation of $5x10^9$ tons of uranium in the oceans (Vernon and Shah, 1983). Most of the uranium in water comes from dissolved uranium through rocks and soil; only a very small part is from the settling of uranium dust out of the air. Activity concentrations of U^{238} and U^{234} in drinking water are between a few tenths of a mBq per liter to a few hundred mBq per liter. The typical activity concentrations of uranium in the air are around 2 μBq per cubic meter (UNSCEAR, 2000).

On average, approximately 90 μg (micrograms) of uranium exist in the human body from normal intakes of water, food, and air; approximately 66% is found in the skeleton, 16% in the liver, 8% in the kidneys, and 10% in other tissues (WHO, 2001).

Uranium in the environment can exist in 5 oxidation states (+2, +3, +4, +5, and +6), where complexation and redox reactions control the mobility of uranium (Langmuir, 1978).

The primary abiotic and biological processes that transform uranium in the soil are oxidation–reduction reactions that convert U (VI) (soluble form) to U (IV) (insoluble form). Under oxidative conditions uranyl, UO_2^{+2}, forms highly mobile compounds, but under the reductive condition U (IV) is oxidized to stable uranyl oxide UO_2. This process determines uranium's varied reaction in soil (Vodyanitskii, 2011). Abiotic and biological processes transform uranium in the environment where these reactions then form complexes with inorganic and organic ligands. The mobility and solubility of uranium in soil is dependent on several factors such as: pH, redox potential, temperature, soil texture, organic and inorganic compounds, moisture and microbial activity (Gavrilescu *et al.*, 2009).

2. Uranium in the Environment - NORM / TENORM

Background radiation dose consists of the radiation doses received from natural and man-made background. Isotopes that contribute to natural radiation can be divided into those materials which come from the ground, rocks, building materials, and drinking water supplies (*terrestrial sources*) and those which are produced as a result of the interaction of atmospheric gases with cosmic rays (*cosmogenic*). Man-made radiation is generated from nuclear and non-nuclear (industrial) activities. Natural background radiation contributes about 81% of the annual dose to the population and man-made background radiation contributes the remaining 19%. (http:// www. doh. wa. Gov /ehp/ rp/ factsheets/factsheets-htm/ fs10bkvsman.htm). It is estimated that the Chernobyl radioactive fallout added about 2% to the natural radioactive background.(Yablokov et al., 2009).

Anthropogenic (man-made) potential sources of uranium could be classified as following (Stojanović and Milojković, 2011):

- Nuclear activities (nuclear industry through enriching uranium and energy production, radioactive waste management, nuclear medicine, accident situation),
- Military activities (production and use of nuclear weapons)
- Conventional (non-nuclear) activities (mining, coal production and combustion, phosphate industry, oil and gas exploitation, water treatment…)

The major contributor of background radiation is NORM - "Naturally Occurring Radioactive Material." NORM is used for all naturally occurring radioactive materials where human activities have increased the potential for exposure compared with the unaltered situation and potentially includes all radioactive elements found in the environment.

Long-lived radioactive elements such as uranium, thorium, and potassium and any of their decay products, such as radium and radon are examples of NORM. These elements have always been present in the Earth's crust and atmosphere. The term NORM exists also to distinguish "natural radioactive material" from anthropogenic sources of radioactive material, such as those produced by nuclear power and used in nuclear medicine, where incidentally the radioactive properties of a material may be what make them useful. However, the perspective of radiation doses on people, this distinction is completely arbitrary. http://world-nuclear.org/info/inf30.html

The acronym TENORM, "Technologically-Enhanced Naturally-Occurring Radioactive Materials," or technologically enhanced NORM is often used to refer to those materials where the amount of radioactivity has actually been increased as a result of industrial processes, or as a result of past or present human practices (http://www.epa.gov/radiation/tenorm). The Committee on Evaluation of EPA Guidelines for Exposure to Naturally Occurring Radioactive Materials, of the National Academy of Sciences and National Academy of Engineering defines TENORM as "*any naturally occurring material that is not subject to regulation under the Atomic Energy Act whose radionuclide concentrations or potential for human exposure have been increased above levels encountered in the natural state by human activities.*" Much TENORM contains only trace amounts of radiation and it is part of our everyday landscape. Some TENORM, however, contains very high concentrations of radionuclides that can produce harmful exposure levels. EPA is concerned about TENORM because of its potential for harmful exposure to humans and the environment. TENORM does not include background radiation or the natural radioactivity in an area (EPA, 1999).

Over the past 20 years, EPA and other federal as well as state government agencies, industries, and other organizations have identified an array of naturally occurring materials that, because of human activity, may present a radiation hazard to people and the environment. Many industrial processes are using NORM as feed materials or have TENORM by-products. It has been recognized that certain materials used or generated in certain industry sectors have higher than average levels of NORM, or may concentrate the NORM into TENORM. The existence of TENORM is ignored by industrial managers and TENORM wastes might accidentally be discovered.

The major uranium sources of TENORM are:

- uranium mining (uranium mining overburden, uranium *in situ* leachate, evaporation pond, solids)
- phosphate industry (phosphor gypsum, phosphate sludge, phosphate fertilizer)
- coal combustion (fly ash, bottom ash, sludge)
- petroleum oil, and natural gas production (sands, sludge, hard scales…)
- metals mining (rare earths, monazite, xenotime, bastnasite), gold and silver, titanium, copper, aluminum, zircon, molybdenum, tin, tungsten, vanadium-tailings, dump and heap leach wastes, waste rock and overburden, solid waste from liquid
- waste and drinking water treatment (treatment sludge, treatment plant filters, treatment plant ash)
- geothermal energy production (scale, brine pond deposit)
- building materials (certain construction materials- gypsum and stone (e.g., granite))

Many of the materials that are technically TENORM have only trace amounts of radiation and are part of our everyday landscape. However, some TENORM has very high concentrations of radionuclides that can result in elevated exposures to radiation. According to the Environmental Protection Agency (EPA) TENORM poses three problems:

- it has the potential to cause elevated exposure to radiation;
- people may not be aware of TENORM materials and need information about them;
- the industries responsible for generating these materials may need additional guidance to help manage and dispose of them in ways that protect people and the environment that are economically sound (http://www.epa.gov/radiation/ tenorm/sources.html#summary-table).

While regulations may vary from state to state, operations involving NORM or TENORM are rarely licensed by the NRC or an equivalent agreement state program. This is due to the fact that workers are unlikely to exceed 10 % of the occupational whole body dose limit, which would require the use of a radiation monitoring device. There is very limited data, if any, on the industrial workers exposure from operations involving the generation of TENORM. In addition, TENORM may be of concern under non-occupational exposure scenarios (ASTSWMO, 2011).

2.1. Sources of NORM and TENORM in Serbia

In Serbia, the main sources for uranium deposits are magmatic, sedimentary, carbonate rocks, and granites. Natural uranium is found in greater amounts (mean 50-100 Bq/kg, up to 200 Bq/kg, due to the mineral content in the soils), in these mountain regions: Bukulja, Janja, Stara Planina, Kucani-Beljanik, Avala-Kosmaj-Babe, Rudnik, Cer, Vranje-Pcinj, Fruska Gora, and Vrsacko brdo. The distribution of uranium correlates with radio ecological areas of Central-Eastern Serbia, Central Serbia, Southern and South-Eastern Serbia. Other important sources of uranium in Serbian area are certain technological production processes, such as production and combustion of coal in the power plants, the production of phosphoric acid, phosphate fertilizers, phosphogypsum and others.

Near the thermal power plant "Kolubara" and "Nikola Tesla," strength of the equivalent dose is in the 1.42 to 4.87 nSv/h range (Djuric and Popovic, 2001). The production and the use of phosphate fertilizers are other important sources of uranium in Serbian environment. The concentration of uranium in phosphate ore ranges from 12 to 180 Bq/kg. Serbia annually imports about 1,000,000 tons of phosphorite for mineral fertilizers production. The average concentration of uranium in phosphorite imported was 150 Bq/kg. This means that Serbia annually imports about 150 tons of uranium or 50 TBq of radioactivity are applied to Serbian agricultural soils (Djuric and Popovic, 2001).

The processing of phosphorite creates a certain amount of phosphor gypsum, phosphate for feeding livestock, etc. The concentration of uranium in these productions is in the 5.6 to 14 Bq/kg range. The production of phosphoric acid annually processes about 120,000 tons of raw phosphate with an average uranium concentration of 90 g/t (about 100 GBq). The activity of U^{238} in the phosphoric acid is 467 Bq/dm^3 and for U^{235} it is 34.4 Bq /dm^3; the sediment in this production contains 360 Bq /dm^3 U^{238} and 20 Bq / dm^3 U^{235}.

Finally, waste rock dumps and uranium mines are important uranium generators: in the southeast of Serbia, in the place of Gabrovnica-Kalna, uranium mine still exists, but it had stopped mining ore and the uranium flotation process more than 40 years ago. Barren soil, from this closed uranium mine in Gabrovnica-Kalna, contained uranium in range from 15,33 mg/kg to 17 mg/kg, which today cover an area of about 0.1 km^2. Water which is springing from the mine, contained 0,053 mg U/dm^3 water (Sarić *et al.*, 1995; Stojanovic *et al.*, 2009). Perennial vegetation experiments in real and controlled conditions of different plant species (crops and vegetables), on tailings from closed uranium mine in Kalna, have indicated that the content of uranium in the plants that were growing on these tailings, have been larger than uptake from plants that were growing on the control, non-contaminated soil, and that degree of uranium adoption strongly depends on the plant species, genotypes, and plant organs (Stojanović and Milojković, 2011).

The facts previously described, indicate that the sources of uranium in Serbia results from natural, geological and geochemical origin of sediments, rocks and soils (NORM) and present "background of natural ionizing level." Technological processes in power plants and the production and import of phosphate fertilizers, as well as their use, has also contributed to the increased concentration of uranium in certain areas, thus presenting the main form of TENORM. According to some estimates the "natural level of ionizing radiation" in Serbia has increased about 30 times in the last 40 years (Djuric and Popovic, 2001).

Thoroughly, in the next section will be explore processed forms of TENORM through the phosphate industry problems, as well as the consequence of depleted uranium use; both as a source of uranium environmental contamination.

3. NORM and TENORM in Phosphate Industry

Phosphate rock is the major source of phosphorus in nature. It exists mainly in the form of hydroxyl and fluoroapatite, $Ca_{10}(PO_4)_6(OH)_2$ and $Ca_{10}(PO_4)_6F_2$, respectively; or a mixture of both that is used mainly for the production of fertilizers and elemental phosphorus.

The uranium in the secondary geochemical environments shows tendency for phosphate binding and it is significantly concentrated in the phosphate rock in the range of 5 – 300g/tU. Uranium in phosphate rock may be present in ionic form U^{4+} or U^{6+} or both. Tetravalent uranium state in isomorphism substitution with calcium ion; the ionic radius of U^{4+} is 0.97Å being very near to that of Ca^{2+} which is 0.99Å (Erdem *et al.*, 1996). Therefore, from the point the dislocation of uranium from the soil into the plant, calcium is considered a metabolic analogue of uranium (Stojanovic, 2006)

There are two main types of phosphate rock:

- *Sedimentary:* Supplies 85% of the phosphate rock for the industry: containing 100-300 ppm of uranium, 2-20 ppm thorium, and about 0.5% lanthanides. North Africa and Florida phosphates are typical for this category.
- *Igneous (eruptive):* Supplies the remaining 15%, and contains < 10 ppm of uranium but also contains appreciable amount of thorium and lanthanides. Kola phosphate (Russia) is typical for this category.

Current world phosphate rock production is about 170 million tons annually, which represents a potential source of about 17,000 tons of uranium per year.

Phosphate rock used for fertilizers is a major NORM with regard to redistribution of uranium, thorium and radium and with significant inventories uranium on the world level before any production occurred (Table 1).

Approximate uranium inventories in world phosphate deposits are about 9 million t U. (IAEA, 2001). Phosphorite deposits represent large uranium resources, but at a very low grade. Uranium can be recovered as a by-product of phosphate production.

The most important use (over 90 %) of phosphate rock is in the production of phosphate fertilizers. Before phosphate ore is turned into fertilizer or other products, it is transformed into either phosphoric acid (through the "wet process"), or elemental phosphorus (through the thermal process). This processing concentrates NORM in the waste products, transforming them into TENORM. The wet-acid process is the widespread method used by the industry and varies little worldwide. For every ton of phosphoric acid produced there are five tons of phosphogypsum as by-product. This process includes reaction of sulphuric acid with phosphate rock which produces phosphoric acid and phosphogypsum according to the reaction 1:

$$Ca_3(PO_4)_2 + 3H_2SO_4 + 6H_2O = 3CaSO_4 \cdot 2H_2O + H_3PO_4 \quad (1)$$

As indicated, sedimentary phosphate rock contains radionuclides of uranium, thorium and actinium series. The activity concentrations of the elements of these series vary with the source of the ore. The rock itself also varies in mineral structure and constituents. There are some identifiable trends in the distribution of radionuclides during processing of phosphate rock and production of phosphoric acid and fertilizers, presented in Figure 1 (Stojanovic, 2006). Generally, "wet acid processes" is in category of NORM and generate TENORM product and by-product. Relationship of uranium content in mineral phosphate fertilizers and phosphate content in phosphate rock is linear dependence (Stojanović, 2006).

Table 1. Content of U in phosphate rock and approximate U invertors in world phosphate deposits

Origin	*COUNTRY*	*U Concentration, Bq/kg (~g/t)*	Inventories in phosphate deposits / million t U
Igneous * sedimentory	USA	259 - 3700 (20-300)	
	USA-FLORIDA	1500 - 1900 (117-147)	1.20
	MOROCCO	1500 - 1700 (110-132)	6.90
	JORDAN	1300 - 1800 (100-140)	0.10
	BRASIL	114 - 880 (~9-68)	NO DATA
	SENEGAL	100 - 1332 (~8-103)	NO DATA
	AUSTRALIA	130 - 900 (~10-70)	NO DATA
	MEXICO	90 - 2500 (70-200)	0.15
	RUSSIA*	< 10 (< 50)	0.12
	OTHER -	0.65	
	TOTAL	- 9.12	

Source: (IAEA, 2001).

The annual production of phosphate rock of 142 million tonnes per year is giving 66 million tonnes of concentrate. Marine phosphorite deposits are 80% of the world output of phosphate based fertilizer products, and 70% of this is converted into wet process phosphoric acid, as the base for the current uranium extraction process. Assuming an average recoverable content of 100 ppm of uranium, this scenario would result in an annual output of 3700 t U/a. [IAEA, 2001]. Eight plants for the recovery of uranium from phosphoric acid have been built and operated in the United States since 1976 (Florida 6, Louisiana 2).

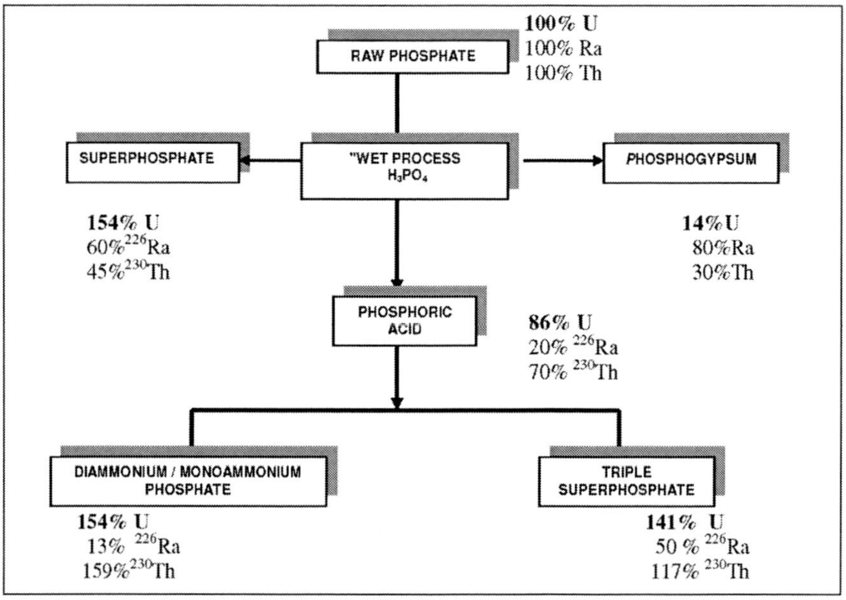

Figure 1. Distribution of the radionuclide in the "wet" process of phosphoric acid production.

Plants have also been built in Canada, Spain, Belgium, Israel, and Taiwan. Unfortunately, this is a very small number when compared to 400 wet-process phosphoric acid plants around the world (Chmielewski, 2008).

Radionuclides from the phosphoric acid contained in the final product, mineral fertilizers, whose implementation leads to contamination of the environment. Table 2 presents the average content of radionuclides in phosphate fertilizers.

Table 2. Concentration of radionuclides in fertilizers (Bq/kg)

Products	U-238	Ra-226	Th-232
Phosphoric acid	1200-1500	300	-
Normal super-phosphate	520-1100	110-960	15-44
Triple superphosphate	800-2160	230-800	44-48
Mono-ammonium phosphate	2000	20	63
Diammonium phosphate	2300	210	<15
Dicalcium phosphate	-	740	<37
PK fertilizer	410	370	<15
NP fertilizer	920	310	<30
NPK fertilizer	440-470	210-270	<15

Source: IAEA Tech Report 419, p.100.

3.1. Effect of Long-Term Application of Phosphorus Fertilizers on Soil

Use of phosphoric fertilizers is the main anthropogenic source of the uranium input in the environment (about 73% of the total uranium input). On the basis of the U concentration in phosphate fertilizers, McBride and Spiers (2001) estimated that 50 years of the application of a specific phosphate fertilizer (e.g., 100 kg/ha/year as P_2O_5) would lead to the addition of 2.4 kg of U per hectare to the topsoil, corresponding to an increase of about 1 mg/kg in the soil U.

In a paddy field in Japan, the increase in soil U due to the application of calcium superphosphate fertilizer (600 kg/ha/year as mixed fertilizer) for 10 years was estimated to be 5.3% of the total U in the soil (Tsumura and Yamasaki, 1993).

Centuries of mining and milling of uranium and other elements have resulted in the generation of significant quantities of radioactive waste materials, considered a menace to public health and environmental quality. Generally, a mine capacity of 100,000 tonnes of uranium ore annually will simultaneously produce 100,000–600,000 t of waste tailings (Gavrilescu *et al.*, 2009). As a consequence, there may be a risk for ecosystems, agro-systems, and health.

Kratz and Schnug (2006) reported that a mean value of uranium from 7 to 23 g/ha (annually) can be loaded onto soil using of 22 kgP/ha from various phosphorus fertilizers.

Effect of long-term application of phosphorus fertilizers on uranium contamination soils was the subject of many studies in the world. The first report on this topic was in the 1960s (Rothbaum *et al.*, 1979).

Increased content of U in the topsoil due to the long term application of chemical P fertilizers ranges from 0.08 to 1.3 mg / kg soil for periods of application ranging from about 20 to over 80 years (Rothbaum *et al.*, 1979; Stojanovic *et al.*, 2006; Takeda *et al.*, 2006; Taylor, 2007; Tunney *et al.*, 2009; Yamaguchi *et al.*, 2009)

Takeda *et al.*, (2006) found that in the soils at 0-35 cm depth, during a 61-year cultivation period, U concentration increased about 200 mg/m^2 by the vertical profile. Also, Taylor (2007) found that mean levels of total U increased by 1.30 ± 0.03 mg/ kg of soils sampled about 40 years ago compared with recent soil samples, an annual increase of 0.033 ± 0.008 mg/ kg yr^{-1}, and these increases are probably associated with phosphate fertilizer application.

On the other hand, there are authors who found no significant accumulation of U after application of fertilizer, suggesting either that less U were applied using a different P fertilizer and / or that soil heterogeneity masked significant effects (Wetterlind et al., 2011). Jones (1992) reported no changes in uranium concentration in the soil during 82 years of application of phosphorus fertilizers, and assumed that the loss of uranium over the years were due to a leak from the topsoil, surface soil erosion, or adoption by the plants.

Schnug (2005) calculated in his report that barely more than 1 g/ha of uranium is removed by crop products, leaching, and erosion. Also, Mortvedt (1994) reported no differences between uranium concentrations of unfertilized and fertilized soil with the triple superphosphate made from Florida phosphate rock, which, according to data collected by Kratz and Schnug (2006), has a uranium concentration in the range of 65-141 mg/kg. Saueia and Mazzilli (2006) evaluated that radiological impact of long term applications of phosphate fertilizers is negligible and below the 1 mSv/year.

Contradictory results were expected and indicate that the effect of long term application of phosphate fertilizers to increase the content of uranium in the soil cannot be generalized, but that is specific to each individual site.

Uranium present in soil can be dissolved in solution, or ion exchanged in the reaction, complexed with soil organics or precipitate as pure or mixed solids with consequence for ecosystems, agro-systems and health, passing through the food chain.

Pathways of uranium determine different factors such as: soil pH, Eh-redox potential, content and types of organic and inorganic ligands, amount of Fe oxide, mineral composition of soil, physico–chemical properties of soil, partial pressure of CO_2, temperature, microbiological activity (Stojanovic, 2006).

Many authors emphasize different factors as dominant in the distribution of uranium in environmental and in the risk assessment. Soil pH should be the focus variable for reduction of the uncertainty associated with the soil Kd value used in environmental risk assessments, even for reducing the uncertainty in site-specific Kd values. Positively charged UO_2^{2+} is adsorbed on the negatively charged sites of soil components, and these sites increase with soil pH. The U sorption capacity of soil, therefore, increases with soil pH. Under a reductive environment, the major U species is insoluble UO_2. In addition to soil composition, the types and concentrations of coexisting ions, soil pH, and redox conditions control the mobility of U in soil. This implies that agricultural practices have significant effects on the mobility of U in soil (Echevarria *et al.*, 2001). However, when carbonate concentration increased with an increase in pH, U became mobile in soil because of the formation of a soluble and negatively charged carbonate-U complex. UO_2^{2+} is also adsorbed on Fe and Al oxides (Catalano and Brown, 2005; Yamaguchi *et al.*, 2009).

Important factors for controlling the accumulation of U in soil are the amount of organic matter and the concentration of Fe oxide (Yamaguchi *et al.*, 2009).

Takeda *et al.* (2006) compared U speciation in agricultural soil with that in non-agricultural soil in the adjacent area and showed that fertilizer-derived U is mainly associated with the organic matter and Fe oxide in the upland field andosols in Japan.

The importance of soil organic matter as a pool of fertilizer-derived U has also been suggested by Rothbaum *et al.* (1979). The solid–liquid distribution coefficient (Kd) of U is significantly correlated with organic matter and amorphous Fe contents of soil (Vandenhove *et al.*, 2007).

3.1.1. Phosphorus Fertilizers as Source of Uranium in Serbian Soils

Around 1500 t of mineral fertilizers based on phosphorus are applied per annum in Serbia. It is estimated that around 210 kg of uranium (30 g/ha) are in this way introduced into the environment (Mitrović *et al.*, 2011; Tunney *et al.*, 2009; Stojanović *et al.*, 2006.).

Due to this fact there is a risk of exposing of population to ionizing radiation. Stojanović *et al.*, (2006) investigated whether long-term application of phosphorus fertilizers causes increase of uranium content in arable soils in Serbia.

Mineral fertilizers based on phosphorus that were distributed in Serbia in 2009 had a uranium content from 0.3mg/kg to 153 mg/kg (Stevanovic *et al.*, 2009).

These investigations were made possible by virtue of number of experimental fields that were set up on three types of soil, different acidity *(chernozem, smonitza and pseudogley)* more than 40 years ago. Same variants of mineral nutrition were used in these experiments which all had parcels without application of phosphate fertilizers (controls). Soil samples for uranium analysis were taken from two soil layers (0-30 and 30-60 cm) over five years, after harvesting of wheat and maize.

Results indicate significant and highly significant differences among pseudogley in relation to smonitza and chernozem, for both sampling depths and both nutrition treatments (Tables 3 and 4).

Table 3. Significance of differences of average values for U content (ppm) for three soil types (controls and treatments) and two depths (after maize grow)

Types of soil	Variant of Nutrition	Average value	D e p t h 0 cm – 30 cm				
Chernozem	Control	1.53±0.08	0.23*	0.41**	0.03	0.12*	0.16
	Treated	1.69±0.10	0.39*	0.57**	0.13	0.28*	–
Smonitza	Control	1.41±0.11	0.11	0.29*	0.15	–	–
	Treated	1.56±0.09	0.26	0.44**	–	–	–
Pseudogley	Control	1.12±0.10	0.22	–	–	–	–
	Treated	1.30±0.10	–	–	–	–	–
	D e p t h 30 cm – 60 cm						
Chernozem	Control	1.62±0.09	0.28**	0.34**	0.04	0.11	0.11
	Treated	1.73±0.06	0.39**	0.45**	15	0.22**	–
Smonitza	Control	1.51±0.09	0.17**	0.23	0.07	–	–
	Treated	1.58±0.10	0.24**	0.30	–	–	–
Pseudogley	Control	1.28±0.11	0.06**	–	–	–	–
	Treated	1.34±0.09	–	–	–	–	–

Meaning of symbols: $^{*}p < 0.05$ $^{**}p < 0.01$.

Table 4. Significance of differences of average values for U content, (ppm) for three soil types (controls and treatments) and two depths (after wheat grow)

Types of soil	Variant of Nutrition	Average Value	D e p t h 0 cm – 30 cm				
Chernozem	Control	1.62±0.09	0.11*	0.17*	0.07	0.02	0.7
	Treated	1.69±0.07	0.18*	0.24*	0.00	0.09*	–
Smonitza	Control	1.60±0.09	0.09	0.15*	0.09*	–	–
	Treated	1.69±0.10	0.18	0.24*	–	–	–
Pseudogley	Control	1.45±0.11	0.06	–	–	–	–
	Treated	1.51±0.10	–	–	–	–	–
D e p t h 30 cm – 60 cm							
Chernozem	Control	1.56±0.11	0.13**	0.21**	0.009	0.01	0.13**
	Treated	1.69±0.09	0.26**	0.34**	0.04	0.12**	–
Smonitza	Control	1.57±0.12	0.14**	0.22**	0.08**	–	–
	Treated	1.65±0.12	0.22**	0.30**	–	–	–
Pseudogley	Control	1.35±0.18	0.05	–	–	–	–
	Treated	1.43±0.10	–	–	–	–	–

Meaning of symbols: $^{*}p < 0.05$ $^{**}p < 0.01$.

The lowest uranium contents were reported in pseudogley soil types and the highest in chernozem soil types. The same trend is noted for both investigated depths. Established differences result from the physical and chemical properties of investigated soil types.

Table 5. Depleted Uranium (^{235}U 0.2%, ^{238}U 99.8%)*

Isotope	Mass portion, U %	Specific activity (Bq/mg OU)
^{238}U	99.799	12.38
^{235}U	0.2000	0.16
^{234}U	0.0010	2.29
		Σ 14.83

* Browne *et al.*, 1986.

Acidified soils, poor with humus as Serbian pseudogleys, are ideal for the acid soluble forms of uranium, available for plant adsorption. Such uranium forms migrate downwards the soil profile up to geochemical barriers, suitable for overturning or fixation of uranium.

Therefore, the uranium content in top soils, of pseudogley type, is the lowest in relation to other investigated soils. Soil type smonitza had the higher uranium content in regard to pseudogley and the lower in regard to chernozem. Smonitza belongs to acid soils with an average clay and dust content of around 70%, while chernozem belongs to the category of neutral soils with less content of clay and dust (50%) and higher content of organic substance. Same trend was confirmed for both plant species. Statistical analysis of the results obtained indicates that significant differences (at $p<0.50$) exist between the control and application of phosphorus fertilizers in the layer from 0-30 cm, while no such differences were found for the layer from 30-60 cm.

Statistical analysis of the results obtained indicates that significant differences ($p<0.5$) exist between the control and application of phosphorus fertilizers in the layer from 0-30 cm, while no such differences were found for the layer from 30-60 cm.

There are significant differences in the average uranium content between pseudogley on one side and the other two soil types (chernozem and smonitza) on the other side. Highest value for uranium content was in chernozem and the lowest in pseudogley. Physicochemical soil properties have effect on process of uranium migration and mobilization, and for this reason fixation of uranium in investigated soil types decreases in the following direction: *chernozem > smonitza >> pseudogley*. Generally, the values obtained for total uranium content in the investigated experimental variants were in the range from 0.65 to 1.94 ppm.

Natural content of uranium in Serbian soils is in the interval from 0.08 to 5.9 ppm, so it can be concluded that the results obtained in this investigation are within natural limits. Manojlović *et al.* (1989) did not determine statistically significant differences in uranium content in non-fertilized Serbian soils and soils treated with phosphate fertilizers.

3.2. Uranium Uptake of Plants due to the Application of Phosphate Fertilizers

There is no doubt that with mineral phosphate fertilizers uranium enters into the environment. Ways of distribution and the risk of inclusion of uranium in the food chain are different for each sample site. The uranium pathway depends on the geomorphology, physical and chemical properties of soils, groundwater, physiological and morphological characteristics of plants that are treated with these fertilizers, and risk assessment cannot be generalized. For this reason, different opinions in the research are logical (Stojanovic *et al.*,

2006; Vukašinović *et al.*, 2010; Stojanovic and Milojkovic, 2011). The literature is generally referred to for the research in the degree of uranium adoption by different plant species grown under controlled conditions in pots with the addition of uranyl nitrate *or in natural condition* on the contaminated soils of anthropogenic origin (Stojanovic *et al.*, 2009; Stojanovic *et al.*, 2010).

There are contrasting reports on the effect of phosphate in the soil solution on uranium plant uptake. Laroche *et al.*, (2005) reported that increasing phosphate concentration, from 0 to 15 μM in nutrient solution in a hydroponics culture, decreased free uranyl concentration about 43%, 68%, and 12% at pH 4.9, 5.8, and 7 respectively. Furthermore, Rivas (2005) reported no significant effect of phosphate on the uranium concentration and uranium uptake of faba bean (*Vicia faba*), and uranium concentration of maize (*Zea mays*) on a contaminated soil. On the other hand, Ebbs *et al.,* (1998) reported that addition of phosphorus had largely overcome the toxic effects of uranium in a hydroponics solution at pH 5.0, most likely due to complexation of the uranium with phosphate.

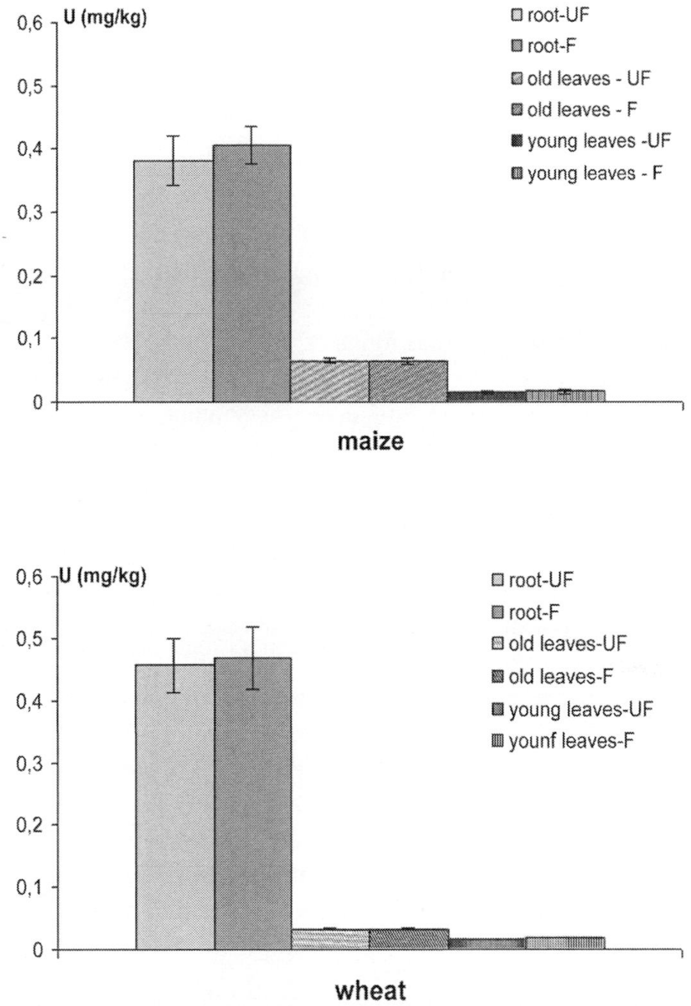

Figure 2. Mean concentrations and standard deviations (*n* =5) of uranium in maize and wheat.

Negative effect of phosphate in soil culture on uranium concentration in plant tissues or total plant uranium uptake has been reported by some authors (Lamas, 2005; Rivas, 2005). The total uranium uptake by the maize decreased when the phosphate fertilizer was used on contaminated soil (Rivas, 2005). However, increasing effects of phosphate (by applying P-fertilizer) on uranium concentration and uptake by hay on weakly acidic soil, with a pH of 5.3-6.7, was reported by Ananyan (1991).

Contrasting effects of phosphate on uranium plant uptake can be explained by different types of uranium phosphate complexes that may form depending on soil or pH solution and types of crops. Phosphate can decrease concentration of free uranyl cation in solution and the type of complexes varies according to kind and amount of electrical charge. Then in a pH higher than 6, the major uranium phosphate complex is $UO_2PO_4^-$ and it can be readily taken up by plants. In the pH range of 4 to 6, and less than 4, the most abundant uranium phosphate complexes are UO_2HPO_4 (aq) and $UO_2H_2PO_4^+$, respectively, and it seems that they are not available to plants. Pulhani *et al.*, (2005) studied the effect of two soil type, unfertilized and fertilized, on the soil–plant (wheat) transfer of uranium and other radionuclides, and concluded that the availability of essential nutrients like calcium and potassium regulates the uptake of nonessential elements, also the soil properties such as clay mineral and soil amendments such as fertilizer application strongly affect the uptake, retention, and distribution of radionuclides in plants.

The presence of phosphate contributes to precipitation of uranyl ions as uraninite, autunite, and uranium phosphite [$U(PO_3)_4$] low solubility, to $Kp = 10^{-49}$ as a basis of their stability in the long geological period under very different conditions, important from the point of view of environmental protection and production of healthy and safe food (Stojanović, 2006).

It is confirmed that the contents and forms of phosphorus in the soil as well as soil pH, redox-potential, plant and mineral composition of the solid phase, concentration and type of inorganic and organic compounds in soil and soil solution, temperature, humidity and pressing, potentially accessible forms of uranium may become bioavailability for plants adoption (Sarić *et al.*, 1995; Vandenhove *et al.*, 2007; Tunney *et al.*, 2009; Stojanović *et al.*, 2009; Stojanović *et al.*, 2010). Through the processing of this perennial results through interaction of the three factors (soil type, plant organs, and nutrition variation) shows that there were no statistical differences between the investigated plant organs of fertilized wheat and maize in relation to controlled one. For both plant species, the largest uranium content was in its root, in wheat more than in corn, then in the old leaves, and finally in young leaves.

Perennial results indicate that the contents of uranium in grain of wheat and maize in all of analyzed samples in all locations were below the detection limit of method (<0.01 mg/kg) for both variants of nutrition (Stojanović, 2006). Figure 2 shows the differences of average content of uranium in the plant organs for both plant species in all three types of soil regarding the nutrition variations (unfertilized-UF and fertilized-F)

It can be assumed that the application of phosphate fertilizers changes the relative ratio of fixed and exchangeable forms of uranium and on that relationship affect: type, amount of fertilizers and soil properties (Rafsanjani, 2009; Yamaguchi *et al.*, 2009).

3.3. Available Forms of Uranium from Phosphate Fertilizer

Interest of many studies was based on the method of determining the available form of uranium for adoption by the plant.

Tessier *et al.*, (1979) defined procedure of the fractional extraction method which determined the contents water-soluble and specifically adsorbed and structurally related forms of uranium in different types of soil.

Yamaguchi *et al.*, (2009) found that long-term application of phosphate fertilizers causes accumulation of U in the Japanese surface soil of agricultural fields, whereby soil organic matter is an important pool of U in upland field and pasture soil, whereas poorly crystalline Fe/Al minerals are important pools of U in paddy soil experiencing alternating changes in redox conditions.

Stojanović *et al.*, (2011) studied the influence of soil type and physical chemical properties on uranium sorption and bioavailability in order to reduce the uncertainty associated with this parameter in risk assessment models. The tests were conducted on three types of Serbian soils, alluvium, chernozem, and gajnjaca, in total there were 67 samples. Determinate the content of total and available form of uranium and phosphorus, the degree of acidity (pH_{KCl}), and humus and their correlation in order to define the dependence of the dominant factors on the mobilization of uranium. Content accessible form of uranium according to the type of soil is declining in number: gajnjaca> alluvium>chernozem. Between pH values and the content available uranium in chernozem and gajnjaca found the mean correlation dependence, statistically significant at the level of significance of 99% and the alluvium at the level of significance of 95%. Correlation coefficients in all cases were negative, indicating that the reduction in pH increases the mobility of uranium and thus its availability for the adoption of the plants. All contents of total uranium (of 1.2 ppm - 2.91 ppm) in 67 samples of tested three types of soil were less than 4 ppm, which is considered as a natural uranium content in the lithosphere (UNSCEAR, 1977), and therefore lower than 5.9 ppm, which is considered the upper limit of the natural uranium content in the soil of Serbia (Jelenković, 1991). According to UNSCEAR (2000), average content of uranium in the soil worldwide is about 3.0 ppm (or 35 Bq / kg for ^{238}U isotope). According to Lamas (2005), natural uranium content in soil varies from 0.1 to 11.2 mg kg^{-1} in different areas.

The contents of accessible forms of uranium were in the range of 0.027 ppm – 0.211 ppm and from 1.88% - 7.25% of the total. Literature data indicate that the contents of mobile forms of natural uranium present in the total uranium geochemical origin, with 0.1% - 3%. All values over this threshold may indicate anthropogenic origin of uranium. The highest and lowest content available form of uranium was determined on gajnjaca which leads to the conclusion that it cannot be attributed to the predominance of one type of soil over another, in the form of available content of uranium (Stojanović *et al.*, 2011a). Soil pH was the only dominant factor that significantly controlled the uranium value without further significant contribution of other soil parameters.

The same conclusions came from Echevarria *et al.*, (2001), thus, soil pH should be the focus variable for reduction of the uncertainty associated with the soil Kd value used in environmental risk assessments, even for reducing the uncertainty in site-specific Kd values, that any other effect of soil properties on U sorption is hidden.

Table 6. Amounts of DU (tons) dispersed into the environment, during previous wars, according to various sources

Country	Period of conflict	DU dispersed into the environment [tons]	Source
Iraq and Kuwait Gulf War	1990–1991	300- 321	Bleise *et al.*, (2003); Hamilton (2001); Handley-Sidhua *et al.*, (2011).
Gulf War	1990–1991	258	The Royal Society, (2001).
Bosnia–Herzegovina	1994–1995	2,750	The Royal Society, (2001).
Bosnia and Herzegovina	1995	3	Hamilton, (2001); Bleise *et al.*, (2003).
Serbia and Montenegro (Kosovo conflict)	1999	10	Bleise *et al.*, (2003); UNEP, (2001); Zunić *et al.*, (2008); Rajković and Đorđević, (2006).
Serbia and Montenegro (Kosovo conflict)	1999	11	Hamilton, (2001); Bleise *et al.*, (2003).
Kosovo conflict	1999	8.5	The Royal Society, (2001).
Iraq War	2003	170-1700	MOD, (2008); UNEP, (2007).
Iraq War	2003	> 30	Jia *et al.*, (2006)
Iraq War	2003	1000	Brown, (2009).

Table 7. Certain properties of the tested Serbian soil: alluvium, gajnjaca, and chernozem

Soil type	Utot (mg/kg)	Uav (mg/kg) Uav/ Utot %	P_2O_5 tot ,%	P_2O_5 av (mg/100g)	pH KCl	Humus (%)
alluvium n=25	1.2–2.50	0.062–0.131 3.6%–6.42%	0.081–0.29	2.5 – >36	3.90–7.40	1.66–5.40
Gajnjaca n = 20	1.40–2.91	0,021–0,211 5.10%–7.25%	0.080–0.210	0.9 – >36	4.10–7.20	1.96–3.69
chernozem n = 22	1.8–2.6	0.09–0.113 3.62%–6.09%	0.100–0.368	0.7 – >36	4.90–7.50	1.86–4.18
n = 67	1.2–2.91	0.021–0.211 3.6%–7.25%	0.080–0.368	0.7 – >36	3.90–7.50	1.66–5.40

4. Depleted Uranium - Characteristics

Depleted Uranium (DU) is a by-product of nuclear fuel cycle that occurs in the process of isotope ^{235}U separation from the natural uranium, whereby its concentration decreases from 0.7% up to 0.2-0.3%. The use of uranium as nuclear fuel requires its enrichment with isotope

^{235}U so that the percentage of the uranium-235 isotope is increased, typically to 3 to 5%. In this way the extracted depleted ^{238}U is no longer economically valuable and has a status of radioactive waste (Rajković, 2001). It is estimated that worldwide over 1.1 million tons of DU lie on dumpsites, a minimum of 46 thousand tons are added to this every year.

The term depleted (that implies safety) is based upon the fact that dangerous γ -and β - radiation has very low intensity and α-radiation, does not take into account its destructive biological effects. Some basic data on DU are given in Table 5.

DU has a specific activity of 14.83 Bq/mg which is approximately 40% less radioactive than natural uranium (25.4 Bq/mg). Besides its low-radioactive material (specific activity of 14.83 Bq/mg), its Gamma radiation rate is also low, -48 keV. The concentration of ^{234}U is very low, according to its small atomic mass (Morgan, 1980).

Chemical properties of natural and depleted uranium are identical and their chemical toxicity, which was discovered two centuries ago, does not depend on isotopic composition. Depleted uranium has been approved by the U.S. Department of Defense for the production of ammunition and missiles containing the 0.2% of the isotopes ^{235}U and 0.0003% of isotope ^{236}U (AEPI, 1995). Penetrator is made of an alloy composed of depleted uranium (99.25%) and titanium (0.75%) (Žakula, 2002).

The following characteristics of depleted uranium make it as a very suitable material for a munitions production:

- Depleted uranium is very dense (19.05 g/cm^3), almost 70% denser than lead (11.35 g/cm^3)
- Kinetic energy of the bullet with depleted uranium does not cause any explosion. Its fragments are passing through the solid target and a flame appears due to the pyrophoric properties of the uranium and extremely high temperatures during the impact (SAIC, 1990).
- Ammunition with depleted uranium has large breakdown strength. Increased impact effect of 30 mm caliber is able to penetrate through 6-9 cm thick steel improving the ballistic missile trajectory.

4.1. Use of DU Ammunition in Past Wars

In all the wars in the past 20 year (1991 Gulf War, the Bosnia and Serbia war, and the 2003 invasion of Iraq) approximately 1.4 million DU missiles were used. During Gulf War I (1990–1991), approximately 320 tons (equivalent to over 1 million 30-mm rounds) of DU munitions were used by the US forces (Bleise *et al.*, 2003), and approximately 1 ton of DU was fired from UK tanks (Bleise *et al.*, 2003). During the Bosnia–Herzegovina conflict (1994–1995), approximately 3 tons of DU were fired in NATO air strikes, and about 10 tons of DU during the 1999 Kosovo conflict (Bleise *et al.*, 2003). During the 2003 Iraq War, approximately 2 tons of DU were fired by the UK MOD (MOD, 2008), the amount of DU by USA forces has not yet been disclosed, but speculative figures range between 170 and 1700 tons (UNEP, 2007).

Generally there is no precise data on the amounts of the used DU ammunition during the past wars. Table 6 presents some information on DU dispersed in the environment found in different sources.

4.1.1. Depleted Uranium on the Territories of Former Federal Republic of Yugoslavia

During the Kosovo conflict, NATO aircrafts had used DU weapons on many locations. According to data from the Pentagon, hundred missions flights of the A-10 "Thunderbolt" aircraft launched about 31,000 rounds only at Kosovo and Metohija (at 112 locations) and 3000-5000 bullets at southern Serbia (12 sites) and at the Lustica Peninsula, Cape Arza near Herceg Novi, Montenegro (2 sites). That means that according to official data, NATO had landed on Yugoslavia around 10 tons of depleted uranium (UNEP, 2001; 2002a, b).

According to estimates of Yugoslav Army at Kosovo and Metohija were launched around 50,000 cartridges or about 13,550 kg of ammunition containing depleted uranium. In the tested soil samples from most endangered sites, areas of Vranje, Bujanovac, and Presevo; near the Macedonian border, during the NATO aggression measured radioactivity was 230 to 1100 times higher than the allowed, in the range of 5580 to 235,000 Bq/kg.

4.2. Conduct of Depleted Uranium Ammunition after Launching

After firing, the projectile with depleted uranium burns for about 10% of its weight during the flight. Confronted with the target, 18-70% of the penetrator body ignites and oxidizes in the flame (US Army Chemical School, 1995).

Kinetic energy of a penetrator does not results in an explosion due to a relatively low melting temperature of uranium (1132 °C) and its pyrophoric properties. When confronted with target, frontal surface of penetrator lights up and releases the extremely high energy which causes an increase in temperature to about 800 C (Morgan, 1980; SAIC, 1990). Because of its extremely high density penetrator goes through 6-9 cm thick steel with a very characteristic small, round entrance hole. Bottom of For Thus, depleted uranium accumulates (deposits) on the surface of land in the form of burnt bullet pieces, fine dust particles (< 5μm) or by combustion exceeds into uranium oxides. Because of its thermo-chemical instability uranium primarily occurs in U (VI) oxide form of UO_x^{2+} (where $0 < x < 0.4$). Further oxidation lead to the formation of mixtures oxides containing insoluble UO_2 and water soluble UO_3 building uranyl ion, UO_2^{2+} (Rostker, 1998; Ebingen *et al.*,1998; Harley *et al.*,1999; AT SDR, 1997).

Under the oxidizing conditions uranium, in the soluble form of uranyl ions, is able to move through the environment and come into contact with all living organisms. Under the reduction conditions, most of the uranium is in the solid state insoluble form of ceramic oxides (Rajkovic, 2001). The main threat to the environment originates from produced aerosols. At the initial phase their dominant presence is in the air, and subsequently in the soil and water. Contamination of the environment, after impact of the missile with depleted uranium into a solid target.

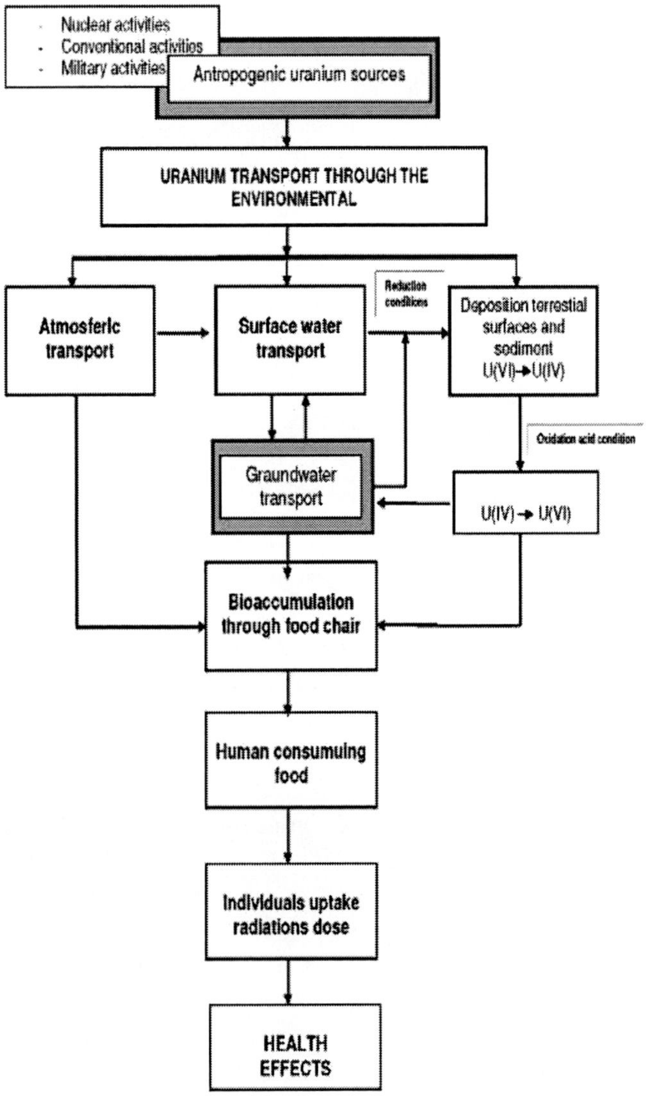

Figure 3. Uranium environmental pathways and potential risk to Human health.

4.2.1. Dept Contamination-Penetration of the Undamaged (Unburned) DU Missiles in the Soil

The behavior of DU penetrators have been studied Erikson *et al.*,(1990), who determined that the uranium from these penetrators mainly corrode into the water soluble hydrated U (VI) oxide. Migration potential of the uranium initially depends on the chemical composition of the local soil and water and DU oxidative product. The mobility of the dissolved uranium product is subject to Eh, pH and the presence of the complex agents at local ground waters.

Uranium (VI) exists in the solution as uranyl ion (UO_2^{2+}) and its more mobile than U (IV) because it easier builds soluble complexes with complex ligands from the water. The presence of carbonates and the phosphates have been found to be very important for these processes. Transportation of soluble uranium may also be accelerated by dilution, as it reduces the concentration of uranium in ground waters and the surface waters. These reactions include ion-exchange reactions and the specific adsorption of uranium with the organic substances, clay minerals, Fe (III) and the oxy hydroxide which are commonly present in soils.

It was determined that land, if it is acidic, can cause an increase mobility of uranium ions, as well as the balance of sorption processes. Authors have also studied adsorption capacity of different soil types and found that soils with high content of carbonates have the lowest absorption capacity, probably by building a soluble uranyl carbonates like UO_2CO_3, $[UO_2(CO_3)_2]$ or $[UO_2(CO_3)_2]$. Under the aerobic conditions, iron also could play a crucial role in the control of uranium movement in soil. Uranium can built complexes with iron which is located in number of soil salts. Furthermore, uranium can adsorb humic acid when present in soil. Complexation with organic compounds affects the reduction rate of uranium migration through the land even for several orders of magnitude. (Duff *et al.*, 2002). Uranium (including depleted uranium) environmental pathways and the potential risk to human health are shown in Figure 3.

4.2.2. Dotted Contamination: Contact with the Burnt Remains of the Projectile

In case of that burnt remains of the projectile (penetrators) with depleted uranium remain on the ground surface, a point contamination directly around the place where the missile fell can occur. On these locations, the surface soil is contaminated with small fragments of depleted uranium oxides or parts of the projectile liner. After the removal of small fragments and uranium oxides, the liner parts of surface layer of soil 3-5 cm thick, speed of impulses falls abruptly. On Cape Arza there were 144 such locations (Vukotic, 2002).

4.2.3. Superficial Contamination: The Origin of Uranium Oxides Aerosol

When the penetrator strikes a solid target and fires oxides of uranium form aerosols comprising of particles microscopic in size, invisible to the naked eye and ordinary microscope, but detectable by radiometric methods. According to some literature data, 50-96% particles in the air have a diameter smaller than 10 microns (usually about 5 microns) and 17-48% of these particles are water soluble. All particles generated when confronted with the target are potentially respiratory toxic (ARDEC, 1991). Depending on the time zone conditions (wind, the type precipitation) depleted uranium aerosols can be scattered from 150m to 5km away from the points of impact (Gršić, 2002; Parkhurst *et al.*,1995).

Monitoring at Cape Arza was performed of the air during cleaning-up in the spring of 2002. Collected air samples were analyzed by high-resolution Alpha Spectrometer. The obtained value for the concentration of uranium was approximately ten times higher than from those referent for the air (Radenković, 2004).

4.3. Monitoring of Radioactivity in Serbian Territory

Keeping in mind that during the NATO bombing campaign in 1999, ammunition with depleted uranium had been used, the investigations regarding the consequences of depleted uranium application followed immediately. The aim was to review all the aspects and impacts of ammunition with depleted uranium: the military, contamination of the environment and negative consequences to human health, influence on local population as well as the ecological and social law aspects. Monitoring of the environmental radioactivity in the system includes a vertical analysis: air - precipitations - land - water - plants - animals - a man (Kathren, 1986).

The results have showed that Uranium, applied to the soil as Uranium oxide, is dissolved and that it can be absorbed by plants by physicochemical and biological processes.

After three years up to 40% of the supplied Uranium was converted into mobile species. Such mobile Uranium species can either be absorbed by plants or leached from the soil to water bodies. In the tests of FAL the Uranium contents supplied by the plants directly depended on the Uranium concentrations in the soils. In respect to the total Uranium content of the soil 0.4 - 0.6 %, or in respect to the available Uranium share 5-6 % went to above ground plants from the soil. Radenković *et al.*, (2008) investigated what happened with released DU corrosion products after three years elapsed for soil contaminated during the military actions (Bratoselce, Southern Serbia), by applying a sequential extraction procedure as the most valuable technique for geochemical fractionation and speciation of radionuclides in soils. Depleted uranium penetrator was found in the soil at 0.5 m depth, covered with oxides.

Table 7. Specific activity of particularly radionuclide in soil samples from locality of Pljačkovica during 2002 (Bq/kg)

Type of sample and location	^{40}K	^{232}Th	^{238}U	^{235}U	^{226}Ra	^{137}Cs
soil from the projectile hole	445±15	41±2	2272±8	19,6±0,8	22±2	4,6±0,4
land within the enclosure area	429±15	39±2	77±26	3,3±0,5	45±8	277±8
land below the relay	588±19	49±3	149±21	4,9±0,5	68±3	29±1
land outside the enclosure area	574±19	39±2	70±30	3,7±0,5	49±8	85±3
lichen below the relay	203±13	9±3	< 48	< 2,9	< 0,4	136±56
moss within the enclosure area	110±4	9,3±0,7	16±0,7	0,7±0,2	15±3	28,3±0,9
moss outside the enclosure area	543±20	42±3	102±20	4,0±0,5	53±10	105±3
vegetation below the relay	172±6	4,5±0,4	23±5	0,8±0,1	11±2	2,0±0,1
cortex of wood outside the enclosure	58±4	< 2,1	< 16	< 1,8	< 12	10,8±0,8
drinking water, Bujanovac	58±4	0,03	<0,08	<0,003	< 0,03	< 0,004
drinking water, Mečkovac	0,11±0,	< 0,02	< 0,11	< 0,07	< 0,11	< 0,006
sludge from the tank, Balinovac	324±20	27±2	48±20	1,0±0,3	24±5	1,1±0,4
rabbit	87±3	< 0,2	< 2,3	< 0,1	< 2,1	0,16±0,05
grapes, Balinovac	34±2	< 0,8	< 4,7	< 0,2	< 2,7	0,19

Soil samples were collected in the surrounding impact zone at the position of the projectile entrance into ground, the path through the subsurface ground, soil layer just next to DU penetrator, and different distances down along the depth profile. The specific activity of ^{238}U in the soil sample taken at the spot of the projectile entrance into the ground was about 90 kBq/kg, in the projectile path it was near 78.5 kBq/kg and in the soil layer next to the penetrator it was about 263 kBq/kg and contamination decrease with the distance from the source through the soil profile.

The samples were collected from the enclosure which the Yugoslav Army designated as a contaminated area and assumed non-contaminated area, outside of the enclosure. During 2001 and 2002, sampling was conducted within monitoring program on the territory of Serbia, partly within the framework of an organized mission of UNEP and former Federal Secretariat for labor, health, and social welfare. Collecting samples of drinking water was carried out four times per year from the water used by local residents. The obtained results are shown in Table 7 (Javorina, 2004).

The results of measurements showed good agreement of natural radionuclides ^{40}K, ^{232}Th, ^{226}Ra and their homogeneous distribution on tested micro-locations. Contamination originating from the depleted uranium was not homogeneous, which is illustrated through the change of specific activities ^{238}U and ^{235}U.

Dependence of the fractionation on the contamination levels and presence of weakly bonded, depleted uranium in the hot spots areas was evident (Radenković *et al.*, 2008). Depleted uranium fractionation indicated an increase of mobile fraction of uranium in contaminated sites than the control, which confirms the rule that the uranium originating from anthropogenic sources is more easily included in the food chain from natural uranium.

Important indicator of the presence of depleted uranium in the soil is a dysfunctional relationship isotopes activities $^{234}U/^{238}U$, $^{235}U/^{238}U$, and $^{236}U/^{238}U$, which is entrenched in the natural uranium. Just this amended isotopic ratio of activity is detected in the soil samples collected in Serbia and Montenegro, which confirms the presence of DU (Radenković *et al.*,2007; 2008).

4.3.1. Examination of the Contents of Radionuclides in Human and Animal Feed

UNEP / UNCHS Balkans Task Force (BTF) have performed a theoretical evaluation of the inhalation of depleted uranium during the utilization of DU - ammunition (BTF 1999; UNEP, 2000):

- *instantaneous inhalation of depleted uranium dust* after the impact into a solid target, with a maximum input of 100 mg of DU near the target and instantly after a hit, may lead to acute chemical toxicity and the total effective radiation dose caused by inhalation of DU of a minimum 10 mSv ;
- *inhalation of depleted uranium aerosols in the immediate vicinity of the target,* under the assumption that approximately 6 mg of DU is in 1 mg of dust in the vicinity of the target and in the area around the target surface of 1,000 m^2. Depending on the concentration in the air, this assumption corresponds to the concentration of DU in the air from 0.3 $\mu g/m^3$ (in the air "normal" density) to 30 $\mu g/m^3$ (in a very "dense" air). During a continuous exposure, this value corresponds to effective dosage for inhalation from 0.3 to 30 mSv / year;
- *ingestion (consumption) of vegetables* (or generally food of plant origin) contaminated with dust with DU which occurs by using the leafy parts of plants, that were on the surface of the contaminated dust, during the first weeks after exposure with 100 mg of depleted uranium. This quantity is very important in terms of chemical risks. The resulting radiation dose is the order of 0.1 mSv, (BTF 1999). The scientists found out that the mobilization of Uranium grows with a decreasing fertility of the soil (minor pH value, less content of mineral plant nutrients, especially phosphorus). In the tests of FAL the uranium contents supplied by the plants directly depended on the uranium concentrations in the soils. In respect to the total uranium

content of the soil 0.4 - 0.6 %, or in respect to the available uranium share 5-6 % went to above ground plants from the soil. (Schnug, 2005)

- *ingestion of contaminated groundwater* as a result of the reactions between soil and groundwater, which leads to the migration of uranium through land. This reaction involves precipitation of secondary uranium minerals, ionic exchange of uranium in clay and specific ionic adsorption of uranium on the surface of minerals. All these types of mass transfer reactions can lead to a decrease in uranium concentration in groundwater across the land (Crancon, 2001).

In utilization of groundwater contaminated with depleted uranium in the areas where the target was located, BTF provides the next assessment:

"The concentration of depleted uranium in groundwater can exceed 1 mg/dm^3. At this concentration of uranium, it can be noticed as chemical toxicity. The current radiation dose caused by using contaminated water is 1 mSv (BTF 1999).

5. Health Aspects of Depleted Uranium

There is a plenty literature that enables the health impact assessment of uranium and its compounds on the health of workers exposed to the uranium impact by its mining, milling or working in the nuclear plants (Polednak and Frome,1981; McGeoghgan and Spinks, 2000). But the negative health effects of uranium in the form of depleted uranium used as armor-piercing has been accentuated after the serious health problems observed both by soldiers – veterans, and at local populations, that has occurred after its application in the military conflicts in the Gulf and the Balkan wars (Durakovic, 2003; Milacic *et al.*, 2004; Bem and Bou-Rabee, 2004). Although many researchers and authors of different reports and studies are diminishing the effects of depleted uranium to human health, there is a well-documented evidence of both chemical and radiological toxic properties of uranium isotopes that has been an area of numerous research studies and scientific reports on its organotoxic, mutagenic, teratogenic, and carcinogenic effects (Durakovic, 2003).

We can talk about two types of health risks associated with uranium exposure: immediate health risks that include kidney diseases and respiratory problems, and the long-term risks which include the lung, bone, or other types of cancer. The chemical toxicity and radiological effect of uranium and its compounds can cause adverse health effects to human population.

The toxicity type is determined by the form of uranium – low soluble aerosols of uranium will predominate in radiological toxicity, while the soluble metal forms will induce the chemical toxicity. The chemical toxicity of this radionuclide element and its compounds is similar to that for other heavy metals, for example cadmium or mercury, due to its chemical affinity to proteins which seriously disturb the process of re-absorption of glucose and amino acids

If we are considering radiological effects of depleted uranium ammunition, we can split it into two parts: the acute risk of external exposure, which is by some author negligible and internal radiological exposure, which comes from intake of depleted uranium, and which cannot be venial (Bem and Bou-Rabee, 2004). Although the depleted uranium can be considered to have less radiological impact than uranium, it has stronger chemical toxicity because of its high linear energetic transfer (LET) irradiation, the long time of body extrusion (5,000 days) and the ability of different tissue deposition (Milacic *et al.*, 2004).

Figure 4 represents the main pathway and the final targets for human body contamination by any form of uranium. The picture is based on many different published reports that have cited consequences of exposure to uranium, or depleted uranium, with reference to different negative effects on human health such as kidney or lung damage, diverse type of organs or tissues cancer, DNA damage (Miller *et al.*, 2002), mutagenicity (Lin *et al.*,1993), and neurological deficits (Pellmar *et al.*, (1999).

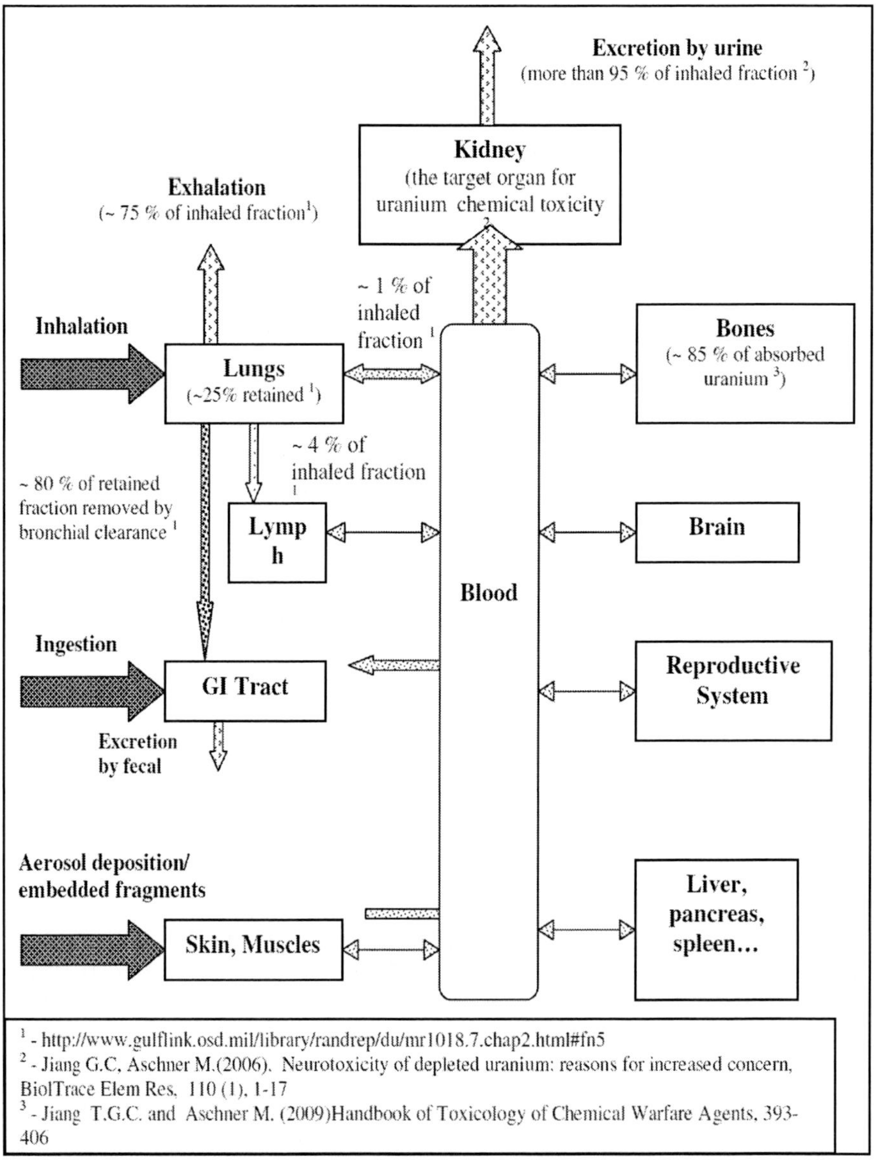

Figure 4. The main pathway of uranium in human body.

The medical effects of internal exposure to uranium are depending on many factors such as: the uranium quantity intake and the impact contamination time, including also the age, sex, genetic predisposition, and the exposure to the other materials, previous health status, radio-sensitivity of the organism, but also the diet and stress (Durakovic, 1999). The different tissue distribution studies have reported DU accumulation in the bone, kidney, reproductive system, brain, and lung, with verified genotoxic, mutagenic and carcinogenic properties, as well as reproductive and teratogenic alterations (Horan et al.,2002).

The laboratory experiments with animals that were conducted in order to assess the health risk connected with depleted uranium, have proved that DU certainly can cause appearance of leukaemia and lymphoma as well as secondary myelodisplastic syndrome (Milacic et al.,2004).

Although there is not clear evidence that uranium affects human central nervous system, the bioaccumulation experiments conducted on rats have showed that after exposure to repeated inhalation, depleted uranium is able to enter the brain, producing strong behavioural changes (Monleau *et al.*,2005).

As a consequence of uranium entry into the organism, numerous health risks appear depending on four main factors: isotopic composition, chemical composition, particle size and solubility. The same emerging pathogenic significance of uranium has chemical toxicity and radioactivity of gamma emission $^{(235}U)$ and alpha emission (^{238}U). Kinetics of uranium in the body is very specific. Blood is a fundamental transport factor in the distribution and redistribution of toxic uranyl (U^{+6}) and uranium (U^{+4}) ions. Active role in the transport of uranium have red blood cells and plasma proteins. For about 60 minutes 95% uranium disappears from the bloodstream. The biological half-life of uranium in the body is 2 to 5,000 days (Rajkovic, 2001).

The common exposure pathways of depleted uranium are as following:

- *Inhalation* (This is the most likely way of uranium intake. During the military impacts by depleted uranium munitions or after a plane crash, certain aerosols are produced. These aerosols can be spread out in the huge areas around the impact sites, and thus be inhaled by civilians and military personnel, leading to their further contamination (the particles smaller than 10 μm, so called repirable fraction, can reach deeper pulmonary regions (bronchiole and alveoli) and deposit there for considerable time). In this case, not more than 1% of inhaled insoluble uranium is finally transferred and retained in the target organ—kidney (ICRP-72, 1995). The re-suspended particles of depleted uranium that are not inhaled are further deposited on the ground soil and vegetables, which then leads either to the contamination of underground water or the DU incorporation into the food chain);
- *Ingestion* (This way of body entering will occur by intake of food and water that are directly contaminated with certain amounts of depleted uranium. It might also come from hands coming in contact with contaminated materials (e.g., skin, clothes, soil…). Considering this way of uranium intake, the special concern should be paid on young children who might play in the contaminated areas and by hand-to-mouth activity ingest contaminated soil);
- *Dermal contact* (Relatively negligible pathway of depleted uranium intake since the amount of depleted uranium that will directly pass through the skin and enters a body

is vanishing, unless if we are not considering the open wounds or embedded parts of ammunition).

5.1. "Invisible Threat "of Depleted Uranium in Serbia after Military Conflict in 1999

The official data about the health conditions of populations directly or indirectly exposed to DU in Serbia after military conflict in 1999 are pretty precautionary and scanty. The data presented in different studies are at least careful and discreet because no one wants to provoke 'uranium panic' especially with the lack of money for meaningful objective international research and continuous and appropriate clean-up methodology.

In the paper presented in TI Conference, Pantelic and Golubovic (2008) have exposed the following: the data published in Serbian medical journal *'Praksis Medika'* presents that in 1997 the number of patients in Kosovska Mitrovica hospital (pneumonic-physiological department) with lung tumors was 2.6%, while in 2000 this number has significant increase to 22%. The same research presents that the urology department of the same hospital marked 1.6% patients with tumors, while in 2000, only one year after the NATO bombing with DU ammunitions, this number was 22%. The same source (Pantelic and Golubovic, 2008) also indicates that in Balkan symposium held in 2005, the medical section of Serbian Academy of Science has presented data concerning pregnancies, births and malformations of new born/dead babies in Serbia, in the period from 1989 to 2001.

All the data collected from 30 Serbian Health centers were elaborated according to appropriate WHO methodology. These results indicated that the report in Serbia presented significant decrease in the number of births while the number of spontaneous miscarriages (abortion) had a big increase. They also highlight that after the 1999 the number of dead newborn babies had drastically increased from 50 to 357, while the number of dead babies born with malformations increased from 121 to 610. In this paper, the authors have also stipulate that each year in Serbia has registered more than 30,000 new cancer patients which is, regarding the overall Serbian populations of ~7.1 million people, much more than average rate in Europe.

The authors of *The Consequences of NATO Bombing on the Environment in Serbia,* Milačić and Simić (http:// irpa11. irpa.net /pdfs/ 1d11.pdf) have published the results of investigations that were conducted in order to assess the consequences of DU use on human health and environment after the NATO bombing. They have compared the health status of the population from Vranje region (South Serbia) and Kosovo that were directly exposed and the nonexposed populations and underlined that one of the examined groups from DU contaminated region have had certain changes in blood cells that were mainly morphological, in cytoplasm, nucleus and chromosomes, while some samples of urine of mentioned group were also contaminated with DU. The authors have concluded that although they could not expect the direct influences on human health, they also could not exclude the later consequences. This is one of the reasons why longer and bigger scientific investigation should be necessary conducted!

The general conclusion of many authors, no matter which emphasis their investigation took: in order to get insight on uranium or DU toxicity on human health, the increased and objective scientific fieldwork with appropriate laboratory measurement without any political or any other interest is necessary!

6. Hybrid Technologies for Remediation of Uranium Polluted Soils

State of the environment and the concept of sustainable development have caused the emergence of new technological solutions that would reduce impact of human activities on the environment. Hybrid, combined, remediation technology for cleanup of contaminated media by uranium, are increasingly being used for environmental and economic reasons. Integrated management of soils contaminated with uranium, based on results obtained in Institute for nuclear and other mineral raw materials in Belgrade, have been effectively applied many times. Conceptual approach is a synergy of physical, chemical, and biological remediation processes and techniques. *In situ* treatment includes a combination of phyto-extraction and phyto-stabilization with by uranium hyper accumulator plants with functional materials based on Serbian (domestic) alumosilicate minerals.

6.1. Methods and Techniques for Uranium Removal

Remediation technologies for treatment of uranium contaminated soils and groundwater could be applied either as *ex situ* or *in situ* techniques. According to Gaverilescu *et al.*, (2009), we can classify methods and techniques for uranium removal as: Natural attenuation, Physical processes, Chemical methods, Biological methods, and Electrocinetic methods. These techniques can be applied individually or in combination (hybridization) as it is presented in Figure 5.

Each one of these fundamental technical choices will direct decision makers to substantially different paths with regard to their subsequent choices, actions, and potential results; making available significantly different technological options for application, within a remediation program, which involves multidisciplinary environmental research on characterization, monitoring, modeling, and technologies for remediation

6.1.1. Phytoremediation

In situ remediation techniques are more suitable for radioactive contamination due to reduced exposure of workers during the construction or transportation. Among the remediation techniques currently used, soil excavation is the most common treatment for radioactively contaminated soils as well as encapsulation size separation, soil washing, electrokinetics and ion exchange (Evans, 2000). However, these *in situ* techniques are expensive compared to phytoremediation techniques. Phytoremediation involves the use of plants to extract, sequester and/or detoxify the pollutants present in soil, water, and air. Plant-assisted remediation of soil can generally occur through one or more of the following mechanisms: phytostabilization, phytodegradation / phytotransformation, phytovolatilization, rhisodegradation, phytohydraulics, phytoextraction.

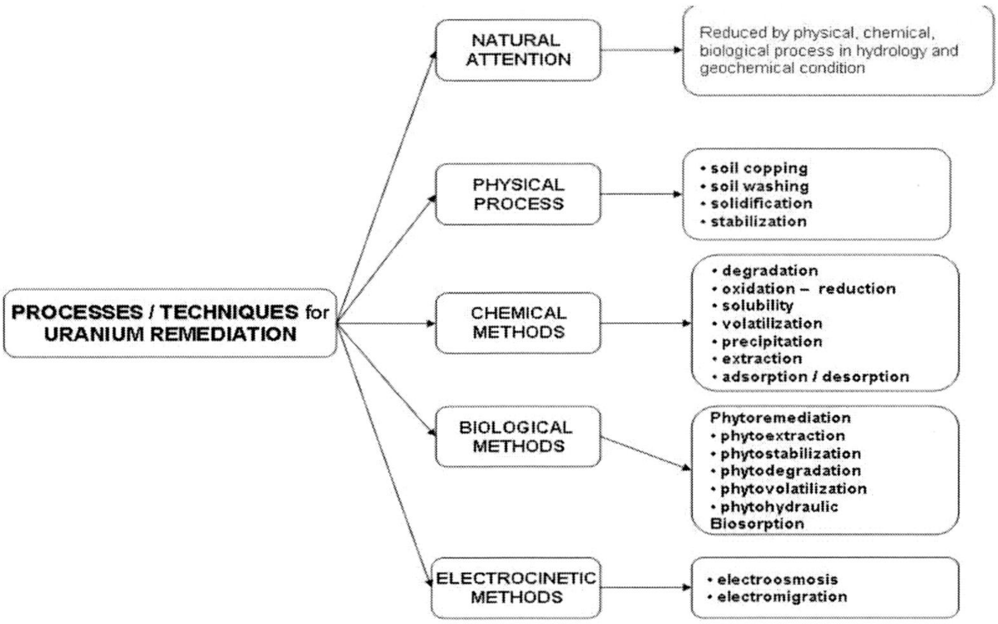

Figure 5. Processes and techniques for uranium removal.

Phytoremediation offers advantages such as: the cheap and simple option, better metal recycling, better public acceptance and less destruction to remediation sites. Phytoextraction efficiency is determined by the metal availability, which is influenced by pH, redox potential and metal complexation. The major disadvantage of the technique is the time of requirement - from 18 to 60 months or even decades. Phytoextraction, a type of phytoremediation, use of metal-accumulating plants (hyperaccumulators) that can transport metals from the soil to the roots and concentrate them in above-ground plant tissue - shoots (AbdEl-Sabour, 2007; Ebbs,1998). Hyperaccumulators plants are those plants that adopt the pollutants from the soil at a much higher rate than other plants (100 to 1000 times).

According to the PHYTOREM data base sunflower is recognized as hyperaccumulator of uranium. PHYTOREM was developed by Environment of Canada and this database consist of 775 plants with capabilities to accumulate or hyperaccumulate one or several of 19 key metallic elements. Species were considered as hyperaccumulators if they took up more than 1000 mg/kg dry weight of most metals. Plants hyperaccumulators like the sunflower had content of uranium more than 15000 mg kg^{-1} dry weight (McIntyre, 2003). There are two general approaches to phytoextraction: continuous and chemically enhanced phytoextraction. The first approach uses naturally hyperaccumulating plants with the ability to accumulate an exceptionally high metal content in the shoots. Another method is the application of synthetic and nature organic agents in order to improve the mobilization of uranium and increase efficiency of phytoextraction. A key to the success of uranium phytoextraction is to increase soil uranium availability to plants (Leštan, 2006).

In literature there have been numerous reports about amendments in phytoremediation compounds that increase the uptake of uranium by various plants. Addition of chelates increases the mobility of metals in the soil and form complexes with metals reducing the positive charges and thus affects the availability of metal to plants. Amendments could be organic compounds: synthetic chelating agents (ethylenediaminetetraacetic acid (EDTA),

N-hydroxyethyl-ethylenediamine-N,N',N'-triacetic acid (HEDTA), diethylenetrinitril opentacetic acid (DTPA)), natural fulvic acid, humic acid and more natural low molecular weight organic acids (citric, malic, oxalic, and acetic acid).

The most frequently used is EDTA, which has been reported as more effective than other synthetic chelators for several heavy metals (Stojanović *et al.*, 2011), Phytoremediation in certain situations, show superior economic, technical, and environmental advantages compared to other remediation techniques. Therefore, a great deal of research focus and investment can be expected into management of the soil–plant system in the next 50 years (Willey, 2007).

6.1.2. In Situ Immobilization of Uranium Using Alumosilicate Minerals as Sequestering Agents

Addition of chelating agents in order to enhance phytoextraction may promote leaching of the pollutants (uranium) into groundwater. During the selection of remediation technologies one of the key factors is the optimal choice of immobilization materials, based on efficiency and the price. Therefore, there is a need for application of sequestering agents (apatite, zeolite, clay, zerovalent iron, bentonite, etc.), that will enable hydrological control, immobilization and transformation of excess uranium, which the plants were unable to accept. Furthermore, sequestering agents can be used as ground cover in perennial phytoremediation, for adsorption of uranium, which can leach from fallen leaves in autumn.

Application of fast - acting adsorbents (diatomite, zeolite modified by organic agents and bentonite) with slow - acting adsorbent (natural phosphate), achieves a synergistic effect of mixtures of reactive materials like a permanent solution for the "*in situ*" stabilization of uranium ions in the remediation of soil contaminated by uranium (Milojković *et al.*, 2009; Stojanović *et al.*, 2011b). Figure 6 shows the effect of immobilization of uranyl ions with different alumosilicate materials on pH 5.5.

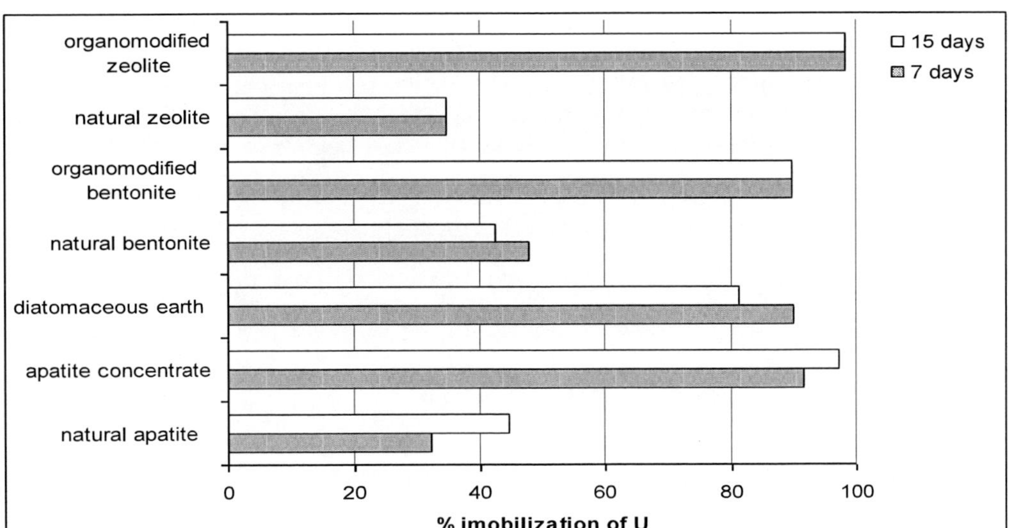

Figure 6. Immobilization of uranyl ions for different alumosilicate materials with pH 5.5.

Table 8. Efficiency mixture of apatite and immobilization of uranium OZ

Type of mixture	% unadsorbed U	% adsorbed U
100% apatite	83.33	16.67
75% apatite 25% OZ	21. 90	78.10
50% apatite 50% OZ	6.70	93.30
25% apatite 75% OZ	4.86	95.14
100% OZ	0.04	99.96

The results of adsorption and immobilization effect of apatite mixtures and organomodified zeolite (OZ) with the reaction time of 7 days are presented in Table 8.

The results show that OZ is more efficient than apatite because within 7 days it absorbs nearly 100% of uranium. With increasing content of OZ in the mixture, adsorption index increases. Thus, mixtures with 75% apatite and 25% OZ showed more than four times better adsorption characteristic than natural apatite in the same period of action.

Mixture of apatite and organomodified zeolite confirmed the synergism of action of these materials which is reflected in the rapid binding of uranium with organomodified zeolite and the formation of stable phase uranium-phosphate-autunite.

Application of this mixture eliminates the risk of desorption of organomodified zeolite due to changes in soil conditions. For the remediation of large areas of contaminated soil with a lower level of radionuclides application of natural apatite is justified (Matijasevic *et al.*, 2006; Stojanović *et al.*, 2008). Generally, remediation of soil contaminated with uranium requires a holistic approach including the use of secure "environmental friendly" materials that are cheap, easily applicable and available locally. The results confirm the applicability of the studied natural and modified aluminosilicate materials (zeolite, apatite, bentonite and diatomite) individually and in mixture.

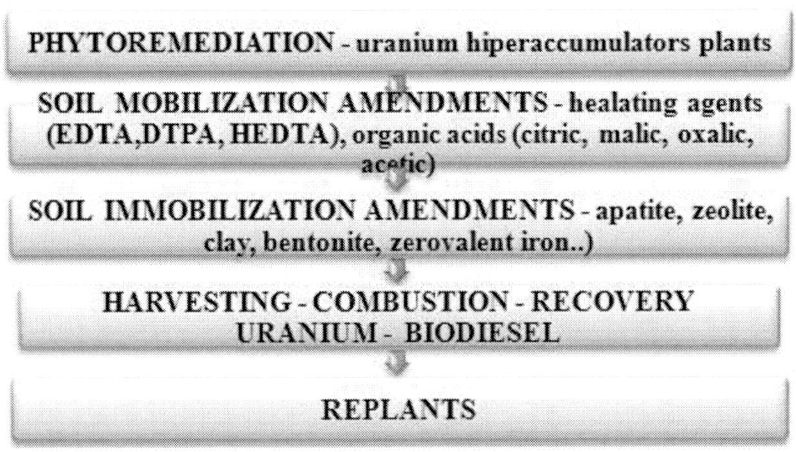

Figure 7. Hybrid technology for remediation of uranium polluted soils.

The use of adsorbents with faster and stable action, together with the materials with slower acting, provide synergistic effect of reactive materials mixtures for *in situ* stabilization of uranium ions. Hybrid technology represents a permanent solution and means for integrated management strategy of contaminated sites which include: proper selection of plants, improving mobility of uranium with amendments (organic agents) and application sequestering agents for immobilization and transformation of excess uranium where plants were unable to accept. Scheme of innovative hybrid technology for remediation of uranium polluted soils is shown in Figure 7 (Stojanović and Milojković, 2011).

CONCLUSION

Environmental contamination by radioactive substances has become one of the important problems of modern civilization. The technological revolution that is achieved by applying nuclear energy for peaceful purposes, despite the enormous progress that has been made for environmental protection in this area, immediately leads to further increase of radio-contamination of the biosphere. The potential danger from the use of nuclear weapons and the risk of accidents at nuclear power plants make this potential problem have even greater significance. On the other hand, the permanent intake of anthropogenic origin radionuclides into the environment is achieved through the products of non-nuclear industry which leads to the so-called TENORM, "Technologically-Enhanced Naturally-Occurring Radioactive Materials." The paper presents some of the aspects of NORM / TENORM sources at the global level with special emphasis on the situation in Serbia. In the past 20 years, Serbia has faced the consequences of economic sanctions, the economy transition and the NATO bombing. Lack of legislation for many decades in the field of environment protection and protection against ionizing radiation has contributed to the increased level of TENORM in the environment. On the other hand, the lack of financial resources has led to no appropriate depleted uranium decontamination of Serbian environment. One of the results of these circumstances is increasing number of people suffering from cancer since 1999 year.

In the belief that when addressing such topic as radiation always makes it necessary for the need to say or do something more, the authors would like to ultimately express this personal view of this topic. The material presented here should be treated as a kind of declaration of technological repentance, sincerely recognition that the path of human development has threatened serious unintended effects. This is also a sort of appeal to the population attitude towards the environment to recover from the abuse by seeking reasonable procedures with a nature use. It is our obligation to the world that we inherit to learn safeguards for the generations that come!

ACKNOWLEDGMENT

The authors are grateful to Serbian Ministry of Education and Science which support this research through its Project N° TR 31003 and TR34013

REFERENCES

AEPI, (1995). Health and environmental consequences of depleted uranium use in the U.S. Army: Technical Report, Army Environmental Policy Institute, Atlanta, Georgia.

Agency for Toxic Substances and Disease Registry (AT SDR) (1997). Public Health Assesment: US Army Materials Technology Laboratory Watertown, Middlesex Country Massachusetts, MA 0213820939, Department of Health and Human Services.

AbdEl-Sabour, M.F. (2007) Remediation and bioremediation of uranium contaminated soils, *Electronic Journal of Environmental, Agricultural and Food Chemistry*, *6* (5), 2009-2023.

Ananyan, V.L., (1991). Effect of fertilizers on the accumulation of uranium by herbaceous plants in various natural zones of Armenia. *Scripta Tecnica Bol.*, *23*, 118-123.

(ARDEC) Development and Engineering Center (1991). Summation of ARDEC test data pertaining to the oxidation of depleted uranium during battefield conditions, US Army Armament Research.

ASTSWMO (2011). Radiation Focus Group, Federal Facilities Research Center, Incidental TENORM: A Guidance for State Solid Waste Managers, Washington.

Bem, H., and Bou-Rabee, F., (2004). Environmental and health consequences of depleted uranium use in the 1991 Gulf War, *Environment International*, *30*, 123-134.

Bleise, A., Danesi, P.R., and Burkart, W. (2003). Properties, use and health effects of depleted uranium (DU): A general overview. *J. Environ. Radioactiv.*, *64*, 93-112.

Browne, E., Firestone, R.B., and Shirley, V. (1986). Table of isotopes. Lawrence Berkeley Laboratory, University of California. John Wiley and Sons Inc., New York.

Brown, P. (2003). Gulf troops face tests for cancer, guardian.co.uk April 25, 2003. Retrieved February 3, 2009.

UNEP/UNCHS Balkans Task Force (BTF) (1999). The potential effects on human health and the environment airsing from possible use of depleted uranium during the 1999 Kosovo conflict. A preliminary assessment, Geneva, p.76.

Boukhenfouf, W., and Boucenna, A. (2011) Uranium content and dose assessment for phosphate fertilizer and soil samples: Comparison of uranium concentration between virgin soil and fertilized soil, *Radiation Protection Dosimetry*, doi: 10.1093/rpd/ncr025.

Catalano, G.J., and Brown, E.G. (2005). Uranyl adsorption onto montmorillonite: Evaluation of binding sites and carbonate complexation. *Geochim. Cosmochim. Acta.*, *69*, 2995-3005.

Chmielewski, G.A. (2008). Nuclear fossile fuels worldwide reserves, *Nukleonika*, *53*, 11-14.

Crancon, P. (2001). Migration de l'uranium dans un Podz. Le Role des Colloides dans la Zone non saturee et la Nappe: Application aux Zlandes de Gascogne. Ph.D thesis, University of Grenoble 1, France.

Duff, M.C., Coughlin, J.U., and Hunter, D.B. (2002). Uranium co-precipitation with iron oxide minerals. *Geochim. at Cosmochim. Acta.*, *66*, 3533-3547.

Durakovic, A. (1999). Medical effects of internal contamination with uranium, *CMJ*, *40* (1), 1-18.

Durakovic, A. (2003). Undiagnosed illnesses and radioactive warfare, *CMJ*, *44* (5), 520-532.

Djuric, G., and Popovic L.D. (2001). Uranium in the environment- "Zero state" in Serbia, *Chemical Industry*. *55* (7-8), 289-294.

Duquène, L., Vandenhove, H., Tack, F., Van der Avoort, E., Wannijn, J., and Van Hees, M. (2006). Phytoavailability of uranium: Influence of plant species and soil characteristics.

Ebbs, S.D., Brady, J.D., and Kochian, V.L. (1998). Role of uranium speciation in the uptake and translocation of uranium by plants, *J. Exp. Bot.*, *49* (324), 1183-1190.

Ebingen, M.H., Essington, E.H., Gladney, E.S., Newman, B.D., and Reynolds, C.L. (1990). Long term fate of depleted uranium at Aberdeen and Yuma proving grounds, Phase I: Geochemical transport and modelling, LA-11790-MS, Los Alamos National Laboratory.

Echevarria, G., Sheppardi, I.M., and Morel, J. (2001). Effect of pH on the sorption of uranium in soils, *Journal of Environmental Radioactivity*, *53*, 2,257-264.

EPA (1999). Evaluation of guidelines for exposures to technologically enhanced naturally occurring radioactive materials, *Committee on Evaluation of EPA Guidelines for Exposure to NOTM*, National Academy Press, Washington.

EPA 402-R-00-01 (2000). National Research Council Committee on Evaluation of EPA Guidelines for Exposure to Naturally Occurring Radioactive Materials, Evaluation of guidelines for exposures to technologically enhanced naturally, Washington.

Erdem, E., Tlnklh, N., Yllmaz, T.W., Uyamk, A., and Olmez, H. (1996). Distribution of uranium in the production of triple superphosphate (TSP) fertilizer and phosphoric acid, *Fertilizer Research. 44*, 129-131. Erikson, R.L., *et al.*, (1990). A review of the environmental behavior of uranium derived from depleted uranium alloy penetrators, Pacific Northwest Laboratory, Richland, Washington, PNL-7213.

Evans, K.G. (2000). Methods for assessing mine site rehabilitation design for erosion impact. *Aust. J. Soil. Res.*, *38*, 231-248.

Gavrilescu, M., Pavel, V.P., and Cretescu, I. (2009). Characterization and remediation of soil contaminated with uranium , *Journal of Hazardous Materials*, *163*,457-510.

Gršić, Z., Jovašević-Stojanović, M., and Milutinović, P. (2002). Procena rasprostiranja aerosola osiromašenog uranijuma kroz atmosferu pri pogotku projektila u tvrdu metu, *Chemical Industry Series*. *2*, 69-72.

Hamilton, E.I. (2001). Depleted uranium (DU): A holistic consideration of DU and related matters. *The Science of the Total Environment*, *281*, 5-21.

Handley-Sidhua, S., Miranda, J., Keith-Roach, J.R., Lloyd, D.J., and Vaughan, A. (2010). Review of the environmental corrosion, fate and bioavailability of munitions grade depleted uranium. *Science of the Total Environment*, *408*, 5690-5700.

Harley, H., Foulkers, E.C., Hilborne, L.H., Hudson, A., and Anthony, C.R. (1999). A review of the scientific literature as it pertains to Gulf War illnesses: Vol.7, *Depleted Uranium*, MR-1018/7-OSD, RAND.

Horan, P., Dietz, L., and Durakovic, A. (2002). The quantitative analysis of depleted uranium isotopes in British, Canadian, and U.S. Gulf War veterans [published erratum appears in *Mil. Med.*, 2003; 168:474]. *Mil. Med.*, 167:620-627.

IAEA (2001). *http://www.wise-uranium.org/purec.html.*

International Atomic Energy Agency, IAEA (2003). Extent of Environmental Contamination by Naturally Occurring Radioactive Material (NORM) and Technological Options for Mitigation, *Technical Reports Series No*. 419, STI/DOC/010/419 (ISBN: 9201125038).

International Commission on Radiological Protection (1995). Annales: Part 4. Inhalation dose coefficients. Age dependent doses to members of the public from intake of radionuclides. *25*, 3-4.

Javorina, L., Pantelić, G., Tanasković, I., Vuletić, V., Eremić-Savković, M. (2001). Radioaktivnost različitih uzoraka iz Pljačkovice, XXI *Simpozijum Jugoslovenskog Društva za Zaštitu od Zračenja*, Kladovo, pp. 161-163.

Jelenković, R. (1991). Uranium mineralization in the Sumadija metallogenic district (Serbia): Genetic and morphostructural types. Republic Fund for Geological Investigation, Belgrade.

Jones, L.R. (1992). Uranium and phosphorous content in Morrow plot soil over 82 years, *Commun. Soil. Sci. Plant Anal.*, *23*, 67-73.

Jia, G., Belli, M., Sansone, U., Rosamilia, S., Gaudino, S. (2006). Concentration and characteristics of depleted uranium in biological and water samples collected in Bosnia and Herzegovina, *Journal of Environmental Radioactivity*, *89*, 172-187.

Kathren, R.L. (1986). *Radioactivity in the Environment: Sources, Distribution, and Surveillance*, Harwood Academic Publishers, Chur.

Kratz, S., and Schnug, E. (2006). Rock phosphate and P-fertilizers as sources of P contamination in agricultural soils. In: B.J. Merkel and A. Hasche-Berger (Eds.), *Uranium in the Environment* (pp. 57-68). Berlin Heidelberg, Springer.

Lamas, M.D.C. (2005). *Factor Affecting the Availability of Uranium in Soils*. Landbauforschung Völkenrode Sonderheft 278, Germany, Braunschweig.

Langmuir, D. (1978). Uranium solution-mineral equilibria at low temperature with applications to sedimentary ore-deposits, Geochim. *Cosmochim. Acta.*. *42*, 547-569.

Leštan, D. (2006). Enhanced heavy metal phytoextraction. In: *Phytoremediation Rhizoremediation*, pp. 115-132, Springer the Netherlands.

Lin, R. H., Wu, L. J., Lee, C. H., and Lin-Shiau, S. Y. (1993). Cytogenetic toxicity of uranyl nitrate in Chinese hamster ovary cells, *Mutat. Res.*, *319*, 197-203.

Manojlović, S., Bikit, I., Slivka, J., Vecković, M., Čonkić, L.J., and Doret B., Kruar, M. (1989). Da li đubriva koja sadrže fosfor zagađuju zemljišta radionukleidima. *VIII Jugoslovenski Simpozijum (1989) Oštećenja zemljišta i problemi njegove zaštite Žabljak,* Zbornik radova, pp.11-28.

Matijasević, S., Daković, A., Tomasevic Canović, M., and Stojanović, M. (2006). Uranium(VI) adsorption on surfactant modified heulandite/clinoptilolite rich tuff, *Journal of the Serbian Chemical Society*, *71*, 1323-1331.

McGeoghgan, D., and Spinks, K. (2000). The mortality and cancer morbidity experience of workers at Springfields uranium production facility, 1946–1995. *J. Radiat. Prot.*, ***20***, 111-138.

McIntyre, T. (2003). Phytoremediation of Heavy Metals from Soils. In: *Advances in Biochemical Engineering/Biotechnology*, *78*, 97-123, New York: Springer-Verlag Berlin Heidelberg.

Mesler, B., The Nation – The Pentagon's Radioactive Bullet. *http:/ /www. thenation. com/issue/961021/1021mesl.htm.*

Milacic, S., Petrovic, D., Jovicic, D., Kovacevic, R., and Simic, J. (2004). Examination of the health status of populations from depleted-uranium-contaminated regions, *Environmental Research*, *95*, 2-10.

Milojković, J., Stojanović, M., Ileš, D., Grubišić, M. (2009). Bentonit i diatomejska zemlja kao sekvencijalni agensi za adsorpciju urana, *Proceedings of 47th Meeting of the Serbian Chem. Society*, March 21, 2009, Belgrade, Serbia, pp.131-135.

Miller, A. C., Stewart, M, Brooks, K., Shi, L., and Page, N. (2002). Depleted uranium-catalyzed oxidative DNA damage: Absence of significant alpha particle decay. *J Inorg. Biochem.*, *91*, 246-252.

Mitrović, B., Vitorović, G., Stojanović, M., and Vitorović, D. (2011). Radioactivity of phosphate mineral product, *Veterinarski glasnik*, *65*, 1-2, 133-140.

MOD. (2008). The UK Ministry of Defence, Depleted Uranium. *http://www.mod.uk/DefenceInternet/AboutDefence/WhatWeDo/HealthandSafety/DepletedUranium/Depleted Uraniumdu. htm 2008.*

McBride, M.B., and Spiers, G.A. (2001). Trace element content of selected fertilizers and dairy manures as determined by ICP-MS, *Commun. Soil Sci. Plant.*, *32* (1-2), 139-156.

Morgan, K.Z. (1980). *Hazards of Low-Level Radiation, Yearbook of Science and the Future*, Supplement of the Encyclopedia Britannica.

Monleau, M., Bussy, C., Lestaevel, H.P., Paquet, F., and Chazel, V. (2005). Bioaccumulation and behavioral effects of depleted uranium in rats exposed to repeated inhalations, *Neuroscience Letters*, *390*, 31-36.

Mortvedt, J.J. (1994). Plant and soil relationships of uranium and thorium decay series radioradionuclides – A review. *Journal of Environmental Quality*. *23*, 643-650.

Pantelić, M., and Golubović, D. (2008). The influence of the depleted uranium on health of the population and the environment, *Tehnika i Informatika u Obrazovanju*, pp.71-89. Parkhurst, M.A., Johnson, J., Mishima, J., Price, J.L. (1995). Evaluation of Depleted Uranium Aerosol Data: Its Adecquancy for Inhalation Modeling, PNL-10903, Richland, WA: Battelle Pacific Northwest laboratory.

Pellmar, T. C., Fuciarelli, A. F., Ejnik, J.W., Hamilton, M., *et al.,* (1999). *Toxicol. Sci.*, *49*, 29-39.

Polednak, P.A., and Frome, L.E. (1981). Mortality among men employed between 1943 and 1947 at a uranium processing plant. *J. Occup. Med.*, *23*, 169-175.

Popovic, D., Djuric, G., and Todorovic, D. (2001). Possible impacts of depleted uranium (DU) ammunition on health and environment, *Danubius*, No. 1-2.

Pulhani, V.A., Dafauti, S., Hegde, A.G., Sharma, R.M., and Mishra, U.C. (2005). Uptake and distribution of natural radioactivity in wheat plants from soil. *Journal of Environmental Radioactivity*, *79*, 331-346.

Radenković, M., Anđelić, T., Kovačević, M., and Vukotić, P. (2004). Depleted uranium in the air during the cleanup operations of Cape Arza, *J. Serb. Chem. Soc., 69* (12), 1153-1155.

Radenkovic, M., Kandic, A., Vukanac, I., Joksic, J., and Djordjevic, D. (2007). Chemical and radiochemical characterization of depleted uranium in contaminated soils, *Russian Journal of Physical Chemistry A*, *81* (9), 1448-1451.

Radenkovic, M., Cupac, S., Joksic, J., and Todorovic, D. (2008). Depleted uranium mobility and fractionation in contaminated soil (Southern Serbia). *Env. Sci. Pollut. Res., 15* (1), 61-67.

Rajković, M.B. (2001). Osiromašeni uranijum – I. Uranijum, radioaktivnost i zakonska regulativa, *Hemijska industrija*, *55* (4), 167-182.

Rajković, M.B., and Vucelić-Radović, B. (2001). Ciklus kruženja uranijuma i kontaminacija biološkog lanca ishrane osiromašenim uranijumom, *EKO-KONFERENCIJA 2001: Zaštita životne sredine gradova i prigradskih naselja*, Septembar 26-29, 2001.god., *Novi Sad*, pp.381-385.

Rajković, B.M., and Đorđević, A. (2006). Possibility of Serbian soil contamination by depleted uranium after NATO bombing 1999, Radionuclide contamination of Serbian soil and remediation possibility. (Ed.) M. Stojanović, Belgrade, pp.167-219.

Rafsanjani, H.M. (2009). Comparative studies on the solubility of uranium and phosphorus in phosphate-fertilizers and their uranium transfer to plants, Julius Kühn-Institut, Federal Research Center for Cultivated Plants, Quedlinburg, Germany.

Rivas, M.D.C. (2005). Interactions between soil uranium contamination and fertilization with N, P, and S on the uranium content and uptake of corn, sunflower, and bean, and soil microbiological parameters, Landbauforschung Völkenrode Sonderheft 287, Germany, Braunschweig.

Rostker, B. (1998). Environmental exposure report: Depleted uranium in the Gulf, Department of Defense.

Rothbaum, P.H., McGaveston, D.A., Wall, T., Johnston, A.E., and Mattingly, G.E.G. (1979). Uranium accumulation in soils from long-continued applications of superphosphate. *Eur. J. Soil. Sci., 30*,147-153.

Sarić, M. R., Stojanović, M., and Babić, M. (1995). Uranium in plant species grown on natural barren soil. *Journal of Plant Nutrition 18* (7), 1509-1518.

Saueia, C.H.R., and Mazzilli, B.P. (2006). Distribution of natural radionuclides in the production and use of phosphate fertilizers in Brazil, *Journal of Environmental Radioactivity*, *89* (3), 229-239.

Singh, S., Malhotra, R., and Bajwa, B.S. (2005). Uranium uptake studies in some plants, *Radiation Measurements 40*, 666-669.

Stevanović, D., Kresojević, M., Stojanović, M., and Grubišić, M. (2009a). Actual assortment of mineral fertilizers in Serbia and their quality and problems of applications, Proceedings of XXIIIst Conference of Agronomist, Veterinarians and Technologiest, Belgrade, 169-177.

Stojanović, M. (2006). Radionuclide contamination of Serbian soil and remediation possibility, Institute for Technology of Nuclear and Other Minerals Raw Materials, Belgrade.

Stojanović, M., Grubišić, M., Stevanović, D., Milojković, J., and Ileš, D. (2008). Remediation of the Serbian soils contaminated by radionuclides in the function of sustainable development, *Chemical Industry and Chemical Engineering Quarterly*, *14*(4), 265-267.

Stojanović, M., and Milojković, J. (2011). *Phytoremediation of Uranium Contaminated Soils, Handbook of Phytoremediation*, Ivan Golubev, (Ed.), Nova Science Publishers Inc., New York, United States of America, ISBN: 978-1-61728-753-4, 93-136.

Stojanović, M., Milojković, M., Lopičić, Z., Mihajlović, M., Radulović, D., and Ileš D. (2011b). *In situ* immobilisation of uranium using alumosilicate minerals as sequestering agents, *The 43rd International October Conference on Mining and Metallurgy*, Kladovo, Srbija, pp.246-249.

Stojanović, M., Mrdaković- Popić, J., Stevanović, D., and Martinović, L. (2006). Phosphorus fertilizers as Ssurce of uranium in Serbian soils, *Agronomy for Sustainable Development*, *26*, 179-183.

Stojanović, M., Stevanović, D., Iles, D., Grubišić, M., and Milojković, J. (2009b). The effect of the uranium content in the tailings on some cultivated plants, *Water Air Soil Poll.*, *200*, 101-108.

Stojanović, M., Stevanović, D., Iles, D., Grubišić, M., and Milojković, J. (2010). Phytotoxic effect of uranium on the growing up and development plant of corn, *Water Air Soil Poll.*, *209* (1-4), 401-410.

Stojanović, M., Stevanović, D., Milojković, J., Mihajlović, M.L., Lopičić, Z., and Šoštarić, T. (2011a). Influence of soil type and physical chemical properties on uranium sorption and bioavailability. *Water, Air, and Soil Pollution, 223* (1), 135-144.

Science Applications International Corporation (SAIC), (1990): *Kinetic Energy Penetrator Environmental and Health Considerations* (Abridged), Vol. 2, pp. 2-4.

Schnug, E. (2005). Metal of Dishonor - munitions from depleted uranium (DU) contaminates soils in crisis areas, *http://www.innovations-report.com.*

Takeda, A., Tsukada, H., Takaku, Y., Hisamatsu, S., and Nanzyo, M. (2006). Accumulation of uranium derived from long-term fertilizer applications in a cultivated Andisol, *Science of the Total Environment, 367* (2-3), 924-931.

Taylor, D.M. (2007). Accumulation of uranium in soils from impurities in phosphate fertilisers, *Landbauforsch Volkenrode, 2,* 133–139.

Tessier, A., Campbell, M.P., and Bisson, M. (1979). Sequential extraction procedure for speciation of particulate trace metals, *Anal.Chem., 54,* 844-851.

Todorov, P.T., and Ilieva, E.N. (2006). Contamination with uranium from natural and anthropological sources, *Applied Physics Rom. J. Phys. 51,* 27-34.

Tsumura, A., and Yamasaki, S. (1993). Behavior of uranium, thorium, and lanthanoids in paddy fields, *Radioisotopes*, (42), 265-272.

The Royal Society, (2001). The health hazards of depleted uranium munitions: Part I. London.*(http://royalsociety.org/uploadedFiles/Royal Society_Content/policy/publications/2001/10023.pdf).*

Tunney, H., Stojanović,, M., Mrdaković-Popi, J., McGrath, D., and Zhang, C. (2009). Relationship of soil phosphorus with uranium in grassland mineral soils in Ireland using soils from a long-term phosphorus experiment and a National soil database, *J. Plant Nutr. Soil Sci., 172* (3), 346-352.

(UNEP/United Nations Centre for Human. Settlements [UNCHS] Balkans Task Force 1999) *http://www.wise-uranium.org/dissbk.html.*

United Nations Scientific Committee on the Effects of Atomic Radiation (1993). Sources and effects of ionizing radiation.) *(http:// www. unscear.org/unscear/en/publications/1993.html).*

UNEP (2000): NATO confirms to the UN use of depleted uranium during the Kosovo Conflict, Press Release, 21. March 2000.

UNEP (2001). Depleted uranium in Kosovo, post-conflict environmental assessment. United Nations Environment Programme, Switzerland. *http://postconflict.unep.ch/publications/uranium.pdf.*

UNEP (2002a). Depleted Uranium in Serbia and Montenegro—Post Conflict Environmental Assessment. UNEP, Geneva.

UNEP (2002b). Discussion of health issues: Appendix P. In: Depleted Uranium in Serbia and Montenegro—Post Conflict Environmental Assessment. UNEP, Geneva, pp. 179–183.

UNEP, (2007). Technical report on capacity-building for the assessment of depleted uranium in Iraq. Switzerland: United Nations Environmental Program; 2007. UNSCEAR (2000).*http://www.unscear.org/.*

United Nations Scientific Committee on the Effects of Atomic Radiation (UNSCEAR) (2000). Report to the general assembly. Annex B: Exposures from natural radiation sources, New York, USA.

US Army Chemical School (1995). Depleted Uranium Training Support Packets: Tier I – General Awareness. *http:// www. thenation. com/issue/961021/1021mesl.htm.*

Vukotic, P., Andjelic, T., Zekic, R., Kovacevic, M., Vasic, V., Ristich, N. (2002): Decontamination of Cape Arza (Montenegro) from depleted uranium, *Proceedings of the European IRPA Congress 2002*, Florence, Italy, 132-R.

Waite, T.D., Davis, J.A., Payne, T.E., Waychunas, G.A., and Xu, N. (1994). Uranium (VI) adsorption o ferrihydrate - application of surface complexation model, *Geochimica et Cosmochimica Acta*, *58*, 5465-5478.

Vandenhove, H., Van Hees, M., Wouters, K., and Wannijn, J. (2007). Can we predict uranium bioavailability based on soil parameters? Part 1: Effect of soil parameters on soil solution uranium concentration, *Environ. Pollut.*, *145* (2), 587-595.

Vernon, F., Shah, T. (1983). The extraction of uranium from seawater by poly(amidoxime)/poly(hydroxamic acid) resins and fiber Reactive Polymers, *Ion Exchangers, Sorbents*, *1* (4), 301-308.

Vodyanitskii, N. Yu. (2011). Chemical aspects of uranium behavior in soils: A review, *Top of Eurasian Soil Science*, *44* (8), 862-873.

Vukašinović, I., Đorđević, A., Rajković, M., Todorović, D., and Pavlović, V. (2010). Distribution of natural radionuclides in anthrosol-type soil. *Turkish Journal of Agriculture and Forestry*, *34*, 539-546.

Wetterlind J., Richer De Forges, C.A., Nicoullaud B., Arrouays, D. (2011). Changes in uranium and thorium contents in topsoil after long-term phosphorus fertilizer application *Soil Use and Management*, doi: 10.1111/j.1475-2743.2011.00376.x.

Willey, N. (2007). Soils Contaminated With Radionuclides: Some Insights for Phytoextraction of Inorganic Contaminants, Methods in Biotechnology, 23 Phytoremediation: *Methods and Reviews*, pp.305-317, Totowa, NJ: Humana.

WISE Uranium Project - FAQ (2004): Depleted Uranium in Urine of Soldiers (*http://www. wise-uranium.org/diss.html*).

WHO- World Health Organization, Department of Protection of the Human Environment Depleted uranium Sources, Exposure and Health Effects, Geneva, 2001. (*http:// www.who. int/ionizing _radiation/ pub_meet /en/Depluraniumintro.pdf*).

Žakula, B. (2002). Municija sa osiromašenim uranijumom, *Chemical Industry*, *2*, 93-97.

Zunić, S.Z., Mietelski, J.W., Blazej, S., Gaća, P., Tomankiewicz, E., Ujić, P., Celiković, I., Cuknić, O., and Demajo, M. (2008). Traces of DU in samples of environmental bio-monitors (non-flowering plants, fungi) and soil from targetsites of the Western Balkan region, *Journal of Environmental Radioactivity*, *99*, 1324-1328.

Yablokov, V.A., Vassily, B., Nesterenko, B.V., and Nesterenk, V.A. (2009). *Chernobyl Consequences of the Catastrophe for People and the Environment*, Published by Blackwell Publishing on behalf of the New York Academy of Sciences, Boston, Massachusetts.

Yamaguchi, N., Kawasaki, A., and Iiyama, I. (2009). Distribution of uranium in soil components of agricultural fields after long-term application of phosphate fertilizers, *Science of the Total Environment*. *407* (4), 1383-1390.

http://irpa11.irpa.net/pdfs/1d11.pdf.

http://www.unscear.org/docs/reports/2008/09-86753_Report_2008_Annex_B.pdf.

In: Uranium: Characteristics, Occurrence and Human Exposure ISBN: 978-1-62081-207-5
Editors: Alik Ya. Vasiliev and Mikhail Sidorov

Chapter 3

URANIUM MINING IN PORTUGAL: FROM UNMINED URANIUM ORES TO MINING WASTE DEPOSITS

Maria João Batista, Cátia Prazeres, Maria Ondina Figueiredo, Jaime Leote, Daniel P. S. de Oliveira, Augusto Filipe and Teresa Pereira da Silva

LNEG - Laboratório Nacional de Energia e Geologia, Unidade de Investigação Recursos Minerais e Geofísica, Alfragide, Portugal

ABSTRACT

In Portugal there are three uranium provinces - Beiras, Trás-os-Montes and Alto Alentejo - from which uranium mining only took place in the first one. Extraction in the Beiras province was undertaken in about 60 mines. The first license dates from 1907 to mine radium, and the interest in uranium only started after the Second World War when the market demand for defense purposes started to grow. The market decreased later due to international treaties of non-dissemination of weapons of mass destruction and the demand turned to energy purposes. Uranium production was never intended for the internal market but rather for export purposes. Known uranium deposits in Portugal mainly occur as veins in granitic rocks and in the metasediments of the contact. There is only one known sedimentary (alluvial) deposit located in the Urgeiriça mining area. Most uranium ore deposits are small (less than 500 t of U_3O_8) but there are some ore deposits of medium size. The generally accepted genesis model of the Portuguese vein-type ore deposits consider the primary ore hydrothermal deposits as resulting from magmatic differentiation and the secondary as leaching processes from the host rocks and from the primary deposits in a long period of enrichment. Ore concentration techniques varied according to the grade, the ore type and the mining process. Natural leaching was the first method used, after that and in related with the grade of the ores, in situ leaching and heap leaching was used. Until 1946 only radium was extracted and treated and about 15-20 kg/t of ore processed to radium was then recovered to uranium. Several experiments to improve the recovery of ore deposits in the poorest ores of Urgeiriça, Rosmaneira and Bica showed the diversity of its compositions and ways to improve the efficiency of ore treatment. The exploration of uranium stopped late in the eighties and mining finished with the closure of mines in the Beiras province in 2001. After this period a remediation program started to solve environmental problems. In this chapter a series of considerations based on the information archived at the Lab oratorio National de Energies e Geologic (LNEG, the Geological Survey of Portugal) will be made now that uranium exploration in Portugal is gaining the interest of mining companies.

INTRODUCTION

Uranium is many times associated with the effects of the exposure to radioactivity; nevertheless, this element has a great economic importance for the world and many applications in both the military and civilian sectors. However, the most relevant uranium use is for highly efficient energy production. Europe has several countries that have considerable resources of uranium. Total European reserves are about 1 196 674 t of U distributed mainly by sandstone, intrusive and vein type deposits (IAEA, 2009). Portugal is still one of the small European countries that contribute to these totals (see Table 1). As an important out-put of Portugal's mining legacy of the 20th Century, uranium was actively explored in approximately 500 prospects and mined in 61 of those, located within three mining districts. Like many European vein-type deposits, the uranium ore-bearing rocks in Portugal are associated to late post-organic granites in the Vermicides (Plant, 2003; Costello Neiva, 2003) mainly two mica, monzonitic and peraluminous granites, all of them from fine to course grained aging between 278 and 366 Ma with various U and Th grades fractionated from different magmas' evolution. U mineralization through secondary minerals is also present in metamorphism aureoles disseminated in metasediments. In the past mining period, environmental effects were not taken into consideration by the mining companies. First, because it was apparently an unknown issue, and second, because legal obligations until the nineties of the 20th century were practically inexistent. European countries were abridged only by the EURATOM Treaty that in its article 35° states that "*each Member State shall establish facilities necessary to carry out continuous monitoring of the levels of radioactivity in air, water and soil and to ensure compliance with the basic safety standards*".

Uranium mining involves preliminary studies dealing with economical aspects of the market demands (product receptors trade options), geological aspects (resource evaluation and cut off), mining (operation aspects), environmental (risk assessment) and all these together with a risk/benefit evaluation. This risk/benefit evaluation includes nowadays simulation scenarios of land occupancy and employment over time. Social aspects of poverty, education level, job options and all health risks involved are also included and balanced in modern mining projects where uranium is no exception. The recovery of a region from the economical and social point of view is a matter of great importance. The health risks involved creates apprehension in populations close to uranium mines. The same is not true for the uranium ore deposits, sometimes with higher radiation exposure than the mined sites.

The legacy of almost a century of mining and milling of uranium ores in Portugal has resulted in a variety of smaller and larger environmental liabilities in the centre and north of the country demanding care and attention. A thorough knowledge of the uranium contents in tailings and other waste sites is essential to understand how these areas should be remediated to avoid new legacies for the country and for the local population.

Table 1. Ore deposits reserves of uranium and the ore-type deposit related by country in Europe (IAEA, 2009)

Country	Tons of U	Ore deposit types
Germany	334 525	sandstone, vein, other
Check Republic	233 425	sandstone, vein, other
Ukraine	174 940	sandstone, quartz-pebble, intrusive, metasomatite, vein, other
Sweden	135 603	volcanic, intrusive, vein, other
France	110 842	sandstone, vein, other
Bulgaria	49 060	sandstone, volcanic, vein, other
Romania	37 071	sandstone, vein
Denmark	27 000	intrusive
Spain	26 398	sandstone, vein
Slovakia	19 065	volcanic
Poland	13 724	sandstone, vein
Finland	9 660	sandstone, intrusive, vein, other
Slovenia	8 500	sandstone
Serbia	6 701	sandstone, vein
Portugal	5 360	vein
Italy	4 800	volcanic

Remediation projects based on ore formation knowledge are usually less aggressive and have the purpose of being less costly, and to generate less immediate impacts and uses natural environment potentialities (Kalin *et al.*, 2005). These techniques have sometimes the disadvantage of showing their effectiveness not in short-time although very stable in long-term periods. When dealing with radioactive materials, remediation objectives need to take care of two main aspects: i) chemical immobilization; ii) radioactivity confinement. This chapter is intended to make known the information archived at LNEG about the studies performed by many authors during exploration, mining and ore treatment of the uranium deposits in Portugal. This is done by characterizing the ore deposits from the unmined deposits in Alto Alentejo and Trás-os-Montes regions to the mining region of Beiras. The post mining actions and management of the wastes in Portugal will also be reported.

Historical Perspective of Mining

The uranium producing regions in Portugal are: Beiras, Trás-os-Montes and Alto Alentejo Provinces. Uranium mining was carried out in Portugal from 1907 to 2001 with the closure of the Sevilha and Quinta do Bispo Mines. A total of 60 mines were active (45 open-cast; 15 underground) during this period. Uranium deposits in Portugal were first discovered in 1907 and the first mining concession, Romaneira, dates from 1909 followed by several others, namely Alto da Várzea (1911), Bica (1912), Coito da Casa Nova (1912) and Urgeiriça (1912). The most important uranium mine was the Urgeiriça mine where approximately 1100 t of U_3O_8 were mined (from 1944 until 1992) in both underground and open pit settings. This mine also produced radium from 1913 until 1944.

During the first years of mining (1909-1912) the ore was sent to France for processing but in 1913 the first ore treatment plant was built at Urgeiriça. Only radium was produced up to 1929 and uranium was rejected as gangue. Starting 1929, the emphasis switched to uranium although radium was produced up to 1944. In 1929 the *Companhia Portuguesa de Radium Lda* (CPR) was formed and acquired some of the concessions and continued to mine for radium that was exported to England under the form of barium and radium concentrates (JEN, after 1963). The company CPR became interested in uranium in 1941 and from 1941 to 1944 slowed radium production until abandoning it completely in 1944. From 1945 exploration, mining and ore treatment of uranium were increased substantially and an ore treatment plant was installed at Urgeiriça with an annual capacity of 125 t U_3O_8. After the Second World War the market demand for defense purposes started to grow, decreasing later due to international treaties of non dissemination of mass destruction weapons and turning the demand to energy purposes. The uranium mining industry was never for the internal market but to export; *Empresa de Desenvolvimento Mineiro* (EDM) reports 4400 t of U_3O_8 of total historic production. Up to 1962, CPR maintained 16 active mines of which Urgeiriça and Bica were the most prominent. These two mines produced 555 000 t of ore with an average grade of 0.267% U_3O_8, which resulted in 1325 t U_3O_8 almost all exported to the USA (JEN, after 1963). In 1954 the *Junta de Energia Nuclear* (JEN) was formed. This company carried out a systematic uranium exploration program from 1955 to 1959 that covered 50000 km^2. As a result, 377 new uranium occurrences were identified, 192 of which were classified as vein-type and 185 classified as impregnation-type. A total calculated resource of 26 000 t U_3O_8 was estimated by the JEN. From 1962 to 1977 JEN produced 9763 t of U_3O_8. *Empresa Nacional de Urânio E. P.* (ENU) was constituted in 1977, inheriting all the concessions, rights and competencies previously attributed to the JEN. From 1977 to 1991 1875 t U_3O_8 were produced. The entire amount was sold to Japan, Germany, Iraq, USA and France. Production was markedly reduced up to 2000 with Portugal producing only 246 t U_3O_8 through the treatment of low-grade ore by leaching processes. After a mere 5 t U_3O_8 produced in the first trimester of 2001, ENU ceased uranium production (Parra and Filipe, 2003). The reduction of uranium production (between 10 and 30 t U) in the last years of mining was due to the closure of the Chemical Treatment Plant at Urgeiriça and the production was totally from static leaching and from *in situ* leaching. In 1999 ENU gave up the Nisa mining project due to low uranium market prices. Mining was solely restricted to Central Portugal and therefore, the uraniferous regions of Trás-os-Montes and Alto Alentejo remain unexploited. The main known reserves were classified as medium size (IAEA, 2009) where the Nisa deposit, from Alto Alentejo uranium province, is the biggest (2440 t U_3O_8), most studied and better known deposit. The estimated reserves of Horta da Vilariça deposit (Trás-os-Montes) amout to 945 t U_3O_8.

Geological Setting of Uranium-Rich Granites

The Hercynian Orogeny that occurred during Devonian to Late Permian formed the latest supercontinent Pangea with direction approximately N-S. During Visean-Westephalian the consolidation of the European part of Hercinian Orogeny occurred and is referred as Variscan orogeny. Meanwhile, during the Late- to post-collision of extensional orogenic collapse (end of Carboniferous to early Permian), large quantities of radiothermal high heat production granites were emplaced including along major fault zones (Plant *et al.*, 1999). In fact, exposure of deformed rocks from the middle crust generally affected by high-temperature, low to medium pressure metamorphism, partial melting resulted in widespread granite emplacement (Malavieille, 1993).

Cameron (1960) suggests that U-rich granites that eventually form vein-type deposits belong to four regions in Europe related with Variscan Orogeny: 1) the Portuguese-Spanish region; 2) the Massif Central in France; 3) Vorges-Black Forest region; 4) Bohemian-Erzegebirge region. The first region, Portuguese and Spanish, resulted from Variscan intrusions elongated in a NW-SE direction presumably intruded along the lines of Variscan folding. The Massif Central granites are at the centre of a NW-SE elonged series of granites from Sardinia and Corsica to Brittany and Cornwall. The Vorges-Black Forest batholit probably connected with the Ardennes indicates another NW-SE series. The Harz-Bohemian region has the same NW-SE elongation. The author suggests that all these NW-SE granite orebodies belong to the Variscan Orogeny and the uranium bearing region crosses an ENE-WSW direction from the Iberian Peninsula until Bohemian-Erzegebirge region.

To have U-rich granites it is necessary to assume that the mantle was originally enriched in U, and by episodes of partial fusion the magma is enriched in U. The felsic character of the rock indicates even higher uranium content. The uranium along with other incompatible elements with large ionic radius do not fit in the silicate structure therefore enriches the magmatic fluids. The enrichment of this incompatible element in crust is continuous (Cuney, 2010).

Among felsic igneous rocks three types may constitute productive U sources: highly fractionated per alkaline; metaluminous high-K calc-alkaline rocks and peraluminous igneous rocks derived from low degree of melting of supracrustals. Another type even lower grade of partial melting corresponds to granitic dikes emplaced in migmatites with weak peraluminous composition. Each of these granitic compositions is characterized by a particular fractionation of U and Th and a specific par agenesis of accessory minerals (Cuney, 2009).

a) Peraluminous leucogranites (S-granites) result from a variety of processes and a series of sub-types were identified. The ones responsible for large uranium deposits have less than 10% biotite a strong peraluminosity index [$Al_2O_3/(CaO+Na_2O+K_2O)>1$] with fractionation, meaning that the remaining Al from feldspars can be included in muscovite or cordierite (Frost *et al.*, 2001). These low temperature peraluminous melts fractionation of low solubility monazite and zircon result in depletion of Zr, Th and REE. Further fractionation of this melt is enriching in U until saturation of uraninite and a Th-poor uraninite rich rock is obtained with the easiest leachable U source (Cuney, 2009). Common examples that can be found in the mid-European Variscan belts are peraluminous leucogranites, muscovite-bearing and high-silica content that result from partial melting of metasediment rocks (Frost *et al.*, 2001).
b) Peralkaline granites (Al-type granite), syenites and volcanic rocks are similarly enriched in U and Th and other large and highly charged elements, and although with high U content are not generally associated with vein-type uranium deposits unless U-Th minerals are metamict. The peralkaline volcanic rocks are a high source of U because it is mainly in glass which can be leached during devitrification (Cuney, 2009).

c) High-K calc-alkaline metaluminous granites can be enriched in U during magmatic fractionation but Th, Zr and REE are either depleted as in peraluminous granites or have similar behavior such as peralkaline granites depending of temperature and the degree of peraluminosity $[Al_2O_3/(CaO+Na_2O+K_2O)]$. These Ca-rich minerals (amphibole, titanite, allanite) incorporates most REE but low contents of Th. Thorium with UO_2, forms uranothorite which can be a refractory source of U for hydrothermal fluids circulating after the granite emplacement. If the Th/U is very low the uraninite can crystallize in highly fractionated high-K calc-alkaline granites (Cuney, 2009).

Most of the Portuguese granites were intruded during the Variscan orogeny in different phases, but especially related with the third deformation phase F3 (third ductile phase) (JEN, 1968; Ribeiro, 2001; Dias *et al.*, 2010) emplaced in a post-collisional stage of the Variscan Orogeny (290-321 Ma). The first emplacement models already relates these granites with important tectonic structures and considered younger granites the ones with circular structures and formed in deeper areas of the crust whereas the older granites ascended through shear zones.

The main granitic groups can also be spatially related with the important shear zones of the Variscan chain. The first domain was considered to NE of the Carboniferous Durico-Beirão belt which corresponds to a major NW-SE shear zone. The second domain occurs between the previous shear zone and the Porto-Tomar-Badajoz-Cordoba Shear zone, where the granites intruded mostly the pre-Ordovician megasequence of metapelites and greywackes of the Schist Greywacke Complex. These granites occur aligned with these belts. The third domain of granites aligned between the Porto-Tomar-Badajoz-Cordoba Shear and the Ferreira-Ficalho Shear zone (JEN, 1968; Ribeiro, 2001). Therefore, the most important granitic emplacements also related with uranium vein-type deposits were intruded taking advantage of shear zone spaces or in the intersection between these shear zones and major faults. In fact, in the Beira region the NNE-SSW to N-S direction faults constitutes the main mineralized fault directions and act as left shear in some cases as Celorico da Beira and Santa Comba Dão mineralizations and right shear in Pinhel, Bica and Vale da Arca mineralizations. In Guarda, parallel sinistral faults to the NNE-SSW oriented Vilariça fault and E-W, N-S filled with basic rocks (JEN, 1968).

In a close relation with the U occurrences the Nisa granites part of an important of Late-Variscan granites alignment (Nisa-Albuquerque- Los Pedroches) along the border of Central Iberian and Ossa Morena Zones. In the western part it has an arc shape and creates a metamorphic belt in metasediments in the surroundings filling the shear zones and confirming its relation with these structures in the last phases of the Variscan ductile phase (Ferreira *et al.* 1987). The late-Variscan folding of the granites in the region is reflected in the N20-30E direction filled by quartz veins and N15-30-50W later, rejected to the right side, some of them with U mineralization (Campos and Pereira 1991).

Most of the Variscan granites related with U deposits in Portugal are peraluminous leucogranites (S-granites) formed with partial fusion of metasediments of pre-Ordovician age (Neiva and Gomes, 2001; Dias *et al.*, 2010). Figure 1 shows the 1:500 000 geological map (Oliveira *et al.,* 1992) overlaid with the known U mineral occurrences.

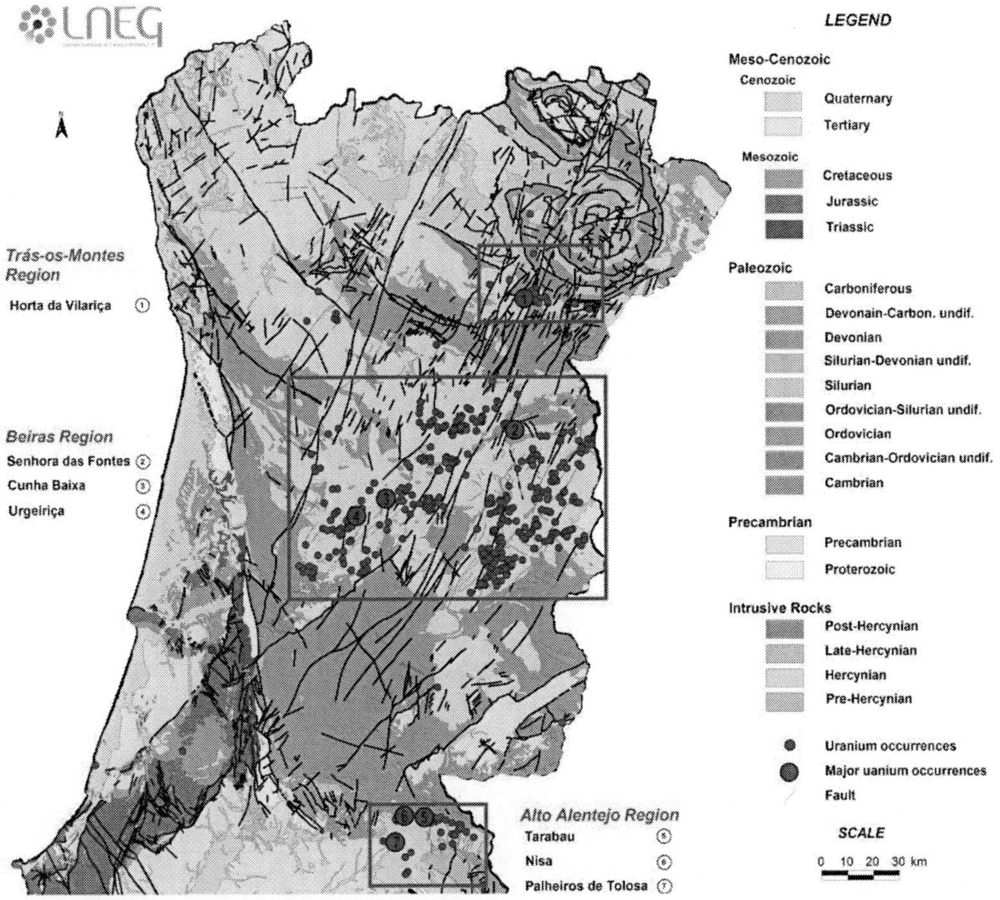

Figure 1. Geological map of northern and central Portugal overlaid with the known uranium mineral occurrences (adapted from Oliveira *et al.*, 1992). 1- Horta da Vilariça; 2- Senhora das Fontes; 3- Cunha Baixa; 4- Urgeiriça; 5- Tarabau; 6- Nisa and 7- Palheiros de Tolosa.

The various Variscan U-rich granites in Portugal are different in age and in U and Th content, are mostly monzonitic granites, two mica or predominantly biotitic, peraluminous with uraninite present as a minor accessory mineral. Whereas granite γ_{2b}^{II} is 297-319 Ma with 5-14 U mg kg^{-1} and Th 12 to 47 mg kg^{-1}; the γ_{3b}^{II} 277-297 Ma with 6-21 mg kg^{-1} U and 8-35 mg kg^{-1} of Th; the γ_{3c}^{II} 278-305 Ma with 6-13 mg kg^{-1} U and 1-43 mg kg^{-1} of Th corresponding from a different magmatic origin. The γ_{3d}^{II} derive from γ_{2b}^{II} magma and contains at least 8 mg kg^{-1} of U and 2 mg kg^{-1} of Th. The γ_3^{I} of 300-366 Ma with 5-23 mg kg^{-1} U and 4-61 mg kg^{-1} of Th are two mica with predominant biotite or muscovite. The γ_3^{III} are of 282-308 Ma with 6-26 mg kg^{-1} of U and 13-52 mg kg^{-1} of Th being slightly more uraniferous than the other granites (Cotelo Neiva, 2003).

The main minerals of the 'fertile' granites are: quartz, feldspar minerals (albite, albite-oligoclase, and rarely andesine), microcline and microcline-perthite. The accessory U-bearing minerals are biotite, muscovite, rutile, zircon, titanite, tourmaline, andalusite, cordierite and sillimanite. The main accessory minerals are chlorite, kaolinite, calcite and iron oxides. Weathering of these minerals by acid hydrothermal fluids or meteoric waters configures a supply of uranium rich waters at the surface and to groundwaters (Plant *et al.*, 1999; Plant *et al.*, 2003; Cabral Pinto *et al.*, 2009).

Late-Variscan-Alpine Folding

The late Variscan Orogeny consists of a post-tectonic uplift with N-S direction stress during the Stefanian and W-E during the Permian. The stress system of Stefanian age produced two sets of sinistral faults with NNE-SSW direction to ENE-WSW and a dextral with NNW-SSE to NW-SE. These systems affect the alkaline granites (300±10 Ma) and the calc- alkaline granites intrusion (280±10 Ma) seems to be partially controlled by this event. The great extension and importance of these faults during the Cenozoic leave its footprint in the geomorphology of the territory with a graben such as Vilariça and a horst such as "Cordilheira Central" and the intense folding in the Beira Alta region. During Permian and in a maximum compression the direction turns to E-W with evidences in structures N-S such as Autunian age in Buçaco that affects the western border of the Iberian Massif. The last late Variscan folding was after the Autunian and prior to Triassic in Coimbra which was not affected by N-S folding (Ribeiro *et al.*, 1980).

During the Mesozoic and Cenozoic age, the arc deformation events form the Cantambrian Mountains, Iberian Chain and Serra Morena and the compression of the Alpine trench resulted in the Pyrenees and the Betic Mountains. The Alpine orogeny is characterised by E-W and NE-SW fracture systems that result in uplift and depressions such as Serra da Estrela, Mortágua, Celorico da Beira and Cova da Beira.

Variscan and Alpine Folding in Relation with the Uranium Mineralizations

Initially, Variscan NW-SE and N-S fractures were reopened and filled with quartz veins and basic rock dikes in N-S and NNW-SSE systems. After that event there was the formation of Pyrenees and Betic Belt by uplift and during since Cretaceous and Oligocene. Then there was the separation in E-W blocks and the filling of veins and basic rock dikes of NE-SW and E-W systems. After wood important shear zones form and also Serra da Estrela, Guadarrama, Credos Mountains rise (JEN, 1968; Ribeiro *et al.*, 1980, Jabaloy *et al.*, 2002). After these events a step fracturing event occurs with the definition of the *Cordilheira Central.* After this phase the NW-SE, N-S and E-W faults were rearranged and filled with veins and basic rock dikes of the NE-SW system and epirogenic movements, after that tension fractures of NE-SW direction were formed and filled with basic and uranium rich fluids. Reopening of the fractures takes place rearranging the E-W and NE-SW blocks resulting in intense installation of uraniferous fluids. After this period a series of small scale movements rejected to SSE the mineralization and the Vilariça fault occurred as a post mineralization event.

Vein Type Portuguese Uranium Deposits

Most of the uranium ore deposits are small (less than 500 t of U_3O_8) but there are some ore deposits of medium size, according to the IAEA (2009) classification. The most important known ones which remain unexploited are Nisa in the Alto Alentejo Region and Horta da Vilariça in the Trás-os-Montes region.

They are coeval with both the Variscan and Alpine orogenies resulting in late deposits of mesoalpine phase of Eocene-Oligocene age. Cerveira (1950) considered the mineralizations within structures orientated NW-SE related with the Variscan Orogeny and the mineralization found in structures NE-SW oriented in accordance with Alpine Orogeny. Vein deposits were later on divided into two types (Matos Dias and Andrade, 1970): from the Variscan Orogeny (Sn, W, Au, Ag, As, Pb, Zn and Fe) and related with the Alpine Orogeny (Pb, Zn, Sb and U). They still grouped pegmatite mineralizations in irregular structures (Sn, W, Ti, Mo, Ta, Li, Be and U) and alluvial deposits (Sn, Ti, Ta, V and U). Ages of mineralizations were determined to Lenteiros, Urgeiriça and Carrasca and the results were 60-100 Ma of the Upper Cretaceous/ Eocene and Triassic.

Cameron (1960) suggests three main categories of vein-type U ores occurring in Central Portugal: red jasper veins, ferruginous white quartz veins and black and white banded quartz veins, the former being considered by the author as the most favorable for U installation. Another interesting result of this study was that when U is related with other metallic minerals in the veins - such as hematite, pyrite, chalcopyrite, galena, sphalerite and arsenopyrite - hematite is by far the better indicator of the presence of U.

In relation with common wall-rock alteration – like hematitization, sericitization, tourmalinization - the first is the most important indicator of uranium presence in a quartz vein; at a regional scale, sericitization is the most important alteration event related with U installation in veins. However, Ávila Martins (1959) suggests that hematitization of feldspars, biotite alteration to chlorite and considerable loss of quartz in relation to unaltered granites. The lost of quartz induced sometimes to consider these rocks as syenites. This author also reports that the alteration phenomena could be result of circulation of Fe-rich solutions acting in granitic zones previously deformed caused by tectonic events. Moreover, the ferruginous aspect of these frequently called red granites of Beiras, due to the presence of Fe oxy-hydroxides, with well known capacity to adsorb UO_2^{2+} ions, could therefore be responsible for the higher radioactivity in these granites.

Cameron (1960) suggests that uranium genesis in unaltered granite at first with high grade of U the biotites with a series of events associated with the Alpine Orogeny would cause the installation of uranium vein deposits. The first phase of events occurred in Tertiary age where movements of uplift and northeastwards thrust of Serra da Estrela, caused the ascending of solutions that attacked the wall rock granites causing the removal of Fe, U, Mg, Na, Ca and Al from the original granite. It was suggested that Fe and U removed was carried to considerable depth and kept in solution. These statements were proved by the presence of pleochroic haloes in the biotites found in the thin sections of the granites of Beiras. A second phase where the movements, with uplift and northeastwards thrust caused the reopen of sections of the original shears and produced tension fractures that facilitated the ascent of hot acidic solutions. These local structures were able to receive silica and hematite removed in the first phase forming jasper with a typical red coloration. A third phase, related with the second and even more restrict, causing only horizontal adjustments received more silica and hematite and uranium from the first phase during the sericitization process. Further silica gels were produced and the U removed from the solutions was retained in the silica gel in colloidal or microbotryoidal forms.

Basham and Matos Dias (1986) admitted that the vein-type mineralizations have different ages and origins from hydrothermally jasper-type originated at medium to high temperatures passing to lower temperature quartz veins and peribatholitic dissemination of what they called the Iberian type; to supergene enrichment in surface related traps such as weathered basic dikes. All of them vein-type mineralizations have close relation with Variscan granites where uraninite with low Th content is the main source of U.

The same authors discuss the characteristics of the mineralization based on the following classification of Portuguese U ore deposits:

1. Hydrothermal, polymetallic pitchblende veins in open structures with jasper or ferruginous quartz gangue.
2. Hydrothermal veins with banded milky white and smoky quartz gangue.
3. Pitchblende-fluorite veinlets in episyenites.
4. Dominantly hexavalent uranium minerals in quartz veins, tectonic breccias and basic dykes.
5. 'Iberian-type' disseminated deposits.

It is also reported by the same authors that all known U-bearing hydrothermal veins are in the younger (post-tectonic) Variscan granites that intruded the pre-Ordovician metapelites and greywackes; these are intragranitic veins and predominantly appear in Guarda-Viseu region and also in some mineralizations of Alto Alentejo. These granites cut across the NW-SE trend of metasediments (Cameron, 1960; Ribeiro 2001). In general the older granites are barren in U although some indications occur in Pinhel the older granites.

1. The mineral paragenesis of these jasper or ferruginous white quartz veins is composed of quartz, jasper or ferruginous quartz, pyrite, sphalerite, galena and fluorite, micro-crystaline quartz, pitchblende and chalcopyrite. The wall rock alteration was described as in Cameron (1960). The age is attributed to the Alpine Orogeny period of faults opening and as a consequence was attributed to Tertiary age. This type of deposit considered hydrothermal appears in Urgeiriça, Guarda and Reboleiro and in Alto Alentejo.
2. The hydrothermal veins with banded milky white and smoky quartz gangue control and paragenesis are similar to the jasper type deposits but the pitchblende tends to be microbotryoidal and is more associated with pyrite that is sometimes the only sulfide present. An example is the Cunha Baixa deposit.
3. The pitchblende-fluorite veinlets in episyenites occur mainly in the transition to unaltered monzonite granites. Within the episyenites occurs mainly autunite formed by supergene remobilization and the primary origin was low temperature hydrothermal fluids installed along shear zones.
4. The ore-type of dominantly hexavalent uranium minerals in quartz veins, tectonic breccias and basic dykes appears only at the surface where pitchblende sometimes appearing below the oxidation surface. These deposits occur both in the Beiras and Alentejo regions and are related with the late Variscan fracture systems. The surface alteration is observed until about 15 m and the possibly primary ores are not observed. These types of veins were described as being successive concentrations of U by long-term weathering throughout Tertiary age by supergene processes

accumulating in fractures and traps. In fact, this explanation is compatible with a combination of supergene processes and the passing of hydrothermal fluids as it was described for the Cunha Baixa mineralization.

5. The 'Iberian-type' disseminated deposits consist of the mineralized area of metamorphism aureole of the metapelites and graywackes of the Schist Greywacke Complex with the younger Variscan granites where U-bearing minerals and intragranitic vein-type deposits are also present. In fact, no mineralization of such type was found in barren granites.

These so called 'Iberian-type' disseminated deposits (Basham and Matos Dias, 1986) were previously described by Limpo de Faria (1966) that classified these mineralizations in two types:

1. Dissemination deposits where the mineralization is dispersed in fractures along the planes of schistosity in the altered surface zone ore concentrated in shear zones and with supergene characteristics. Examples of these are Nisa, Horta da Vilariça and Ázere deposits.
2. This type of deposits are veined epithermal mineralizations with a surface zone identical to the dissemination-type being the extension of the uraniferous intragranitic veins and the examples pointed out are Cunha Baixa, Tarabau and Póvoa e Meadas.

The Douro region mineralizations and the Vilariça region from which the most important one is Horta da Vilariça, the Ervedosa region up north and the Valongo-Esposende area are all in the vicinity of the granite. In Horta da Vilariça just as in Nisa the mineralised rock is the quartz schist with cordierite, but also aplite-pegmatite veins occur with smoked quartz. The mineralization was identified as autunite and occurs similarly in metasediments and in granites. In the Ervedosa area the uranium mineralization identified is torbernite and autunite and occurs in clay schists and micaschists that are heavily oxidized. In the last northern area the occurrences were identified in graphitic, carbonific schists in Fe oxides altered areas.

Uranium Exploration in Portugal

History of Exploration

Regional geological mapping made it possible to view both local and regional anomalies which were the starting point for the systematic studies initiated in 1955. The use of aerial photo interpretation permitted detection of regional lineaments and consequently allowed for selection of favorable areas for exploration. The first phase of exploration studies used to define the regions and geological structures preferential for the uranium exploration was concluded in 1959. A lengthy regional radio-prospecting campaign started with the installation of scintilometers in airplanes, vehicles and manual. By airborne methods, a grid of 400 meters spacing and 100 meters mean high above the ground was undertaken. A mean density of 1.22 measures per sq. km was sufficient to identify that the Variscan granites were the most important lithologies for the uranium ore occurrence. This led to the conclusion that, not only the granites, but also the metamorphic aureole of the contact of the Ante-Ordovician metasediments, were mineralized with uranium (JEN after 1963).

Junta de Energia Nuclear (JEN) implemented general, regional and local programs of uranium exploration. The general program was developed along with detailed geological mapping in an irregular grid of 200 per 300 m. The regional program was based on the anomaly areas obtained in the general program and consisted of further detailed geological and radiometric studies in grids of 25 per 25 m or 25 to 50 m with recording of the readings of radiometry every 5 m. In mineralized areas (local program) the studies were carried out with high sensitivity Geiger-Müller counters or high sensitivity scintilometers.

In these studies the counting plan was made to obtain the geometry of the mineralized body and the grid was about 1 per 5 m. Isoradic maps of 1:200 m scale were obtained. Further studies in local programs to obtain the 3D geological characteristics electrical methods were used through the resistivity charts, spontaneous polarization added to the electrical measurements made possible to establish better interpretation of the geological mineralized structures (JEN, after 1963).

After these non-invasive methods an accurate evaluation of resources was carried out and other methods such as trenches perpendicular to the mineralized veins separated from 10 to 20 m apart, drilling with or without recovery of the drilling core were made. At this stage, JEN also collected samples for chemical and radiometric analysis to obtain the uranium concentrations and ore grades. Drilling was made and samples for mineralogical and petrological analyses were obtained from the cores to evaluate the ore reserves. Drilling was carried out in plans of 30 to 40 m apart.

After many years of exploration, the JEN classified the Portuguese uranium deposits as vein type deposits and impregnation (secondary minerals type). The company *Empresa Nacional de Urânio* (ENU), that succeeded JEN, used similar methods of exploration but abandoned the trenches opening and the studies passed from geophysical methods to drilling. With the closing of JEN, some experts moved to ENU and those specialized in the nuclear field to *Instituto Tecnológico Nuclear* (ITN) while other specialized in chemical processing and U beneficiation moved to the *Laboratório Nacional de Engenharia, Tecnologia Industrial* (LNETI) and part of the exploration group that remained in the public service was integrated in the *Serviço de Fomento Mineiro* (SFM) later integrated in the *Instituto Geológico e Mineiro* (IGM) and after that, *Laboratório Nacional de Energia e Geologia* (LNEG). In the 1980's SFM developed a program to recover the radiometric surveys collected from JEN, ENU and SFM and recalibrated the surveys. In a later stage of uranium exploration, the sedimentary rocks were investigated. In fact, the western belt of Meso-Cenozoic age was investigated in a series of campaigns of radiometry and geochemistry by the SFM between 1982 and 1984. Exploration techniques evolved to sophisticated methods and an example of that was the Rebolia Sediments near Figueira da Foz (Prudêncio *et al.*, 1987). In this study a combination of radioemanometry, gamma ray spectrometry in the field in profiles distant of 5 metres using SPP2 scintilometer and integral alpha activity was measured by EDA RD-200 emanometer. Soil samples taken in a tight grid and submitted to chemical analysis by Instrumental Neutron Activation Analysis (INAA) in the Portuguese Research Nuclear Reactor of Sacavém at a neutron flux of 1.6×10^{11} n cm^{-2} s^{-1} where Na, K, Sc, Cr, Mn, Fe, Co, Zn, As, Rb, Cs, Ba, La, Ce, Nd, Eu, Tb, Yb, Lu, Ta, Th and U were determined (Cabral *et al.*, 1979, Gouveia *et al.*, 1986). Leachable and total U was determined by fluorimetry using a reflection fluorimeter (Cajão, 1969). Due to the specificities of uranium mineralizations, an internal report on occurrences in Portugal was elaborated at LNEG within an Information System on Mineral Occurrences and Resources, SIORMINP (www.lneg.pt), combining the information contained in several previous JEN reports, especially the one dated 1968, where a classification according to mining districts was introduced. This report complements the uranium occurrences inventory that is now overlaying the maps of geology and radiometry.

Calibration of Old Detectors to Obtain Consistent Country Coverage

Most of the country was surveyed by vehicle-mounted and ground-based radiometric equipment in the 1950s and 1960s. Under an IAEA project, a study was undertaken on the recalibration of these old surveys (Torres, 1992; Torres and Grasty, 1993). It was found that much of this old data could be reprocessed to produce a map of the country's natural radioactive environment. This information was combined with results of airborne gamma spectrometric surveys. A truck-mounted 256-channel gamma spectrometric system was acquired to carry out this project. The natural radiation thus depends, as noted above, from the concentration of uranium in rocks. The uranium concentration in the Variscan granites of Central Portugal is reflected by the elevated values of radiation, measured in counts per second and transformed in exposure rate, in μR/h (Torres, 1992; Torres and Grasty, 1993). To obtain this information at a national level, a data calibration was performed using data acquired with old equipments and more recent ones (Grasty, 1993). Exposure rate was obtained from old campaigns of radiometric exploration due to its ability to distinguish between different background levels in different lithologies and in different geological settings. Radiometric properties were used not only for exploration purposes but also for definition of different levels of intervention in remediation processes (Saraiva *et al.*, 1998). A calibrated Exploranium 256 channel spectrometer with an ANI (T1) crystal, energetic range spectra of 0 to 3 MeV, window total count (0.41 to 2.81 MeV), potassium (1.37 to 1.57 MeV), uranium (1.66 to 1.86 MeV), thorium (2.41 to 2.81 MeV) and cosmic radiation >3 MeV and two Saphymo SRAT SPP2 scintilometers of ground measurements were used for calibration. 32 measurements in 14 places were made and verified that the scintilometers (the counts per second) and spectrometer (converted to exposure rate) measurements fit into a linear regression. A regression equation was obtained where E (exposure rate) = 0.128xcps (counts per second)-2.47 (Torres, 1992). The airborne calibration was made by taking out the cosmic radiation, the aircraft radiation and the radiation from the radon. After that the Compton correction was made, and the altitude smoothed and stabilized at 90 m high. In ground level surveys Exploranium GR-650 with GPX-256 of 4.2L, GPX-21 of 0.35L and Saphymo SRAT SPP2 were used. The data processing and homogenization of data was made using homemade and Exploranium softwares (Saraiva *et al.*, 1998).

Principles of the Exploration Tools Used

Geology related with the events of the late-Variscan events and Alpine orogeny combined with granitic composition and the degree of fractionation are key factors of investigation and were explained before. Other important factors base themselves in natural radioactivity directly intensified when in the presence of uranium mineralizations and the identification of hexavalent uranium minerals at surface, easily detected by gamma ray measurements and good indicators of mineralization and origin of the uranium.

Natural Radioactivity

All elements that have an atomic number above 80 have radioactive isotopes and isotopes with atomic number above 82 are radioactive, potassium being an exception with atomic number 40.

The radioactive elements are grouped into three groups of families: descendants of ^{238}U, descendants of ^{235}U and descendants of ^{232}Th from which the most important families of ^{238}U in the case of natural.

All uranium isotopes are radioactive and the three natural uranium isotopes found in the environment, ^{234}U, ^{235}U, and ^{238}U have radioactive decay by emission of an alpha particle accompanied by weak gamma radiation.

The dominant isotope, 238U, forms a long series of decay products that includes the key radionuclides ^{226}Ra, and ^{222}Rn. The decay process continues until a stable, non-radioactive decay product is formed. The annual dose of external radiation according to the UNSCEAR (1958) is presented in Table 2.

Other naturally occurring isotopes in the Earth's crust come from cosmic radiation (e. g. ^{14}C that is formed in the higher atmosphere). In the atmosphere most of the radioactivity is a result of diffusion across the Earth's crust of radon and thoron descendent gases of uranium and thorium decay chain.

Radon has a bigger half-life (3.8 days) then thoron (52 seconds). Uranium has a more significant concentration whereas thoron quickly decays to its stable non-gas descendants.

After irradiation of gases into the atmosphere and starting decay they bind in the air, in colloidal particles of sizes (aerosols) along with other metals, in water molecules. These tend to acquire the equilibrium and precipitate, stable isotopes are the radium D (^{210}Pb) in the case radon, and thorium D (^{208}Pb) in the case of thoron progeny. After that the dilution in surface and groundwater bodies occurs.

Table 2. Annual dose of external radiation per lithology

Igneous rocks				Sandstones sedimentary rocks			
^{226}Ra	**^{238}U**	**^{232}Th**	**^{40}K**	**^{226}Ra**	**^{238}U**	**^{232}Th**	**^{40}K**
24 (mrad/y)	25.8 (mrad/y)	36.8 (mrad/y)	34.6 (mrad/y)	13 (mrad/y)	7.7 (mrad/y)	18.4 (mrad/y)	14.6 (mrad/y)
48 Bq/kg	48 Bq/kg	48 Bq/kg	810 Bq/kg	26 Bq/kg	15 Bq/kg	24 Bq/kg	330 Bq/kg
Shales sedimentary rocks				Limenstones sedimentary rocks.			
^{226}Ra	**^{238}U**	**^{232}Th**	**^{40}K**	**^{226}Ra**	**^{238}U**	**^{232}Th**	**^{40}K**
20 (mrad/y)	7.7 (mrad/y)	30.6 (mrad/y)	30 (mrad/y)	7.7 (mrad/y)	8.4 (mrad/y)	4 (mrad/y)	3.6 (mrad/y)
40 Bq/kg	15 Bq/kg	41 Bq/kg	810 Bq/kg	16 Bq/kg	15 Bq/kg	5 Bq/kg	81 Bq/kg

Uranium and Uranium Minerals

The most insoluble forms of uranium minerals are those containing the reduced species of uranium (tetravalent). These minerals in general display a regular polyhedral coordination around U^{4+}cations and tend to be isostructural with other minerals containing no uranium. Thus, U^{+4} ions substitutes for other cations in many minerals; conversely, minerals containing the oxidized species U^{6+} are not isostructural with other natural compounds due to their complex structures and unusual crystal chemistry (Burns and Finch, 1999). Uranium occurs in a number of different primary minerals in igneous rocks, especially in pegmatites and is abundant in accessory minerals of these rocks (e.g., uraninite). Uranium is a lithophile chemical element with a great affinity to silica; it is mobile under acidic conditions and also under alkaline conditions in an oxidizing environment, and thus, forms a wide variety of mineral complexes with anions (sulfate, phosphate, carbonate, silicate, vanadate, arsenate and others), plus cations (IAEA, 1980). In general, the global average concentration of uranium in rocks ranges from 1 to 10 mg/kg (Wanty *et al.*, 2000). Rocks of late magmatic differentiation facies, richer in silica, present higher U concentrations. Thus granites and rhyolites tend to have higher concentrations of uranium than the more mafic rocks. Deposits formed from residual fluids differentiation of these rocks (more felsic) tend to be enriched in U. Generally, also potassium and thorium contents decrease in igneous rocks as they become less felsic in composition (Pagel, 1979). Uranium has the ability to substitute for Th, Ca, Zr and REE in the structure of accessory minerals. These minerals can grant radioactive properties to the rocks without the presence of uranium deposits; in some cases they can even be highly radioactive, example of these are xenotime, monazite, pyrochlore, zircon, allanite and titanite. The above mentioned minerals can give rise to uranium sedimentary deposits if concentrated in a sedimentary basin, because these accessory minerals are also very resistant to alteration and in certain conditions they concentrate as placer deposits.

Figure 2. Present field work equipments (scintillometer SHAPHYMO SRAT SPP2 and Exploranium GR-256 Spectrometer) used during the Nisa region studies.

Figure 3. The experimental open pit in Nisa (Photo taken in November 2010).

The most representative primary mineral found in the Portuguese vein deposits is pitchblende (essentially UO_{2+x}), which alteration gives rise to a variety of oxihydroxides like clarkeite $[(Na,Ca,Pb)_2U_2(O,OH)_7]$ and fourmarierite $[PbU_4O_{13}.4H_2O]$ and to silicates such as uranophane $[Ca(UO_2)_2Si_2O_7.6H_2O]$, plus the poorly defined "gummite". The crystalline variety of pitchblende (uraninite, UO_2) is less frequent. The so-called black products of uranium - very common in these deposits - are essentially formed by an amorphous variety of pitchblende. Hexavalent uranium is therefore the dominant ion in the minerals occurring in Portuguese uranium ores. The autunite phosphate group, $[(Ca,Ba,Cu,Mg)(UO_2)_2(PO_4)_2.8\text{-}12H_2O]$, results in a number of specific minerals: autunite (the calcium term), uranocircite (the barium term), torbernite (the copper term), saleeite (the magnesium term) or the aluminous phosphate sabugalite, $HAl(UO_2)_4(PO_4)_4.16H_2O$. The precipitation of saleeite at the surface of Fe minerals can be responsible for U retention within the quartz veins (Cabral Pinto *et al.*, 2008).

New Developments in Exploration

Nowadays, LNEG continues to carry out the radiometric mapping and a 1:500 000 map of Portugal with about 800 000 calibrated measurements of radiometry transformed from airplanes, vehicles and ground measurements obtained for exploration purposes (presented at both Gy/h units and μR/h units) is about to be finished. Studies in the Nisa region included mineralization reconnaissance of the previous studies in the region from Nisa to Castelo de Vide. In this region soil sampling and a preliminary qualitative mapping of U, Th, Rb, Zr by X-ray fluorescence spectrometry (XRF) and radiometric measurement were carried out at regional scale and combined with the country wide mapping (Batista *et al.*, 2008). Particular focus was given to richest area in Nisa where the experimental open pit is located, with detailed 1:100 scale geological and radiometric studies in today scenario (Figure 3). Detailed study updates of geological and radiometric surveys were also carried out at Tarabau and Melriça mineralizations (Prazeres, 2011).

Trás-os-Montes Area

In Trás-os-Montes area uranium mineralizations were identified in Horta da Vilariça and Eiras da Horta and up north in the Ervedosa region. Some of these mineralizations are dessiminations in schist such as the case of Horta da Vilariça.

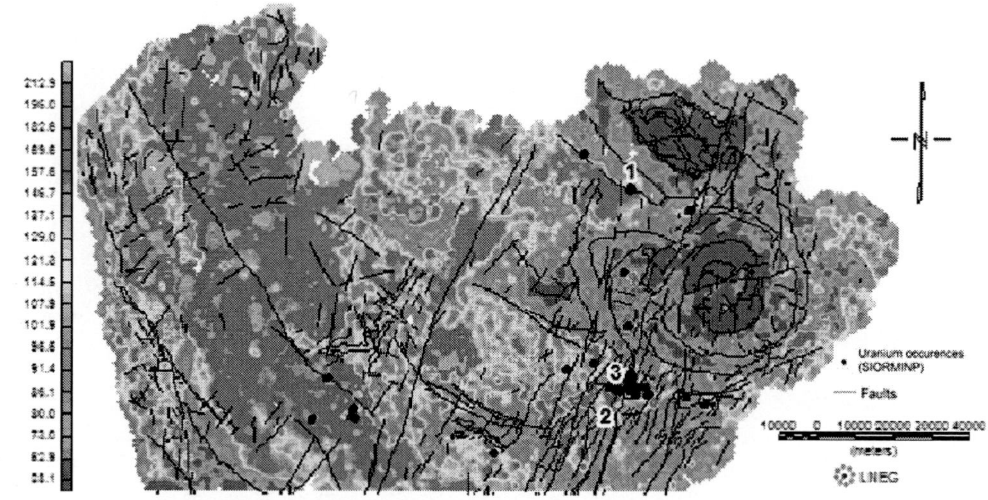

Figure 4. Radiometry map of northern Portugal with the main fault systems and location of known U occurrences (SIORMINP database) 1 – Ervedosa, 2 - Eiras da Horta, 3 - Horta da Vilariça.

Geology and radiometry combined with the SIORMINP database of occurrences indicates the mineralizations alignments at regional scale. The evident reject that is observed even in the mineralizations show the posterior implantation of the Vilariça fault in relation with the mineralizations and that the preference fault system is the NE-SW direction specially in the intersection with E-W direction or even ENE-WSW (Figure 4).

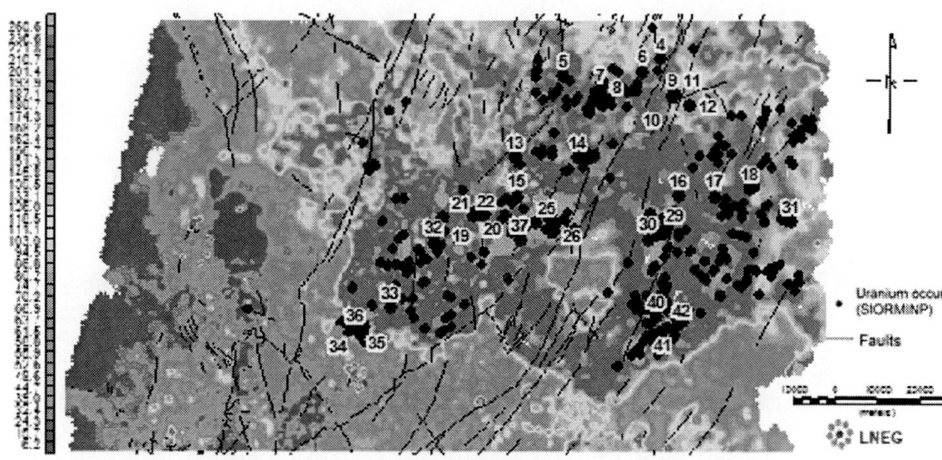

Figure 5. Radiometry map of Central Portugal (Beiras region) with the main fault systems and location of known U occurrences (SIORMINP database) 5 - Fontaínhas de Gradiz, Ervideira, Luz e Mestras; 6 - A-do-Cavalo; 7 – Lenteiros; 8 – Ferreiros; 9 - Alto da Rasa; 10 – Cótimos; 11 - Quinta das Seixas; 12 - Senhora das Fontes; 13 - Corga do Valbom; 14 – Vales; 15 - Pinhal do Souto; 16 - Pêra do Moço; 17 - Prado Velho; 18 -Barroco D. Frango; 19 - Quinta do Bispo; 20 - Cunha Baixa; 21 - Carvalhas, Picoto e Montes Fuscos; 22 – Espinho; 23 – Valdante; 24 – Formiga; 25 - Canto do Lagar; 26 - Castelejo (I e II); 27 – Tentinolho; 28 - Barroca Funda; 29 - Alto da Várzea e Quinta do Pinheiro; 30 - Forte Velho; 31 - Ribeira do Ferro; 32 – Urgeiriça; 33 – Sevilha; 34 - Mondego Sul; 35 – Ázere; 36 - Vale da Abrutiga; 37 - Ribeira do Boco; 38 – Rosmaneira; 39 - Vale da Arca; 40 – Coitos; 41 – Carrasca; 42 - Pedreiros e Bica.

The northern region had a low investment in exploration hence, is a not so well known region and is a non mined region where the potential of new discoveries can be higher than in the other regions.

Beiras Area

Radiometry of Beiras region shows very well the favorable composition of granitic intrusions where uraninite was possibly an important accessory mineral with poor content in Th. It is also possible to observe alignments of mineralizations in relation with the NE-SW direction (Figure 5).

Alto Alentejo Region

Nisa granite is evidenced by the radiometry and uranium mineral occurrences. Although, at the present scale it is not possible to observe the intense fault system with NE-SW and almost N-S direction within the granite that is present especially in the eastern side of the batholith (Figure 6).

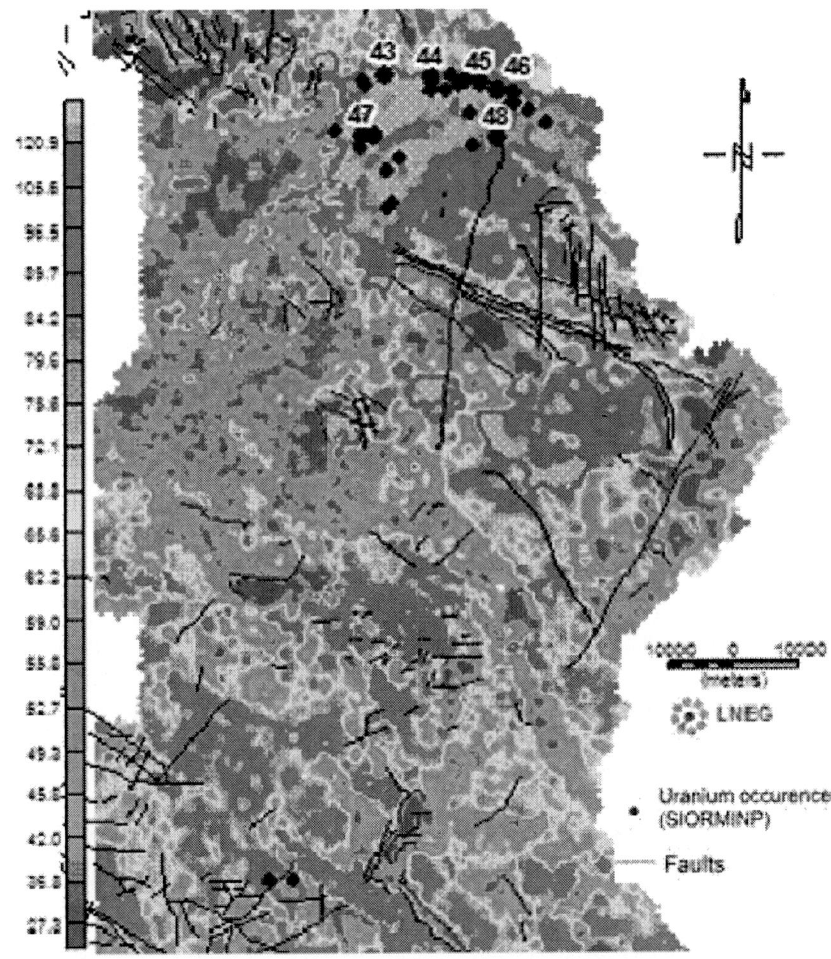

Figure 6. Radiometry map of Alto Alentejo with the main fault systems and location of known U occurrences (SIORMINP database) 43 – Nisa; 44 – Tarabau; 45 - Póvoa e Meadas; 46 – Meada; 47 - Palheiros de Tolosa; 48 – Melriça.

The fault systems where U installation occurred were the WNW-ESSE when intersecting metasediments and NNW-SSE and NE-SW when intersecting granites (Campos and Pereira, 1991). Exploration was intense during the mining period in the Beiras region, medium intensity of exploration was made in Alto Alentejo region and low in Trás-os-Montes region (Matos Dias and Correia Marques, 1980).

Uranium Ore Processing History in Portugal

The classical treatment of uranium ores was natural uranium leaching. In 1963 JEN reported 10 years of experience with this kind of ore treatment from *Companhia Portuguesa de Radium, Lda* (Ramos, 1963). Natural leaching was used in radium recovery until 1946 afterwards uranium. About 15-20kg/ton of ore processed to radium was then recovered to uranium. Initially the ore was transported to be concentrated in the Chemical Treatment Plant at Urgeiriça. The ore was crushed to the granulometry of 5x8 inches to a daily treatment from 90 to 100 t of ore.

Leaching of uranium with water was always associated with the presence of sulfides. Several observations were appointed (Pinch, 1966), such as:

- several types of ore leaching tests;
- secondary minerals were less resistant than primary ones;
- pyrite is essential in the process;
- leaching is proportional to the degree of acidity of the liquor, and alkaline liquors may result from the presence of carbonates;
- washing the ore cannot be continuous to facilitate the oxidation;
- reducing the grain size raises the leaching properties.

Experiments to improve the recovery of ores in the poorest deposit at Urgeiriça, Rosmaneira, Bica showed that these mines contained minerals with different compositions: at Urgeiriça there was essentially a low concentration primary ore containing appreciable amounts of visible pyrite; at Rosmaneira, with an initial content of U_3O_8 higher than at Urgeiriça, the ore exhibited a similar rate of leaching and a similar pyrite content; Bica ore proved to be not easily leached although with some visible pyrite. It emerged from the experience that the leaching was slow at first, faster in the interim stage and returned to be slower in the third phase that also decrease the concentration of uranium in the liquor also increased the concentration of iron which required greater consumption of magnesium per kg of uranium precipitate.

Extraction tests proceeded with the construction of stoneware pipes in which the ore was placed and where it was watered daily. Pyrite, pyrolusite and iron sulfate were added to the ore to turn the leaching more efficient. The addition of finely calibrated sand and pyrite from the Chemical Treatment Plant not only enhanced the permeability of the ore but also facilitated the leaching (Pinch, 1966). The facilities have undergone improvements in 1968 in order to reduce the grain size of the ore as an attempt to increase uranium recovery (Conceição, 1972).

In the Chemical Treatment Plant at Urgeiriça, after the ore was milled, subjected to a solid-liquid separation and filtrated, uranium extraction with organic solvents or ion exchange resins allowed a recovery of about 90% U_3O_8. The pursuing procedure was the precipitation with ammonia, filtration, drying and packaging of the ore.

To meet the international demands of uranium, the processing capacity achieved was 155 t of ore per day. The pilot plant updated was installed in the mine of Sra. das Fontes at Pinhel and batteries of resins and solvents in Bica mine (Matos Dias, 1971).

The poor ore treatment facilities in Rosmaneira worked from 1953 until 1960. In Urgeiriça the first ore treatment phase started in 1954 and ended in 1961. In Bica the ore treatment started in 1954 whereas in Valinhos was in 1957 and Vale da Arca in 1959. In 1958 was built the called Urgeiriça II and closed in 1962 when it ended the contract of *Companhia Portuguesa Radio Ltd* with the Portuguese State. These facilities consisted of a series of tanks where the mineral was deposited, collecting tanks where the liquor obtained was stored and a facility where the precipitation process occurred. The various containers were arranged whenever possible in cascade to avoid hydraulic pumping. The bottom of the tanks was coated with cement and Flintkote. Each container of ore had an independent irrigation system and liquors were collected to the collecting tank. The precipitation was done in Urgeiriça Chemical Treatment Plant, in Bica, Rosmaneira and Vale da Arca mines. The precipitation was made after passing the liquors from the collecting tank to the precipitation tank, where agitation of about 30 minutes with compressed air and addition of magnesium until the pH reaches 6.3 was made. After this process a sample of the suspended precipitate was collected to check the pH and the flocculant was added and agitated about 5 minutes. After this the air was let out to let the precipitate settle. After the precipitate is deposited the liquid supernatant was extracted and the precipitate was dried and packaged. These ore processing facilities treated more than 60,000 t of poor ore and 47 t of U_3O_8 was produced until 1962 (Elias, 1971). Grosso and Urbano (1972) determined the U concentrations in the ores dividing the U regions in Urgeiriça, Guarda and Alto Alentejo. Calculations were made to increase the ore treatment capacity of Urgeiriça plant to 300 t /day or construct more facilities and transport the ore to those other facilities as well. In Guarda the balance between transport of poor ore from other mines to the Urgeiriça treatment plant or make treatment of the ore in mobile and semi-mobile installations on the sites was considered ineffective so the ore was then transported to Urgeiriça. In Alentejo an investigation pit was opened in late 1969 to study the ore treatment and if transport to Urgeiriça would be necessary. It was observed that this ore is different from the rest treated in Portuguese plants. The results showed that the Nisa ores can be leached with acid and that due to the low grade of the ore, compared to the usually treated in Urgeiriça, roasting at 350°C during one hour after the leaching, increased in 10% the ore recovered. The grain size for treatment of Nisa ore would be lower than 3.36 mm and milled dry (Elias *et al.*, 1973). It was also observed that the other smaller Alto Alentejo ore deposits would be transported to the projected treatment plant of Nisa but that further recover tests were needed because there are differences between them (S.P.E.M., 1973). *In situ* leaching was another of the methods used for the extraction of the ore in Urgeiriça. This method consisted of leaching the ore without taking it out of the deposit. Thus the dissolution of the uranium deposit was made after adding sulfuric acid at pH 1.8 to the ore where reagent adsorption on the mineral surface, desorption and reaction with the transfer from the mineral to the fluid took place. This method was used when the content of the ore did not allow the removal of the deposit by conventional methods or where mining was difficult. This method was made in all situations after termination of the Chemical Treatment Plant. To have better contact between the solution and the deposit, fractures were induced. The fluids were injected into holes and sometimes it was necessary to modify the route of the fluid. Some open pits were also used for static leaching after collection of liquors from leaching *in situ* within the mine (Matos Dias, 1976). This procedure has become common practice. The static leaching was the dissolution of the ore at a certain granulometry with a concentration of sulfuric acid. In the case of Cunha Baixa the ore was extracted and the radiometry was measured and divided into three categories that were indicative of the ore grade: 1) above the cut-off (0.02% U_3O_8) the ore grade was taken to the Chemical Treatment Plant of Urgeiriça; 2) less than 0.02% U_3O_8 and below the cut-off grade the ore was stored for static leaching for which the opencast of the mine was used; 3) less than the previous grades was stored as sterile. In this case tests on stoneware pipes were also carried out that led to the conclusion that the ore from the Cunha Baixa due to their mineralogical and chemical characteristics presents a low recovery with natural leaching, i.e. with neutral water recovery is low. With the irrigation of the ore with a solution pH 1 of sulfuric acid there is a high recovery although a little slow. With the use of effluent Chemical Treatment Plant of Urgeiriça the recovery is also high and is faster. Static leaching was then adopted for the treatment of poor ores and considered efficient and economically viable. Thus, mining operations in 1996 at Quinta do Bispo and Bica had ore leaching facilities using immersion in acid solutions where ore was then submitted to ion exchange processes. The operation of removing the uranium from the resin was made in Urgeiriça. In the final processing only Quinta do Bispo had the leaching by immersion process in the mine open pit followed by the extraction of uranium concentrate from the leached solutions. This was done with ion exchange resins that were regenerated in order to remove the uranium concentrate in Urgeiriça. The static leaching had already been abandoned. The latest activities of processing and production of ore were undertaken in 2001 where the treatment of ore by means of ion exchange resins was made in Quinta do Bispo from ore from the Sevilha mine. The ores of the Sevilha mine were entirely extracted and deposited in tanks in Quinta do Bispo and Urgeiriça being leached afterword in Quinta do Bispo and the product subsequently recovered from the resins in Urgeiriça. In order to minimize the environmental impacts caused by the treatment of the ore held in Cunha Baixa, whose operation ceased in 1991, ENU proceeded to the neutralization of water and soil surrounding the mine. The waters of the open pits passed by a circuit to which was added calcium hydroxide and barium chloride to increase the pH and precipitate the radium. Simultaneously a monitoring of seepage water from tailings was undertaken through the analysis of water from two piezometers. In 1992 the leaching of marginal ores started which over time had been deposited in the surroundings of Quinta do Bispo. To put the waste from leaching in the open pit its bottom was covered with 30 cm thick clay layer. The solutions were recycled without the need of adding new water to the circuit, when in summer there was a lack of clean water the circuit was fed by a borehole. After that the excess water was treated with lime and barium chloride and launched in precipitation ponds. For the treatment of Sevilha mine ore a leaching tank was built with a capacity of 2000 t in the base with a layer of sand on which geotextile was deposited. After that layer a film of high density PVC of 1.5 cm, and again geotextile; a drainage layer of 0.5 m thick coarse gravel, a drainage system and perforated pipes upon which the ore is deposited. The uranium concentrate was stored in Urgeiriça before being exported. Storage was made according to international standards applied to these products.

Environmental Issues Related with Radioactive Deposits

Uranium transport generally occurs in oxidizing surface water and in groundwater under the form of uranyl ion, as uranyl fluoride or carbonate complexes. Uranyl fluoride complexes dominate in acidic oxidizing waters, whereas carbonate complexes dominate in near-neutral and alkaline oxidizing waters (Langmuir, 1978; Littleton, 2006). Uranium migration in soils can be reduced when UO_2^{2+} is reduced to U^{4+} by humus, peat, or other organic matter under anaerobic conditions; this knowledge is presently used for non invasive *in situ* remediation processes. Also phosphates and sulfides usually precipitate uranium and hence stop migration, a behavior that can be exploited in remedial operations as well. Hydroxyl, silicate, sulfate and organic complexes might also be important, specially sulfate in mining and milling operations that use sulfuric acid as a leaching agent. Maximum sorption of uranyl ions in natural environments occurs at a pH of 5.0-8.5. The sorption of uranyl ions by such natural media appears to be reversible; to immobilize the uranium permanently it is necessary to reduce it from the hexavalent form to the tetravalent form by means for instance of hydrogen sulfide (H_2S). The tetravalent uranium ion is essentially insoluble (Langmuir, 1978). In the case of *in situ* non removable techniques it is necessary to bear in mind the other hazardous property of uranium: radioactivity, that is not taken care only by immobilization but also by isolation where coverage is necessary as well (Pereira *et al.*, 2004). Radon emanation from rocks is different from lithology to lithology and should not be ignored or unknown in remediation processes because sometimes natural levels are higher than intervention levels (Neves and Pereira, 2004). Efficient remediation techniques apply comparison between pre- and post-mining operations, but this information is difficult to obtain generally because mining was started before modern diagnosis techniques (Pereira *et al.*, 2006). Different soil types also influence very much the distribution and mobility of uranium. Lateritic soils can mobilize uranium easier than for instance hydromorphic podzols where uranium can accumulate in reduced organic matter (Barbier, 1974; Rachkova and Shuktomova, 2009). Clays have an important part in fixing uranium in soils, being a natural trapper of uranium (Zaidan *et al.*, 2003; Jeffrey *et al.*, 2005; Figueiredo *et al.*, 2011); not less important fixing role is played by organic matter in reduced environment and by iron oxy-hydroxides. The mechanisms of ore forming in sedimentary environments also support itself in these properties such as organic matter reduction (Spirakis, 1996; Kyser and Cuney, 2008). Important factors that result in presence, solubility and immobilization of uranium in waters and in superficial environments that are important to retain for remediation purposes are: 1) uranium content in rocks, soils and sediments and its leaching ability; 2) proximity to uranium-bearing rocks and mineralizations; 3) climatic effects in the media and seasonal variability especially evapo-transpiration; 4) pH, oxidation state; 5) the presence of anions such as SO_4^{2-}, F^-, Cl^-, CO_3^{2-} to form sulfates, fluorides, chlorides, carbonates; 6) chlorides to form insoluble uranium minerals 7) highly sorptive materials such as organic matter, ferric, manganese, titanium oxy-hydroxides and clays (Langmuir, 1978). The nature of uranium species depends on the nature of the solutions, redox conditions, pH, temperature and uranium concentration in the source media (Cuney, 2010). At 25°C and atmospheric pressure in acid aqueous and oxidant the principal stable uranium ion is uranyl, when pH rises the hydroxides appear. The uranium mobility increases especially in the groundwater where soluble uranium species complexes with anions to form carbonates, sulfates, phosphates, fluorides and chlorides (Langmuir, 1978; Kyser and Cuney, 2008). Chemical transformation, dilution, sorption, and radioactive decay, are responsible for natural attenuation regarding the presence of metals and radionuclides. The health and environmental risks may be reduced by changing the amount of exposure, the exposure pathway, or the toxicity (Vuong, 1998). Fine grained uranium minerals can be trapped in sediments if organic matter or reduction conditions immobilize uranium. The knowledge of these ore formation conditions can also play an important role in remediation because it is part of the sedimentary ore processes at low temperature (Langmuir, 1978).

Exposure in Uranium Regions

EPA information and according to the USA's National Council on Radiation Protection and Measurement, the average annual radiation dose per person in the U.S. is 620 milliRem (6.2 millisieverts) and that the common dose of radiation from radon and thoron emanation, intra and extraterrestrial background represents 50% of the average dose per person which is significant. Primary contributors of radiation from the natural environment are soil gamma radiation and inhaled radon. Annual doses for people living in brick homes may increase up to 10 milliRems/year (mRem/yr) due to naturally occurring thorium, uranium, and radium found in clays often used to make bricks (NCRP, 1987). A study was conducted as a joint venture between Health, Environment and Economy Ministries to confirm or reject the idea that workers from Urgeiriça underground mine were, due to labour, exposed to radioactivity during the living-period of the mine. The project was called MinUrar (Falcão *et al.*, 2005). The main results indicate a significant probability that the workers were exposed to radioactivity. Following this study, EDM made it a priority to decommission the radioactive mines (www.edm.pt). The main issue of management of uranium wastes in a uranium rich region such as Beira is to define the reference level and remediate considering this reference. Background level is much higher in this region than in other regions of the country. These differences can be observed through the natural radioactivity obtained by the radiometry measurements. Furthermore, it is vital to monitor long-term decommissioning actions and the biological community exposed to radiation, even from natural radiation.

Overview of the Uranium Waste Management in Portugal

The Central Portugal was the only region where uranium was mined, thus it is the region with the greatest number of environmental problems and waste management needs. EDM is the national company with authority to decommission the abandoned mine areas called "orphan sites". Because radioactive sites could configure a risk to the public in the uranium region (Falcão *et al.,* 2007) they were considered a priority of intervention.

A national plan was carried out and the intervention was divided according to is radiological impact of 60 mines and one factory, whereas 45 were open pits and 15 underground. Therefore, group 1 corresponds to the mines where only mining occurred and that the ore was transported all to other mines where treatment facilities were installed. This group included Barroco D. Frango, Pera do Moço, Ribeira do Ferro, Corguinha/Prazos, Quinta das Seixas, Sentinela, Barroca Funda, Cruz da Faia, Cótimos, Alto da Rasa, Reboleiro, Formiga, Ervideira, Ferreiros, Lenteiros, Carril, Luz, Fontainhas/Gradiz, Coitos, Pedreiro and Carrasca.

The second group corresponds to the group of mines where mining occurred, and where by means of radiometric measurement and classification rich ores were transported to treatment facilities elsewhere and the poor ores were stocked in piles on the site. This group included Mondego Sul, Valdante, A do Cavalo, Espinho, Tentinolho, Mestra, Picoto, Vale Covo, Fonte Velha, Vales and Alto da Várzea.

The third group corresponds to mines where heap leaching was carried out in stockpiles and where remaining wastes of this activity were still on site. This group included Castelejo, Vale de Abrutiga, Sra das Fontes, Barracão (Fábrica), Prado Velho, Forte Velho, Rosmaneira, Vale D'Arca, Mortórios, Ribeira do Boco, Fontinha, Barroco I, Maria Dónis, Canto do Lagar and Corga de Valbom. The fourth group corresponds to mines that involved leaching and chemical treatment (ENU, 2000): Urgeiriça, Quinta do Bispo, Cunha Baixa and Bica.

At the time of this study, the sites of Sevilha, Vale da Videira, Vale do Tamão, Tapada dos Mercados, Freixinho, Póvoa de Cervães mines were already remediated and Vale de Abrutiga was starting to be remediated. Monitoring was already implemented at Urgeiriça, Quinta do Bispo, Cunha Baixa, Bica, Vale de Abrutiga and Castelejo where acid mine wastes occurred. Recently, a wide study was undertaken in order to manage the wastes from the 61 mines; the mines were divided into groups according to accessibility and natural characteristics with the objective of economically transporting the wastes to the main mine of the group where they will be stored and confined. The groups are: Castelejo, Mortórios, Prado Velho and Quinta do Bispo. Complementary studies will be made in these mines in order to evaluate the natural characteristics has a storage place for the wastes (http://www.edm.pt). A plan was previously made to Barragem Velha from the Urgeiriça project (RNT, 2004). In 2008 EDM decommissioned the mines by removing the wastes placed in S. Barbara poor ore waste pile and other poor ore dump into the "Barragem Velha" and sealed the site with several layers, ranging from clay up to a cover of geotextile material. Monitoring spots were also emplaced and periodic monitoring is made. Also the main Chemical Treatment Plant of the uranium mining period was decommissioned and used as heritage and possibly will be used as a study centre. All uranium occurrences in Alentejo and Trás-os-Montes provinces are unexploited and wastes are practically inexistent. A minor waste production was carried out during prospective mining at Nisa and Palheiros de Tolosa and the ore was taken to Urgeiriça for mineralurgical tests.

CONCLUSION

The uranium producing regions in Portugal are Beiras, Trás-os-Montes and Alto Alentejo Provinces. The known mineral deposits are mainly vein-type deposits although disseminated deposits in the contact metamorphic aureole in metasediments also occur. Few occurrences are reported in sedimentary rocks. An alluvial type deposit was only identified in the Urgeiriça mining field.

Uranium mining was carried out in Portugal from 1907 (Rosmaneira) to 2001 with the closure of the Sevilha and Quinta do Bispo Mines. The most important uranium mine was the Urgeiriça mine where approximately 1100 t of U_3O_8 were mined (1944 until 1992) in both underground and open pit settings. This mine also produced radium from 1913 until 1944. A total of nearly 500 mineral occurrences were identified and evaluated during exploration work in the 50's and 60 mines were exploited (45 open-cast; 15 underground). Mining was solely restricted to Central Portugal and therefore, the uraniferous regions of Trás-os-Montes and Alto Alentejo remain unexploited. The main known reserves were classified as medium size where the Nisa deposit, from the Alto Alentejo uranium province, is the biggest (2440 t of U_3O_8), most studied and better known deposit. The Horta da Vilariça deposit has an estimated reserve of 945 t of U_3O_8 but is considered a less studied region where the investment in exploration was small.

The considered "fertile" fault systems were the ones resulting from the Alpine Orogeny that in Beira region (NE-SW and ENE-WSW orientation) and in the Alto Alentejo region (WNW-ESSE orientation) when intersecting metasediments as in the Nisa deposit and NNW-SSE and NE-SW when intersecting granites. The considered "fertile" granites in Portugal are peraluminous leucogranites (Variscan S-granites), formed by partial fusion of metasediments of pre-Ordovician age.

In times of great economic importance and low raw material supply within Europe, uranium exploration gained new interest for mining companies. Portugal still has unmined areas with a relatively high degree of geological knowledge such as Alto Alentejo region and the less known Trás-os-Montes region which are the most attractive areas of the country. Remediation projects already developed in the Beira region makes this region less attractive, although exploration cannot be excluded in this part of the country where it is geologically favorable and the history of mining showed a great potential in the past.

REFERENCES

Ávila Martins, J., 1959. *Hematitization and ferruginous alteration of Beiras granites and their relation to uranium occurrences (in Portuguese).*Memória nº16, Junta de Energia Nuclear.

Barbier, M..J., 1974. Continental weathering as possible origin of vein-type uranium deposits. *Mineral. Deposita*, 9, 271-288.

Basham, I.R., Matos Dias, J.M., 1986. Uranium veins in Portugal. Vein type uranium deposits. *Report of the Working Group on Uranium Geology,* Internat. Atomic Energy Agency (IAEA).

Batista, M.J., Figueiredo, M.O., Silva, T.P., Costa, A., Lourenço, C., Midões, C., Ramalho, E., Solá, R., Prazeres, C., Fernandes, J., Ferreira, M.L., Leote, J., Ribeiro, L., Martins, L., 2009. *Geoenvironmental caracterization of Nisa uranium* deposit (in Portuguese). Technical Report for EDM, 28pp.

Burns, P.C., Finch, R.,1999. Uranium: Mineralogy, geochemistry and the environment. *Reviews in Mineralogy,* Mineral. Soc. Amer. 38, 679p.

Cabral, J.M.P, Gouveia, M.A., Prudêncio, M.I., 1979. Determination of rare earths and other oligoelements in rocks by activation with thermal neutrons (in Portuguese). *Boletim Mus. Lab. Miner. Geol. Fac. Ciênc. Lisboa*, 16 , 205-218.

Cabral Pinto, M.S., Silva, M.V.G., Neiva, A.M.R., 2008. Geochemistry of U-bearing minerals from the Vale de Abrutiga uranium mine area, Central Portugal. *Neues Jahrbuch für Mineralogie - Abhandlungen,* 185, 183-198.

Cajão, M.M. 1969. *Internal Report,* Junta de Energia Nuclear.

Campos, A.B.A., Pereira, L.C.G., 1991. Architecture and fracturation of Nisa granite, Alto Alentejo-Portugal (in Portuguese). *Memórias e Notícias, Publ. Mus. Lab. Mineral. Geol., Univ. Coimbra*, 112, 121-133.

Cameron, J., 1960. The ore controls of some uranium vein deposits in Portugal, Technical paper, 26. Junta de Energia Nuclear. 62p.

Cerveira, A., 1950. About uranium metallogenesis (in Portuguese). *Bol. da Soc. Geol. de Portugal*, 8 (3), 141-179.

Conceição, A., 1972. Preliminary study on the reduction of the grain size section of Oficina de Tratamento de Minérios da Urgeiriça (in Portuguese) – Relatório da Direcção-Geral dos Serviços de Prospecção e Exploração Mineira, Junta de Energia Nuclear.

Cotelo Neiva, J.M., 2003. Portuguese deposits of uranium ores and their geneses (in Portuguese). In: A Geologia de Engenharia e os Recursos Geológicos, Imprensa da Univ. Coimbra, vol. I, pp. 15-76.

Cuney, M., 2009. The extreme diversity of uranium deposits. *Minerallium Deposita*, 44, 3-9.

Cuney, M., 2010. L'uranium dans la croûte terrestre et ses propriétés geochimiques. *Géochronique*, 113, 18-20.

Dias, G., Noronha, F., Almeida, A., Simões, P.P., Martins, H., Ferreira, N., 2010. Geochronology and petrogenesis of tardi-varisc plutonism (NW of Portugal): Synthesis and inferences about accretion processes and crustal recycling in Centro-Iberian zone (in Portuguese). In: Ciências Geológicas-Ensino e investigação e a sua História, Vol. I, Chapt. II, 143-160.

Elias, J.T., 1971. Uranium ore concentration industry in the Metropolis (in Portuguese). In: Minérios Nacionais e sua valorização através da indústria metalúrgica. Publication of Laboratório de Física e Engenharia Nucleares, C1-C20.

Elias, J.T., Conceição, E., Santos, C.S., Alves, M., Silva, L.A., Sá, C., Ramos, M.A., 1973. Study on the Nisa ore treatment (in Portuguese). Relatório IQ (TM)-3/73. Serviço de Investigação de Química. Grupo de Tratamento de Minérios. *Laboratório de Física e Engenharia Nucleares*, 45p.

ENU- Empresa Nacional de Urânio, S.A., 2000. Preliminary Project of Environmental Recovery of Uranium Mines: Barroco I, Ribeira de Boco, Canto do Lagar, Corga de Valbom, Vale de Abrutiga, and ex-factory from Barracão – a "Proposition of Methodologies".27p.

Falcão, J.M., Carvalho, F.P., Machado Leite, M., Alarcão, M., Cordeiro, E., Ribeiro, J., 2005. MinUrar – Uranium mines and wastes. Effects in Health of the population. Relatório científico I. Public report.

Falcão, J.M., Carvalho, F.P., Machado Leite, M., Alarcão, M., Cordeiro, E., Ribeiro, J., 2007. MinUrar - Minas de urânio e seus resíduos. Efeitos na saúde da população. Relatório científico II. Public report.

Ferreira, N., Iglesias, M., Noronha, F., Pereira, E., Ribeiro, A., Ribeiro, M.L., 1987. Granites of Central Iberian Zone and ints geodinamic setting (in Portuguese). *In* : Libro Homenaje a L.C. Garcia de Figueirola, 37-51, Ed. Rueda, Madrid.

Figueiredo, M.O., Silva T.P., Batista, M.J., Leote, J., Ferreira, M.L., Limpo, V., 2010. Uranium in surface soils: An easy-and-quick assay combining X-ray diffraction and X-ray fluorescence qualitative data. *J. Geochem. Exploration*, 109, 134-138.

Frost R. B., Barnes, C. G., Collins, W. J., Arculus, R. J, Elis, D. J., Frost, C. D., 2001. Geochemical Classification for Granitic Rocks. *J. Petrology,* 42, 2033-2048.

Gouveia, M.A., Seabra e Barros, J., Tavares, M.J., Freitas, M., 1986. Combination of neutron activation, atomic absoption spectroscopy and conventional methods in elemental analysis of silicate rocks.*J. Radioanal. Nuclear Chemistry*, 97, 265-270.

Grasty, R.L., Tauchid, M., Torres L., 1993. Standardization of old gamma ray survey data. Application of uranium exploration data and techniques in environmental studies. Proceedings of a Technical Committee meeting held in Vienna, 9-12 November, IAEA-TECDOC-827, 35-46.

Grosso, J.C., Urbano, M.F.N., 1972. Determination of ore grade from Urgeiriça, Guarda e Alto Alentejo from which the treatment is apropriate in the classical plantor static leaching and transport quantities estimate (in Portuguese). *Publication from Gabinete de Estudos e Planeamento. Junta de Energia Nuclear*. 12p. Http://www.edm.pt.

IAEA, 1980. Uranium evaluation and mining techniques. International Atomic Energy Agency, Viena, 551p.

IAEA, 2009. World Distribution of Uranium Deposits with Uranium Deposit Classification. IAEA-TECDOC-1629.

Jabaloy, A., Galindo-Zaldívar, J., González-Lodeiro, F., 2002. Palaeostress evolution of the Iberian Peninsula (Late Carboniferous to present-day). *Tectonophys.,* 357, 159-186.

Jeffery, A., Greathouse, H.R., Stellalevinsohn, M.A., Denecke, A.B., Pabalan, R.T., 2005. Uranyl Surface Complexes in a Mixed-Charge Montmorillonite: Monte Carlo Computer Simulation and Polarized XAFS Results. *Clays and Clay Minerals,* 53, 278-286.

Junta de Energia Nuclear (JEN), 1968. The uraniferous province of Central Portugal: structural and metallogenetic characteristics (in Portuguese). 131p.

Junta de Energia Nuclear (JEN), after 1963. Uranium mining industry in Portugal (in Portuguese), 45p.

Kyser, K., Cuney, M., 2008. Chapter 11: Implications for exploration strategies. In: Recent and not-so-recent developments in uranium deposits and implications for exploration. *Short Course Series*. Volume 39. Mineralogical Association of Canada.

Kalin, M., Wheeler, W.N., Meinrath, G., 2005. The removal of uranium from mining waste water using algal/microbial biomass.*J. Environm. Radioactivity*, 78, 151–177.

Langmuir, D., 1978. Uranium solution-mineral, equilibria at low temperatures with applications to sedimentary deposits. In: Kimberley, M. M. Uranium Deposits, their mineralogy and origin. *Short-Course Handbook*, Volume 3. Mineralogical Association of Canada.

Littleton, B., 2006. Depleted uranium. Technical Brief. U.S. Environmental Protection Agency. EPA-402-R-06-011.

Malavieille, J., 1993. Late Orogenic extension in mountain belts: Insights from the basin and range and the Late Paleozoic Variscan Belt, *Tectonics*, 12, 1115–1130, doi:10.1029/93TC01129.

Matos Dias, J. M., 1971. Geoeconomical perspectives of Portuguese uranium (in Portuguese). *Memórias e Notícias, Museu e Lab. Min. Geol. Univ. Coimbra*, 71, 36-47.

Matos Dias, J., Andrade, A. S., 1970. Uranium deposits in Portugal. *I.A.E.A Uranium Exploration Geology*, 129-142.

Matos Dias, J.M., 1976. Perspectives of uranium mining industry in Portugal (in Portuguese). *Memórias e Notícias, Museu e Lab. Min. Geol. Univ. Coimbra*, 81, 51-74.

Matos Dias, J.M., Correia Marques, J.A., 1980. Potencialidades uraníferas de Portugal. *Comum. Serv. Geol. Portugal*, t. 66, 177-184.

Neiva, A.M.R., Gomes, M.E.P., 2001. Diferentes tipos de Granitos e seus Processos Petrogenéticos: in Granitos Hercínicos Portugueses. *Memórias da Academia de Ciências de Lisboa* 39, 53–95.

NCRP, 1987. National Council on Radiation Protection and Measurements. Exposure of the Population in the United States and Canada from Natural Background Radiation. NCRP Report No. 94. Washington, DC: NCRP, December 1987.

Neves, L.F., Pereira, A.C., 2004. Radioactividade natural e ordenamento do território:o contributo das Ciências da Terra. *Geonovas*, 18. 103-114. Associação Portuguesa de Geólogos Magazine.

Oliveira, J.T., Pereira, E., Ramalho, M., Antunes, M.T., Monteiro, J.H., 1992. Geological map of Portugal 1: 500 000. *Serviços Geológicos de Portugal, Lisboa.*

Pagel, M., 1979. Granites et gisements d'uranium du Portugal: Typologies, altération et minéralisations. Technical Commitee Meeting on the Geology of vein and Similar Type uranium deposits.

Parra, A., Filipe, A., 2003. Portuguese uranium occurrences (in Portuguese). Instituto Geológico e Mineiro (Unpublished report).

Pereira, A.C., Neves, L., Dias, J.M., 2006. Estimating Radiological Impacts of Uranium Mining: The Evaluation of Natural Background in the Beiras Metallogenic Province (Central Portugal). TA6 – Radiation Protection of the Public and the Environment. *Proceedings of the Second European IRPA Congress on Radiation Protection.*

Pereira, A.C., Neves, L., Dias, J.M., Campos, A.B., Barbosa, S., 2004. Evaluation of the radiological hazards from uranium mining and milling wastes (Urgeiriça – Central Portugal). 11th International Congress of the International Radiation Association.

Pinch, C., 1966. Influence of ore granulometry on static lixiviation (in Portuguese). *Public. Lab. Phys. Nuclear Eng.* LFEN-N1.29.

Plant, J.A., Simpson, P.R., Smith, B., Windley, B.F., 1999. Uranium ore deposits; products of the radioactive Earth. *Reviews in Mineralogy and Geochemistry,* 38, 255-319.

Plant, J. A., Reeder, S., Salminen, R., Smith, D. B., Tarvainen, T., De Vivo, B., Petterson, M. G., 2003. The distribution of uranium over Europe: geological and environmental significance. *Applied Earth Science: IMM Transactions section B,*112 , 221-238.

Prazeres, C.M., 2011. Geochemical, radiometric and mineralogical characterization of some uranium mineralizations of the Nisa region (in Portuguese). MSc thesis, Univ. Lisbon, 147pp (Unpublished).

Prudêncio, M.I, Gouveia, M.A., Marçalo, J., Albuquerque, Barroso, E., 1987. Radioemanometry, Gamma-spectrometry, Fluorimetry, Neutron Activation and Numerical Taxonomy Methods in anuranium prospection program. A study of Rebolia Sediments (Portugal). *Comunic. Serv. Geol. Portugal*, 73, 11-19.

Rachkova, N. G., Shuktomova, I. I., 2009. Changes in the Mobility of Uranium, Radium, and Thorium Compounds in the Plow Layer of Podzolic Soil. *Eurasian Soil Science,* Vol. 42, No. 2, 194–200.

Ramos, A., 1963. Natural lixiviation of uranium Portuguese ores, 10 years of practice (in Portuguese). Report, Direcção-Geral dos Serviços de Prospecção e Exploração Mineira, Junta de Energia Nuclear.

Ribeiro, M.L., 2001. Modelos de implantação dos granitos variscos Portugueses. *Memórias da Academia das Ciências* 39, 33–52.

Ribeiro, A., Antunes, M.T., Ferreira, M.P., Rocha, R.B., Soares, A.F., Szyszewski, G., Moitinho de Almeida, F., Carvalho, D., Monteiro, J.H., 1980. Introduction à Geologie General du Portugal. 26éme Congrés Geologique Internacional, Paris, France. 114p.

RNT 2004. Environmental remediation in mining area of Urgeiriça (in Portuguese). Plano director de intervenção e projecto de selagem e recuperação ambiental da Barragem Velha. Relatório não Técnico do Estudo de Impacto Ambiental. http://www2. apambiente.pt/IPAMB_DPP/ docs/RNT1228.pdf

Saraiva, J.M., Torres, L., Leote, J., 1998. Aplicações da cartografia radiométrica na protecção ambiental e na prospecção de recursos, IX Asamblea Nacional de Geodesia y Geofísica, I Asemblea Hispano-Portuguesa de Geodesia y Geofísia, Aguadulce (Almeria), Espanha, 9-13 Febrero.

SIORMINP- Base de Dados de Concessões Mineiras (www.lneg.pt).

Spirakis, C.S., 1996. The roles of organic matter in the formation of uranium deposits in sedimentary rocks.*Ore Geology Reviews* Vol. 11, Issue 1-3, 53-69.

S.P.E.M., 1973. Projecto da oficina de tratamento químico de minérios de urânio de Nisa. Junta de Energia Nuclear. Direcção Geral dos Serviços de Prospecção e Exploração Mineira.

Torres, L., 1992. Radiometric map of Portugal. Internal report. *Direcção Geral de Geologia e Minas, Serviço de Fomento Mineiro,* 20p.

Torres, L.M., Grasty, R.L, 1993. The natural radioactivity map of Portugal. In: Application of uranium exploration data and techniques in environmental studies. *Proceedings of a Technical Committee meeting held in Vienna,* IAEA-TECDOC-827, 127-133.

UNSCEAR, 1958. Report of the United Nations Scientific Committee of the effects of Atomic Radiation. Supplement nº 17 (A/3838), 230p.

Vuong L., 1998. Natural attenuation of inorganic contaminants. Center for Public Environmental Oversight. http://www.cpeo.org/pubs/fsna3i.html.

Wanty, R., Miller, W., Briggs, P., McHugh, J., 2000. Geochemical Processes Controlling Uranium Mobility in Mine Drainages, In: The Environmental Geochemistry of Mineral Deposits, Part A: Processes, Techniques, and Health Issues, Reviews in Economic Geology, 201-213.

Zaidan, O.F., Greathouse, J.A., Pabalan, R.T., 2003. Monte Carlo and Molecular Dynamics Simulation of Uranyl Adsorption on Montmorillonite Clay. *Clays and Clay Min.,* 51, 372-381.

In: Uranium: Characteristics, Occurrence and Human Exposure ISBN: 978-1-62081-207-5
Editors: Alik Ya. Vasiliev and Mikhail Sidorov

Chapter 4

LASER-INDUCED NUCLEAR DECAYS IN URANIUM ISOTOPES

G. A. Shafeev
Wave Research Center of A.M. Prokhorov
General Physics Institute of the Russian Academy of Sciences, Russia

ABSTRACT

Experimental results are reviewed on laser-induced accelerated nuclear decays in Uranium isotopes and nuclides of its branching under laser exposure of metallic nanoparticles in aqueous solutions of Uranium salts. Laser exposure of metallic targets (Au, Ag, Pd) in aqueous solutions of $^{238}UO_2Cl_2$ leads to generation of nanoparticles of corresponding materials in the bulk of the liquid and simultaneously shifts the secular equilibrium in the activity of nuclides of its branching, such as ^{234}Th and ^{234m}Pa. Real-time acquisitions of gamma-spectra of solutions indicate the presence of non-equilibrium gamma-activity in the range of photon energies between 50 and 60 keV. This activity is attributed to the laser-assisted generation of short-living isotopes ^{214}Pb and ^{214}Bi that stand at the end of ^{238}U branching. The experimental results are presented on laser-induced accelerated alpha-decay of ^{232}U under exposure of aqueous solution of its salt in presence of metallic nanoparticles. The results on the shift of secular equilibrium under laser action are compared with similar results on electro-explosion of metallic wires in aqueous solutions of Uranium salt. Possible mechanisms of laser-induced decays are discussed on the basis of amplification of the laser wave on metallic nanoparticles due to plasmon resonance.

INTRODUCTION

Since their advent more than 50 years ago lasers became versatile tools widely used both in variety of scientific researches and technological processes. Lasers produce a flow of highly organized photons, both in space and spectral range. This allows focusing laser beams to very small spatial dimensions that are, however, limited to the wavelength of laser light. The energy of laser photons is usually of several eV if no methods of frequency up-conversion are used. Free electron laser systems are capable of generation of X-ray photons. Still, however, photon energies of laser photons are much smaller than characteristic energies of nuclear levels. Recent theoretical work shows the capability of laser radiation to directly excite nuclear levels of energy [1]. However, exciting a nuclear transition would require an x-ray or gamma-ray lasers with an intensity greater than 10^{20} Watts/cm^2.

At the same time, laser generated photons can indirectly influence the nuclear levels and alter or induce nuclear reactions. Laser beams at intensity level of 10^{18} W/cm^2 are capable of inducing nuclear transformations under laser exposure of solid targets in vacuum. This can be achieved at pico- and femtosecond laser pulse durations [2-4].

The emission of alpha-particles was observed under exposure of ^{11}B target to picosecond laser pulses. Experiments were carried out on the "Neodymium" laser facility at the pulse energy of 10–12 J and pulse duration of 1.5 ps. The composite targets $^{11}B + CH_{2n}$placed in vacuum were used. The yield of 10^3 alpha particles per pulse has been detected at peak laser intensity on the target of 2×10^{18}W/cm^2 [5]. The mechanism of initiation of a neutron-less fusion reaction is due to the acceleration of electrons and protons in the plasma separated by strong laser field.

Another approach to laser initiation of nuclear reactions is based on the laser exposure of nanoparticles suspended in a liquid (colloidal solution). This process becomes possible at much lower level of intensities of laser radiation in the medium, of order of 10^{12}-10^{13} W/cm^2. The crucial role in laser-induced acceleration of nuclear decays belongs to nanoparticles (NPs) that interact with laser beam in presence of unstable isotopes. Each nanoparticle can be considered as a target for laser exposure. Unlike bulk targets, metal nanoparticles are optically thin. The critical density is reached from the onset of laser pulse; this is the density of free electrons in the nanoparticle but owing to the small size of the target the reflection is negligible. In this case the whole nanoparticle is heated up by the laser beam homogeneously. The same applies to the late stages of laser heating – the material of the target is ionized and expands in the shape of spherical cavity but the laser radiation is still absorbed by the plasma due to the small size of the cavity. Small size of NPs leads to local enhancement of the field of laser wave, and the efficient intensity in the vicinity of nanoparticle can be several orders of magnitude higher due to plasmon resonance of charge carriers. Such electric field is already comparable with electric fields inside the atoms and this causes significant distortion of electronic shells. This distortion, in turn, alters the probability of spontaneous nuclear decays, either alpha- or beta ones.

Since the acceleration of nuclear decays occurs under exposure of a solid target immersed into liquid to laser radiation, an introduction section is needed to outline main phenomena that take place in this experimental setup.

Generation of Nanoparticles under Laser Ablation of Solids in Liquids

Formation of nanoparticles under laser ablation of solids either in gas or in vacuum has been extensively explored during the last decade. Understanding of the mechanisms of cluster formation is needed to control the process of Pulsed Laser Deposition (PLD) that is widely used now for deposition of a large variety of compounds. Formation of nanoparticles under laser ablation of solids in liquid environment has been studied to a lower extent. A number of wet chemical methods of nanoparticles (NPs) preparation are known up to date. Laser ablation of solids in liquid environment emerged as an alternative technique that is capable of producing “pure” NPs without counter-ions and surface-active substances. Primary motivation for generation of NPs by laser exposure of solid targets in liquids was the hope that pure NPs would be more efficient in the process of Surface Enhanced Raman Scattering (SERS), since no other compounds can obscure the surface of metallic NPs [6-10].

Formation of both large varieties of NPs under laser ablation of corresponding solids has been reported during the last decade [11-23].

Further studies showed that the laser ablation in liquids is a complex process with a number of experimental parameters. In particular, interaction of ejected nano-sized species with hot liquid vapors at high transient pressure results in large diversity of final particles that remain in the liquid. Also, surface tension at the interfaces is dominant force that governs the dynamics of interaction of molten target material with vapors of surrounding liquid on a nanometer scale. Under pulsed laser exposure the liquid also undergoes chemical changes that may be stipulated by ejected NPs that act in this case as a catalyst. Initially liquid substance may become supercritical in the vicinity of the target, which alters its reactivity and opens new channels of chemical reactions especially in case of aqueous solutions.

Interaction of metal nanoparticles with laser light proceeds via its absorption by free electrons. Free electrons exhibit plasmon resonance, whose position is determined by both their concentration, their effective mass (that depends on the solid) and particle size. NPs of noble metals, e.g., Au or Ag, show a strong selective absorption in the visible, so the suspensions of these NPs in liquids are colored. Plasmon resonances of other metallic NPs are situated in the UV region. Though if the liquid is transparent at the onset of laser exposure of the target, the appearance of NPs may change this situation, and the laser radiation is absorbed by the colloidal solution. This may lead to poor control of the laser spot size on the target due to either thermal lensing inside the solution or different non-linear phenomena induced by laser beam in NPs. The position of plasmon resonance of metallic NPs virtually coincides with that of NS generated on bulk targets under their laser ablation in liquids (see above). In other words, metallic NPs in the bulk of the solution are of the same origin as NS left after solidification of the melt on the target. These NPs represent the part of the target material that has been sputtered into the liquid under the action of the pressure wave of the liquid that surrounds the target. The well-studied NPs generated by laser ablation in liquids are gold NPs. This is due to their perfect chemical stability, which guarantees the metallic nature of generated NPs. Synthesis of a stable colloidal solution of Au NPs in water has been reported by several groups [31 – 35].

Optical and chemical properties of Au NPs are thoroughly explored owing to well-established chemical methods of their synthesis. Spherical Au NPs are characterized by a single (degenerated) band of plasmon resonance situated in the visible range of spectrum. Elongated Au NPs are characterized by two plasmon bands situated in the visible, and their position varies with the aspect ratio of NPs. This offers a good coupling of the visible laser radiation to Au NPs. Moreover, tuning the laser wavelength in various parts of their spectrum allows shape-selective fragmentation of Au NPs [36].

Typical view of Au NPs generated by the laser exposure of a bulk Au target in water radiation of a Cu vapor laser in D_2O is presented in Figure 1.

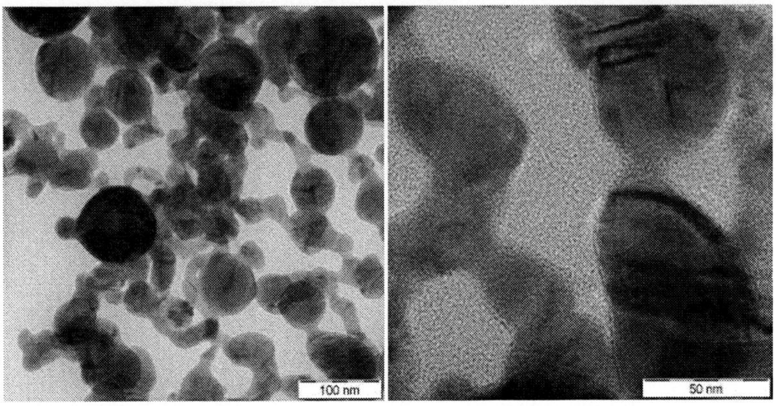

Figure 1. Transmission Electron Microscope (TEM) view of Au NPs generated by ablation of Au NPs in D_2O by radiation of a Cu vapor laser, wavelength of 510 nm, pulse duration of 15 ns.

Average size of generated NPs is about 50 nm and in general depends on the laser fluence on the target. Generated NPs remain in the solution and can be again exposed to the laser beam. This secondary exposure of NPs causes the change of their average size, and the size distribution function of NPs shifts to lower dimensions. This is so called fragmentation process. Fragmentation of NPs under their laser exposure in liquids has been reported as early as the synthesis of NPs itself [6, 11, 19]. The process manifests itself as a gradual decrease of the size of NPs upon laser irradiation. As a result, the distribution function of particle size shifts to lower dimensions. Optical constants of metallic targets do not vary largely in the wavelength range of lasers used for NPs synthesis, so fragmentation of NPs should be ascribed to interaction of synthesized NPs themselves with the laser beam inside the liquid.

The physical processes that lead to fragmentation of NPs in the laser beam are still under discussion. The most probable reason leading to reduction of NPs size are the instabilities that develop at the interface of molten NPs with the vapors of surrounding liquid [24]. Indeed, NPs may absorb enough energy from the laser beam to undergo the phase transition into a liquid state. The temperature of NPs exceeds at this moment the boiling temperature of surrounding liquid. The latter is therefore vaporized and forms a shell around the NPs. The pressure of vapors in this shell is around 10^8 Pa, and its possible asymmetry would result in the break of molten NPs inside it into smaller parts. The most probable process is splitting of the molten NPs into two equal particles, since, first, this corresponds to lowest value of total surface energy, and second, is the lowest mode of perturbation of the surface of molten NPs. This hypothesis is in qualitative agreement with the fact that the average size of NPs exposed to laser radiation in various liquids decreases with the decrease of boiling temperature of the liquid. The balance is gained when the pressure inside the NPs of radius R is of order of the pressure of the surrounding liquid p_{liq}: $2\sigma/R \sim p_{liq}$, where σ stands for the surface tension of molten NPs.

Fragmentation of the NPs under their laser exposure introduces feedbacks into the system “laser radiation – NPs”. Indeed, the average size of NPs under their laser exposure is inversely proportional to the peak intensity of the laser beam inside the liquid. The size of NPs decreases upon exposure till they become so small that the energy absorbed from the laser pulse is not sufficient for melting. The system therefore demonstrates a negative feedback and auto-stabilization. On the other hand, melting temperature of NPs decreases with its size, which is a positive feedback. Finally, small NPs formed via fragmentation from bigger ones do not absorb enough energy from the laser beam to be molten, and the ensemble of NPs acquires its lower limit of their size.

NANO-STRUCTURES FORMATION UNDER LASER EXPOSURE IN LIQUIDS

Laser exposure of a metallic target in liquid environment leads not only to ejection of NPs into the surrounding liquid. The rate of NP generation depends not only on the laser fluence, but also of experimental parameters such as the metal reflectivity at the laser wavelength, the nature of the liquid, etc. The rate of NP production decreases with the laser fluence. At lower fluence the rate of NP generation drops, while the surface of the target undergoes eye-visible changes. The coloration of the exposed areas is well pronounced while the target remains in the liquid, whereas it fades if the target is removed from the liquid and dried. In case of the Ag target the color of laser-treated areas is yellow, though the maximum of absorption for freshly-exposed Ag surface lies in UV region (370 nm, [25]). Exposed areas of Au sample look like metallic copper with corresponding maximum of absorption around 570 – 590 nm. The coloration is due to plasmon oscillations of free electrons in nano-protrusions of the surface exposed to laser radiation in water.

The formation of nanostructures on a rough metal surface subjected to laser ablation in water is due to the modification of the topology of the molten area upon decreasing the laser fluence. At high laser fluence (~ 1 J/cm^2 in our case), the molten area is continuous, and this melt is effectively dispersed as nanoparticles by the recoil pressure of the surrounding liquid. The recoil pressure of the vapor surrounding the target pushes these molten areas from the target and generates a nano-relief.

In other words, formation of nano-spikes is a result of a hydrodynamic instability at the interface "liquid vapor – melt" characterized by very small period of order of hundreds of nanometers. The described mechanism is independent on the target material, and similar nanostructures may be grown on any other material that absorb at the laser wavelength.

These data are corroborated with the images taken with a. The surface of either Ag or Au target is covered by a dense array of nanostructures (Figure 2). Of course, the protruding NS will be molten by the next laser pulse since they weak thermal coupling with the substrate. However, new NS will arise as soon as the laser pulse is over, and each subsequent pulse will be absorbed by nano-structured surface of the target. Therefore, interaction of pulsed laser radiation with metals immersed into liquid is characterized by formation of different nano-objects: nanoparticles in the liquid and nanostructures left on the target surface. The technique of laser ablation in liquids is very flexible and allows generating NPs of numerous metals and compounds in virtually any liquid. New phenomena arise if metallic NPs are dispersed in the aqueous solution of salts of unstable isotopes, e.g., Uranium. Laser exposure of such solution leads to irreversible changes of its activity and activities of its daughter nuclides. As it is demonstrated below, laser exposure accelerates both alpha- and beta-decays of all nuclides of ^{238}U branching. The effect of laser exposure of NPs on the activity on unstable isotopes was firstly observed in aqueous solutions of ^{232}Th salts [26].

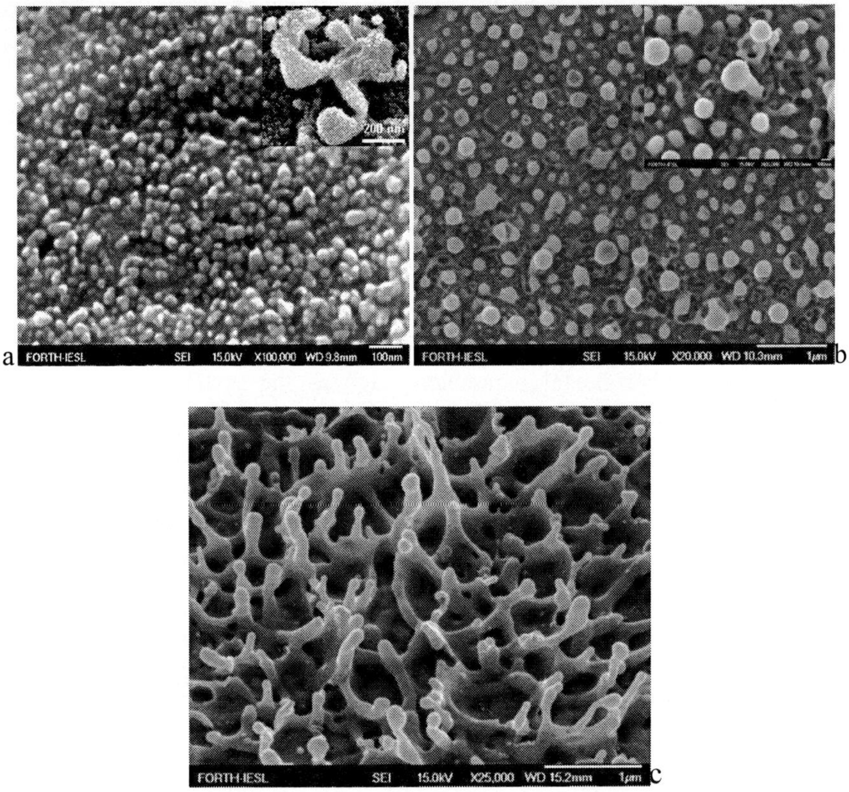

Figure 2. Scanning Electron Microscope with Field Emission (FE SEM) view of NS on Ag (a) and Au (b) targets produced by their ablation in water with radiation of a 5 psKrF laser, wavelength of 248 nm. Insets show the view at different scale. NS on bulk Au produced by femtosecond laser exposure in water, wavelength of 800 nm, pulse duration of 100 femtosecond (c).

It was experimentally found that the activity of daughter nuclides can be affected by laser exposure of Au NPs in solutions of $Th(NO_3)_4$. Also, laser exposure altered the activity of ^{137}Cs presented in the solutions as impurity. Further investigations showed that this phenomenon is rather general and is also observed in solutions of Uranium salts subjected to laser pulses of high intensity of order of 10^{12} W/cm^2, though this intensity is several orders of magnitude smaller than used in above mentioned experiments on laser exposure of various targets in vacuum.

The results of laser-induced accelerated decay of Uranium isotopes are compared with experimental results obtained as early as in 2002 on deviation of secular equilibrium in Uranium nuclides under electro-explosion of metallic foils in aqueous solutions of Uranil.

Aqueous Solutions of Uranil-Ions

Aqueous solutions of Uranil salts of natural isotope composition were used in all experiments, either $UO_2(NO_3)_2$ or UO_2Cl_2. Both salts dissociate in aqueous medium to UO_2^{-2} ion and corresponding anion. Uranium demonstrates 6+ valency in these compounds. In order to affect the activity of nuclides, the laser radiation should becoupled to the solution. This can be achieved using metallic nanoparticles suspended in the solution of Uranium salt. Figure 3 shows absorption spectra of both colloidal solutions of Au NPs and aqueous solutions of UO_2Cl_2. Radiation of both Cu vapor laser and the second harmonics of a Nd:YAG laser is efficiently absorbed by Au NPs owing to plasmon resonance of electrons in NPs situated for Au around 530 nm.

The absorption of the solution increases in UV range of spectrum, and the radiation of the third harmonics of a Nd:YAG laser with wavelength of 355 nm is strongly absorbed by Au NPs due to inter-band transitions. Moreover, this radiation is absorbed by Uranil-ion itself causing its strong green luminescence.

The fundamental output of a Nd:YAG laser at wavelength of 1064 nm lies outside the absorption range of the above solution. However, under sufficiently high intensity of laser radiation NPs serve as centers for ionization, and laser radiation couples to the colloidal solution through absorption by plasma shells around NPs. The density of electrons in these shells varies during the laser pulse but at each moment before recombination of plasma they can be considered as nanoparticles of plasma that are characterized with their own plasmon resonance.

That is why as it will be demonstrated below the influence of laser exposure on the activity of radionuclides in the solution weakly depends on the nature of NPs. The initial absorption due to plasmon resonance of electrons in specific metallic NPs serves as a precursor for further development of plasma cloud around NPs.

The decay of ^{238}U proceeds through well known sequence of alpha- and beta-decays of these isotopes and their daughter nuclides:

$$^{238}U \rightarrow {}^{234}Th \rightarrow {}^{234m}Pa \rightarrow {}^{234}U \rightarrow \ldots {}^{214}Pb$$

Typically, beta-decay leads to the formation of daughter nuclei in the excited state, and gamma-emission of this daughter nuclide allows tracing its activity.

In the equilibrium conditions (so called secular equilibrium) the ratio of activities of neighboring nuclides is inversely proportional to their half-life times.

This equilibrium ratio shifts under laser exposure of metallic NPs in aqueous solution of salts of Uranium. As it will be seen from results presented below the deviation from secular equilibrium is observed for both isotopes ^{238}U and ^{235}U and their daughter nuclides. High isotope selectivity is observed in certain range of laser parameters.

Namely, the activity of ^{238}U branching is strongly affected by laser exposure of the solution, while the activity of ^{235}U in the same solution remains constant within the accuracy of measurements. Similar results were experimentally studied for another Uranium isotope, namely, ^{232}U. The influence of laser exposure on the activity of nuclides of ^{238}U branching was estimated on the basis of gamma-spectra of solutions before and after exposure. Typical gamma-spectrum of the solution of UO_2Cl_2 used in the work is presented in Figure 4. The emission of gamma-photons accompanies the beta-decay of corresponding nuclides.

The activities of isotopes ^{234}Th, ^{234m}Pa, and ^{235}U before and after laser exposure were measured in the following series of experiments. The gamma-activity of ^{235}U was determined form the gamma-activity of its daughter nuclide ^{231}Th [27]. The activity was calculated as the area under the corresponding peaks of gamma-spectrum.

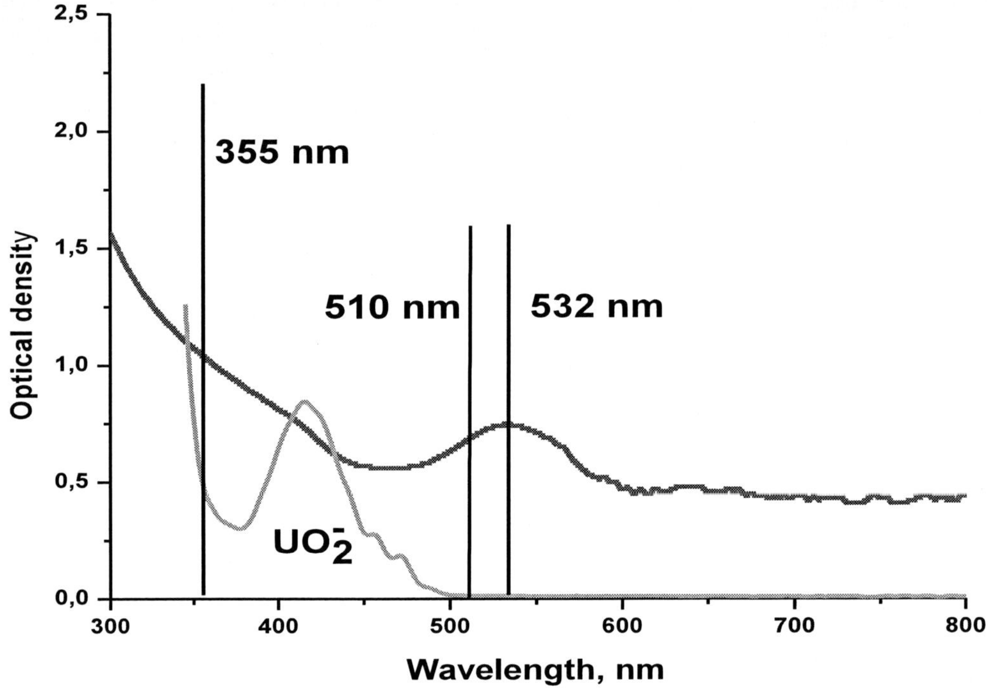

Figure 3. Absorption spectra of an aqueous solution of UO_2Cl_2 (orange) and of a colloidal solution of Au NPs (red). Vertical lines indicate the position of the second and third harmonics of a Nd:YAG laser as well as the position of the emission line of a Cu vapor laser. The position of the first harmonics of a Nd:YAG laser at 1064 nm is not shown.

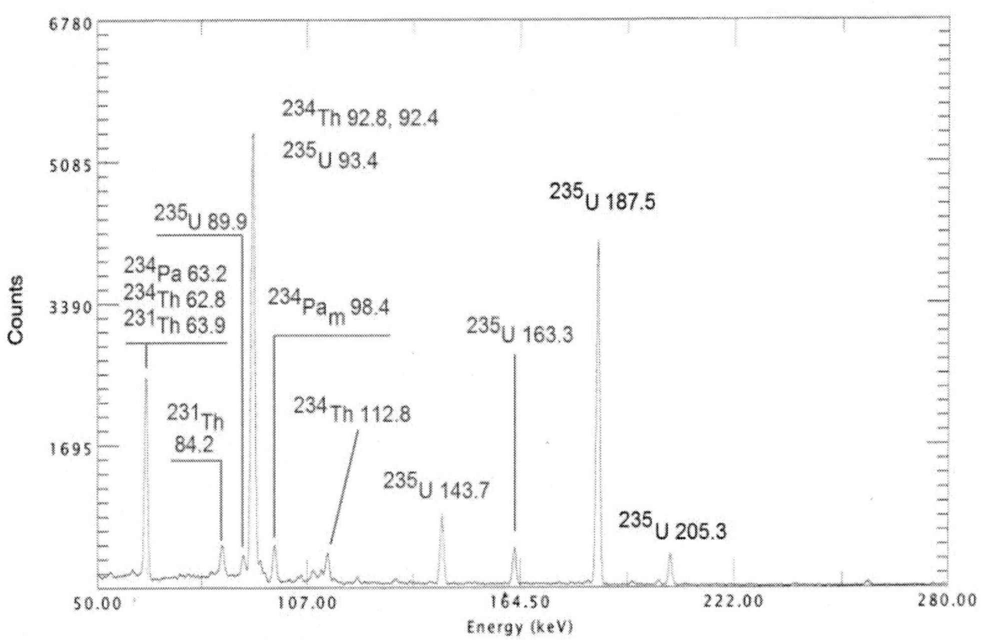

Figure 4. Gamma-spectrum of the initial solution of UO_2Cl_2 in H_2O. The spectrum obtained with the help of a semiconductor gamma-spectrometer Ortec-65195-P.

LASER EXPOSURE OF AQUEOUS SOLUTIONS OF URANIL-CHLORIDE WITH AU NANOPARTICLES

Gold is the most stable metal against oxidation, which stipulates its frequent use for laser generation of NPs. Colloids of Au NPs in pure water are perfectly stable without addition of any surface active substances. First experiments with laser-induced nuclear decays have been conducted exposing the mixture of colloidal solutions of Au NPs generated by laser ablation of a bulk Au target with aqueous solutions of UO_2Cl_2.

However, the presence of Cl^- ions in the solution leads to rapid sedimentation of Au NPs. The NPs agglomerate and sediment on the cell walls. Therefore, the absorption of laser radiation in the medium decreases with time and its coupling becomes less efficient. New approach has been elaborated to overcome this problem.

Namely, the solution of Uranium salt initially does not contain NPs. Instead, a bulk Au target is placed inside the solution, and NPs are generated by laser ablation of this target by laser beam focused through the layer of solution. In this way NPs are generated during laser exposure of the solution, and their concentration varies during exposure. In this case the presence of anions is not that critical for the stability of the colloid. Indeed, if NPs agglomerate, new NPs are generated at their place. This approach is schematically illustrated in Figure 5.

One more advantage of this scheme is the generation of nanostructures (NS) on the target surface, which is especially pronounced under ablation of metal using sub-nanosecond laser pulses.

These NS may contribute to the amplification of the laser wave in their vicinity. The temperature rise in the cell during laser exposure depends on the average power of laser radiation and in typical conditions does not exceed 10 K. At higher average power of laser radiation of several Watts cooling of the cell with water flow can be used. The following sections describe the results of laser exposure of Au target in aqueous solutions of UO_2Cl_2 to different laser sources [28].

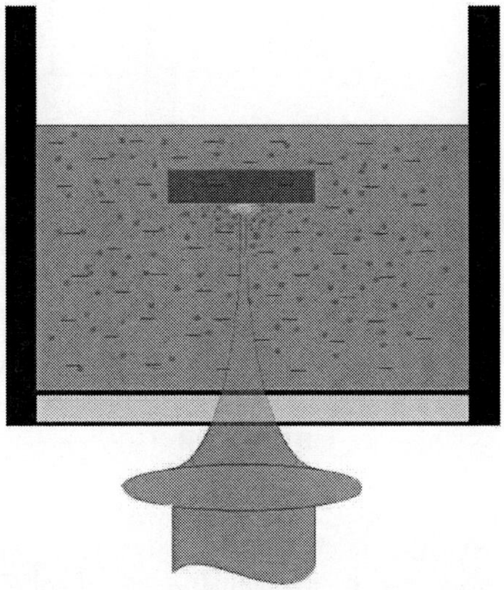

Figure 5. The scheme of laser exposure of metallic target in aqueous solution of Uranil-chloride.

Femtosecond Ti: Sapphire Laser

Laser exposure of the target has a pronounced effect on the activity of the nuclides present in the solution. Changes occur at all the gamma photon energies characteristic of a given nuclide. As an example, Figure 6 illustrates the effect of laser exposure on the ^{234}Th activity. The activity is seen to markedly decrease. At the same time, it depends little on whether the exposure was performed in heavy or light water. The half-life of ^{234}Th is 24 days, so its activity just after the exposure was even lower than that in Figure 6.

Figure 7 illustrates the influence of laser exposure on the activity of the nuclides in question. After the laser exposure, the activity was measured twice, which allowed us to examine its relaxation to the equilibrium state. The activities of the three nuclides are seen to decrease.

After the exposure, the ^{234}Th and ^{234m}Pa activities gradually approach their initial level, whereas the ^{235}U activity varies insignificantly. This is due to the continuing spontaneous alpha decay of ^{238}U to ^{234}Th and then to ^{234m}Pa. Note that the above data qualitatively correlate with results for aqueous uranium salt solutions exposed to 150-ps 1064-nm Nd : YAG pulses [29].

When there are no external influences, each nuclide of the ^{238}U series has an equilibrium activity, determined by its half-life. ^{235}U belongs to another decay series and, when alpha-decaying to ^{231}Th, it has an equilibrium activity. The ^{234}Th to ^{235}U activity ratio in uranium of natural isotopic composition is constant under secular equilibrium conditions. The A_{234Th}/A_{235U} activity ratio in solutions after 800-nm femtosecond laser exposure is presented in Figure 8.

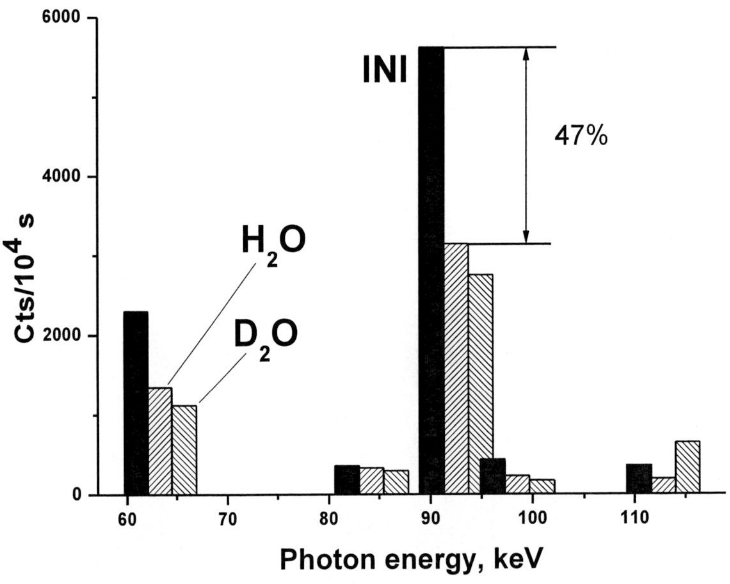

Figure 6. Activity of Thorium-234 before (INI) and after laser exposure of solutions of UO_2Cl_2 in H_2O и D_2O. Laser wavelength of 800 nm, pulse duration of 180 fs. Exposure time of 1 hour at repetition rate of laser pulses of 1 kHz. Gamma-spectra were acquired 14 days after laser exposure, accuracy of activity measurements is ±5%.

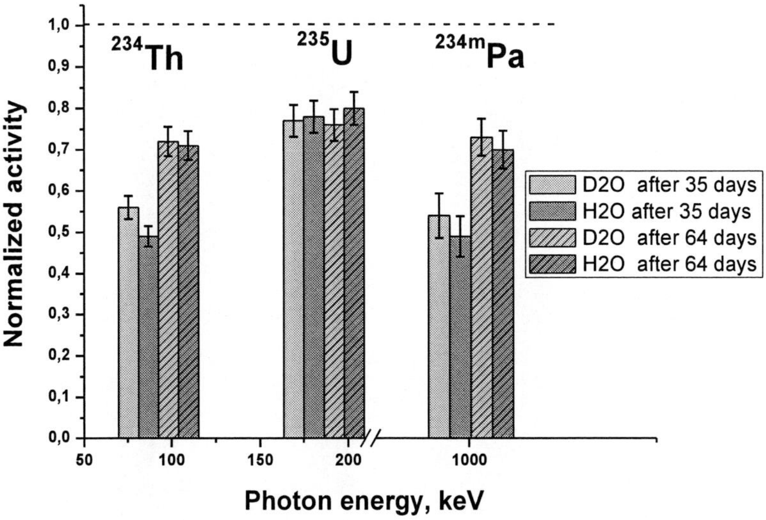

Figure 7. Activities of ^{234}Th (92.5 keV), ^{235}U (186.5 keV), and ^{234m}Pa (1001 keV) normalized on their activity in the initial samples. The activities of laser-exposed samples either in H_2O or in D_2O were measured 35 and 64 days after laser exposure. Dashed line denotes the ratio of activities in the initial sample. Ti:sapphire laser, wavelength of 800 nm, pulse duration of 180 fs. Exposure during 1 hour at repetition rate of 1 kHz.

Laser exposure of solutions results in the decrease of activities of all of three nuclides. The activities of both ^{234}Th and ^{234m}Pa recover to their initial value with time after laser exposure. This is due to the continuing alpha-decay of ^{238}U to ^{234}Th, which in turn decays to ^{234m}Pa. Note that the activity of ^{235}U does not change after laser exposure and remains constant within the accuracy of measurements being still smaller than in the initial sample. This behavior is evident since there is no parent nuclide for ^{235}U.

In the conditions of secular equilibrium the activity of all nuclides of ^{238}U branching has its equilibrium value. ^{235}U belongs to another branching, and its activity also has its equilibrium value that is established during its alpha-decay to ^{231}Th. The ratio of activities of ^{234}Th and ^{235}U in Uranium of natural isotopic composition is constant in secular equilibrium. Deviation of this ratio from its equilibrium implies the presence of non-equilibrium processes induced by laser radiation in respective branches. The ratio of activities of ^{234}Th and ^{235}U after exposure of Au target in either H_2O or D_2O to femtosecond laser radiation is plotted in Figure 8. One can see that the activity ratio undergoes strong changes after laser exposure and then recovers to its initial equilibrium value.

Since the variation activity of ^{235}U after laser exposure is relatively small, one may ascribe the restoration of the activity ratio to recovery of the activity of ^{234}Th. The activity ratio of Thorium-234 and Uranium-235 in the initial sample was measured twice. One can see that this value is the same in these two independent measurements. This confirms the reliability of measurements.

Laser exposure markedly reduces this ratio, which then gradually reverts back to its initial level. The deviation of the activity ratio from its equilibrium value is essentially independent of water composition (H_2O or D_2O). Similar deviation of the activity ratio of these nuclides is observed under electro-explosion of Ti foils in Uranil solutions (see corresponding section below).

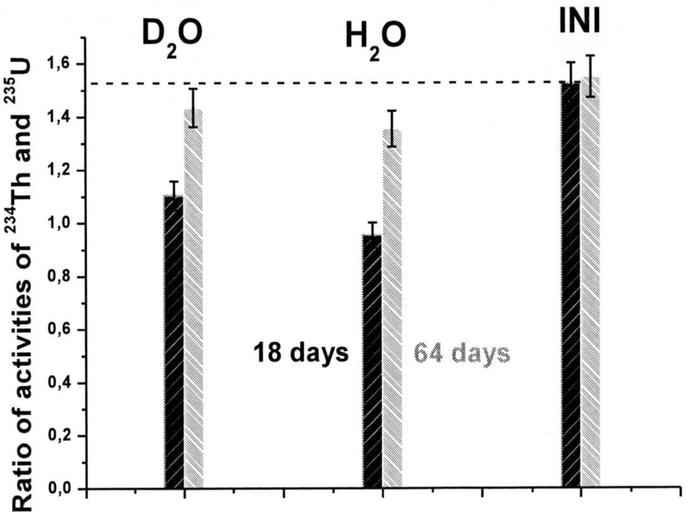

Figure 8. Ratio of activities of Thorium-234 and Uranium-235 measured at 92.5 and 185 keV, respectively, under exposure of a bulk Au target in aqueous solutions of Uranil-chloride to femtosecond laser radiation. Measurements were carried out 18 (black columns) and 64 days (light columns) after laser exposure. Dashed line shows the activities ratio in the initial sample.

PICOSECOND ND: YAG LASER

Nd:YAG lasers are currently the most popular laser sources. Frequency up-conversion using non-linear crystals allows covering wide spectral range. In particular, frequency up-conversion of the fundamental output at 1064 nm of these lasers may couple radiation to different parts of absorption spectra of Au NPs and Uranil ion (see Figure 3) [30]. Initiation of nuclear decays using the first harmonics of a picosecond Nd:YAG laser is presented in details elsewhere [29]. Experimental results on laser initiation of decays in Uranium branching under exposure to harmonics of a Nd:YAG laser are presented in this section.

SECOND HARMONICS (532 NM)

The exposure of a gold target to the second harmonic, which lies within the plasmon resonance of gold nanoparticles, in aqueous Uranyl-chloride solutions is also accompanied by a considerable deviation of nuclide activities from their equilibrium values. Figure 9 plots the activities of the three main nuclides in irradiated solutions against UO_2Cl_2 concentration. As seen, the ^{234}Th and ^{234m}Pa activities in the irradiated solutions markedly exceed those in the non-exposed solution, and the deviation of the activities increases with increasing concentration.

At the same time, the ^{235}U activity in the same irradiated solutions coincides with its equilibrium value to within the present measurement accuracy. This means the efficient enrichment of the solution by ^{235}U isotope. One can see from the plots above that the activities of both ^{234}Th and ^{234m}Pa are much higher than that in non-exposed solutions, and this deviation increases with concentration of UO_2Cl_2.

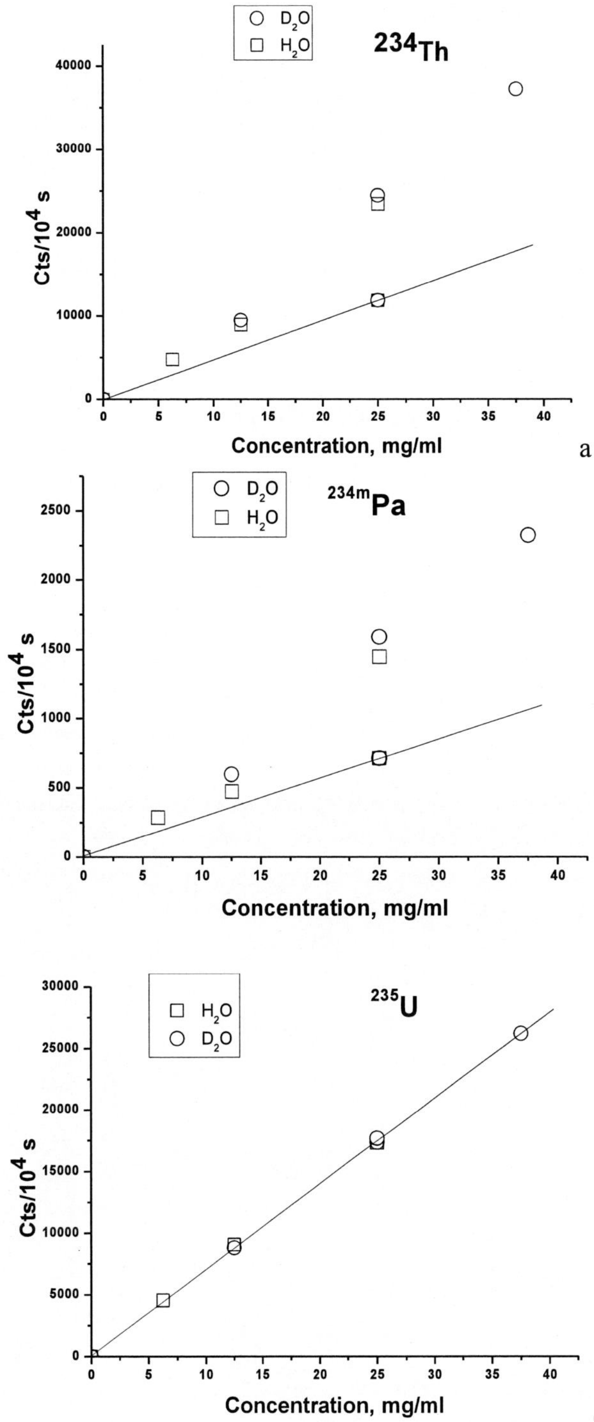

Figure 9. (a) ^{234}Th, (b) ^{234m}Pa, and (c) ^{235}U activities (counts over acquisition time of 10^4 s) as functions of concentration for Uranyl-chloride solutions in D_2O (circles) and H_2O (squares) containing gold nanoparticles and exposed to the second harmonic of a neodymium laser at 532 nm and pulse duration of 150 ps. The straight lines are drawn through the nuclide activities before the exposure. The size of the symbols represents the experimental error.

At the same time, the activity of ^{235}U in the same exposed probes remains at its equilibrium values. Therefore, very high isotopic selectivity is observed under exposure of a bulk Au target in aqueous solutions of UO_2Cl_2 to picosecond radiation at wavelength of 532 nm. The increase of activity of Thorium-234 indicates accelerated alpha-decay of its parent nuclide Uranium-238. In turn, the increase of activity of Protactinium-234m indicates that Thorium-234 undergoes accelerated beta-decay.

This observation is in good agreement with results of previous studies on electro-explosion of metallic wires in aqueous solutions of Uranil salts, as will be described in the corresponding section below.

THIRD HARMONICS (355 NM)

UV laser exposure of a gold target in an aqueous Uranyl-chloride solution at pulse duration of 150 ps leads to a considerable deviation of the A_{234Th} /A_{235U} ratio from its equilibrium value and in some range of parameters this deviations reaches a 5-fold decrease. This activity ratio is shown in Figure 10 as a function of uranium salt concentration in solution.

Laser exposure markedly reduces the ^{234}Th activity, whereas the ^{235}U activity changes little. That the activity ratio approaches its equilibrium value at high Uranium salt concentrations seems to be caused by the fact that the solution absorbs the laser radiation while it travels towards the target. As a result, the rate of gold nanoparticle generation decreases with increasing salt concentration, and laser exposure has a weaker effect on the nuclide activities.

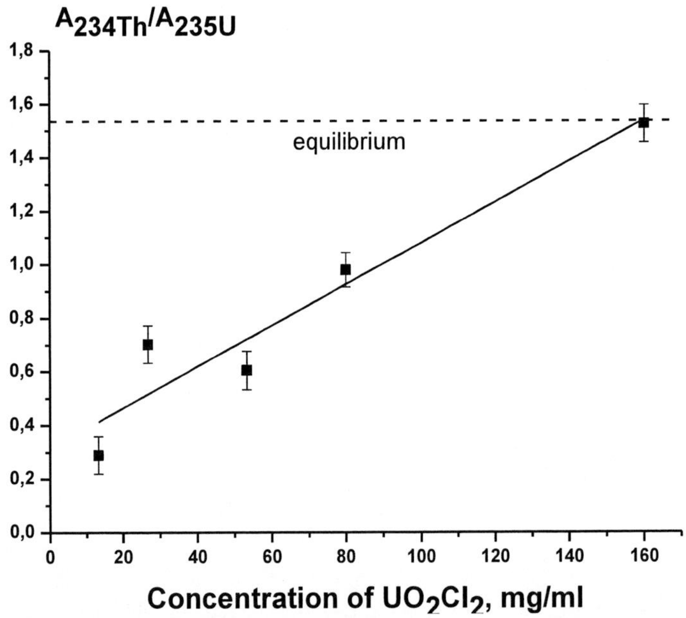

Figure 10. $^{234}Th/^{235}U$ activity ratio as a function of UO_2Cl_2 concentration in D_2O after 36 000 pulses of the third harmonics of a Nd:YAG laser at wavelength of 355 nm and pulse duration of 150 ps. The dashed line shows the equilibrium activity ratio.

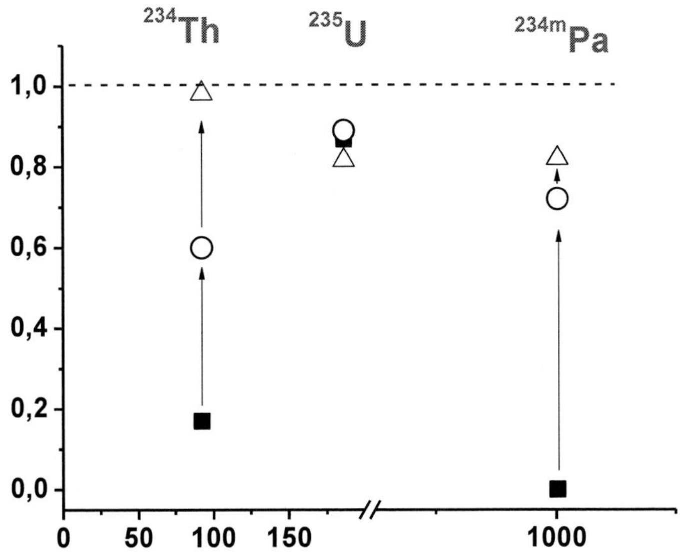

Figure 11. Normalised ^{234}Th (92.5 keV), ^{235}U (186.5 keV) and ^{234m}Pa (1001 keV) activities after laser exposure of a gold target. The activities are normalized to their values before the exposure. The measurements were made 28 (solid squares), 66 (open circles) and 131 days (open triangles) after laser exposure (36 000 pulses) in H_2O at the Uranyl-chloride concentration of 13.3 mg/mL. The dashed line indicates the activity before the exposure. The vertical arrows show the change in activity between the measurements. The size of the symbols represents the experimental error.

Figure 11 shows normalized activities of the three nuclides in irradiated solutions at the lowest salt concentration. As in the case of 800-nm Ti:sapphire laser exposure, the Nd : YAG third harmonic reduces the nuclide activities, including that of ^{235}U. The ^{234m}Pa activity 77in the irradiated sample was zero to within the present experimental uncertainty. It reached a measurable level only two months after the laser exposure. Moreover, the gamma spectra of irradiated samples contained peaks of nuclides (e.g. ^{214}Pb) that had not been detected in the solution before the laser exposure.

One can see that the activity of ^{234m}Pa is 0 within the accuracy of gamma-measurements even 28 days after laser exposure. Here again, the activity of ^{235}U remains constant after laser exposure being slightly lower than that in the initial sample. This means very large drop of activities ratio of ^{234}Th and ^{235}U induced by laser exposure and indicates large deviation from secular equilibrium.

Laser Exposure of Uranil-Chloride Solutions with NPs of Palladium

In another series of experiments the influence of laser exposure on the activity of radio-nuclides was tested using NPs of Pd dispersed in the solution of Uranil-chloride [30]. The plasmon resonance of Pd NPs lies in UV range near 250 nm. The position of this resonance is red-shifted with the increase of NPs size making them suitable for coupling UV laser radiation to the solution.

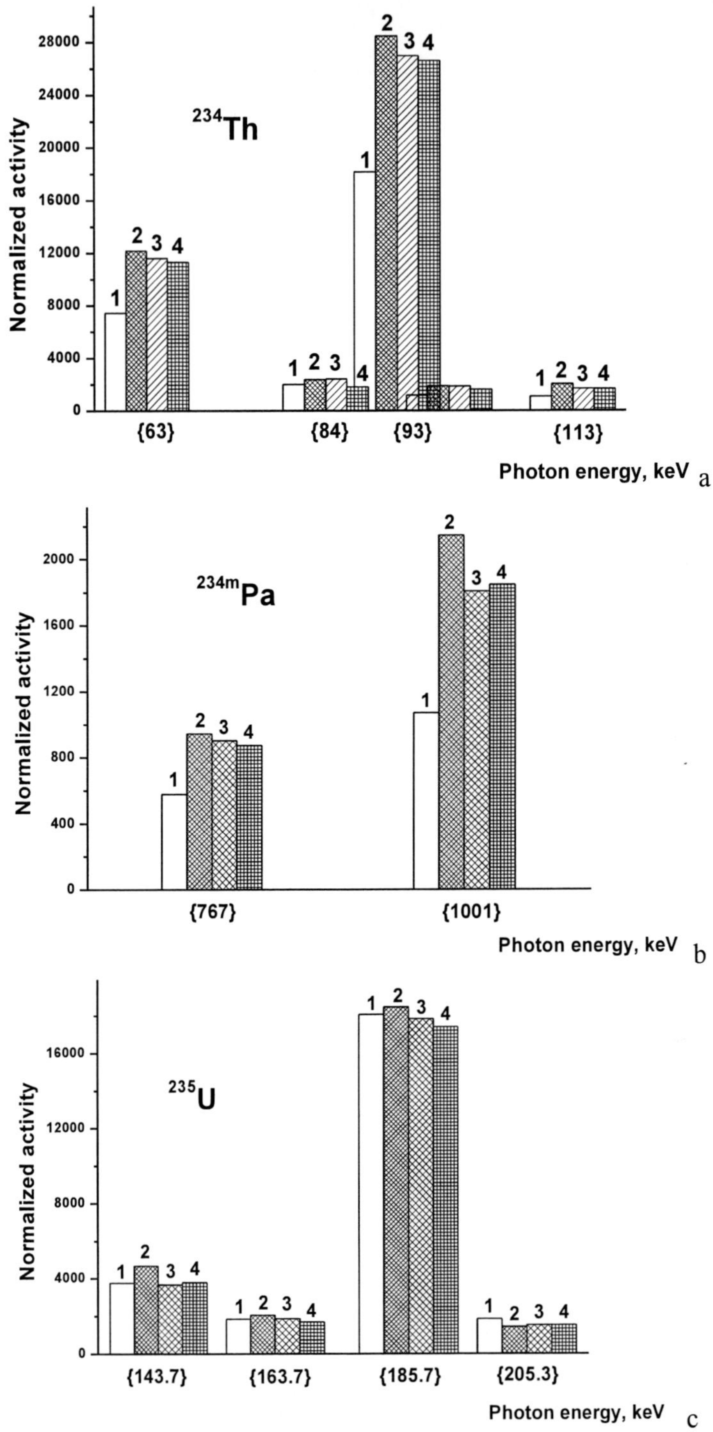

Figure 12. Normalized activities of ^{234}Th (a), ^{234m}Pa (b), and ^{235}U (c) after laser exposure of aqueous solutions of UO_2Cl_2 with Pd NPs. Changes of activities are shown on main photon energies of each nuclide. Laser wavelength of 355 nm, pulse duration of 150 ps, 1 hour exposure at repetition rate of 10 Hz. Concentration of UO_2Cl_2: 1 (initial solution) – 93 (in H_2O), 2 – 47 (in H_2O), 3 – 47 (in H_2O), 4- 23 (in D_2O) mM. Estimated peak power in the solution of 10^{11} W/cm^2.

Pd NPs were generated by ablation of a bulk Pd target in either H_2O or D_2O. Laser exposure of UO_2Cl_2 solutions with Pd NPs was carried out using the same scheme as for Au NPs. Third harmonics of a Nd:YAG laser was used at wavelength of 355 nm and pulse duration of 150 ps. Comparison of nuclides activities before and after laser exposure is presented in Figure 12.

Changes of nuclide activities are symbatical for all energies of gamma-photons of the same nuclide. The activity of ^{234}Th increases after laser exposure, which can be explained by accelerated alpha-decay of its parent nuclide ^{238}U. The activity of ^{234m}Pa changes in a similar way. On the contrary, the activity of ^{235}U after laser exposure remains the same within the accuracy of measurements, similarly to results obtained with Au NPs. Note that the deviation of nuclide activities is not sensitive to the type of aqueous medium.

Laser Exposure of Uranil-Chloride Solutions with Ag NPs

Next series of experiments dealt with laser exposure of aqueous solution of Uranil-chloride in presence of Ag NPs. Plasmon resonance of Ag NPs in water is centered at 400 nm, as shown in Figure 13.

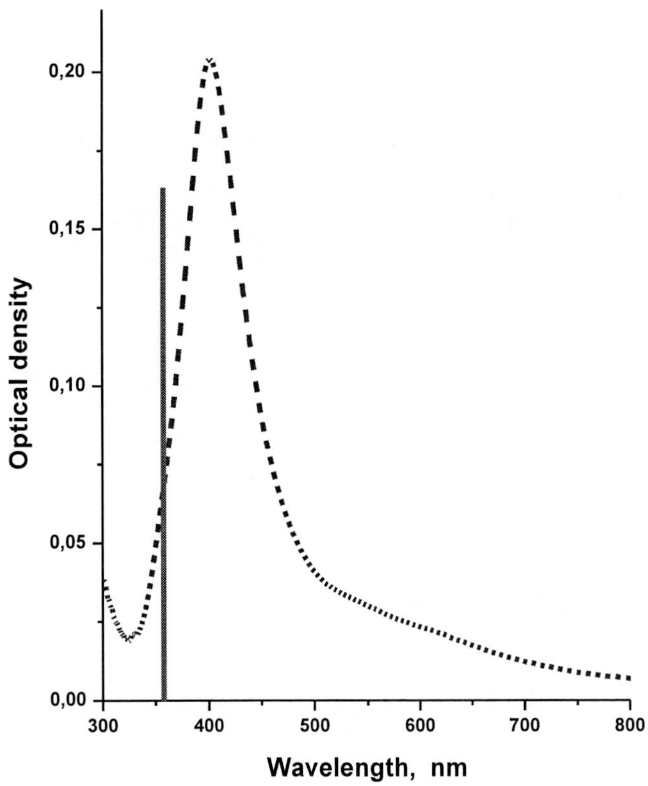

Figure 13. Absorption spectrum of Ag NPs in water. Vertical line indicates the position of the third harmonics of a Nd:YAG laser at 355 nm.

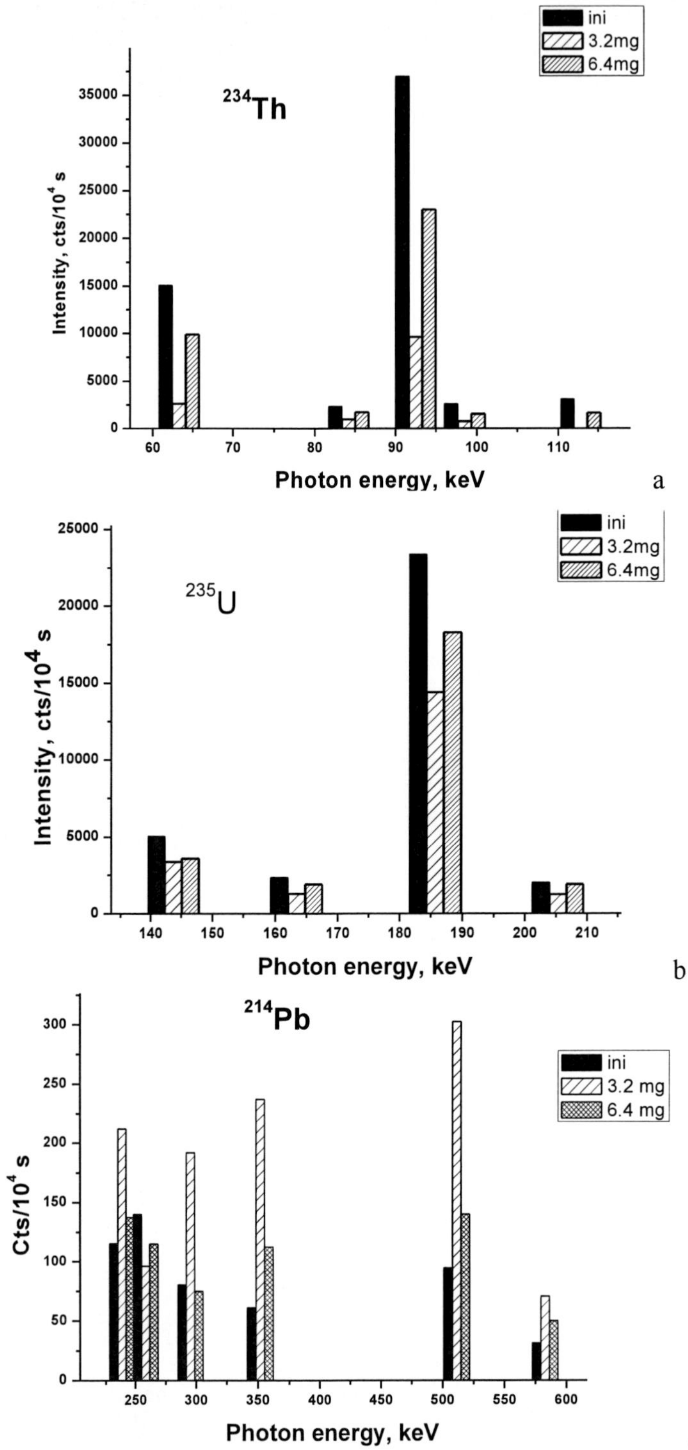

Figure 14. Activity change of ^{234}Th (a), ^{235}U (b), and ^{214}Pb (c) for two values of concentration of $UO_2(NO_3)_2$ in H_2O (3.2 and 6.4 mg/ml) after exposure to the third harmonics of a Nd:YAG laser at estimated peak power of 10^{12} W/cm^2. The most intense peaks of each nuclide are shown. 1 hour exposure at repetition rate of 10 Hz. Activities of nuclides in the non-exposed solution are labeled as INI. Gamma-measurements were carried out 11 days after laser exposure.

Plasmon resonance of Ag NPs is closer to 355 nm than that of Au NPs. Moreover, Ag NPs and NS is one of the most popular nano-objects that demonstrate the effect of Surface Enhanced Raman Scattering (SERS). Ag nano-entities demonstrate highest amplification of the light waves, and one may expect that their influence on the activity of radio-nuclides will be especially pronounced.

Laser ablation of a bulk Ag target in an aqueous medium containing Cl- ions, e.g., UO_2Cl_2, is accompanied by formation of AgCl compound instead of metallic species. For this reason laser exposure of Ag target was carried out in an aqueous solution of $UO_2(NO_3)_2$. The target was exposed to laser radiation at two values of salt concentration. The resulting change of activities is shown in Figure 14 [30].

The presented diagrams show that the activity of Thorium-234 is reduced by a factor of 3 at lower concentration of Uranil and to lesser extent at high concentration. The dependence of deviation of activity on the concentration of Uranil-chloride is typical for exposure to UV laser radiation. This radiation is absorbed by the Uranil-ion itself. The absorption of laser radiation leads to reduction of energy fluence on the target and, consequently, to lower concentration of Ag NPs.

The activity of Uranium-235 also decreases after laser exposure unlike the case of Au and Pd NPs. The increase of activity of ^{214}Pb in exposed solutions is of special interest. This nuclide stands at the end of ^{238}U branching. Its half-life is around 26 minutes (see Real-time gamma-emission section below). Figure 15 shows the fragment of gamma-spectrum of the solution in the vicinity of several peaks. One can see that their intensity in laser-exposed sample is much higher than in the initial one. Its increased activity 11 days after laser exposure indicates strong deviation from equilibrium of its parent nuclides. ^{226}Ra is the closest long-living nuclide to ^{214}Pb.

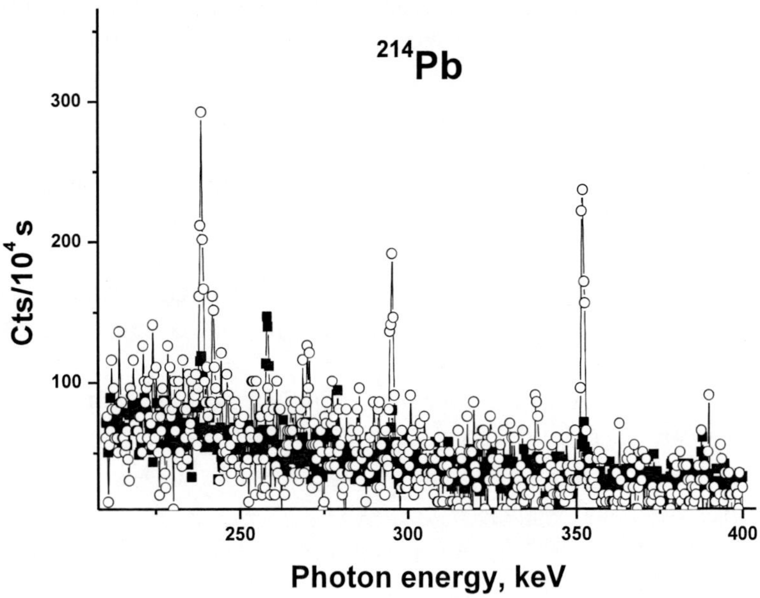

Figure 15. Fragment of gamma-spectrum with peaks of ^{214}Pb. Exposed solution (open circles), initial solution (solid squares). $UO_2(NO_3)_2$ concentration in H_2O of 3.2 mg/ml.

In turn, ^{226}Ra is the daughter nuclide to another long-living nuclide ^{234}U. Therefore, laser exposure causes strong deviation of equilibrium in the whole ^{238}U branching.

The increase of ^{214}Pb activity in solutions exposed to laser radiation is in good agreement with the results on non-equilibrium gamma-emission of Uranil solutions detected in real time during laser exposure (see below).

Real Time Gamma-Activity of Solutions under Laser Exposure

Similarly to laser exposure of aqua-ions of Thorium [26], laser exposure of aqueous solutions of Uranil-chloride is accompanied by non-equilibrium emission of gamma-photons. This section gathers the results on gamma-emission of solutions of Uranil in real time, during the laser exposure. Gamma-emission from samples under their irradiation with a Cu vapor laser was recorded in real time using a scintillation NaI gamma-spectrometer that was fixed near the working cell. Since the initial level of gamma-emission is comparable with the natural gamma-background, both the spectrometer and the working cell were placed into a 5 cm thick lead housing. The laser beam was directed to the working cell through a small aperture in the housing. Acquisition of gamma-photons was carried out through the wall of glass cell with the solution. The energy resolution of the scintillation spectrometer was 3 keV per channel [31]. The samples were the solutions of UO_2Cl_2 of natural isotope composition either in H_2O or D_2O. The typical concentration of Uranium salt was 50 mg/ml (0.15 M/L). Prior to the laser exposure, 1 ml of this solution was mixed with 1 ml of the colloidal solution of Au nanoparticles (NPs) obtained by laser ablation of a bulk gold target in either in H_2O or D_2O [31]. The average size of Au NPs was about 20 nm at concentration of $10^{14}cm^{-3}$. The cell with the solution was sealed during the laser exposure to avoid the evaporation of the liquid. The average stationary temperature of the liquid during the laser exposure did not exceed 45°C. Two laser sources were used, either a visible range Cu vapor laser (wavelength of 510.6 and 578.2 nm and pulse duration of 10 ns) and the third harmonics of a Nd:YAG laser (wavelength of 1064 nm and pulse duration of 350 ps). The estimated peak power in the solution was 6×10^{11} and 10^{12} W/cm^2, respectively. The results obtained with these two laser sources are virtually the same. In order to avoid possible influence of the power supply of a Cu vapor laser on the gamma-spectrometer, shutting down of laser radiation was performed using shutting down the laser beam itself keeping the power supply running. Also, the spectrometer was fed from autonomous battery.

The general view of gamma-spectrum of the solution under laser exposure is presented in Figure 16. The main feature of the spectrum is the peak at small energy of photons between 50 and 60 keV. One can also see several peaks that correspond to Thorium-234 and Uranum-235 (90-100 keV range) and single peak of Uranium-235 at 186 keV.

Low-energy gamma-spectrum of the solution under laser exposure shows strong maximum in the 54-57 keV (see Figure 17). The real-time variation of activity of both ^{234}Th and ^{235}U are close to zerowithin the accuracy of measurements. The signal rapidly decays after shutting the laser radiation, and 16 hours after the exposure the signal is at the level of background (Figure 18).

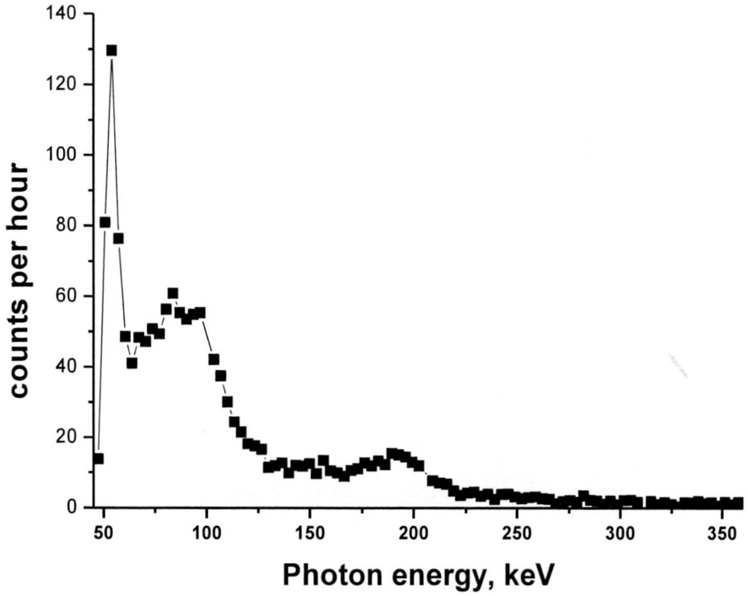

Figure 16. General view of gamma-spectrum of Uranil-chloride solution under exposure to radiation of a Cu vapor laser. Points are channels of the spectrometer.

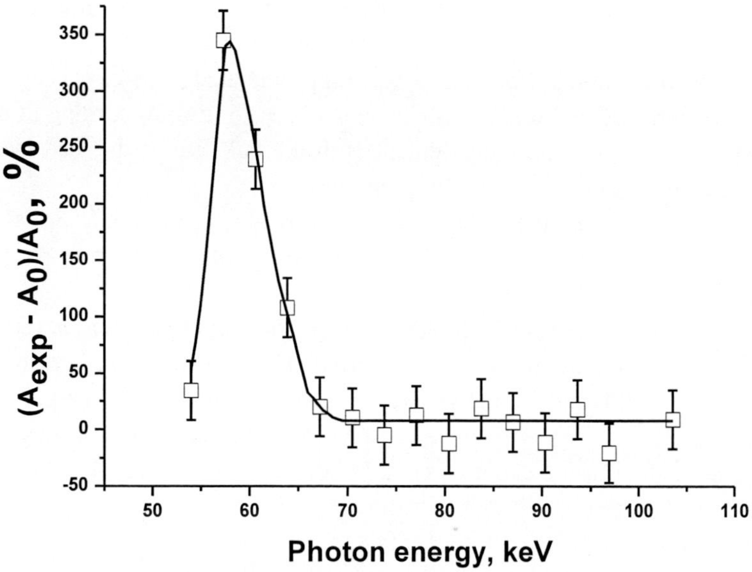

Figure 17. Normalized activity of the solution during laser exposure A_{exp} and before laser exposure A_0 in low-energy part of spectrum of the solution. Cu vapor laser, solution of UO_2Cl_2 in H_2O. Points are channels of the spectrometer.

The increase of gamma-activity in 50-60 keV is higher than before exposure by a factor of 3. The gamma-signal in this range increases under laser exposure during 1.5 – 2 hours and then reaches certain plateau value. The decay of this signal after shutting down the laser beam is shown in Figure 19. The activity rapidly decreases. Determination of the characteristic time of decay may stem the light on the nuclides that emit these gamma-photons.

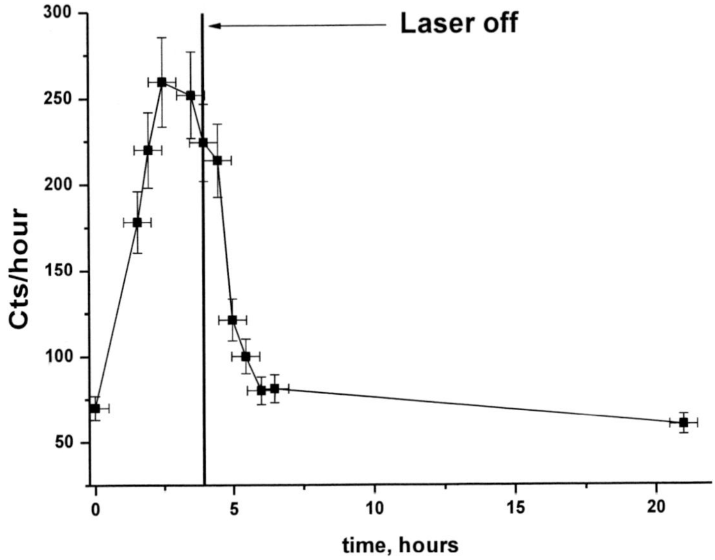

Figure 18. Kinetics of the gamma-emission of the solution of UO_2Cl_2 in D_2O with Au NPs at 53.4 keV photon energy. The acquisition time in each data point was 30 min. Cu vapor laser, the laser exposure starts at t = 0.

It was found that the time constants of decays at 54 and 57 keV are slightly different. This is illustrated in Figure 19, where these decays are plotted.

The decay of the gamma-emission is well approximated with exponential dependence with half- life of either 20 or 26 min with accuracy of ±3.5 min. According to reference data [33], this time constant is in good agreement with the half-life of the lead isotope ^{214}Pb of 26.8 min and bismuth isotope ^{214}Bi of 19.9 min.

These isotopes stand near the end of ^{238}U branching. Other lines of ^{214}Pb (at 241.9, 295.2, and 351.9 keV) are situated in the region of low sensitivity of the spectrometer used and cannot be detected with sufficient accuracy.

The activity of the samples at 54-57 keV reaches its maximum after the onset of the laser exposure (Figure 18). Apparently, this time is needed to produce sufficient amount of short-living isotopes of both Pb and Bi. However, the real-time variation of the gamma-activity of other isotopes, e.g., ^{235}U, ^{234}Th is small and lies within the accuracy of measurements of ±5%.

Cu vapor laser used in the experiments on real-time gamma-emission of Uranil-chloride solutions is relatively weak leaser source with its 10^{10} W/cm^2 intensity in the medium.

Exposure of solutions to radiation of this laser does not lead to any measurable changes in the activity of main nuclides presented in the solution, such as ^{235}U or ^{234}Th. That is why the change of activity in real time is observed only for short-living isotopes of lead and bismuth. However, the observed non-equilibrium gamma-emission confirms the presence of strong deviation from secular equilibrium in ^{238}U branching. The non-equilibrium gamma-emission of ^{214}Pb is in good agreement with results presented in previous sections on the increased activity of this isotope in Uranil-chloride solutions exposed to 150 picosecond laser radiation.

It is pertinent to note that the possibility to observe non-equilibrium gamma-emission of ^{214}Pb and ^{214}Bi is due to the local equilibrium between the rate of laser-induced generation of this isotope from its parent nuclides and laser-induced decay of these nuclides.

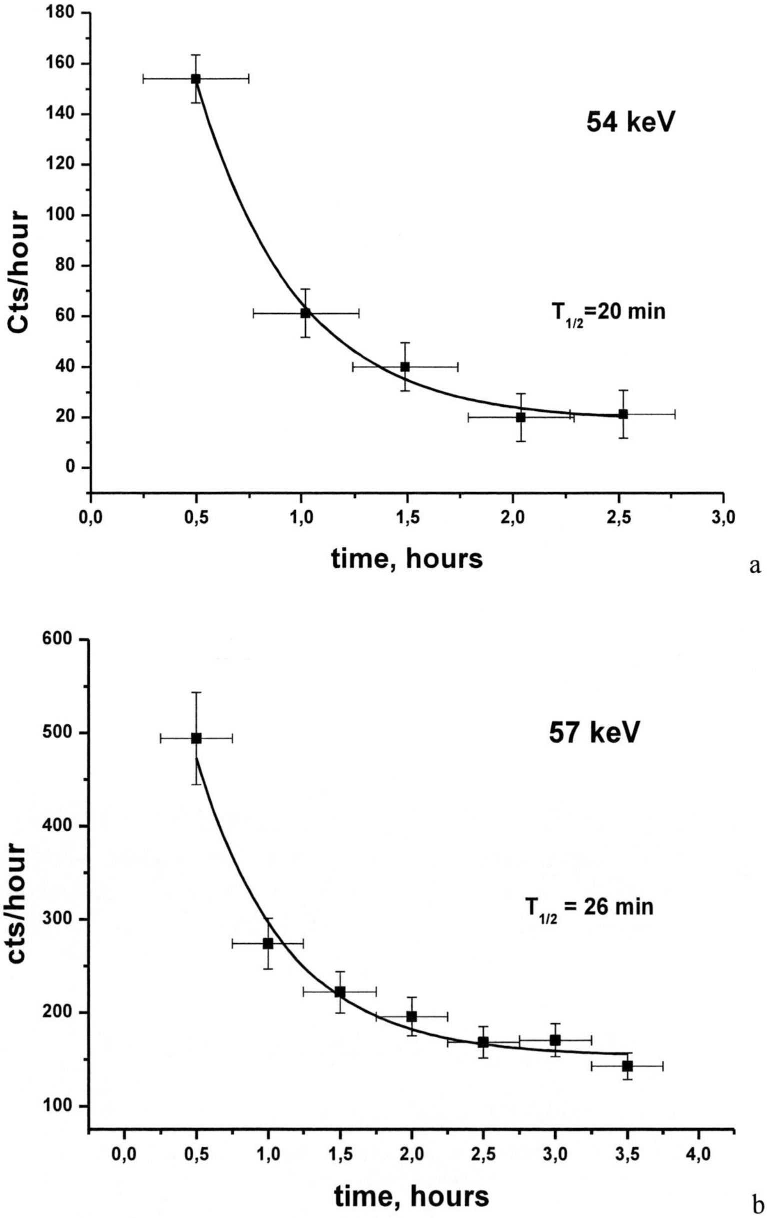

Figure 19. The decay of the gamma-emission at 54 keV (a) and at 57 keV (b) after the end of laser exposure. Exponential fit to the experimental points.

The effect of laser exposure on the activity of nuclides depends on the relative concentration of Uranium salt and metallic NPs.

The fraction of the latter should be sufficiently high to alter the activity of nuclides of Uranium branching. This is illustrated in Figure 20, where the normalized activities of various nuclides are plotted against the volume of the solution of UO_2Cl_2. One can see that the activity of both ^{214}Pb and ^{214}Bi increases after laser exposure by a factor of 20 at low concentration of Uranium salt [30].

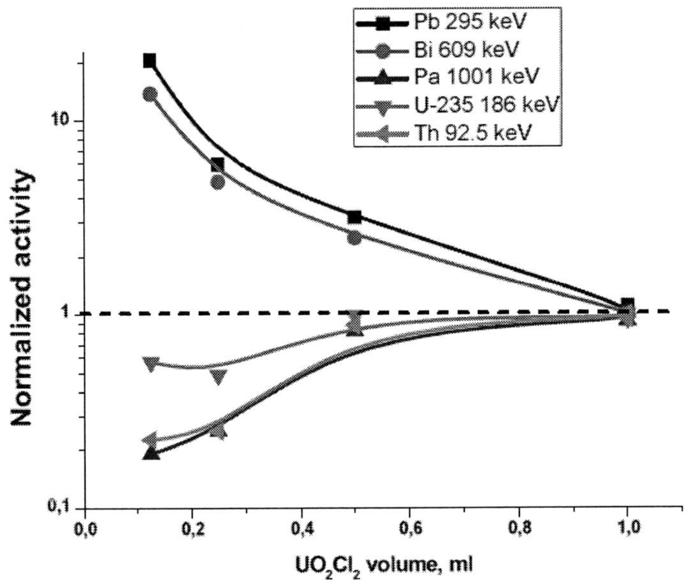

Figure 20. The activities of various nuclides normalized on their value in non-exposed solution as the function of the volume of UO_2Cl_2 solution in H_2O at concentration of 0.2 g/ml in the colloidal solution of Au NPs. Total volume of the solution is 2 ml, 1 hour exposure to the radiation of a Nd:YAG laser at 1064 nm at repetition rate of 10 kHz, pulse duration of 3 ns.

Other nuclides (^{234m}Pa, ^{235}U, ^{234}Th) demonstrate the decrease of activity after laser exposure, and this decrease is also highest at low concentration of Uranium salt. At high concentration of Uranium salt in the solution laser exposure causes only negligible changes of nuclide activities. This indicates that the effect of the laser field enhancement on nano-entities present in the solution plays a key role in laser initiation of nuclear decays.

Laser-Induced Accelerated Decay of ^{232}U

The conclusion on the variation of alpha-activity of ^{238}U under laser exposure was made on the basis of indirect measurements of gamma-activity of its daughter nuclides, ^{234}Th and ^{234m}Pa [28, 29]. The half-life time of alpha-decay of ^{238}U exceeds 10^9 years. It is expected that the influence of laser exposure on the alpha-activity will be more noticeable in case of a nuclide with shorter time of alpha-decay. ^{232}U may serve as such model nuclide. This is an artificial isotope with half-life of alpha-decay of 69 years. The sequence of decays of ^{232}U is as follows:

$$^{232}U \xrightarrow{\alpha} {}^{228}Th \xrightarrow{\alpha} {}^{224}Ra \xrightarrow{\alpha} {}^{220}Rn \xrightarrow{\alpha} {}^{216}Po \xrightarrow{\alpha} {}^{212}Pb \xrightarrow{\beta} {}^{212}Bi \xrightarrow{\alpha} {}^{208}Tl \xrightarrow{\beta} {}^{208}Pb$$

This section describes the results of laser exposure of nanoparticles in aqueous solution of ^{232}U salt on its alpha-activity. The real-time gamma-emission of daughter nuclides ^{212}Pb and ^{208}Tl can be used for tracing the deviation from equilibrium under laser exposure [34].

The alpha-activity of the samples before and after laser exposure was measured using Ortec-65195-P alpha-spectrometer with accuracy of ± 5 %. Aqueous solution of $^{232}UO_2(NO_3)_2$ was divided into several parts. The first part with activity of 53.4 Bq/ml was used as the initial non-exposed solution.

This value of activity corresponds to 60 pg of ^{232}U per ml. Laser exposure of the solution was carried out by focusing the laser beam on a bulk gold target placed inside the solution with the help of an aspherical lens.

Initially the solution did not contain nanoparticles, the y were generated during the laser exposure of the target and dispersed in the solution due to convective motion. This scheme provides the constant level of nanoparticles density in the solution since they may agglomerate and sediment during laser exposure.

Four laser sources with different wavelengths and pulse duration were used for the exposure of solutions.

1. A Nd:YAG laser with wavelength of 1064 nm, pulse duration of 150 ps, estimated peak power in the solution of 10^{13} W/cm^2. Repetition rate was of 10 Hz, which corresponds to 3.6×10^4 laser pulses during 1 hour exposure.
2. The second harmonics of a Nd:YAG laser with wavelength of 532 nm, pulse duration of 150 ps, estimated peak power in the solution of 10^{12} W/cm^2. Repetition rate was of 10 Hz, which corresponds to 3.6×10^4 laser pulses during 1 hour exposure.
3. A copper vapor laser with wavelength of 510.6 nm, pulse width of 20 ns, estimated peak power in the medium of 10^{11} W/cm^2. Repetition rate of this laser was of 15 kHz, which corresponds to 2.16×10^9 pulses at 4 hour exposure.
4. A Nd:YAG laser with wavelength of 1064 nm, pulse duration of 350 ps, estimated peak power in the solution of 10^{12} W/cm^2. Repetition rate was of 300 Hz, which corresponds to 4.32×10^7 laser pulses during 4 hour exposure.

The laser sources have different pulse duration and wavelength of emission. The radiation of both copper vapor laser and the second harmonics of a Nd:YAG laser fits well the position of plasmon resonance of Au nanoparticles in water (around 530 nm).

Only approximate estimations of the peak power inside the solution with nanoparticles can be done. This is due to initiation of various non-linear effects inside the liquid at intensity level 10^{11} – 10^{13} W/cm^2 that may lead to defocusing of laser radiation. This can be, for example, the dynamic Kerr effect. On the contrary, self-focusing of the laser beam may lead to its filamentation and to the increase of laser intensity in the medium. Filamentation consists in the formation of several thin channels instead of the initial single beam due to dependence of the refractive index of medium on the laser intensity.

Finally, the presence of nanoparticles may alter the intensity level due to the absorption of laser radiation via plasmon resonance of charge carriers in nanoparticles or in the plasma produced around nanoparticles. Alpha-particles that are emitted by ^{232}U nuclei have higher energy than those emitted by ^{238}U. This is the reason of much shorter half-life of its alpha-decay.

Change of alpha-activity of the sample produced by exposure of a gold target in aqueous solution of ^{232}U salt in water to radiation of a copper vapor laser is accompanied by modification of gamma-activity of its daughter nuclides, such as ^{212}Pb and ^{208}Tl. This is illustrated by the following diagrams (Figure 21).

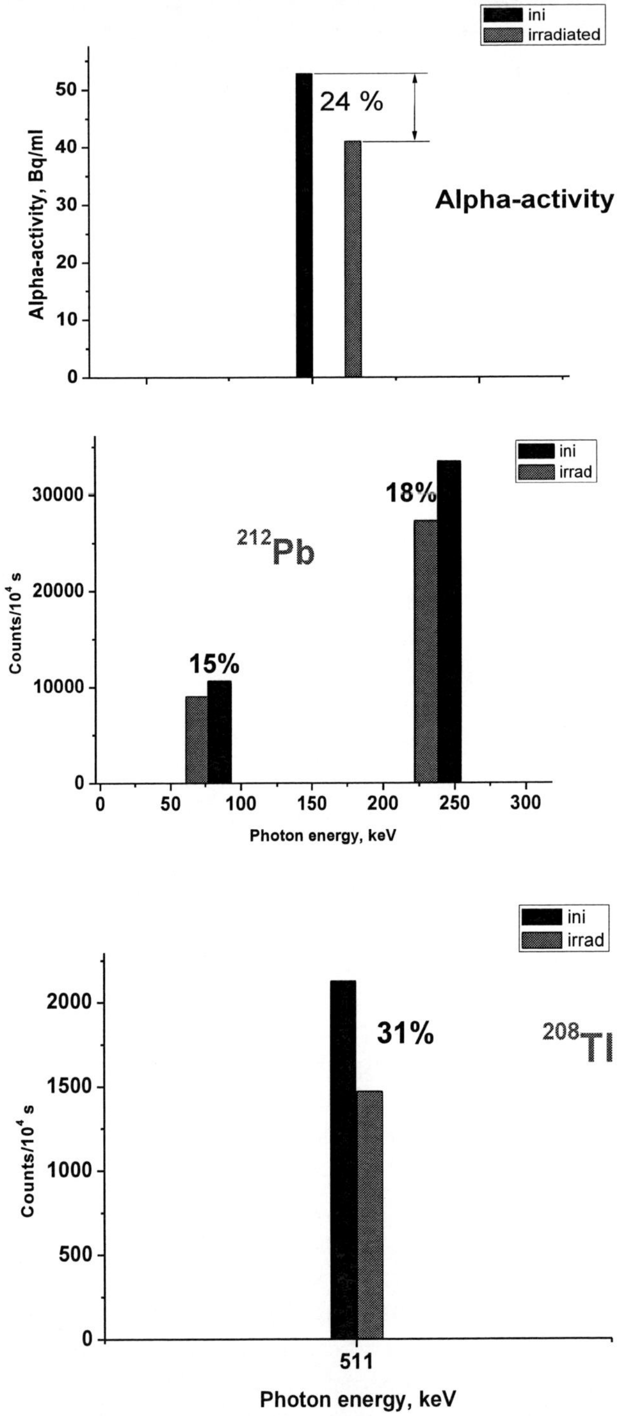

a

b

c

Figure 21. Change of alpha-activity of ^{232}U sample (a) and gamma-activities of ^{212}Pb (b) and ^{208}Tl (c) after exposure of a bulk gold target in aqueous solution of $^{232}UO_2(NO_3)_2$ to radiation of a Cu vapor laser during 4 hours (2.16×10^9 laser pulses).

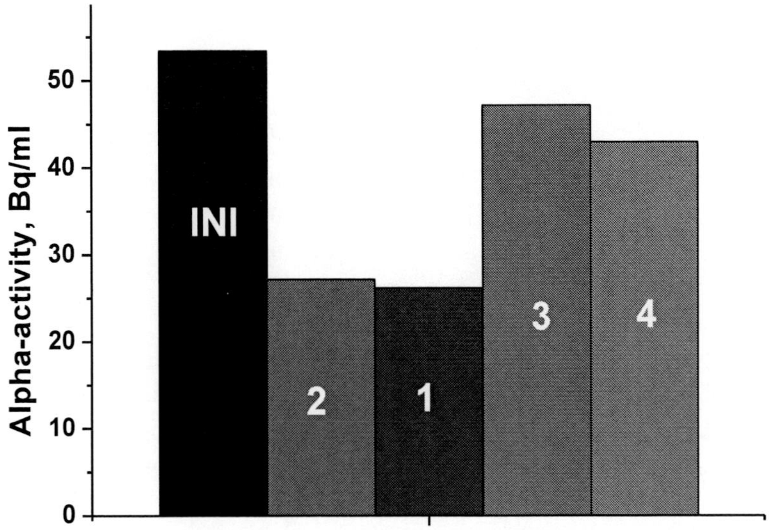

Figure 22. Diagram of comparison of relative effect of laser exposure on alpha-activity of the same solution to different laser sources. 1 and 2 corresponds to the second and first harmonics of a Nd:YAG laser with pulse duration of 150 ps, respectively. Column 3 indicates the alpha-activity of the solution after the exposure to radiation of a Cu vapor laser. Column 4 stands for the activity after exposure to the first harmonics of a Nd:YAG laser with pulse duration of 350 ps. Initial activity is designated by INI.

Similarly to laser exposure of ^{238}U salts, the final change of gamma-activity of daughter nuclides of ^{232}U may have different sign due to different rate of de-activation of gamma-activities of nuclides.

The effect of laser exposure sources is shown in Figure 22. One can see that the activity is decreased by a factor of 2 after 1 hour exposure to laser beam at intensity level of 10^{12} - 10^{13} W/cm^2. It is reasonable to suggest that the alpha-decay proceeds during the laser pulse, while the spontaneous decrease of alpha-activity during exposure is negligible.

This means that alpha-activity drops down by a factor of 2 during 5 μs, which is the total duration of all 150 ps laser pulses. In other words, the half-life of ^{232}U in the laser field is 5 μs instead of 69 years. About 10^{10} nuclei of ^{232}U decay during laser exposure. Another infrared laser radiation with pulse duration of 350 ps affects the alpha-activity of the solution to lesser extent despite to much higher number of laser pulses.

Finally, exposure of the solution of ^{232}U salt to radiation of a nanosecond Cu vapor laser produces quite small effect on the activity, though this effect is still higher than the accuracy of measurements. This relatively low effect is gained despite to much higher number of laser pulses absorbed in the solution (by a factor of 10^5) compared to a 150 ps laser source.

The above results one may conclude that the process of laser-induced alpha-decay has no threshold at least down to 10^{11} W/cm^2.

Laser-induced acceleration of alpha-decay can be observed not only with ^{232}U isotope. At low concentration of Uranil-chloride of ^{238}U (~ 10 mg/ml) laser exposure of its solution with Au NPs to radiation of a Nd:YAG laser with pulse duration of 3 ns during 1 hour at 10 kHz results in the decrease of alpha-activity of ^{238}U by a factor of 1.6. Gamma-activity of daughter nuclides of this isotope at this concentration is too weak for reliable measurements.

Deviation from Secular Equilibrium under Electro-Explosion of Metallic Foils in Solutions of Uranil Salt

Another example of influencing the secular equilibrium in the ^{238}U branching is the electro-explosion of metallic conductors in aqueous solutions of Uranil salts.

The experimental setup used in this series of experiments is completely different from the setup of laser-induced decays. However, the results are quite similar to what has been reported above on the laser-assisted acceleration of nuclear decays.

Electro-explosion of foils consists in discharge of a capacitor through a (thin) metal foil, either in air or in liquid. The electro-explosion of a metallic foil, e.g., Ti foil, immersed into the solution of Uranium salt in water also leads to significant deviation of activities of nuclides from their equilibrium levels. The experimental setup for foils electro-explosion was as follows [35]. Typical volume of aqueous $UO_2(NO_3)_2$ solution was of 20 ml. The battery of capacitances could store up to 22 kJ.

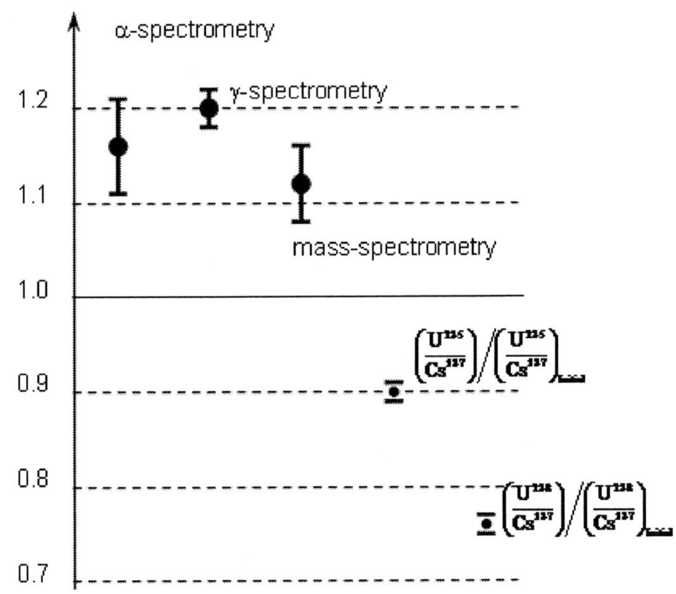

Figure 23. Ratio of activities $^{235}U/^{238}U$ after electro-explosion of a Ti foil in aqueous solution of $UO_2(NO_3)_2$ normalized on its value in the initial solution (courtesy of Prof. L.I. Urutskoev). ^{137}Cs was used as a label.

The stored energy at voltage of 4kV was discharged through a Ti foil placed into a sealed PMMA chamber. The peak value of current during the discharge was 120 kA, duration of the discharge was around 100 μs. After the discharge the metallic foil is dispersed into surrounding solution as molten micro- and nanoparticles of Titanium that interacts with water producing Ti hydroxides.

The deviation of equilibrium activity of nuclides of ^{238}U branching after electric discharge is illustrated in Figure 23. The authors also observed the restoration of ^{234}Th activity several months after the electric discharge. The perturbation of the equilibrium between ^{234}Th and its daughter ^{234m}Pa nucleus detected by γ-spectrometry can be a result of the change in the β^--decay probabilities along different channels. This item was discussed in more detail previously [36].

Unlike the experiments on laser-induced decay described above, no noticeable excess over the background γ-radiation during the discharge was detected. This is probably due to the fact that electro-explosion of a foil is essentially single-pulse process. The discharge can be carried out only once, and then the chamber should be replaced for a new one.

The authors indicate that the ^{238}U concentration decreases more appreciably after the discharge than the ^{235}U concentration, and this is perceived as "effective" enrichment. The alpha-activity of ^{238}U demonstrates almost twofold decrease after the electric discharge. The authors also report the shift in the ratio of ^{238}U/^{137}Cs activity after electro-explsion of a Ti foil in aqueous solution of Uranil ion (see Fig. 23). In general, the shift of secular equilibrium observed under electro-explosion of metallic foils in Uranil solutions are qualitatively close to that observed under laser exposure of NPs in this solution.

Possible Mechanisms of Laser-Induced Accelerated Decays

As it follows from the above experimental data, the decay of ^{238}U under laser exposure of its solutions proceeds according to a classical sequence of alpha- and beta-decays of nuclides of its branching. However, the rate of this decay is many orders of magnitude higher than that of spontaneous decay (more than 10^9 years for ^{238}U). The sequence of nuclear transformations promoted via laser exposure of metallic NPs in aqueous solutions of Uranium salt do not deviate from their natural pathways, through natural sequence of decays of radio-nuclides, and the only parameter which is affected by laser radiation is the rate of the decays.

For any daughter nuclide of ^{238}U branching, e.g., ^{234}Th, the variation of its content in the probe can be written as a kinetic equation:

$$d[^{234}Th]/dt = [^{238}U]\exp(-t/t_{238U}) + k_1[^{238}U] - k_2[^{234}Th] - [^{234}Th]\exp(-t/t_{234Th}),$$

where $[^{23x}N]$ is the content of the corresponding nuclide, and at given volume of the solution is its concentration. In the conditions of secular equilibrium $k_1=k_2=0$, and the above equation describes the equilibrium activity of the corresponding nuclide. The increase of $[^{234}Th]$ is due to spontaneous alpha-decay of ^{238}U with half-life of t_{238U} (negligible during the laser exposure) and laser-promoted alpha-decay of ^{238}U at rate constant of k_1. The decrease of $[^{234}Th]$ is due to the laser-promoted decay constant k_2 and its spontaneous decay with time constant of t_{234Th} (also negligible). Since the duration of laser exposure is much shorter than the time of spontaneous decays mentioned above, the sign of the final variation of $[^{234}Th]$ is determined by the relative values of both k_1 and k_2. Experimental data allow concluding that the values of these constants is the function of experimental parameters, such as laser wavelength, laser peak power, the concentration of ^{238}U, etc. That is why the variation of $[^{234}Th]$ may have different sign after the laser exposure depending on the experimental parameters. For short-living isotopes of Uranium branching, such as ^{214}Pb and ^{214}Bi, their spontaneous decay has to be taken into account, since their half-life is comparable with the duration of laser exposure.

It is known that alpha-decay proceeds via tunneling of alpha-particles through an energetic barrier, and the probability of tunneling is determined by its height E_b and width. Outside the nuclei the potential is self-consistent Colombian and is determined by the sum of electric fields of the nuclei and all electrons in the atom. The energy of laser photons is negligible compared with energy levels of nuclei, and direct excitation of nuclei energy levels is impossible. However, laser exposure of NPs may alter the configuration of the electric field outside the nuclei. The change of beta-decay time of nuclei under multiple ionization of the ^{187}Re atom has been reported [37]. Ionization of the atom alters the electric field around the nuclei and affects the life-time of its decay. Similar process may take place in case of laser exposure of metallic NPs in the solution that contains the nuclei of Uranium. This is due, first, to the high peak intensity of the laser radiation, which results in the high electric field E_{las} oscillating with optical frequency ω. Moreover, this field is enhanced due to the small size of nanoparticles, especially if the laser radiation is tuned into their plasmon resonance. Uranium atom (or aqua-ion in aqueous environment) in the vicinity of nanoparticle is subjected to high electric field of the laser wave E_{las}, and the barrier energy in this case E_{bexp} is modulated by the electric field of the laser wave: $E_{bexp} = E_b + E_{las}\exp(i\omega t)$. Since the probability of alpha-decay depends on E_b exponentially, even small variations of E_b can largely enhance it. Oscillations of alpha-particles inside the nuclei are several orders of magnitude higher, and therefore modulation of potential barrier induced by laser wave can be considered as quasi-stationary.

The mechanism of amplification of a weak Raman signal on either nano-protrusion of the metallic surface or on an isolated metallic nanoparticle is as follows [38]. Let us consider a particle (or protrusions) of radius R and dielectric function $\varepsilon_{met}(\omega)$. The flat incident monochromatic wave with field strength of E_0 can be written as follows:

$$E(r,t) = E_0 \exp(-i\omega t + i\vec{k}\vec{r})$$

In the electrostatic approximation the solution for the local field E_{loc} inside the particle is well-known:

$$E_{Loc}(\omega) = L(\omega)E_0 = \frac{3}{\varepsilon_{met}(\omega)+2}E_0$$

In a dielectric medium, the local factor L(w) is small if the light frequency is less than the bandgap value. However, for metals there is always an optical range in $-\varepsilon_{met}(\omega) < 0$. This means that the external field of certain frequency cannot penetrate inside the metallic sphere. In the resonant conditions , and the field strength at this frequency $E_{Loc}(\omega_{rec})$ is largely increased, while the absolute value remains finite. The precise calculations result in the following expression for the local factor:

$$|L(\omega_{rec})| = \frac{|\varepsilon'_{met}(\omega_{rec})|}{\varepsilon''_{met}(\omega_{rec})},$$

where $\varepsilon'_{met}(\omega_{rec})$ and $\varepsilon''_{met}(\omega_{rec})$ are the real and imageniary parts of dielectric function of a sphere at resonance frequency, respectively. For noble metals, such as Au or Ag, the absorption is small ($\varepsilon''_{met} \ll 1$), and for Ag the local factor L(ω) may reach the value of 10-20. It is evident that the amplification occurs not only on the nano-entity itself but also in its vicinity.

In case of Raman scattering, both fields are amplified by the metallic objects, the incident light wave and the scattered Stokes signal, since their frequencies are very close to each other. As the results, the amplified Raman signal is proportional to $L^4(\omega)$ can be 10^5-10^6 times higher than out of resonance. The amplification of the incident wave is of order of $L^2(\omega)$.

The amplification factor G decreases with the distance from the particle as $G \sim (r_o/R)^{1/2}$, where r_0 stands for the characteristic size of nano-entity. This means that the field enhancement of the incident laser wave should still be significant at some distance from nano-structure.

According to known literature data, the field enhancement on NPs can be as high as 10^4 - 10^5. This gives the effective peak intensity near the nanoparticle of $I \sim 10^{17} - 10^{18}$ W/cm^2. The electric field that corresponds to this intensity is already comparable with internal atomic fields. Therefore, the electronic shell is largely perturbated. The advantage of laser exposure of NPs is that this field is gained inside the solution in their vicinity, otherwise a laser beam of this intensity can propagate neither in liquids nor in air due to various non-linear processes it induces in these media. High electrical field oscillating at laser frequency may alter both the height and the width of energetic barrier for alpha-particles that facilitates their tunneling and the accelerated alpha-decay of Uranium nuclei, either of ^{238}U or ^{232}U. The half-life time of the latter is only 69 yrs compared to 10^9yrs for ^{238}U. Therefore, the potential barrier for alpha-particles in ^{232}U nuclei is lower, and laser exposure of metallic NPs in aqueous solutions of this nuclide results in more pronounced effect. The same oscillating electric field causes the accelerated decay of both ^{234}Th and ^{234m}Pa nuclei through promotion of their beta-decay. The influence of laser exposure on the rate of beta-decay is higher than that on alpha-decay at otherwise equal conditions.

Estimations show that the alpha-decay of 10^6 nuclei of ^{238}U is promoted during one laser shot of 150 ps duration in the exposure conditions shown in Figure 9. The total number of alpha-decays of ^{238}U during the laser exposure is of order of 10^{10}, which represents roughly to 10^{-10} fraction of the total number of Uranium nuclei in the exposed solution. This is much below the accuracy of measurements of the alpha-activity of ^{238}U, which is typically ± 5%.

The enhancement of the electric field near the bulk target subjected to laser exposure can also be due to the presence of nanostructures on the gold target generated by short laser pulses of picosecond range of duration similar to that shown in Figure 5. The enhancement factor of 10^4 in SERS experiments has been realized in case of Au target nanostructured with picosecond laser pulses [39].

The selectivity of the laser-induced alpha-decay of ^{235}U and ^{238}U isotopes is not clearly understood at this moment. One might suggest that the frequency of oscillations of electronic shells around the nuclei is different for these two isotopes, so that the oscillating laser field is far from resonance frequency for one isotope and closer to the other. This point requires further studies.

The highest deviations from equilibrium values of activities of ^{234}Th and ^{234m}Pa are observed under laser exposure of a bulk Au target in aqueous solution of UO_2Cl_2. This scheme is characterized by high concentration of Au NPs inside the laser beam. It is known that the nanoparticles generated by subsequent laser shots are then homogeneously distributed inside the working liquid. Let us estimate the concentration of NPs inside the laser beam n_b before the NPs leave it. The concentration of NPs in the whole volume of the cell V is n, and it was generated with N laser shots. Let us introduce the effective volume of the laser beam inside the colloidal solution, V_b. Then the following relation should be valid: $V_b n_b = nV$, and one can easily obtain:

$$n_b = nV/V_b.$$

The ratio V/V_b depends on the geometry of the particular experiment, and in typical conditions is of order of 10^{-4}, which means that the focused laser beam occupies only a small fraction of the volume of solution [40]. Therefore, the concentration of NPs inside the laser beam is by factor of 10^4 higher than the concentration of NPs in the bulk of the solution. The scheme of laser exposure of a metallic target in the aqueous solution of radio-nuclides (Figure 15) has a mere advantage over the laser exposure of the mixture of the colloidal solutions of NPs with the solution of radionuclides. Indeed, in that scheme the amplification of the laser field takes place both on NPs in the vicinity of the metallic target and on NS left on the target by previous laser pulse. Also, dense ensemble of NPs similar to that shown in Figure 1 or in Figure 2 can contain so called "hot points" that is closely placed nano-protrusions. These hot points can lead to the amplification of the light wave by a factor 10^{16} [41], which do not seem completely credible.

The most probable reason of the laser-assisted accelerated decays in nuclei is large perturbation of the electronic shells that surround Uranium nuclei. Let us estimate typical values of both the electric and magnetic field of the laser wave inside the liquid taking into account the amplification of both fields on metallic nano-entities, either NPs or NS.

At typical peak power of laser radiation in the medium of 10^{12} W/cm^2, amplification factor on either NPs or Ns is of order of 10^4, which is a moderate estimation. The electric field of the electromagnetic wave can be expressed as follows:

$$E(t) = E_0 \cos(\omega t) = \frac{1}{2} E_0 \left[\exp(i\omega t) + \exp(-i\omega t)\right]$$

In case of linear polarization of the radiation, the electric field in V/m can be expressed through its intensity in W/m^2 as follows:

$$E_0 = 27\sqrt{I}$$

For laser intensity I ~ 10^{16} W/m^2 the corresponding electric field E_0 ~ 10^{11} V/m, which exceeds the internal electric field. The corresponding magnetic field of the laser wave can be estimated as 4×10^4 G.

High electric field can produce large perturbations in the electronic sub-system and alter the potential barrier for the tunneling of alpha-particles from ^{238}U nuclei. Also, the electric field of the laser radiation amplified on nano-entities may significantly increase the rate of beta-decays for nuclei in which they are permitted. This possibility is clearly seen from the experimental results presented above.

The possibility of excitation of nuclear levels of energy via generation of high-energy particles, similar to the experiments on laser ablation of targets in vacuum for laser exposure in liquids seems improbable. Indeed, despite to high values of electric field gained in the vicinity of nano-entities presented in the solution, the mean free path of particles in the condensed medium is too short to acquire the necessary energy.

Strong dependence of the acceleration of alpha-decay on the peak power of laser radiation in the medium should be related to the strength of fields of the laser wave. The natural measure of the electrical field is its value inside the atom or ion. The electric field of laser wave becomes comparable with inter-atomic field at intensity level of 10^{16} W/cm^2. Possible mechanism of laser-induced acceleration of alpha-decay can be illustrated as follows (Figure 24). Exposure of NPs to laser radiation leads to its amplification in the vicinity of NPs. If an ion of Uranil is situated near the exposed nanoparticle, then strong electric field of the laser wave disturbes its electronic shells. This perturbation causes the oscillations of the potential near its equilibrium value with the frequency of laser radiation. So do the width and the hieght of the potential barrier for tunneling alpha-particle.

Since the probability of tunelling depends on the barrier widt in an exponential way, so even its small variations can noticeably increase the rate of alpha-decay.

Owing to high concentration of NPs, the reduction of the width of potential barrier occurs near numerous NPs that are inside the focused laser beam. Also, the effect of laser-induced alpha-decay requires numerous laser pulses and has no apparent threshold down to intensities as low as 10^{10} W/cm^2.

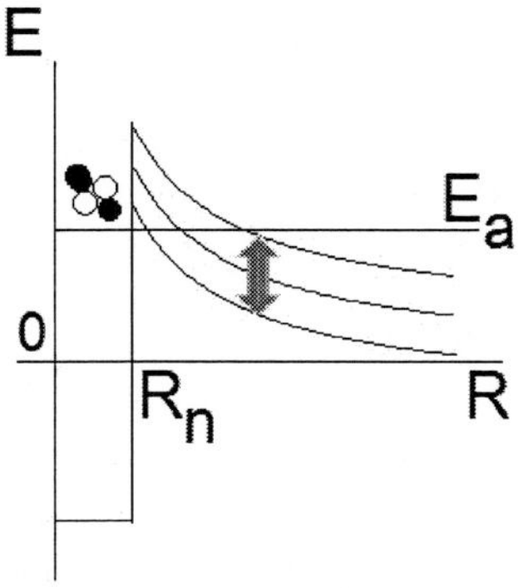

Figure 24. Qualitative illustration of the possible mechanism of laser-induced acceleration of alpha-decay. Open and closed circles symbolize protons and neutrons of the alpha-particle. E_a stands for the energy of tunneling alpha-particle, R_n is the radius of nuclei.

As it has been demonstrated above, the deviation of secular equilibrium in ^{238}U branching occurs in very different experimental conditions, such as laser exposure of NPs in the solution of Uranil-chloride and electro-explosion of a Ti foil in this solution. The common point between these two experimental setups is strong electric and magnetic fields generated in the solution of Uranium salt. It is known that high-current electric explosion of metal wires in a liquid induces strong magnetic fields ($H \sim 1$ MG) and high pulse pressures ($p \sim 10^5$ atm) [42].

The main parameter that is responsible for the deviation from secular equilibrium in experiments with electro-explosion is the magnetic field generated in the solution during the discharge through the foil.

Strong magnetic field changes the energy of the atomic electron shell and, hence, changes also the energy of any nuclear decay, as the decay energy is equal to the difference between the total energies of the initial and final systems with allowance for ionization energies of atoms or ions. For α-decay, the presence of strong magnetic field results in the increase in the decay energy and, hence, in the increase in the α-decay probability [43]. Since the ^{238}U α-decay energy is lower than the energies of the principal channels of ^{235}U α-decay, then the effect of the discharge on the alpha-decay of ^{238}U should be more pronounced. Also, the magnetic field changes the geometry of the problem: the spherical symmetry is replaced by the preferential direction along the magnetic field. This may by be considered as possible explanation of the selectivity of the deviation observed both in laser experiments and in experiments with the electro-explosion of foils in the Uranil solution. One more possible mechanism for $^{238}U/^{235}U$ selectivity is the parity of initial and final states of corresponding nuclei during alpha-decay, which is different for these two isotopes.

Theoretical investigations show that strong magnetic field can significantly accelerate permitted beta-decays [44] and can make possible forbidden beta-decays [45, 46]. Splitting of electronic energy levels in the magnetic field create new states to which the beta-decay becomes probable.

The difference between laser-induced decay and decay induced by electro-explosion is that measurable deviations of activities of nuclides can be observed after only one passage of current through a foil immersed into solution. This indicates that the Uranium nuclei are affected in the whole vessel containing the solution and not only in the microscopic area around the foil. The estimations presented above show that the magnetic field that is generated during the passage of the current through the foil is several orders of magnitude higher than that of the laser wave. The presence of sufficiently strong electric field during electro-explosion of the foil in Uranil solution seems doubtful. Indeed, the electric field may appear at the moment when the metallic foil is broken into pieces. However, the value of applied potential is unknown, since at this moment the battery of capacitors is partially discharged. Also, this field is applied locally just between the isolated pieces of Ti foil and cannot account for the equilibrium shift in the whole solution.

Laser action on the solution mediated by metallic NPs is essentially multi-pulse process. Lower value of both electric and magnetic fields compared to electro-explosion is compensated by the number of laser shots, which is typically of 10^3-10^4. Up to 10^7nanoparticles are inside the laser beam at each laser pulse, and this high number of active sites increases the efficiency of nuclear decays. The increased gamma-activity of Uranil solutions under laser exposure are observed due to accumulation of ^{214}Pb during exposure. No accumulation is possible in case of electro-explosion of foils, since the duration of current is too short.

Conclusion

Thus, it is demonstrated that exposure of aqueous solutions of Uranil ions with metallic nanoparticles to laser radiation at peak intensity of 10^{12}-10^{13} W/cm^2 leads to significant shift of secular equilibrium in the activity of ^{238}U, ^{235}U, and ^{232}U and their daughter nuclides. Laser exposure of the solutions is accompanied by non-equilibrium gamma-emission assigned to the accumulation of ^{214}Pb and ^{214}Bi during laser exposure. Laser influence on the activity of nuclides has no apparent threshold on laser intensity at least down to 10^{10} W/cm^2.

Both approaches, laser exposure of Uranil solution with nanoparticles, and electro-explosion of metallic foils in this solution, lead to qualitatively similar results. The activity of nuclides of ^{238}U branching and that of ^{235}U undergoes irreversible change. Apparently, there is no need to explain this change addressing to unknown physical phenomena. Both setups are characterized by generation of intense electromagnetic fields in the solution that causes the distortion of electronic shells of Uranil ion. This, in turn, alters the probability of nuclear decays.

In both cases, neither fission products nor neutrons generation is detected. From practical point of view it is important that the decay of Uranium and its daughter nuclides can be accelerated with the use of commercially available laser and not only with unique laser systems. These commercial laser sources have high repetition rate of laser pulses, which allows accelerating the decay of radionuclides in reasonably short time. The presented results show that these lasers are efficient tools for acceleration of decay of heavy Uranium nuclei. The efficiency of laser-based technique for lighter nuclei remains to be explored for deactivation of nuclear waste.

Acknowledgments

I thank my colleagues A.V. Simakin, L.I. Urutskoev, A. V. Goulynin, and A. A. Rukhadze for their help in experiments and useful discussions.

References

[1] Wong, A. Grigoriu, J. Roslund, T.-S. Ho, and H. Rabitz, *Phys. Rev.* A84, 053429 (2011).

[2] Andreev A.V., Gordienko V.M., Dykhne A.M., Saveliev A.B., Tkalya E.V., 1997, *JETP Lett.* 66 (5) 312.

[3] Andreev A.V.; Volkov R.V.; Gordienko V.M.; Dykhne A.M.; Mikheev P.M.; Tkalya E.V.; Chutko O.V.; Shashkov A.A.; 1999, *JETP Lett.* 69 (5) 343.

[4] Schwoerer H., Gibbon P., Düsterer S., Behrens R., Ziener C., Reich C., Sauerbrey R. *Phys. Rev. Lett.*, 86 (11), 2317 (2001).

[5] V. S. Belyaev, A. P. Matafonov, V. I. Vinogradov, V. P. Krainov, V. S. Lisitsa, A. S. Roussetsky, G. N. Ignatyev, V. P. Andrianov, *Phys. Rev. E* 72 (2005), 026406.
[6] J. Nedersen, G. Chumanov, and T.M. Cotton, *Appl. Spectrosc.*, (1993) *47*, 1959.
[7] M. S. Sibbald, G. Chumanov, and T.M. Cotton, *J. Phys. Chem.*, (1996) *100*, 4672.
[8] Fojtik, A. Henglein, *Ber. Busenges. Phys. Chem.*, (1993) *97*, 252.
[9] Henglein, *J. Phys. Chem.*, (1993) *97*, 5457.
[10] M. Procházka, Štepánek, B. Vickova, I. Srnova, P. Maly, *J. Mol. Struct.* (1997) *410/411*, 213.
[11] M. Procházka, P. Mojzeš, J. Štepánek, B. Vléková, P.-Y. Turpin, *Anal. Chem.*, (1997) *69*, 5103.
[12] M. Procházka, J. Štepánek, B. Vléková, I. Sronová, P. Malŷ, *J. Mol. Struc.*, (1997) *410*, 213.
[13] Srnová, M. Procházka, B. Vléková, J. Štepánek, P. Malŷ, *Langmuir,* (1998) *14*, 4666.
[14] J.S. Jeon, C.H. Yeh, *J. Chin. Chem. Soc.*, (1998) *45*, 721.
[15] M.-S. Yeh, Y.-S. Yang, Y.-P. Lee, H.-F. Lee, Y.-H. Yeh, and C.-S. Yeh, *J. Phys. Chem.*, (1999) *B 103*, 6851.
[16] Takami, H. Kurita, and S. Koda, *J. Phys. Chem.,* (1999) *B103*, 1226.
[17] S. Link, C. Burda, B. Nikoobakht, and M.A. El-Sayed, *J. Phys. Chem.*, (2000) *B104*, 6152.
[18] F. Mafuné, J. Kohno, Y. Takeda, T. Kondow, H. Sawabe, *J. Phys. Chem.*, (2000) *B 104*, 9111.
[19] F. Mafuné, J. Kohno, Y. Takeda, T. Kondow, H. Sawabe, *J. Phys. Chem.*, (2001)*B 105*, 5114.
[20] A.V. Simakin, V.V. Voronov, G.A. Shafeev, R. Brayner, and F. Bozon-Verduraz, *Chem. Phys. Lett.*, (2001) *348*, 182.
[21] S.I. Dolgaev,A.V. Simakin, V.V. Voronov, G.A. Shafeev, F. Bozon-Verduraz, *Appl. Surf. Sci.*, (2002) *186*, 546.
[22] P. V. Kamat, M. Flumiani, and G.V. Hartland, *J. Phys. Chem.*, (1998) *B 102*, 3123.
[23] G.W. Yang, *Progress in Materials Science*, (2007) 52, 648–698.
[24] G.A. Shafeev, Formation of nanoparticles under laser ablation of solids in liquids, in: *Nanoparticles: New Research*, editor Simone Luca Lombardi, pp. 1 -37, 2008, Nova Science Publishers, Inc., New York
[25] E. V. Zavedeev, A. V. Petrovskaya, A. V. Simakin, G. A. Shafeev, *Quantum Electronics*36(10) (2006), 978-980.
[26] A.V. Simakin, G.A. Shafeev, arXiv:0906.4268; A.V. Simakin, G.A. Shafeev, *Physics of Wave Phenomena,* 16 (4) (2008) 268-274; A.V. Simakin, G.A. Shafeev, *Journal of Optoelectronics and Advanced Materials*, (2010) 12(3), 432 – 436.
[27] http://isotopes.lbl.gov/isotopes/isoexpl.html.
[28] A.V. Simakin, G.A. Shafeev, *Quantum Electronics* 41 (7) 614 – 618 (2011).
[29] A.V. Simakin, G.A. Shafeev, *Physics of Wave Phenomena* (2011) 19(1), 30–38.
[30] G.A. Shafeev, *Applications of Nd:YAG Lasers: From Nano-Structures to Nuclei*, 2011, an open access book at www.novapublishers.com.
[31] Alexandr V. Simakin and Georgy A. Shafeev, *Journal of Materials Science and Engineering B* 1 (2011) 618-625.
[32] G.A. Shafeev, Laser-Based Formation of Nanoparticles, in: *Lasers in Chemistry*, Wiley VCH, Wienheim, 2008, pp. 713-741.

[33] http://nucleardata.nuclear.lu.se/nucleardata.
[34] A.V. Simakin, G.A. Shafeev, arxiv.org/abs/1112.6276 at www.arxiv.org.
[35] G. Volkovich, A. P. Govorun, et al., *Bull. ofLebedev Phys. Inst.* (2002) No 8, 45; *Ann. Fond. L. de Broglie*30 (2005) 63.
[36] F. Cardone, R. Mignani, and A. Petrucci, *Phys. Lett.* A 373 (2009) 1956.
[37] F. Bosch, T. Faestermann, J. Friese, F. Heine, P. Kienle, E. Wefers, K. Zeitelhack, K. Beckert, B. Franzke, O. Klepper, C. Kozhuharov, G. Menzel, R. Moshammer, F. Nolden, H. Reich, B. Schlitt, M. Steck, T. Stöhlker, T. Winkler, K. Takahashi, *Phys. Rev. Lett.* (1996) 77, 5190.
[38] O.A. Aktsipetrov, *Soros Educational Journal*, (2000), vol. 6.
[39] S. Lau Truong, G. Lévi, F. Bozon-Verduraz, A. V. Petrovskaya, A. V. Simakin, G. A. Shafeev, *Applied Surface Science* 254 (2007), 1236-1239.
[40] A.V. Simakin, V.V. Voronov, N.A. Kirichenko, and G.A. Shafeev, *Appl. Phys.,* A79(2004) 1127-1132; S. Lau Truong, G. Levi, F. Bozon-Verduraz, A. V. Petrovskaya, A. V. Simakin, G. A. Shafeev, *Appl. Phys.*A89(2) (2007), 373-376.
[41] SurbhiLal, Stephan Link, Naomi J. Halas, *Nature Photonics*, (2007) Vol. 1, No. 11. (01 November), 641-648.
[42] *Exploding wires* (Eds. W. G. Chace and H. K. Moore), Plenum Press, New York 1962; Y. Bakshaev et al., *Plasma Phys. Rep.* 27 (2001) 1039; A. Velikovich et al., *Phys. Plasmas* 14 (2007) 022706.
[43] L. I. Urutskoev and D. V. Filippov, *Physics-Uspekhi* 47 (2004) 1257.
[44] Philippov D.V., *Physics of Atomic Nuclei*, 70(2), 280–287 (2007).
[45] Philippov D.V., *Physics of Atomic Nuclei*, 70(12) 2068–2076 (2007).
[46] Yu.P. Gangrsky, F.F. Karpeshin, M.B. Trzhaskovskaya, Yu.E. Penionzhkevitch, *Physics of Atomic Nuclei*, 71(6), 979-983 (2008).

In: Uranium: Characteristics, Occurrence and Human Exposure ISBN: 978-1-62081-207-5
Editors: Alik Ya. Vasiliev and Mikhail Sidorov

Chapter 5

URANIUM OCCURRENCE AND BEHAVIOR IN GROUNDWATER (THE EXAMPLE OF NORTH-WEST RUSSIA)

*A. I. Malov**
Institute of the Ecologic Problems of the Northern Region, Ural Division, Russian Academy of Sciences, Arkhangel'sk, Russia

ABSTRACT

In the northern-west part of Russia, the main aquifers are joined to the sedimentary basin of the Mezen syneclise, filled predominantly with sandy-argillaceous sediments. These deposits were formed at the expense of drifting of products of the Baltic Shield breaking. Transfer of uranium in suspended and dissolved states was carried out by surface water and groundwater. During of the continental periods within raised blocks of the syneclise there was redistribution of uranium with the partial removal beyond the uplifts. Elevated (up to 2.6 Bq/L) values of alpha-activity of U isotopes were found in groundwater of the waterbearing complex of sandstones of the Padun suite of the Vend and high (up to 20) $^{234}U/^{238}U$ activity ratios. Study of the distribution of uranium isotopes was made in the North Dvina basin. Also the groundwater residence time in aquifers was determined. To do this was used the generic parameter of *p*: probability of ^{234}U atom transition into water in the decay of ^{238}U, obtained on the basis of geological dating, hydrodynamic calculations and paleo-hydrogeological reconstructions. As a result, eight types of groundwater with ages ranging from 160 to 130,000 years were marked out. Analysis of the calculated dependences showed that high solubility of rocks with Clarke concentrations of uranium, even at the time of contact of water with the rock of 75 000 years it will exceed the maximum allowable level of its content. In almost all samples uranium content is largely determined by its chemical dissolution. The most intense it is exerted on the sides of the basin, where is the development of the youngest water. As ground water moves to the discharge zone and increases its salinity, water's aggressiveness reduces and radiation decay begins to dominate by the leaching of uranium. Approximate estimation of radiation safety of fresh and mineral waters of the

* E-mail: malovai@yandex.ru.

North Dvina basin has shown that they can be used for drinking. At the same time requires the constant monitoring of their radionuclides content.

Keywords: Isotopes of uranium, the age of groundwater, radiation safety, the north-west Russia

GENERAL INFORMATION

Aquifers are usually joined to sedimentary basins filled with permeable sediments. These deposits are formed for account of drifting of products of destruction adjacent to the basins of mountain-folded structures. In the northern-west part of Russia this "pair" is represented by folded massive lifting of the Baltic Shield and located eastward from the Mezen syneclise (northern part of the Moscow syncline) (Figure 1A)

Mesen syneclise is filled predominantly with sandy-argillaceous sediments of marine basins that existed here in Riphean, Vendian, Carboniferous, Permian, partly - in the Devonian, Triassic, Jurassic, Pleistocene and Holocene. Layers of carbonate rocks of the Carboniferous and halogen - Permian are of a secondary importance. The total thickness of sediments reaches up to 3 km (*Malov*, 2004).

In the rocks of the Baltic Shield elevated uranium contents (3.2 ppm) are found in metamorphic rocks, migmatites and granite-gneisses of Kola and Karelian cratons, as well as in the White Sea mobile belt. In intrusive rocks, the highest U concentrations were recorded in the agpaite nepheline syenites of the alkaline suite of the Karelian–Kola alkaline province (10.3 ppm); they are also high in the granites of the White Sea belt (4–7 ppm). In effusive rocks, the highest U concentrations (4.7 ppm) occur in liparites (analogs of granites) (*Smyslov et al.,* 1979).

Also, eight small U deposits, more than 200 ore manifestations (~20 of which are nearly deposits in terms of their characteristics), and several thousands of points of radioactive mineralization U occurred in the form of uraninite (hydrous pitch blende and uranium oxide minerals), coffinite, carnotite, and others (*Metalogeny...*, 1999).

Transfer of uranium in suspended and dissolved states from the Baltic Shield in the Mezen syneclise was carried out mainly by surface water and groundwater. It can be assumed that its distribution in marine sediments was relatively even.

During Cambrian-Devonian and Triassic-the Holocene continental conditions dominated in the Mezen syneclise. The uplift of sedimentary strata at a considerable height above a sea level has contributed to the penetration of water of atmospheric origin to a depth of 650 m (*Malov*, 2004). This process was at most intense in the Mesozoic era and especially in the Pliocene period, when that area was dissected with paleo-incisions down to 300 m (*Malov,* 2003).

Uranium redistribution of the flowing water, saturated with CO_2, O_2 and organic acids was within the raised blocks. The oxidation and transition of U into solution were accompanied by its transfer from groundwater recharge areas into its discharge areas and the partial removal beyond the raised blocks.

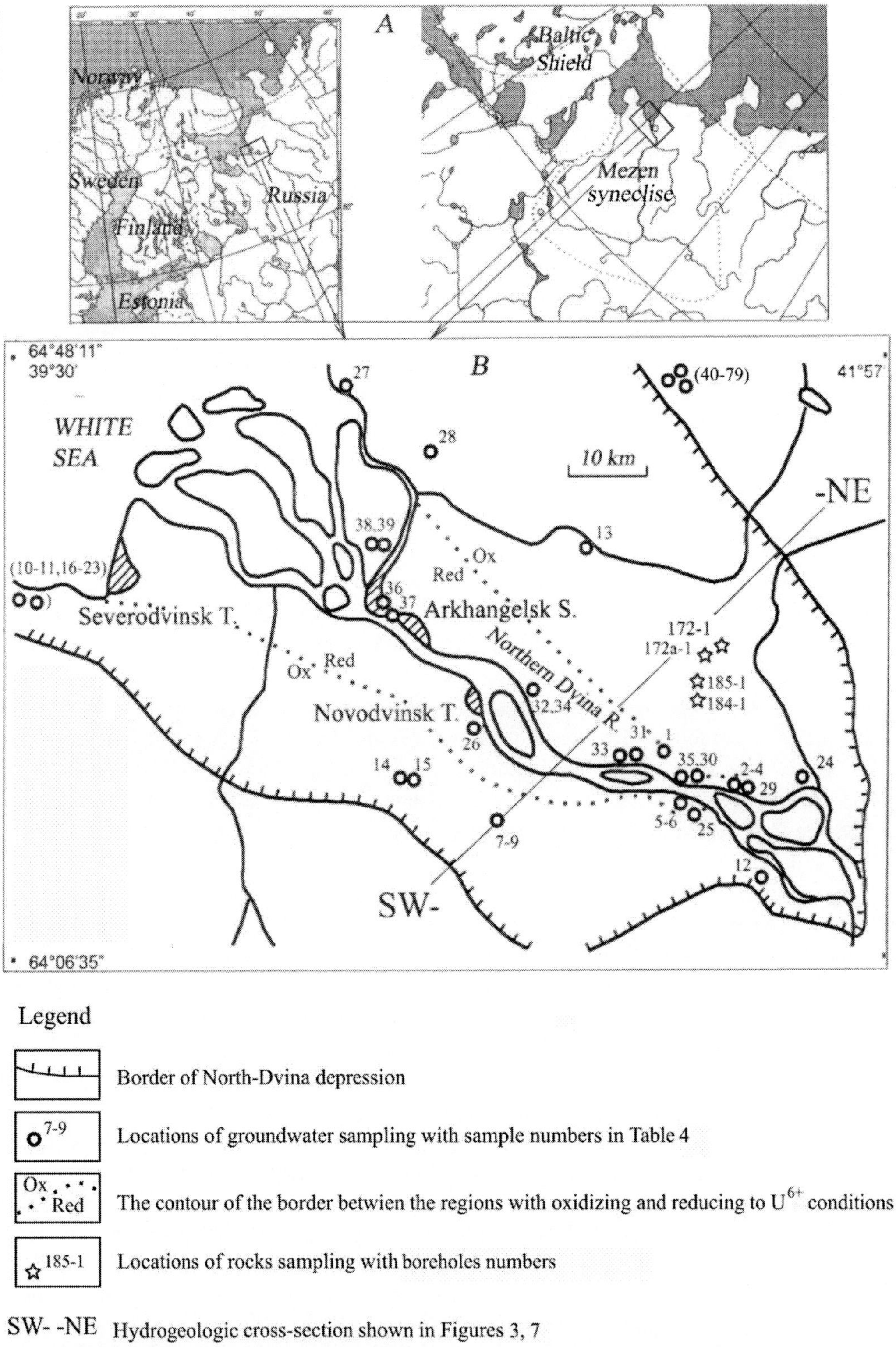

Figure 1. Study area: the North Dvina basin and Mezen syneclise on a map (http://kontur-map.ru/548608.html) (A) and sampling sites (B).

It should be noticed that at present the discharge of U from the territory of the Kola Peninsula and Karelia is about an order of magnitude smaller than that from the Mezen syneclise. According to (*Salminen et al.,* 2004), averages U concentrations in surface water in those two regions are 0.05 and 0.5 μg/L, respectively.

There is no correlation between U concentrations in surface water and the distribution of the mean annual river runoff over the area. For example, the module of river runoff within the Baltic shield is 10–12 L/(s km^2), whereas that in the northern part of the Russian platform and in the Pechorskaya platform is 8–10 L/(s km^2) (*Vodnye…*, 1980).

There is some correlation between U distribution in surface water and the formation conditions of groundwater discharge into rivers. The groundwater runoff accounts for 10–20% of the total river runoff within the Baltic shield and from 20–30% on average in the northern part of the Russian platform to 50% or more in karst regions.

Even closer correlation was found to exist between U concentrations and the underground dissolved solids runoff. According to *(Zverev, 2006)*, the mean discharges of chemical elements with groundwater from the Baltic shield and the Russian platform are 0.14 and 1.5 t/(month km^2), respectively. This means that the specific discharge of chemical elements (including U) from the fracture water basin of the Baltic shield is on average an order of magnitude smaller than that from the Mezen syneclise area.

The underground dissolved solids runoff in mountain folded zones younger than the Baltic shield is much greater. In the Ural mountain folded area, it is 1.4 t/(month km^2), i.e., it is approximately equal to that from the Mezen syneclise area. In the youngest mountain folded areas (the Alpine and the Dzungarian–Tien Shan folded areas), the underground dissolved solids runoff increases up to 5.5 t/(month km^2); and in the areas of the most soluble carbonate rocks of the Southwestern Caucasus it reaches 13.5 t/(month km^2) (*Zverev*, 2006).

Hence, we can conclude that the underground dissolved solids runoff within the Baltic shield in the Proterozoic era was much more intense than that at present; the bulk of U discharge from the Baltic sheet crystal mountain rocks fell just on that time.

It should be noted that at transfer of uranium into groundwater under certain conditions, U, precipitated from solution, forms zones of secondary U enrichment, including the formation of the so-called infiltration deposits of economic importance and wide distribution. They are confined to the permeable layers of variegated sand-clay sequences. Such deposits are known in Germany, France, Great Britain, Italy, Holland, Belgium, Australia, India and many other countries. This type includes numerous of uranium deposits in the Colorado Plateau in the USA. At junction of the Baltic Shield and Mezen syneclise is known deposit Karhu, timed to the contact zone of sedimentary deposits of Riphean with the underlying Archean-Early Proterozoic metamorphic rocks.

The major processes causing U sedimentation were hydrolysis, adsorption on natural sorbents, and the alteration of the environmental conditions from oxidation to reduction ones.

In the course of hydrolytic reactions, Fe hydroxides precipitate among the first dying sandstones red. The sandstones have significant sorption capacity. The precipitation of Fe hydroxide gel causes the formation of a sorption barrier and the co-deposition of a substantial portion of U at this sorption barrier. The zone of red sandstones occurs in Vend deposits (predominantly to a depth of 300 m) in the zone of joining the Mezen syneclise with the Baltic shield.

Among natural sorbents of U, the most important are various forms of organic matter (OM): oxidizable bitumen, coalified organic remains, peat, OM of silts, and so on. The capacity of organic sorbents to concentrate U is directly proportional to their content of oxygen-containing components (that is to their oxidation degree) (*Germanov et al.,* 1975). Organic remains (pellicles of algae) and bitumen widely occur in the Vend sediments deeperthan 800 m.

The zone of anaerobic microbiological processes, creating the reduction conditions for U^{6+} and contributing to U precipitation, occurs in Vend sediments mostly below 300 m. However, it is also present in higher horizons, isolated from the earth's surface, where groundwater contains sulfates and OM. The processes of sulfate-reduction contribute to the formation of the reduction medium, which is favorable for U precipitation.U^{6+} reduces to U^{4+} which precipitates in the form of U oxides and partially in the form of silicates.

Such conditions took place under the strata of accumulating silts in the coastal zones of sea basins that were situated in the territory of the Mezen syneclise in the Riphean, Vend, Paleozoic, Mesozoic and Cenozoic periods in succession. U also sequentially redistributed from more ancient coastal zones into younger ones. The most recent large-scale area of reducing conditions occurred under the Upper-Cenozoic Mikulian Sea sedimentary cover, in particular, in the territory of the North Dvina basin (Figure 1B), which is described below. Then it was "preserved" under the Valdai glacier. Evidently, the U remains that had been earlier carried out from the Baltic shieldt into the Mezen syneclise area were some concentrated in the Mikulian Sea coastal zone, at sea basin sides. After the glacier melting in Valdai and formation of the Northern Dvina erosion depression, hydrocarbonate water flowed from the basin sides toward the river valley. The carbonate-containing water is a solvent of U. In the zones into which this water penetrates, U occurs in the dissolved state and its concentrations are maximal. These zones of oxidation («Ox» in Figure 1) for U^{6+} conditions can be identified by high values of Eh +200 – +300 mV. When the values of Eh are below +100 - 50 mV, concentration of uranium in solution decreases. In these zones, with reducing («Red» in Figure 1) for U^{6+} conditions, an increased content of uranium in sediments is observed.

According to our studies of groundwater isotope composition in the Mezen syneclise in 2002–2010, elevated (up to 2.6 Bq/L) values of the alpha-activity of U isotopes were found in groundwater of the waterbearing complex of sandstones of the Padun suite of the Vend (Vpd in Fig. 2) in the northwestern marginal part of the syneclise bordering with the Baltic sheet. This water is also characterized by high (up to 20) values of $^{234}U/^{238}U$ activity ratio (*Malov et al.,* 2008).

Figure 2 shows variations of ^{234}U and ^{238}U activities in Mezen syneclise groundwater.

The lowest activity of U isotopes and the lowest values of $^{234}U/^{238}U$ activity ratio are characteristic of the upper parts of the active water-exchange zone, whose host rocks feature high hydraulic conductivity and solubility. Such behavior of U isotopes can be explained as follows. In the uranium-containing rocks, the $^{234}U/^{238}U$ activity ratio ≈ 1. Upon the complete decomposition of those rocks, U isotopes pass into solution also in equilibrium proportions. This is characteristic of highly soluble carbonate and sulfate rocks of the Carboniferous and Permian periods (C and P in Figure 2). However, due to the high permeability of these rocks, the water exchange is carried out so quickly that water does not have time to transfer into solution a significant amount of the U. If the solution of rocks is slowed down due to the interaction of water with the aluminosilicates of the Vend period (which takes place at depths of about 100 m or lower), then an excess of ^{234}U in groundwater is observed. This is due to the premature passing of ^{234}U in a solution due to radioactive decay of ^{238}U. The water exchange is slow at these depths, as a result the water becomes more U.

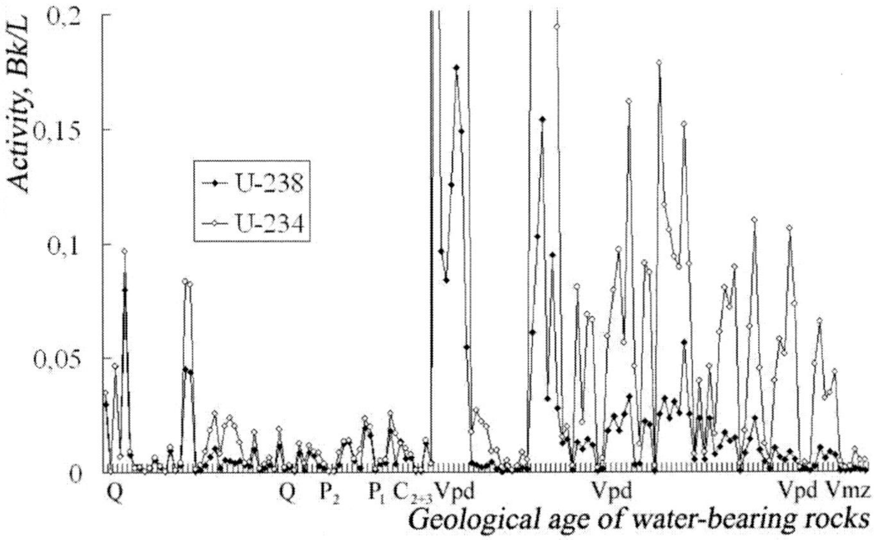

Figure 2. Variations in the activities of ^{238}U and ^{234}U in the Mezen syneclise groundwater.

The distribution of U isotopes in the Vend groundwater is uneven and follows some regularity. The author has studied those regularities within key area: the North Dvina basin.

Description of Hydrogeological Conditions of the North Dvina Depression

North Dvina basin (39°30'-41°57'E, 64°06'-64°48'N) - is a continuation on the land of the Dvina Bay of the White Sea; it stretches from SW to NE up to the mouth of the river Pinega (Figure 1). Currently, the depression is open to the sea bowl-shaped depression in the carbonate-terrigenous deposits of the Middle Carboniferous (C_2) and terrigenous - Lower Carboniferous - Upper Devonian (C_1-D_3), Padun (Vpd) and Mesen (Vmz) formations of the Vendian (Figure 3).

The length of the basin is 120 km, the maximum cross-sectional dimensions (along the shore of the Dvina Bay): the width on top - 100 - 110 km, the width of the bottom - 30 - 40 km, the altitude of the bottom - -60 m. The slopes have a slope of 2^0 - 3^0.

The cavity was filled mainly with clay deposits of Boreal Sea of Mikulian interglacial (mIIImk), which existed there from 140-145 to 70 thousand years ago (*Molodkov, Bolikhovskaya,* 2006). But (*The White Sea system,* 2010) Mikulian sea actually was there 130-117 thousand years ago and in in the period 117-20 thousand years ago in marine sediments accumulated other transgressions and alluvial-lacustrine deposits. In this paper, in figures 3, 7b, all they are shown under the symbol mIIImk. Area of their distribution within the basin is more than 7000 km^2, the thickness reaches the central parts of the basin 50-70 m.

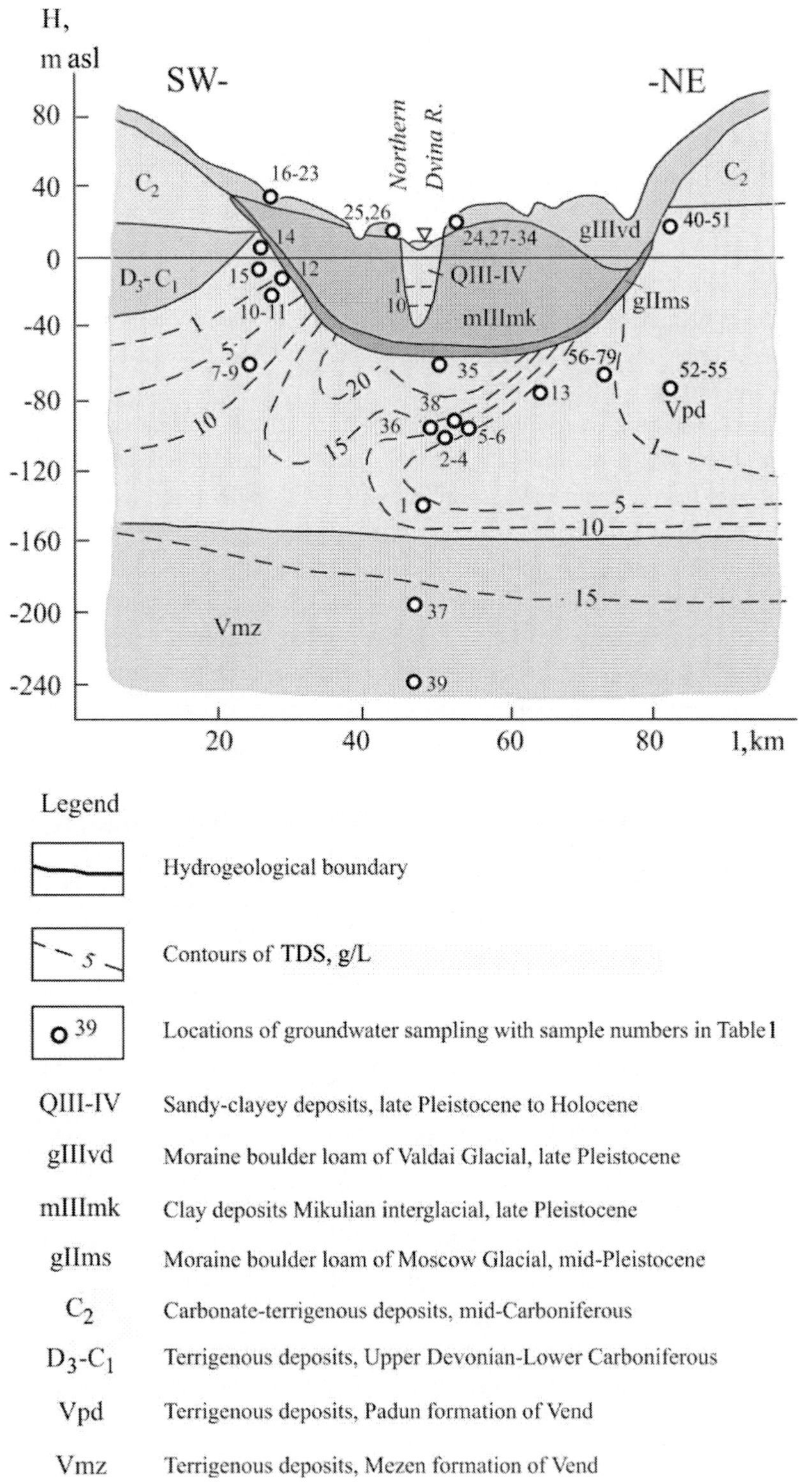

Figure 3. Modern hydrogeologic cross-section along the transect SW-NE in Figure 1. On the cross-section projected all points of groundwater sampling, as shown in Figure 1.

The clay deposits are characterized by large density and the addition of high-strength, which may be associated with compaction under its own weight, as well as under the weight of overlying 70-11.5 thousand years ago (*Molodkov, Bolikhovskaya,* 2006) or 20-17 thousand years ago (*The White Sea system,* 2010) the Valdai glacier, the load from which in his thickness to 1000 meter could reach significant values.

Bottom clay is underlaid by 10-15-meter layer of loam Moscow moraine (gIIms). Age of the Moscow glaciations is > 130-140 thousand years (*The White Sea system,* 2010; *Molodkov, Bolikhovskaya,* 2006).

Below it lies a thick sequence of clastic sediments of the Vendian provided interbedded sandstones, siltstones and mudstones Padun, Mezen and Ust-Pinega formations. Their total thickness is 600-700 m.

The Padun Formation, overlying of the absolute mark of -150 m, consists in the territory of the North Dvina basin mainly of siltstones, rarely – sandstones separated by mudstone interlayers. Rocks have mostly reddish brown color with pale green lenses and patches. The siltstones are dominanted by the coarse-grained (0.01 - 0.05 mm) fraction: 70%, almost always present pelitic particles: up to 30%. The clastic grains consist of quartz (up to 98%), feldspars (up to 10%), and micas (near 1%). The cement has mainly clayey-ferruginous composition. Clay minerals are observed as hydromicas, kaolinite, and chlorite. Up of the section is marked of increase in the content of kaolinite and chlorite impurity reduction up to complete disappearance. In the upper part of the sequence (thickness ~ 50 m), the siltstones are poorly cemented. The chemical composition of rocks is given in table 1.

Table 1. Average chemical composition of siltstones of Padun suite of 18 definitions, %

SiO_2	Al_2O_3	CaO	MgO	Na_2O	K_2O	Fe_2O_3	FeO	TiO_2	LOI	Cr_2O_3	MnO	P_2O_5
72	12,8	0,48	1,24	0,14	2,94	5,3	0,3	0,81	3,36	0,02	0,14	0,07

Deposits of Mezen Formation in the upper part of the section (-150 - -300 m) includes thin intercalation of mudstones (35 - 40%) with siltstones (35 - 40%) and sandstones (20-30%). Painting is from varicolored to completely red in some sections. Siltstones of the Mezen Formation are fine- and coarse-grained. The clastic particles consist of quartz (80-95%), feldspar (3-8%), mica (1-2%) and sometimes there are multimineral rock fragments (up to 3-4%). The cement has usually a mixed composition (hydromicas, chlorite, carbonate, kaolinite and iron hydroxides). The chemical composition of rocks is given in table 2.

The Mikulian clay is overlapped by 10-15-meter layer of Valdai Glacial sediment (gIIIvd) at the top represented mainly moraine boulder loam. The local areas Valdai Glacial deposits are represented by fluvioglacial (fIIIvd) and lacustrine-glacial (lgIIIvd) sands. In limited spaces Valdai and at least - directly Mikulian deposits overlap by 3-5-meter layer of modern sandy-clayey sediments: alluvial (aIV), lake (lIV), marsh (pIV), marine (mIV).

Table 2. Average chemical composition of siltstones of the upper part of the Mezen suite of 66 definitions, %

SiO_2	Al_2O_3	CaO	MgO	Na_2O	K_2O	Fe_2O_3	FeO	TiO_2	LOI	Cr_2O_3	MnO	P_2O_5
69,73	12,99	0,96	1,63	0,25	3,85	5,09	0,37	0,7	3,71	0,02	0,06	0,1

Lower valley of the Northern Dvina is about the thalweg of the North Dvina basin. Its width is from 15 km at the mouth up to 2 km at the narrowest part. Valley is of the erosion origin and it is formed by water currents of the Valdai glacier and cuts through almost to the sole thickness of Mikulian interglacial clay (Figure 3) and in the upper part of the North Dvina basin cuts through their wholly or partially cut into the moraine loam Moscow.

The valley of the Northern Dvina is filled by sandy-clayey deposits (QIII-IV): fluvioglacial, marine, alluvial, lacustrine-swamp. Their total thickness is 40-50 m.

The upper 90-m thick section of the aquifer complex of siltstone and sandstone of Padun Formation, located under the central part of the North Dvina basin is characterized by reversed zoning with decreasing TDS in groundwater from 29 to 5 ppt (Figure 3). It primarily originated due to desalination of groundwater in the long process of continental break in the Mesozoic and Cenozoic. The main reason for desalination was a draining effect of paleovalleys with virtually no overlying aquitard. Desalination of groundwater under the North Dvina basin ceased ~ 130 thousand years ago after the transgression of the Mikulian Sea.

Next to this zone of freshened waters occurred extraction of salt sedimentary pore waters of marine clays Mikulino interglacial under its own weight and the weight of the overlying Valdai glacier. This process is confirmed by reverse zoning on the iodine content is lowered it from 38 to 1.7 ppm (Table 3).

As the table shows, the only source of iodine in this area is the pore water of clay Mikulino interglacial. Elevated concentrations of iodine in them were appeared due to the fact that marine silts contained a large amount (up to 10%) of iodine-containing organic remains of algae. After the decomposition of buried algae, contained in them iodine is partially transferred to the pore solution. Decomposition of organic matter associated with the process of sulfate reduction, as evidenced by the reduction of SO_4^{2-} and increasing - HCO_3^- in the pore water compared to sea water.

These waters are pushed desalinated water down and sideways, and - because of the variance - partly mixed with them. The depth of penetration of the squeezed water was about 90 m, that is - to the roof compared to the impermeable sediments of the Padun Formation of the Vendian deposits of the Mezen Formation (middle filtration coefficients, respectively, 0.56 and 0.03 m/day).

Table 3. The chemical composition of sea water and groundwater

Interval of sampling	Sample location	TDS	Cl^-	SO_4^{2-}	HCO_3^-	Na^+	K^+	Ca^{2+}	Mg^{2+}	I^-
m. asl		ppt				ppm				
0	White sea	35	19353	2701	143	10764	388	408	1297	0.06
- 40	mIIImk	31.3	17582	1653	897	9358	112	201	1492	40
-57 - -68	Vpd	27.7	15231	2312	354	7230	91	1571	858	38
-69 - -80		19.3	10885	1309	196	4425	59	1885	524	22
-90 - -96		12	6360	1087	187	2792	43	1184	239	12
-105 - -145		5	2378	747	127	973	20	634	105	1.7
-150 - 175	Vmz	15.2	7654	2140	40	3788	31	1122	427	0.6
-175 - -250		17.6	9000	2110	61	4500	48	1360	410	1
-250 - -350		25.2	14124	1828	40	6300	39	2211	575	0.6

The Mezen Formation aquifer system is characterized by normal zoning in TDS of groundwaters, rising from the bottom of Padun suite to the regional aquitard, represented by argillites Ust-Pinega Formation of the Vendian, which lie at a depth of 400 meters, from 10-15 to 25 ppt.

As seen from the Table 3 analysis, in groundwater, compared with Sea water and pore solutions Mikulian clay significantly increases the proportion of calcium and significantly reduced the share of bicarbonate ion, which is associated mainly with the hydrolysis of calcium aluminosilicates equilibrium with this type of water. This process is slow, which shows substantial age of groundwater, in particular – Mezen Formation (*Malov* 2004).

Next, no less substantial process of chemical composition of groundwater formation in the upper part of complex of clastic sediments of the Vendian was inflow of fresh water from source areas on the sides of the basin.

This inflow started after the erosion activity of streams from melting glacier and had resulted in the formation of the Northern Dvina River valley, cutting through the thick of the Mikulian clay almost at full capacity (Figure 3). Salty water from the aquifer system of Padun Formation began to flow into the river valley and fresh water from source areas began to leak and replaced them.

On the right bank of the Northern Dvina River, where the aquifer system of the Vendian terrigenous deposits lie directly under carbonate-terrigenous deposits of the Middle Carboniferous, the thickness of the zone of fresh water is maximum (Figure 3). As we approach the central part of the the North Dvina basin, TDS is increased to 1-5 ppt; HCO_3 Mg-Ca composition changes in TDS 1-2 ppt to HCO_3-Cl Na, rarely - Cl-HCO_3 Na, and at 2 - 5 ppt - Cl Na, rarely - Ca-Na. However, where the Vendian aquifer system is covered by sulfate-terrigenous deposits of the Early Carboniferous - Late Devonian (on the left bank of the Northern Dvina River) (Figure 3), underground water at TDS 1-3 ppt have SO_4-Cl Ca-Na or Cl-SO_4 Na-Ca , rarely of SO_4 Na composition, at TDS 3-5 ppt - Cl-SO_4 Ca-Na, rarely - Mg-Ca-Na, Na-Ca, Ca or SO_4 Ca-Na, rarely - Na, Mg-Ca.

Fresh waters of Quaternary deposits, opened with tested wells are characterized by the values of TDS from 0.2 to 1.3 ppt and the composition of HCO_3 Mg-Ca to HCO_3 Na and Cl Na.

Tested sources of mineral waters have 1.6-4.9 ppt TDS and HCO_3-Cl Na composition.

Distribution of Uranium in Groundwater

Figure 4 plots the activities of U isotopes versus the parameters of groundwater chemical composition.

The analysis of the above results shows that the lowest U activity prevails in the most mineralized water (15–27 ppt) which occurs under the deepest part of the Northern Dvina depression, near its thalweg (Figure 3, 4a). This water features somewhat decreased values of pH and Eh and hydrocarbonate-ion concentrations and higher concentrations of Fe (Table 3), (Figs. 4b, 4c, 4e, 4h).

In fresh water, the activity of U isotopes somewhat increases and exceeds the maximum allowable values of alpha-activity for drinking water (Figure 7a). This water is also remarkable for the higher values of pH (up to 9), Eh (up to 300 mv), concentrations of HCO_3^-, O_2, CO_2, and Fe concentrations being rather low (Figs. 4b, 4c–4f, 4h).

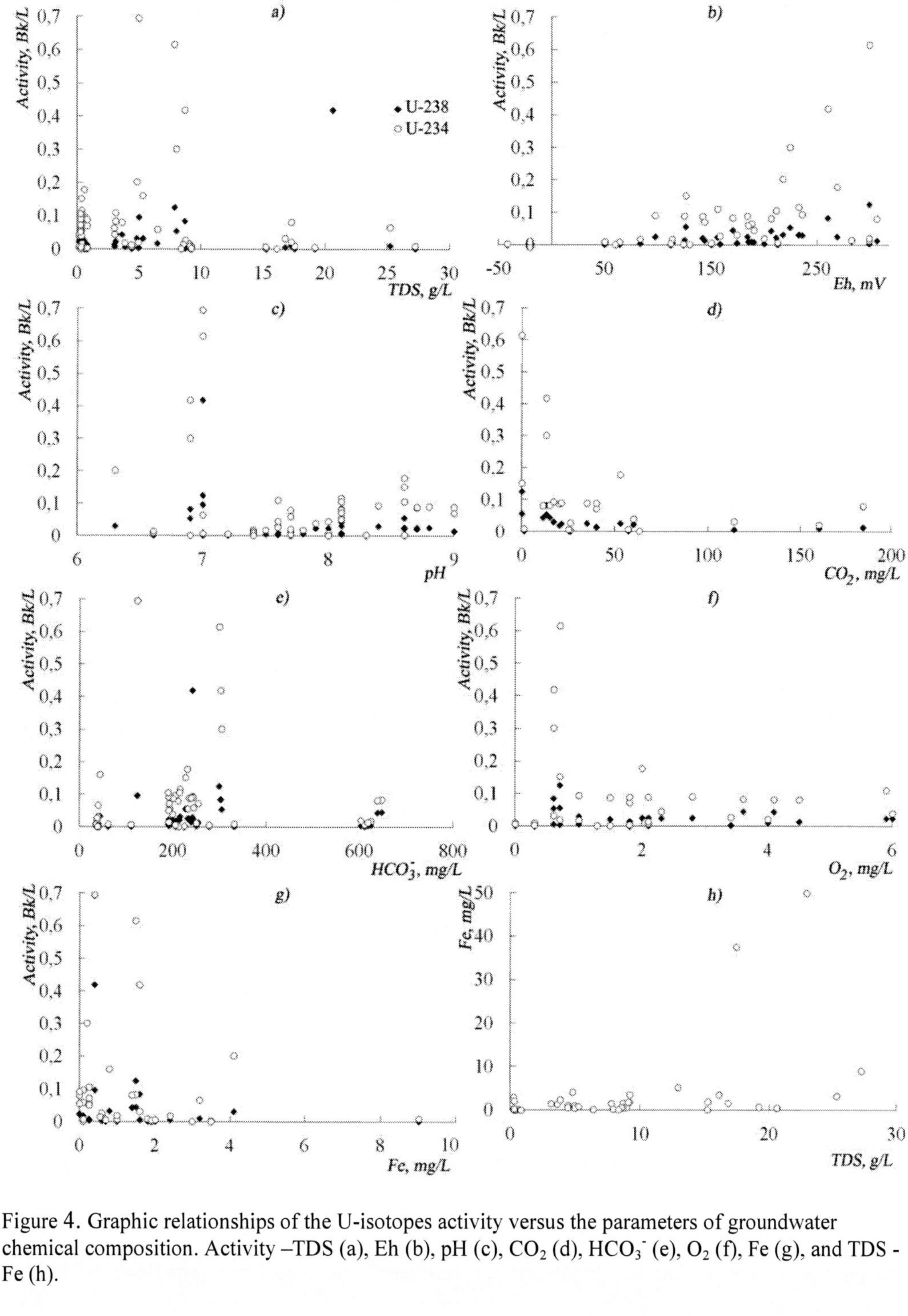

Figure 4. Graphic relationships of the U-isotopes activity versus the parameters of groundwater chemical composition. Activity –TDS (a), Eh (b), pH (c), CO_2 (d), HCO_3^- (e), O_2 (f), Fe (g), and TDS - Fe (h).

The highest activity of U isotopes is recorded in the zones of intrusion of fresh hydrocarbon water into saline water, that is, at the front of "the zones of stratal oxidation", where TDS is 5–10 ppt (Figure 3, 4a). In these zones, the Eh values are up to 300 mV and pH ~7; the concentrations of O_2 (0.5 ppm) и CO_2 (10 ppm), are minimal and those of Fe are low (Figure 4h). An approximate inversely proportional relationship between Fe concentrations and U isotope activities (Figure 4g) can be supposed to exist. The area of maximum U activity coincides with the "hydrogeological window" in the confining layer of Mikulian-Sea clay sediments. The groundwater of the Padun suite of the Vend discharges through this zone into the Northern Dvina valley (*Malov,* 2003) (Figure 1, 3).

For a more detailed analysis of the distribution of uranium isotopes in groundwater is reasonable estimate of age of these types of groundwater, that is timing of their location in the aquifer and the interaction with uranium-bearing rocks.

RATIONALE FOR THE PROPOSED MODIFICATION OF A METHOD OF ASSESSMENT RESIDENCE TIME OF GROUNDWATER IN AQUIFER

In geology are used widely different methods to determine the absolute age of rocks by radioactive isotopes. Despite the questionable accuracy, it is possible to organize and systematize the results of field observations and create a harmonious hypothetical history of the Earth stretching for several billion years.

The use of similar methods in hydrogeology is difficult due to the greater dynamism of groundwater as compared to geological formations and their active interaction with rocks, gases and organic matter, which leads to relatively rapid changes in isotopic composition.

So basically the relatively short-lived isotopes are used that are suitable only for young water with residence time in the aquifer to 30 thousand years (*Ferronsky, Polyakov*, 2009, *Clark and Fritz*, 1997; *Geyh et al.*, 2001; *Kazemi et al.*, 2006; *Solomon et al.,* 1995, etc.).

Attempts to dating older water on even isotopes of uranium were made by many researchers (*Kronfield et al., 1975*; *Andrews and Kay*, 1978; *Barr et* al., 1979, *Andrews et al., 1982, Fröhlich and Gellermann,* 1987) after 1954 when VV Cherdyntsev and PI Chalov discovered "the phenomenon of the natural separation of uranium-234 and uranium-238, which consists in the fact that the transition of uranium from solid natural uranium-containing formations to the liquid, which does not dissolve these structures and isotopic mixture enrich with uranium-234" (wording, introduced in Discoveries of the USSR State Register under the number 163 with a priority of 27 March, 1954).

Calculated dependence, as a rule, included the parameters which characterize the rate of dissolution and precipitation processes of the first-order kinetics, sorption and radioactive decay of ^{234}U and ^{238}U (*Fröhlich and Gellermann,* 1987), the rate of chemical dissolution of zero-order kinetics of ^{234}U and ^{238}U, the recoil range of ^{234}Th atom, the surface area of water-rock interaction and the decay of excess ^{234}U activity, formed by entering the recoil atom in water (*Andrews et al., 1982*).

However, due to inability to determine the exact values of these parameters over long periods of time, the use of the obtained dependencies mainly consisted in analyzing of the impact in changes of values, occurring in theĭn parameters on the isotopic composition of ground water, or pointed to the possibility of their use for dating of groundwater in special cases (*Fröhlich and Gellermann,* 1987).

In the second half of the 80-ies under the aegis of the IAEA the international team of researchers conducted a joint project to assess the possibility of using different isotopes for dating ancient groundwater. Samples were taken from Milk River aquifer in Canada from 16 wells to determine in them ^{2}H, ^{3}H, ^{13}C, ^{18}O, ^{14}C, ^{39}Ar, ^{81}Kr, ^{85}Kr, ^{222}Rn, ^{36}Cl, ^{129}I, ^{234}U and ^{238}U. According to research (*Frohlich et al., 1991*) was concluded that the possibility of using these methods for solving hydrogeological problems are very high, but only provide their joint application.

Thus, the available methods of dating ground water by isotope of uranium, having undoubted theoretical merits, is not quite suitable for practical use. Therefore, this chapter attempts to adapt them for such purposes by introducing a generalized estimated parameter (the probability of passing atom ^{234}U into water in the decay of ^{238}U), obtained through the use of geological dating, hydrodynamic calculations and paleo-hydrogeological reconstructions. It is assumed that this probability is more or less constant for aquifers composed of homogeneous rocks of a certain lithology and oxidizing conditions for uranium.

In the uranium-bearing rocks there is usually an equilibrium ratio between isotopes ^{234}U and ^{238}U: 1 to 18274, which corresponds to one relationship of their activities. Groundwater contained in these rocks, is characterized by an excess of ^{234}U. The natural process of fractionation of ^{234}U and ^{238}U is explained in terms of physics of radiation effects (*Rosholt et al.,* 1963, *Chalov,* 1969, *Kigoshi*, 1971).

In accordance with (*Chalov,* 1969), in alpha decay of ^{238}U atoms in an unbroken part of the mineral part of rocks the breaked (disordered) region appears which was formed by atom recoil ^{234}Th. Among disordered one may be the atom itself, that formed the area. The probability of this event, p: $0<p<1$. Then there is a transition ^{234}Th β (T 24.1 d) → ^{234}Pa β (T 1.18 min) → ^{234}U (T - a half-life).

Excess of ^{234}U I can simply determine by measuring of $^{234}U/^{238}U$ activity ratio - γ:

$$\gamma = \frac{\lambda_2 \cdot N_2}{\lambda_1 \cdot N_1}, \tag{1}$$

where N_1 and N_2 - the number of atoms, respectively ^{238}U and ^{234}U, $\lambda_1 = 1.551 \cdot 10^{-10}$ years^{-1} and $\lambda_2 = 2.835 \cdot 10^{-6}$ years^{-1} - decay constant of these isotopes.

If in undisturbed areas of mineral rocks there is radioactive equilibrium:

$$\gamma = \frac{\lambda_2 \cdot N_{2rock}}{\lambda_1 \cdot N_{1rock}} = \gamma_{rock} = 1, \tag{2}$$

and for the disordered region (dr) there is the probability p that the recoil atom will remain in it and

$$\gamma_{dr} = \frac{\lambda_2 \cdot (N_{2rock} + p)}{\lambda_1 \cdot N_{1rock}} = \gamma_{rock} + \frac{\lambda_2 \cdot p}{\lambda_1 \cdot N_{1rock}}. \tag{3}$$

Thus, $^{234}U/^{238}U$ activity ratio in the disordered region, compared with the undisturbed area, will be more on the value $\frac{\lambda_2 \cdot p}{\lambda_1 \cdot N_{1rocks}}$, which corresponds to the excess alpha-activity of ^{234}U.

Uranium atoms located in the disordered region, primarily go into the water, resulting in the latest by violation of the radioactive equilibrium.

Chemical dissolution of equilibrium uranium, contained in areas undisturbed by the radioactive decay of rocks, reduces the value of excess alpha-activity of ^{234}U in the water.

For long- (geological) term periods of time can be adopted an even transfer of uranium from rocks into the water and we assume that in (3) N_{1rock} - is the total amount of passing into water and remain there ^{238}U atoms for today, which accounts for one ^{238}U decay, $N_{1rock} = N_{1rock\,\Sigma}$:

$$N_{1rock\sum} = N_{dr} + N_{cd}, \tag{4}$$

where N_{dr} - number of ^{238}U atoms, passing into water in the radioactive decay from the disordered region in non-equilibrium relation to ^{234}U;

N_{cd} .– number of ^{238}U atoms, passing into water during the chemical dissolution of rocks from outside the disordered region in equilibrium ratio to ^{234}U.

Accordingly, γ_{dr} – it is the initial activity ratio of passing into water isotopes and remain there for today, $\gamma_{dr} = \gamma_o$, and p - probability of passing into water in the decay of ^{238}U ^{234}U atom. Then the expression (3) can be rewritten as follows:

$$\gamma_o = \gamma_{rock} + \frac{\lambda_2 \cdot p}{\lambda_1 \cdot N_{1rock\sum}}. \tag{5}$$

Knowing the number of ^{238}U atoms, located in the water, contained in a unit volume of rock V_{unit} currently $N_{1unit..water}$, and the activity of ^{238}U of the mineral part in the same volume $A_{1unit.rocks}$, we can roughly estimate the average time of water-rock interaction t number of ^{238}U atoms - $N_{1rock\,\Sigma}$ goes into the water and remain there today:

$$N_{1rocks\sum} \approx \frac{N_{1unit.water}}{A_{1unit.rocks} \cdot t} \tag{6}$$

$$N_{1unit.water} = \frac{n \cdot V_{unit} \cdot c_{1water}}{m_{1atom}} \tag{7}$$

$$A_{1unit.rocks} = (1-n) \cdot V_{unit} \cdot \rho_m \cdot A_{1rocks}, \tag{8}$$

where n - porosity; c_{1water} - concentration of uranium in water; m_{1atom} - mass of one ^{238}U atom, which is equal to $3.952 \cdot 10^{-16}$ µg; ρ_m - mineral density of the rock.

From (5-8) follows:

$$t = \frac{\lambda_1 \cdot n \cdot c_{1water} \cdot (\gamma_0 - \gamma_{rock})}{\lambda_2 \cdot (1-n) \cdot A_{1rock} \cdot \rho_m \cdot m_{1atom} \cdot p}, \quad (9)$$

If the groundwater is constantly supplied by uranium with non-equilibrium isotope ratios γ_o, change in number of ^{238}U and ^{234}U atoms with the time is determined by the differential equations of the form:

$$\frac{dN_1}{dt} = V(t) - \lambda_1 N_1,$$
$$\frac{dN_2}{dt} = K(t) - \lambda_2 N_2 + \lambda_1 N_1, \quad (10)$$

where *V(t)* and *K(t)* - velocity introduces ^{238}U and ^{234}U in water.

The ratio of entering in water uranium isotopes can be assumed as constant, that is $\frac{K(t)}{V(t)} = const$. Then equation (10) can be written as:

$$\frac{dN_1}{dt} = \varphi(t) - \lambda_1 N_1,$$
$$\frac{dN_2}{dt} = b\varphi(t) - \lambda_2 N_2 + \lambda_1 N_1, \quad (11)$$

where *b* - a constant, $\varphi(t)$ - a function, describing the addition in system of U-element instead of its isotopes.

For a homogeneous composition of water-bearing rocks of aquifer one can take $\varphi(t) = const$, and then the solution of equations (11) will have the form (*Cherdyntsev,* 1955):

$$\frac{\gamma_{twater} - 1}{\gamma_o - 1} = \frac{1 - e^{-\lambda_2 t}}{\lambda_2 t}, \quad (12)$$

where $\gamma_{t.water}$ - $^{234}U/^{238}U$ activity ratio at time *t* from the beginning of receipt of uranium in water from water-bearing rocks.

If $\gamma_{rock} = 1$, by solving system of equations (9 and 12) we find the dependence:

$$t = \frac{\ln(k^{-1})}{\lambda_2}, \quad (13)$$

where (14)

The obtained dependences are valid for a model in which the water flowing into the aquifer, does not contain the non-equilibrium uranium. If it contains the non-equilibrium uranium, that is its excess (with subsequent decay of ^{234}U) should be subtracted from the value of γ_{twater}.

Besides, for the use of these equations it is necessary that in the aquifer were oxidative conditions for uranium: Eh> 100 mV, the presence in sufficient amounts dissolved carbonate (*Garrels and Christ,* 1965; *Lisitsin*, 1997).

The key point in estimating the residence time of water in the aquifer is the choice of the calculated value of *p*. It is anticipated that in aquifer, composed of homogeneous rocks, $p = const$.

Assessment of the value *p* is given for the aquifer sands, sandstones and siltstones (for example, the North Dvina basin).

THE RESULTS OF DETERMINING THE RESIDENCE TIME OF GROUNDWATER IN AQUIFERS AND THEIR DISCUSSION

Initial data from the points of groundwater sampling, presented in Figure 1 and 3 are shown in table 4.

Table 4. Characteristics of selected types of groundwaters

Tipe of water	Sample No.	Type of water source	Index of aquifer	Average Uranium content, c_{Iwater}, µg/L	Average $^{234}U/^{238}U$ acnivity ratio, γ_{twater}	TDS, ppm	Average Eh, mV
1	1 - 9	Borehole	Vpd	8.3	5.6	5-10	+238
2	10 - 13	Borehole	Vpd	2.47	3	1-5	+201
3	16 - 23	Spring	QIIIvd	0.35	3.3	1.6-4.9	+188
4	14 - 15	Borehole	Vpd	0.61	2.4	0.1-0.3	+305
5	24 - 34	Well	QIIIvd	0.36	1.7	0.2-1.3	+230
6	35 - 39	Borehole	Vpd, Vmz	0.13	8.4	10-27	-6
7	40 - 51	Borehole	Vpd	0.26	3.2	0.1-0.3	+216
8	52 - 55	Borehole	Vpd	0.51	11.36	0.3	+66
9	56 - 79	Borehole	Vpd	1.81	4.1	0.3-5	+213

According to the content of uranium, the $^{234}U/^{238}U$ activity ratio to values of Eh and types of water sources, 9 types of underground waters are identified.

7-9 types of groundwater tests were sampled from the drainage and observation wells in the Lomonosov diamond deposit, located on the right side of the North Dvina basin. The rest was sampled from holes, springs and wells, confined mainly to the central parts of the basin. 9th water type is similar in composition to water of the 2nd type and the 7th type - to water of the 4th type.

Values of Eh, pH, TDS, the U content in water (c_{Iwater}) and the $^{234}U/^{238}U$ activity ratio (γ_{twater}) in the groundwater aquifer siltstones and sandstones of Vendian are taken on the basis of data, presented in previous publications (*Malov et al*, 2008, 2009, 2010a). In addition to the

Vendian aquifer system was also carried out calculations for a number of springs and wells, draining Pleistocene aquifers.

Calculated density of the mineral part (ρ_m) and porosity (n) of siltstones are defined as average by 52 values: $\rho_m = 2.75$ g/cm^3; n = 0.23 (Figure 5).

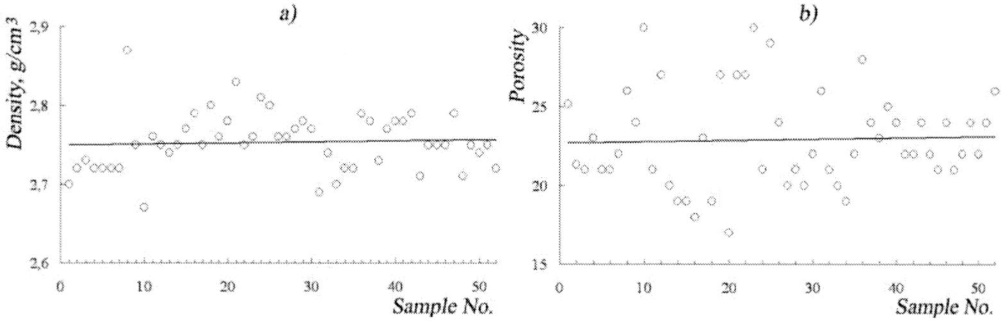

Figure 5. Density (a) and porosity (b) siltstones of the North Dvina basin.

The calculated values for content of U (c_{1rock}) and the $^{234}U/^{238}U$ activity ratio (γ_{rock}) in rocks are defined as average of 22 values: $c_{1rock} = 3$ ppm; $\gamma_{rock} = 1$ (Figure 6).

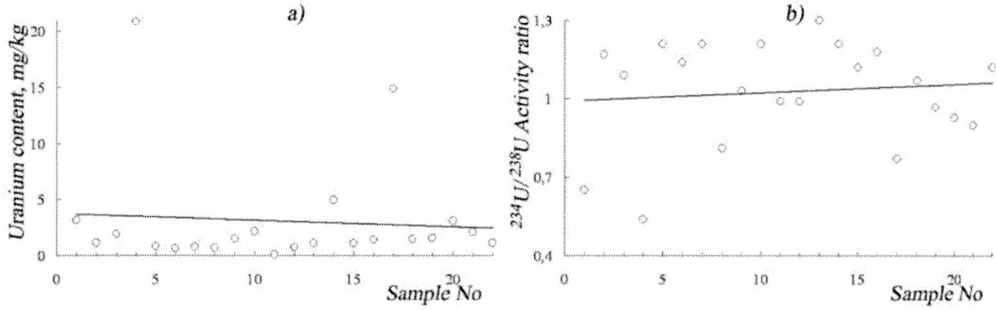

Figure 6. Uranium content (c_{1rock}) (a) and the $^{234}U/^{238}U$ activity ratio (γ_{rock}) (b) in siltstones of the North Dvina basin.

Estimates of residence time of groundwater in the aquifer were carried out on these types of waters.

1 type. Boreholes with high concentrations of uranium, located at the front of "the zones of stratal oxidation" (Samples 1-9 in Table 4)

Under the front of "zones of stratal oxidation» (*Lisitsin,* 1997) is understood the boundary between regions with oxidizing and reducing conditions for U^{6+} (Figure 1)..

These samples were selected from the boreholes, revealed the groundwater with TDS of about 5 - 10 ppt. They are characterized by the values of Eh +142 - +300 mV, indicating the presence of oxidizing conditions for uranium in the aquifer, that is, uranium must be in solution. U concentration in this type of water is maximum: average 8.3 µg/L.

Desalination of saline groundwater in the Permian period, confined to the upper strata of sandy-clayey sediments of the Vendian, which began in Mesozoic in connection with the raising of this territory and emergence of the continental regime, could persist until the end of melting of the Moscow glacier (130 - 140 thousand years ago) when apparently there was a final stage of water exchange in this part of the strata. Then the transgression of the Mikulino Sea isolated groundwater under the most submerged parts of the North Dvina basin (*Malov*, 2004). Extraction of saline sedimentary pore waters from marine clays of the Mikulian interglacial by the last glacier led to marginalization of freshened water downward and sideways. After the Valdai glacier had melted and the Northern Dvina River erosion valley had formed, these waters were returned to its original position, because melted glacial water flowed from the depression sides toward the river valley. As a result of prolonged interaction with the water-bearing rocks, they have acquired its present chemical composition.

Given the existence of maximum concentrations of uranium, the age of water in the aquifer of the Padun Formation of Vendian sediments in the first approximation can be estimated as the maximum, i.e., corresponding to the beginning of the transgression of the Mikulino Sea ~ 130 thousand years. This age, in the obtained for determine time the residence of water in the aquifer sediments of the Vendian equations (13, 14) corresponds the value of $p = 0.45\%$.

Using this value of p, we estimate the residence time of groundwater in the aquifer for the next group of samples.

2 type. Boreholes with average concentrations of uranium, located on the slopes of the *North Dvina* depression (Samples 10-13 in Table 4)

These samples are selected from the boreholes, revealed the groundwater with salinity of ~ 1 - 5 ppt. They are characterized by values of Eh +145 - +283 mV, which also indicates the presence of oxidizing conditions for uranium in the aquifer.

Calculating the age of these waters by (13, 14) yields 14300 years.

That is, this type of waters could be formed in the process of melting the last glacier. According to studies of Quaternary deposits of the Arkhangelsk region, made by the international team of geologists (*Larsen et al.,* 1999, *The White Sea system,* 2010) the North Dvina basin were covered by the last glacier about 20 - 17 thousand years ago. After that, the glacier retreated to the west, and in this place had raised a proglacial water pool. It has been a source of fresh groundwater, which subsequently increased its mineralization up to 1 - 5 ppt and enriched its chemical composition due to interaction with the ion-salt complex of marine sediments of Mikulian interglacial (Na^+, Cl^-), hydrolysis of aluminosilicate (Na^+, HCO_3^-), dissolution of carbonates (Ca^{2+}, Mg^{2+}) and gypsum (SO_4^{2-}).

We estimate the possible range (l) of the water transfer from the recharge area on the right bank of the Northern Dvina River toward to its discharge zone in the river valley by convective transport over time t = 14.3 thousand years:

$$l = \frac{k \cdot I \cdot t}{n_0}, \tag{15}$$

where k - filtration coefficient, m/day, I - gradient of flow, n_0 - active porosity.

The movement of water occurs on the lower half of the non-uniform vertical thickness of the Padun Formation sediment (*Malov et al.,* 2010b). When the calculated values of k = 1.12

m/day, $I = 1.25 \cdot 10^{-3}$ and $n_0 = 0.2$ (*Malov,* 2003), fresh water should pass over 14.3 thousand years in the way of 37 km that conforms to the situation shown in Figure 2.

In addition, we can estimate the reliability of the assumed value of $p = 0.45\%$ for the corresponding for it value of ^{234}Th-recoil range of L (*Andrews et al.,* 1982).

Estimate amount of mineral component of V_{out} in volume of rock $V_{unit} = 1$ cm^3, from which originally recoil atoms may go in the water:

$$V_{out} = (1 - n_0) \cdot V_{unit} \cdot p. \tag{16}$$

The estimated thickness of the exit b:

$$b = \frac{V_{out}}{S_{unit}}, \tag{17}$$

where S_{unit} - the area of pores in volume of rock $V_{unit} = 1$ cm^3 for siltstone with porosity of 0.23 can be evaluated in 393 cm^2.

$$L = 4.255b \text{ ((Andrews et al., 1982)).} \tag{18}$$

For $p = 0.45\%$, $n_0 = 0.23$, we get $L = 1.12 \cdot 10^{-5}$ cm

It should be noted that for granites in (*Andreas et al., 1982*) assumed value of $L = 3 \cdot 10^{-6}$ cm. In the paper (*Chalov and Kucherenko,* 1998) based on a computer modeling, the values of L for crystals of two modifications of zircon and cinnabar is on an order of $0.25 \cdot 10^{-6}$ – $0.4 \cdot 10^{-6}$ cm. For sediment this is probably of a higher value. In any case, for $L = 10^{-6}$ cm, i.e., $p \approx 0.045\%$, age of groundwater of the North Dvina basin will rise to 1.3 million years, which is not consistent with the geological dating.

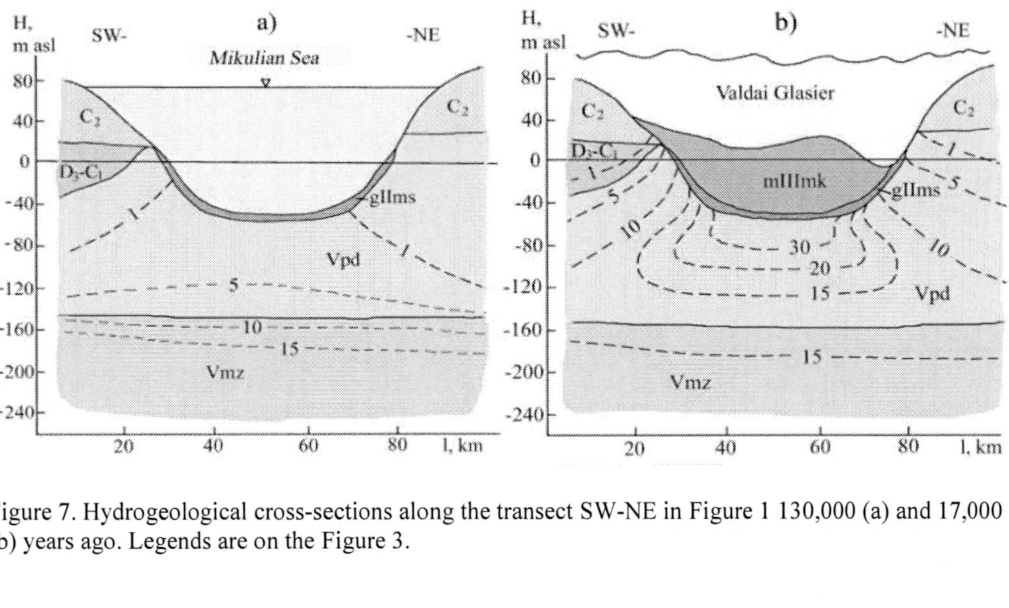

Figure 7. Hydrogeological cross-sections along the transect SW-NE in Figure 1 130,000 (a) and 17,000 (b) years ago. Legends are on the Figure 3.

Thus, the value of probability of passing ^{234}U atom into water under the decay of ^{238}U - p = 0.45%, which we adopted for the calculations corresponds of geological datings, hydrodynamic and balance calculations.

Figure 7 shows the approximate paleo-hydrogeological reconstruction that took place in the estimated time of 130,000 and 17,000 years ago.

3 type. Sources of mineral water (Samples 16-23 in Table 4)

4 type. Boreholes with low concentrations of uranium, located on the sides of the Northern Dvina depression (Samples 14-15 in Table 4)

These samples were selected from mineral springs with TDS of 1.6-4.9 ppt and wells, penetrated underground water with TDS of up to 0.1-0.3 ppt. They are characterized by values of Eh 130 - 340 mV.

Calculating of age of these waters by (13, 14) gives 2300 - 2400 years.

This water has already been formed in the late Holocene due to infiltration of rainfall.

5 type. Wells with fresh water (Samples 24-34 in Table 4).

They are characterized by the youngest waters: 700 years.

6 type. Boreholes with minimal concentrations of uranium, located under the deepest part of the Northern Dvina depression, near its thalweg, are in a region with reducing to U^{6+} conditions (Samples 35-39 in Table 4).

This are the most saline water (10 - 29 ppt). Samples 35, 36, 38 are selected from the boreholes, uncovered the underground water, which were formed during the extraction of saline sedimentary pore waters from marine clays of the Mikulian interglacial by the last glacier, that is, their formation began ~ 130,000 years ago. Samples 37, 39 are confined to the sediments of the Mezen Formation, lying substantially below a zone of penetration of squeezed from clay of the Mikulian interglacial pore solutions (Figure 3, 7b). Hence, it is the most ancient water from tested, their age is more than 130 thousand years. At the same time calculations showed that the age of the groundwater of a sample is 2700 years. This is due to reducing atmosphere, typical for the location of the intervals of tested boreholes. Typical values of Eh 64 - -110 mV, an average of -6 mV, that indicate the presence of reducing conditions for uranium in the aquifer, that is, almost all uranium is transferred to the solid phase. As a result, underground water of this type is characterized by minimal concentrations of uranium to the Vendian aquifers, an average of 0.13 μg/L at the maximum value of activity ratio of even isotopes of uranium: 8.4

Consequently, for aquifers with reducing to U^{6+} conditions: Eh <100 mV, the lack of sufficient amounts of dissolved carbonates, the proposed method of estimating the duration of contact of groundwater with rocks not suitable. Age of groundwater in these conditions can be assessed only by a change in $^{234/238}$ activity ratio on the dependence (12), if we know the initial activity ratio of isotopes ^{234}U and ^{238}U, entering in the water only by the radioactive decay of ^{238}U - γ_o.

7 type. Boreholes with low concentrations of uranium, located on the Lomonosov diamond deposit (Samples 40-51 in Table 4).

These samples were collected in June and July (flood period) from wells in which the uranium content in water depends on seasons: during the flood they are minimal, in the period of low water they increase significantly. This is due to the receipt of surface waters in the deeper horizons in the period of snow melting and diluting it with the more ancient waters. By chemical dissolution of uranium aggressive meteoric waters, the ratio of activities is relatively small. For waters of this type is characterized the minimal activity ^{238}U (<0.01 Bq/L), respectively, the minimum concentration of uranium (average 0.26 μg/L), with an average $^{234}U/^{238}U$ activity ratio 3.2. The average TDS of the mixture of water is 280 ppm, average residence time in the aquifer (age) – 1500 years.

8 type. Boreholes with low concentrations of uranium and high $^{234}U/^{238}U$ activity ratio, located on the Lomonosov diamond deposit (Samples 52-55 in Table 4).

For four samples is characterized the same mineralization - 300 ppm - and low uranium content (average 0.51 μg/L). However, the $^{234}U/^{238}U$ activity ratio is very high: 9 - 14. Waters of this type were taken from the deepest parts of the aquifer complex siltstones and sandstones of Padun formation, where meteoric waters have long periods of time without access. Here, chemical dissolution of uranium is almost absent, and its entrance into the water is determined only by radiative processes. Average time of presence in the aquifer waters of this type is 15,400 years, that is the last glacier meltwater, which existed in this area.

9 type. Boreholes with average concentrations of uranium, located on the Lomonosov diamond deposit (Samples 56-69 in Table 4)

The waters of this type are characterized by slightly elevated salinity (on average 1300 ppm) and maximum concentrations of uranium (on average 1.86 μg/L), with an average $^{234}U/^{238}U$ activity ratio 4.1. These samples were collected from wells located in most cases directly by a relatively small paleovalley Zolotitsa river, that is, conditions for the formation of alpha-activity of these waters are similar to conditions which have taken place in the North Dvina Basin (see above: "2 type"). The age of this water is on an average of 16,300 years. They also formed from melted glacier water, but having a higher mobility compared to 8th type of water, involved in the chemical dissolution of uranium

SOME FEATURES OF URANIUM BEHAVIOR IN GROUNDWATER, OBTTAINED BY USING WITH PROPOSED METHOD

From the expression (9) is, that uranium content in groundwater depends on the residence time of water in the aquifer (for uranium in oxidizing conditions):

$$c_{1water} = \frac{\lambda_2 \cdot (1-n) \cdot A_{1rock} \cdot \rho_m \cdot m_{1atom} \cdot t \cdot p}{\lambda_1 \cdot n \cdot (\gamma_0 - \gamma_{rock})}, \qquad (19)$$

where γ_o - initial activity ratio of isotopes ^{234}U and ^{238}U, entering in water and remaining in it today, from (12):

$$\gamma_o = \frac{(\gamma_{twater} - 1)\cdot \lambda_2 t}{1 - e^{-\lambda_2 t}} + 1. \qquad (20)$$

When $\rho_m = 2.75$ g/cm^3; $n = 0.23$ (Figure 5), $c_{1rock} = 3$ ppm; $\gamma_{rock} = 1$ (Figure 6):

$$c_{1water} = \frac{0.3496\cdot t}{(\gamma_0 - 1)}, \text{ where } t \text{ - a thousand years.}$$

Figure 8 shows the general trends in the isotopic composition of groundwater in sandy aquifers in the course of geological time.

Figure 8 shows that the uranium-bearing rocks of high solubility (γ_o = 1.1) with the Clarke concentrations of uranium to the level of 3 ppm, even at the time of contact of water with the rock of 75,000 years, it will exceed the so-called "intervention level" of ^{238}U, amounting to radiation safety standards, adopted in Russia, 3.1 Bq/L or 0.25 mg/L. At low solubility of the same rocks (γ_o = 8), when the supply of uranium in water is predominantly due to radiation effects, this level is almost never exceeded. But it is clear that in this case may exceed the intervention level for ^{234}U (2.9 Bq/L) at the time of contact of water with the rock about 600 thousand years. Activity ratio of uranium isotopes (Figure 8b) falls naturally within the geological time.

For the specific conditions of the North Dvina basin we shall try to estimate the maximum γo, based in table 4 values γ_{twater} for wells, located in a region with reducing to U^{6+} conditions: on average γ_{twater}= 8.4. Assuming that such activity ratio of uranium isotopes formed here only due to radiative processes, without the participation of the chemical dissolution of uranium, in a time ≈ 130,000 years, from (20) is that γ_o = 9.8.

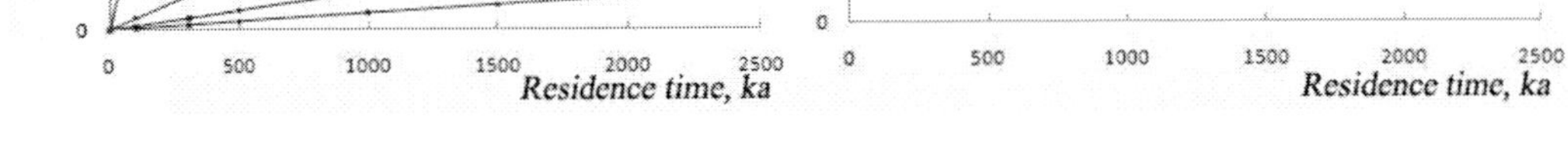

Figure 8. Changing the ^{238}U content in water at different initial activity ratio of isotopes ^{234}U and ^{238}U, entering the water and remaining in it today γ_o (γ_o = 1.1, 2, 4 and 8) (a) and $^{234}U/^{238}U$ activity ratio γ_{twater} (b) depending on the residence time of water in the aquifer t.

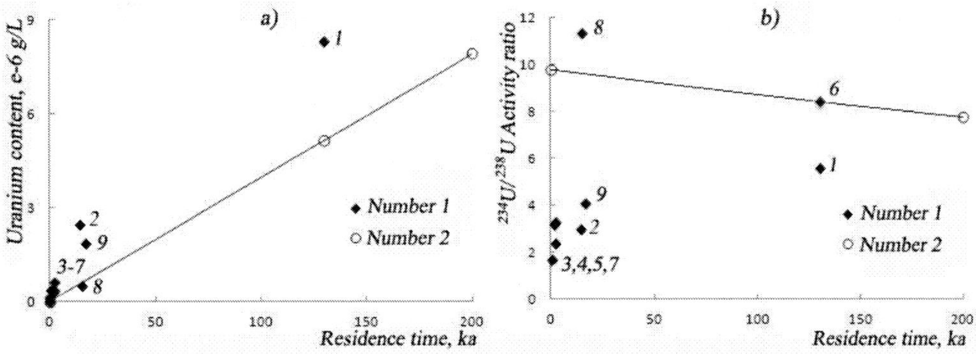

Figure 9. Changing of ^{238}U content in the water (a) and $^{234}U/^{238}U$ activity ratio (b) depending on residence time of water *t* in aquifers of the North Dvina basin. Row 1 - on the basis of calculations by formulas 13, 14 at values c_{1water}, γ_{twater} on the table 4, row 2 - on the basis of calculations of these parameters only through radiative processes in equations 19, 20 at $\gamma_o = 9.8$. The numbers 1 - 9 designated types of water, according to the table 4.

Figure 9 shows the real nature changes in concentrations of uranium and its isotope activity ratio in water in the course of geological time, established on the basis, described in the previous section and calculations by formulas 13, 14 at values c_{1water}, γ_{twater} in the table 4 in comparison with the nature of these parameters only through radiative processes (at γ_o = 9.8), without the participation of the chemical dissolution and sorption of uranium (number 2).

Figure 9a shows that almost all samples are above the line, which characterizes the supply of uranium in the water only by radiative processes, i.e. the content of uranium in them is determined largely by its chemical dissolution. The most intensive is the dissolution, compared with radiation leaching, exerted on the sides of the depression, where the development of the youngest water of 3-5 and 7 types. This is mainly fresh water; the most non-equilibrium is with surrounding rocks and, therefore, the most aggressive towards them. Aggressiveness of water decreases and processes of chemical dissolution becomes weaker compared with radiation leaching as advance groundwater in the direction of the discharge zone and their mineralization (water 2, 9, and 1 type). Later a series of trend lines should merge with trend lines of number 2.

Water of the 8th type is now close to the lines of this trend, and the 6th type is at it.

Use of Groundwater with Increased Radioactivity

In Russia, the degree of danger of radioactive water is measured by "Norms of Radiation Safety", developed in accordance with the recommendations of the World Health Organization (WHO). In order to preliminarily assess whether the drinking and mineral waters meet the requirements of radiation safety, environmentalists use the measured values of the specific total alpha- and beta-activity of water samples (A_α and A_β, respectively) and the absolute errors of their measurements (U_α and U_β).

For the water from the underground sources of water supply, simultaneously with the above measurements, it is necessary to determine the Rn content of this water. The results of measurements are the specific activity of Rn (A_{Rn}) and the absolute uncertainty of Rn measurement (U_{Rn}).

Water is considered to meet the requirements of radiation safety, if the following conditions are simultaneously fulfilled:

$$A_\alpha + U_\alpha <= 0.2 \text{ Bq/L}; \quad (21)$$

$$A_\beta + U_\beta <= 1.0 \text{ Bq/L}; \quad (22)$$

$$A_{Rn} + U_{Rn} <= 60 \text{ Bq/L}. \quad (23)$$

However, if the value of the total A_α or A_β, is exceeded the following additional operations are allowed. The concentrations of individual radionuclides in water are determined and condition (24) is verified. According to the latter, the sum of the ratios of the specific activities of radionuclides to the respective intervention level should not excess 1:

$$SUM_i\,(A_i\,/IL_i\,) <= 1, \quad (24)$$

where Ai is the specific activity of the i-th radionuclide in water, Bq/L; ILi is the respective intervention level for the i-th radio-nuclide, Bq/L.

If condition (24) is not fulfilled it is necessary to assess the radiation doses to population at the expense of drinking water consumption, using formula (25) obtained for the standard water consumption of 730 L/year:

$$E = 7.3\cdot10^5(4.5\cdot10^{-8}A^{238}U+4.9\cdot10^{-8}A^{234}U+6.9\cdot10^{-7}A^{210}Pb+1.2\cdot10^{-6}A^{210}Po+ +2.8\cdot10^{-7}A^{226}Ra+6.9\cdot10^{-7}A^{228}Ra+7.2\cdot10^{-8}A^{228}Th+2.1\cdot10^{-7}A^{230}Th+2.3\cdot10^{-7}A^{232}Th), \text{ mSv/yr.} \quad (25)$$

At a dose >0.1 mSv/yr, the feasibility of mining and respective protection measures should be considered.

Let us roughly asses the radiation safety of the "Bobrovskaya" mineral water occurring in an area within the North Dvina depression, using the above method. TDS of this water is about 5 ppt; the activity of individual radio-nuclides in water is:

^{238}U – 0.1 Bq/L; ^{226}Ra – 0.05 Bq/L;
^{234}U – 0.7 Bq/L; ^{232}Th – 0.04 Bq/L;
^{210}Pb – 0.14 Bq/L; ^{210}Po – 0.01 Bq/L.

At such activities, condition (21) is not fulfilled, because ^{238}U, ^{234}U, ^{210}Po, ^{226}Ra, and ^{232}Th are alpha-active. The total A_α of this water is 0.1 + 0.7 + 0.01 + 0.05 + 0.04 = 0.9 Bq/L.

Condition (24) is not fulfilled as well:

$$^{238}U/IL_{238U}+^{234}U/IL_{234U}+^{210}Pb/IL_{210Pb}+^{210}Po/IL_{210Po}+^{226}Ra/IL_{226Ra}+^{232}Th/IL_{232Th}= =0.1/3.1+0.7/2.9+0.14/0.2+0.01/0.12+0.05/0.5+0.04/0.6=1.24.$$

The calculated dose of radiation E at the expense of consumption of this water for drinking purposes at a norm of 730 L/yr is $E = 0.12$ mSv/yr. Consequently, this water can be consumed in the amount that is about 17% less than the norm, i.e., up to 610 L/yr.

At the North Dvina depression sides, the total activity of U isotopes in the fresh drinking water from the water-bearing complex of sandstones of the Padun suite of the Vend period sometimes exceeds 0.2 Bq/L, that is, condition (21) is not fulfilled. However, at the above activities of ^{210}Pb and ^{210}Po isotopes, condition (24) is fulfilled. The radiation dose at the expense of water consumption at a norm of 730 L/yr is $E < 0.1$ mSv/yr. Thus, this water can be used for drinking practically without limit.

At the same time, both fresh and mineral groundwater requires its radionuclide content to be regularly controlled: twice a year for mineral water and once a year for fresh water.

CONCLUSION

Uranium is a mobile element that is stable in aqueous solutions, containing dissolved carbonates and oxygen. Its main source is the rock. Therefore, the uranium contents in water are determined in most cases the area and duration of water-rock interaction and the concentration of uranium in the rock.

In the northern-west Russia, the main aquifers are confined to the sedimentary basin of the Mezen syneclise, filled predominantly by sandy-clayey sediments. These deposits were formed by the demolition of the products of destruction of the Baltic Shield. The acid and alkaline rocks in the latter are characterized by higher uranium content (3.2-10.3 ppm). Transfer of uranium in suspended and dissolved states from the Baltic Shield in the Mezen syneclise was carried out mainly by surface water and groundwater. It can be assumed that the distribution of it in marine sediments was relatively uniform.

During periods of continental breaks, the uplift of sedimentary strata at a considerable height above sea level has contributed to penetration of water of atmospheric origin to a depth of 650 m. This process was the most intense in the Mesozoic era and especially in the Pliocene, when that area was dissected with paleo-incisions down to 300 m. Within the raised blocks the redistribute of uranium by flowing water, saturated with CO_2, O_2, and organic acids. The oxidation and transition of U into solution were accompanied by its transfer from groundwater recharge areas into its discharge areas and the partial removal beyond the raised blocks.

When uranium is transferred groundwater under certain conditions it precipitated from solution, forming zones of secondary U enrichment, including the formation of deposits. The main processes that led to the deposition of uranium, is the hydrolysis, adsorption on natural sorbents and oxidative changes in environmental conditions for recovery.

According to our studies of groundwater in the Mezen syneclise, elevated (up to 2.6 Bq/L) values of the alpha-activity of U isotopes were found in groundwater of the sandstones of the Padun suite of the Vend and high (up to 20) values of $^{234}U/^{238}U$ activity ratio. Study of the regularities of their distribution was performed on the experimental test site - the North Dvina basin.

It is established that the lowest U activity prevails in the most mineralized water (15–27 ppt) which occurs under the deepest part of the Northern Dvina depression, near its thalweg.

This water features somewhat decreased values of pH and Eh and hydrocarbonate-ion concentrations and higher concentrations of Fe. In fresh water, the activity of U isotopes somewhat increases and exceeds the maximum allowable values of alpha-activity for drinking water. This water is also remarkable for the higher values of pH (up to 9), Eh (up to 300 mV), concentrations of HCO_3^-, O_2, CO_2, and Fe concentrations being rather low. The highest activity of U isotopes is recorded in zones of intrusion of fresh hydro-carbon water into saline water, that is, at the front of “zones of stratal oxidation”, where TDS is 5–10 ppt. In these zones, the Eh values are up to 300 mv and pH ~7; the concentrations of O_2 (0.5 ppm) и CO_2 (10 ppm), are minimal and those of Fe are low. An approximate inversely proportional relationship between Fe concentrations and U isotope activities can be supposed to exist.

For a more detailed analysis of the distribution of uranium isotopes in groundwater was determined age of these types of groundwater, that is, the groundwater residence time in the aquifer and the interaction with uranium-bearing rocks. To do this was used the generic parameter *p*: of ^{234}U atom transition into water in the decay of ^{238}U probability, obtained on the basis of geological dating, hydrodynamic calculations and paleo-hydrogeological reconstructions. It is anticipated that in aquifer, composed of homogeneous rocks and oxidizing conditions for uranium $p = const$.

As a result, found that the maximum age of 130,000 years, have water with TDS 5 - 10 ppt and maximum concentrations of uranium. They are selected from boreholes, located at the front of “zones of stratal oxidation”. These waters have been buried under a layer of clay deposits as a result of the transgression of the Mikulian Sea and then Valdai glaciation. Water with average concentrations of uranium, located on the slopes of the basin, has TDS 1-5 ppt and age of 14,000 – 16,000 years. This type of water formed during the melting of the last glacier. Fresh water and brackish water with low concentrations of uranium, located on sides of paleovalley are between the ages 2300 - 2400 years. These waters have been formed in the late Holocene due to infiltration of rainfall. For wells with fresh water is characterized the youngest water: 700 - 160 years.

The proposed method of estimating the duration of contact of groundwater with rocks is not suitable for aquifers with reducing to uranium conditions. Age of groundwater in these conditions can be assessed only by a change in $^{234}U/^{238}U$ activity ratio, if we know initial activity ratio.

Analysis of calculated dependences shows that the uranium-bearing rocks of high solubility ($\gamma_o = 1.1$) with the Clarke concentrations of uranium to the level of 3 mg/kg, even at the time of contact of water with the rock of 75,000 years, it will exceed the so-called "intervention level" of ^{238}U, amounting to radiation safety standards, adopted in Russia, 3.1 Bq/L or 0.25 mg/L. At low solubility of the same rocks ($\gamma_o = 8$), when the supply of uranium in water is predominantly due to radiation effects, this level is almost never exceeded. But it is clear that in this case may be exceed the intervention level for ^{234}U (2.9 Bq/L) at the time of contact of water with the rock about 600 thousand years. Activity ratio of uranium isotopes falls naturally within geological time.

Analysis of results obtained for the conditions of the North Dvina basin showed that in almost all samples of the uranium content is largely determined by its chemical dissolution. The most intensive is the dissolution, compared with radiation leaching, exerted on the sides of the depression, where the development of the youngest water. The most intensive is the dissolution, compared with radiation leaching, exerted on the sides of the depression, where the development of the youngest water. Aggressiveness of water decreases and processes of chemical dissolution becomes weaker compared with radiation leaching as advance groundwater in the direction of the discharge zone and their mineralization. In the end, with restoration of conditions for uranium, they almost completely stopped.

According to current regulations and standards for risk assessment of radioactive water is necessary to calculate the total radiation doses due to its consumption. It should not exceed 0.1 mSv/year. This requires the addition of uranium to determine the number of other isotopes of uranium and thorium series. Approximate estimation of the radiation safety of fresh and mineral waters of the North Dvina basin has shown that they can be used for drinking. At the same time requires constant monitoring of their content of radionuclides. For mineral water it must be 2 times a year. for fresh - 1 times.

ACKNOWLEDGEMENT

This work was financially supported by the Russian Fund for Basic Research (RFBR grant No. 10-05-00618a).

REFERENCES

Andrews J.N. and Kay R.L.F. The evolution of enhanced $^{234}U/^{238}U$ activity ratios for dissolved uranium and groundwater dating // In: *Short Paper of the Fourth International Conference, Geochronology, Cosmochronology, Isotope Geology*. U.S. Geological Survey Open-file Report. 78-701. 1978. P.11-13.

Andrews J.N., Giles I.S., Kay R.L.F., Lee D.J., Osmond J.K., Cowart J.B., Fritz P., Barker J.F. and Gale J. Radioelements, radiogenic helium and age relationships of groundwaters from the granites at Stripa, Sweden // *Geochimica et Cosmochimica Acta*. 1982. Vol. 46. № 9. P. 1533-1543.

Barr G. E., Lambert S. J. and Carter J. A. Uranium isotope disequilibrium in groundwaters of southeastern New Mexico and implications regarding age-dating of water // In: *Proceedings of the International Symposium on Isotope Hydrology*. STI/PUB/493. Vienna: I.A.E.A. 1979. Vol. 2. P. 645-660.

Chalov P.I. On the mechanism of formation of non-equilibrium relationships between the natural radioactive isotopes of uranium- and thorium-containing natural compounds // *Atomic Energy*. 1969. Vol. 27. No. 1. P. 26-32

Cherdyntsev V.V. On the isotopic composition of radioactive elements in natural materials in connection with the matters geochronology // In *Proceedings of the Third Session of the Commission to determine the absolute age of geological formations*. Moscow: USSR Academy of Sciences. 1955. P. 175-233.

Clark I.D. and Fritz P. *Environmental isotopes in hydrogeology*. Boca Raton: CRC Press. 1997. 328 p.

Ferronsky V.I. and Polyakov V.A. Isotopy gidrosfery Zemli (*Isotopy of the Earth hydrosphere*). Moscow: Scientific World. 2009. 632 p.

Fröhlich K. and Gellermann R. On the potential use of uranium isotopes for groundwater dating // *Chemical Geology* (Isotope Geoscience Section). 1987. Vol.65. № 1. P. 67–77.

Fröhlich K., Ivanovich M., Hendry M.J., Andrews J.N., Davis S.N., Drimmie R.J., Fabryka-Martin J., Florkowski T., Fritz P., Lehmann B., Loosli H.H. and Nolte E. Application of isotopic methods to dating of very old groundwaters: Milk River aquifer, Alberta, Canada // *Applied Geochemistry*. 1991. Vol. 6. № 4. P. 465–472.

Garrels R. and Christ C. *Solutions, Minerals, and Equilibria*. New York: Harper and Row. 1965. 368 p.

Germanov A.I., Panteleev V.M. and Shvets V.M. Geneticheskie svyazi organicheskogo veshchestva i mikrokomponentov podzemnykh vod (*Genetic Relationships between Organic Matter and Microcomponents of Groundwater*). Moscow: Nedra. 1975. 136 p.

Geyh M, Amore F.D., Darling G., Paces T., Pang Z., and Silar J. Groundwater Saturated and Unsaturated Zone // In: *Environmental Isotopes in the Hydrological Cycle – Principals and Application*. W. G. Mook editor. Vol. IV. Vienna: I. A. E. A. 2001. 117 p.

Kazemi G.A., Lehr J.H. and Perrochet P. *Groundwater Age*. Hoboken, NJ: Wiley Interscience. 2006. 325 p.

Kigoshi K. Alpha-recoil ^{234}Th: Dissolution into water and the ^{234}U/^{238}U disequilibrium in nature // *Science*. 1971. Vol. 173. P. 47-48.

Kronfeld J., Gradsztajn E., Müller H.W., Radin J., Yaniv A. and Zach R. Excess ^{234}U: An aging effect in confined waters // *Earth and Planetary Science Letters*. 1975. Vol.27. № 2. P. 342-345.

Larsen E., Lyså A., Demidov I., Funder S., Houmark-Nielsen M., Kjœr K.H. and Murray A.S. Age extent of the Scandinavian ice sheet in northwest Russia // *Boreas*. 1999. Vol.28. №1 P. 115–132

Lisitsin A.K. A Hydrogeochemical Model of an Infiltrative Ore-forming System. // Geokhimiya. 1997. No. 3. P. 294-304 [*Geochem. Int.* (Engl. Transl.). No. 3. P. 250–259].

Malov A.I. Podzemnye vody Yugo-Vostochnogo Belomor'ya: formirovanie, rol' v geologicheskikh protsessakh (*Groundwater of Southeastern White Sea Area: The Formation and the Role in Geological Processes*). Yekaterinburg: UD RAS. 2003. 234 p.

Malov A.I. Water-Rock Interaction in Vendian Sandy-Clayey Rocks of the Mezen Syneclise // *Lithologi and Mineral Resources*. 2004. Vol. 39. No. 4. P. 345–356.

Malov A.I., Kiselev G.P. and Rudik G.P. Uranium in the Groundwater of the Mezen Syneclise // *Doklady Earth Sciences*. 2008. Vol. 421A. No. 6. P. 965–968.

Malov A.I., Kiselev G.P., Rudik G.P. and Zykov S.B. Uranium isotopes in the Groundwater of the Vend of the Mezen Syneclise // *Water Resources*. 2009. Vol. 36. No. 6. P. 682–691.

Malov A.I., Kiselev G.P. and Zykov S.B. Features of the separation of uranium isotopes in the Groundwater of the Vend of the Mezen Syneclise. In: *Proceedings of the Scientific Conference*. Tomsk: SRTPU. 2010a.

Malov A.I, Titenko A.P. and Georgiev A.G. Geological factors of formation of industrial iodine waters in the interglacial marine sediments. In: *Proceedings of 8 meeting of the Urals lithological "Topical issues of lithology*." Ekaterinburg: IGG UD RAS. 2010b.

Metallogeniya Karelii (Metalogeny of Karelia) / Rybakov S.I. and Golubev A.I. editors. Petrozavodsk: Kar. SC RAS, 1999.

Molodkov A.N. and Bolikhovskaya N.S. Long-term palaeoenvironmental changes recorded in palynologically studied loess-palaeosol and ESR-dated marine deposits of Northern Eurasia: implication for sea-land correlation // *Quaternary International*. 2006. Vol.152-153. P. 37–47.

Rosholt J.N., Shields W.R. and Garner E.L Isotopic fractionation of uranium in sandstone. *Science*. 1963. Vol.139. №3551. P. 224-226.

Salminen R., Chekushin V., Tenhola M. et al. *Geochemical Atlas of the Eastern Barents Region*. Amsterdam: Elsevier. 2004. 548 p.

Sistema Belogo morja. Tom 1. Prirodnaja sreda vodosbora Belogo morja (The White Sea system. Vol. I. *Natural environment of the catchment area of the White Sea)* / Lisitsin A.P. editor. Moscow: Scientific World. 2010.

Smyslov A.A., Moiseenko U.I. and Chadovich T.Z. Teplovoi rezhim i radioaktivnost' Zemli (*Thermal Regime and Radioactivity of the Earth*). Leningrad: Nedra. 1979.

Solomon D.K., Poreda R.J., Cook P.G. and Hunt A. Site Characterization Using $^{3}H/^{3}He$ Groundwater Ages, Cape Cod, MA // *Ground Water*. 1995. Vol.33. № 6. P. 988-996.

Vodnye resursy Nechernozemnoi zony RSFSR (*Water Resources of the Nechernozem Zone of RSFSR*). Leningrad: Gidrometeoizdat. 1980. 214 p.

Zverev V.P. Podzemnye vody zemnoi kory i geologicheskie protsessy (*Groundwater of the Earth's Crust and Geological Processes*). Moscow: Scientific World. 2006. 256 p.

In: Uranium: Characteristics, Occurrence and Human Exposure ISBN: 978-1-62081-207-5
Editors: Alik Ya. Vasiliev and Mikhail Sidorov

Chapter 6

URANIUM: NATURAL OCCURRENCE, PHYSICAL AND CHEMICAL CHARACTERISTICS - ENVIRONMENTAL CONTAMINATION AND HUMAN EXPOSURE

Marcos Moisés Gonçalves[1], Marli Gerenutti[2]* and Marta Maria Duarte Carvalho Vila[2]

[1]Institute of Chemistry, University of Campinas, São Paulo, Brazil

[2]Post graduation programs in Pharmaceutical Sciences, University of Sorocaba, Sorocaba, São Paulo, Brasil

ABSTRACT

Uranium has both radioactive and chemical hazards and risk assessments must address both types. Its environmental distribution depends on natural occurrence and has been influenced by human activities such as energy production, coal and phosphate fertilizer exploration which may impact freshwater ecosystems. In terms of radiological exposition, focusing chemical toxicity, uranium that is ingested or inhaled in sufficient amounts can be harmful. Radiological hazards associated with uranium isotopes are usually less severe than those of plutonium. The industrial activities involved in the nuclear fuel cycle should be considered including the uranium ore mining, milling, refining, conversion, fuel production, nuclear power generation, and nuclear waste management that is subjected to licensing and control processes. According to the International Atomic Energy Agency, the waste generated by these activities require appropriate procedures for its treatment, before disposing them in the environment. The use of nuclear fuels based on uranium is closely related to energy production by nuclear power plants, design and production of atomic weapons. Uranium-238 and uranium-235 are the two most abundant isotopes with half-lives respectively with order of magnitude of couple of billion and hundreds of million years. Thus, the different uranium salts used in fuel cycle production represent a permanent risk of environmental exposure as nuclear accidents can spread them out of the nuclear facilities and, depending on the accident, these salts can be chemically dispersed as they dissolve in water sources. When the

* Correspondent author: E-mail: marli.gerenutti@prof.uniso.br.

uranium chemical species are subjected to extremely high temperatures of fire they can merge to form a residue with only a part of it oxidizing providing airborne contamination risk. The fissile nuclide uranium-235 has been used as nuclear fuel due to the fact that nuclear fission reactions take place when slowed thermal neutrons hit ^{235}U nucleus releasing two lighter nuclides, three free neutrons and a tremendous amount of energy. However, in terms of weight, the naturally occurring uranium isotopes has respectively 99.284% of ^{238}U, 0.711 % of ^{235}U and traces of ^{234}U requiring physicochemical separation process that is able to increase the content of ^{235}U. Natural uranium is the term referred to the refined uranium with the same isotopic ratio found in nature and despite the fact that natural uranium can be used to fuel both low- and high-power reactors, but the enrichment processes based on gaseous diffusion and gas centrifuge using uranium hexafluoride UF_6 are very common. Unfortunately, these enrichment processes give rise to large amounts of uranium with a lower content of the fissile isotope ^{235}U named depleted uranium. About 95 % of the depleted uranium produced to date is stored as UF_6, in steel cylinders in open air yards close to enrichment plants causing a permanent environment risk contamination. The long-term storage of depleted uranium as UF_6 represents safety risks and potential danger to health and environment because of its chemical instability. When exposed to moist air, UF_6 reacts with the water in the air to produce uranyl fluoride UO_2F_2 and hydrogen fluoride both of which are highly soluble and toxic. Storage cylinders must be regularly inspected for signs of corrosion and leaks. The estimated lifetime of the steel cylinders is measured in decades. There have been several accidents involving uranium hexafluoride in the United States and the government has been converting depleted uranium UF_6 to solid uranium oxides for disposal. Such disposal of the entire depleted UF_6 inventory could cost from tens to hundreds of thousands of U.S. dollars.

INTRODUCTION

Uranium is a notable element that is directly related with the discovery of nuclear fission because researchers used neutrons to bombard uranium nuclei. Nuclear fissions have become crucial for electricity production and elaboration of nuclear weapons since the process splits the uranium nuclei into lighter elements releasing even more neutrons that can feed a chain reaction and generates large amounts of energy. Thereafter there was great interest in exploiting its natural reserves all over the world and it's mainly currently used in the energy production. Uranium-based power plants produce approximately one tenth of the world's electricity. However, several difficulties must be overcome such as complicated management of wastes that keep their radioactivity for extended periods, human exposure at all stages of energy production and the impacts to the environment and living organisms caused by nuclear accidents.

Uranium Occurrence

Uranium is a naturally-occurring element, its redistribution into the environment has been influenced by several anthropogenic activities such as nuclear fuel cycle, military use of depleted uranium, coal and phosphate fertilizer use may impact freshwater ecosystems [1]. The average concentration of natural uranium in soil is about 2 ppm and in granite ranging from 2 to 20 ppm. Table 1 provides tabulated data by countries regarding quantities of uranium recoverable from ore, production and demand:

Table 1. Uranium resources, production and demand

	Uranium Resources (tU) In 2005 [UDEPO}	Uranium Resources (%) In 2005 [UDEPO]	Uranium Production (tU) In 2005	Uranium Production (%) In 2005	Nat. Uranium Demand for 2007 (tU)	Nat. Uranium Demand for 2007 (%)
Country	RAR	RAR %				
Australia	725,000	21.72	7,600	17.54	-	-
Belgium	-	-	-	-	1,065	1.47
Brazil	157,400	4.72	-	-	-	-
Canada	329,200	9.86	9,850	22.73	1,900	2.62
China	48,800	1.46	-	-	1,500	2.07
France	-	-	-	-	9,000	12.40
Germany	-	-	-	-	3,490	4.81
India	48,900	1.46	-	-	-	-
Japan	-	-	-	-	8,790	12.11
Kazakhstan	378,000	11.32	7245	16.72	-	-
Korea, Republic	-	-	-	-	3,420	4.71
Namibia	176,400	5.28	3,800	8.77	-	-
Niger	243,100	7.28	3,633	8.39	-	-
Russian Federartion	172,400	5.16	3,381	7.80	4,100	5.65
South Africa	284,400	8.52	750	1.73	-	-
UK	-	-	-	-	1,500	2.07
USA	339,000	10.15	2,000	4.62	22,890	31.53
Ukraine	-	-	-	-	2,480	3.42
Uzbekistan	72,400	2.17	2300	5.31	-	-
Others	363,300	10.9	2769	6.39	12,465	17.17
Total	3,338,300	100.00	43 328	100.00	72,600	100.03

Notes: Data collected from '2007 Red Book' [2]. Reasonably assured resources (RAR) include uranium that occurs in known mineral deposits of delineated size, grade and configuration that could be recovered within the given production cost ranges, with currently proven mining and processing technology. RAR express quantities of uranium recoverable from ore. Uranium occurrence worldwide may also be assessed by UDEPO database which contains information per country, per facility on technical and geological characteristics and general information of uranium deposits.

Uraninite, commonly in the form of pitchblende is the dominant uranium mineral in the monometallic deposit and the main mineral in the polymetallic deposit that includes several elements. Uranium deposits can be classified as fracture controlled and clay bounded, two sub-types of unconformity related deposits that reflect the stratigraphic and structural controls. Unconformity related deposits comprise massive pods, veins and/or disseminations of uraninite spatially associated with major unconformities that separate two rock masses or strata of different ages, indicating that sediment deposition was not continuous. Fracture controlled is dominantly basement hosted whereas clay bounded is a massive ore developed along and just above or immediately below the unconformity in the overlying cover sandstones.

Uranium can also be found in another kind of deposit named phosphorite. Uraniferous phosphorite deposits consist of syn-sedimentary, stratiform, disseminated uranium in marine phosphate-rich rocks or phosphorite deposits formed in continental shelf environments. Phosphorite deposits constitute large uranium resources, but normally at a very low grade (25–150 ppm) [3]. The main subtypes of phosphorite deposits are bedded deposits and nodular deposits.

Bedded deposits are layerlike deposits of replacement origin, usually they are found parallel with the stratification of sedimentary rocks and of contemporaneous origin. Bedded phosphorites are deposits in distal continental shelf environments where phosphorus raised to the surface waters. This upwelling P-enriched seawater provided a renewable source of P and where upwelling U-enriched water flooded the phosphorite accumulations. These environments were characterized by very slow rates of deposition. In Idaho, USA, in the Phosphoria Formation there is a representative bedded phosphorite deposit [3].

Nodular phosphorite are deposits nearer shore in shallower water than their bedded counterparts. Land-pebble phosphate mineralization is composed of U enriched apatite within nodular phosphorite beds interbedded with sand, clay fine to medium-grained and carbonate beds. The marine transgressions leads to a variation of nodular phosphorite. Nodular phosphorite deposits typically average 10 to 20 % P_2O_5 and 20 to 80 ppm U.

Land pebble phosphorites in the Pliocene Bone Valley Formation in Florida contain fluor-carbonate apatite nodules and grains that are enriched up to 30 % P_2O_5 and 500 ppm U [4].

Nuclear Fission - A Discovery Related to Uranium

The discovery of nuclear fission took place precisely with uranium when Otto Hahn and colleagues in 1938 have preceded the research initiated by Enrico Fermi in 1934 when they bombarded uranium with neutrons. Before the discovery of nuclear fission by Otto Hahn and colleagues in 1938, more than a century had passed since the discovery of uranium by Martin Heinrich Klaproth in 1789 and this rare element had had a quiet existence and it hasn't called attention. At that time, uranium occupied the highest place in the table of the elements. Uranium is such an important element in terms of its remarkable radiation phenomena discovered by Henri Becquerel in the spring of 1896 which were at that time grouped together under the name of radioactivity. The observations of Becquerel on uranium radioactivity acquired a great importance when the Curies succeeded in separating the active substances, polonium and radium, from uranium minerals. In his Nobel lecture held in 1946 entitled

“From the natural transmutations of uranium to its artificial fission” [5], Otto Hahn drew attention to fact that uranium had the highest place of the then known chemical elements justified the fact that it has important properties by which it was distinguished from all other elements. According to him, it was especially Fermi who first realized the great importance of neutrons for the production of nuclear reactions, he and his co-workers irradiated the elements of the periodic system with neutrons.

Especially when working with slow neutrons, the course of the reaction was such that these neutrons were captured by the nucleus. They found that in many experiments an unstable isotope of the irradiated element was produced possessing higher atomic number and emitting β-rays giving rise to a stable atom occurring among the usual chemical elements, as follows (Figure 1):

They discovered that when the heaviest natural element, uranium, is bombarded by neutrons, several radioactive elements similar to uranium are formed. Fermi inferred the production of so-called "transuraniums," and then elements not known naturally, i.e. higher than 92, would become known. But, unfortunately Fermi’s findings were not accepted because the possibility of the radioactive species detected with the greatest certainty, so-called 13-minute element, be an isotope of element 91, i.e. protactinium, could not be rejected. As the chemical properties of protactinium were familiar to then, Otto Hahn and Lise Meitner decided to repeat Fermi’s experiments in order to test 13-minute element and they were able to prove without doubt that the 13-minute element of Fermi was neither a protactinium isotope, nor a uranium, actinium, or thorium and thus Fermi’s assertion should be correct.

At that time the possibility of a breakdown of heavy atomic nuclei into various light ones was considered as completely excluded. Otto and Meitner in co-operation with F. Strassmann found that these phenomena were much more complicated than had originally been supposed. After many years of work, they obtained a great number of artificial active kinds of atoms, which appeared to be formed directly or indirectly by β-radiation from the supposed short-living uranium isotopes, and which therefore must represent the so-called transuraniums elements higher than uranium. The fact of the so-called 3.5 hour substance obtained by Curie and Savitch in 1937 and 1938 by irradiation of uranium with neutrons did not have its chemical properties readily determined intrigued Otto Hahn in such a way that he with Strassman have tried to get it. After they had done their experiments carefully and considering chemical properties involved, they concluded that only barium and radium could be considered. Thus, they mixed the pure natural radium isotopes with their artificial isotopes, previously freed from their decay products, and fractionated chemically the mixture.

$$^{127}_{53}I + {}^{1}_{0}n \longrightarrow {}^{128}_{53}I^{*} + \gamma$$

$$^{128}_{53}I \xrightarrow[25\ min]{\beta} {}^{128}_{54}Xe$$

Figure 1. Iodine bombardment by a neutron and radioactive decay.

The result showed that natural radium isotopes could be separated from barium, but the artificial ones could not. This finding clearly demonstrated that radium should be excluded and the element might be barium. To reinforce such evidence, they checked the results considering the type of radioactive decay of radium and barium. If the artificial alkaline earth isotopes were radium, then the decay products produced directly through β-emission should consist of actinium, i.e., from the element 88 should be produced the element 89. If on the other hand it was barium-56, then lanthanum-57 should be formed. Firstly, they mixed an isolated actinium radionuclide, ^{228}Ac with the known primary decay products of artificial radium isotopes formed from the reaction of uranium with neutrons. Then they performed the chemical separation of actinium and lanthanum by the consecrated method of Madame Curie. During the fractionation of lanthanum oxalate with actinium-228, the latter accumulates in the final fractions. However the primary decay product remained with the lanthanum and it was established that the alkaline earth isotope, was an artificial active barium.

Once the idea of fission was too drastic at that time, they tested the most stable of active isotopes, now identified as barium, in a series of precipitation assays. This stable isotope was freed from active decay products and other impurities by recrystallization with inactive barium. One quarter of the amount was reserved and three quarters underwent the following cycle of six successive precipitation steps with many of the compounds crystallizing beautifully and resulting in barium chloride.

$$^{1}_{0}n + ^{235}_{92}U \longrightarrow ^{142}_{56}Ba + ^{91}_{36}Kr + 3\,^{1}_{0}n$$

Figure 2. Fission reaction of uranium – 235.

The reserved and barium chloride preparations were measured alternately using the same counter, with equal weights and equal thicknesses of layers. Within the limits of error, the initial activity and the increased activity due to the formation of active lantanum were the same for both preparations.

They concluded that, the crystallization of so many and such different salts had produced no separation of the active barium from the carrier. As the first communication on these tests in January 6, 1939 were in opposition to all the phenomena observed up to the nuclear physics, the tests mentioned had not been entirely completed and they had expressed the results cautiously. Finally they proceeded to a direct test mixing pure natural radium isotopes with their artificial "radium" isotopes, also previously freed from their decay products and fractionated the mixture in the same way as before.

They fractionated the mixture in the same way as before. The result was that the natural radium isotopes could be separated from barium, but the artificial ones could not. This evidence demonstrates that the fission product was not radium. If the artificial alkaline earth isotopes were radium, then the decay products produced directly through β-emission should consist of actinium: from the element 88 should be produced the element 89. If on the other hand it was barium, then lanthanum should be formed: from element 56 the next higher element 57. Lanthanum could have been produced only from barium and not from radium.

Immediately after this first publication on the production of barium from uranium, there appeared as a first communication an article by Lise Meitner and O. R. Frisch in which the possibility of a breakdown of heavy atomic nuclei into two lighter ones, with total charges equal to that of the original nucleus, was explained with the aid of Bohr's model of the original nucleus.

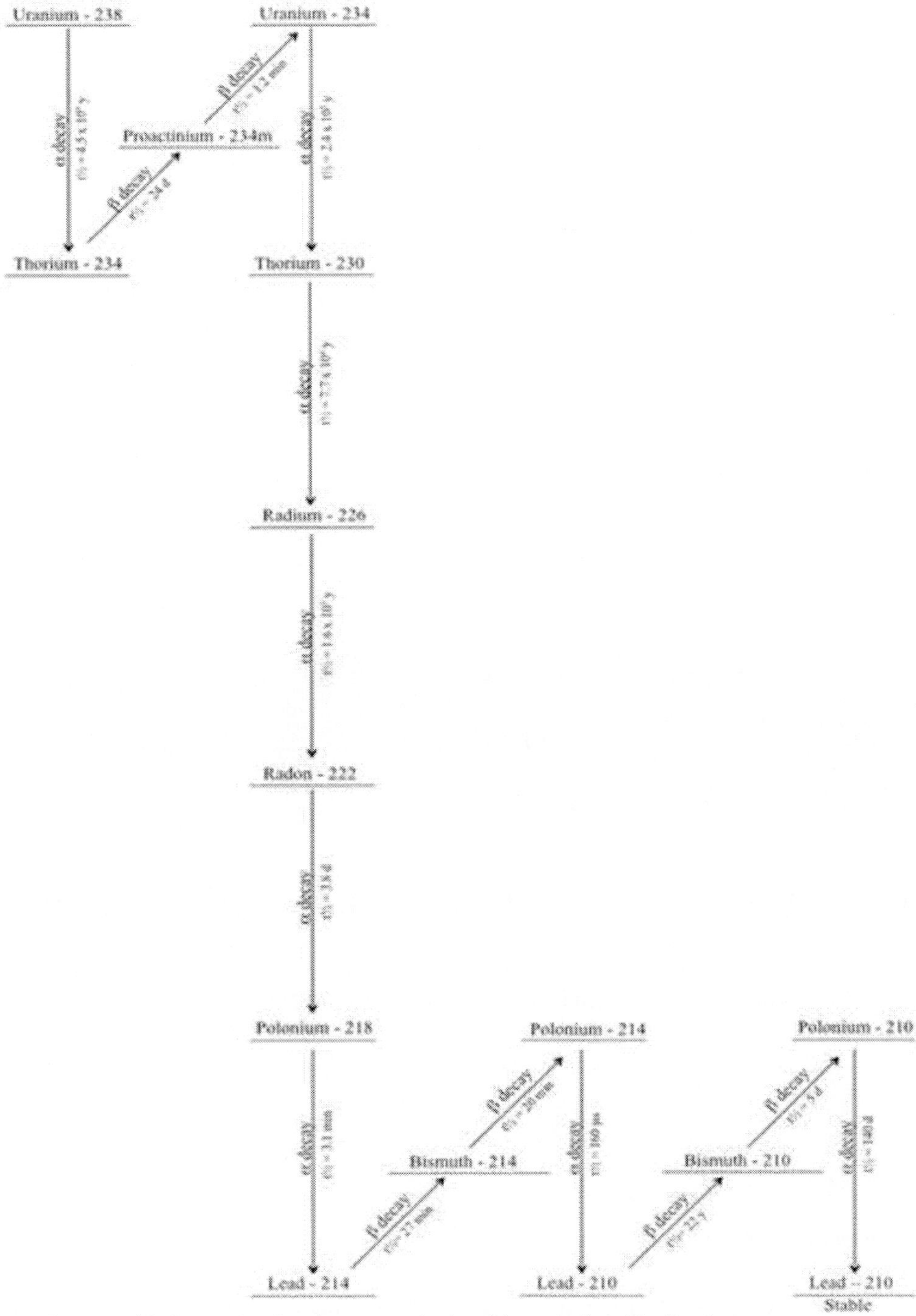

Note: That α and β decays are drawing respectively with vertical and inclined arrows.

Figure 3. Decay scheme of uranium – 238.

To this plausible explanation Hahn sent an account of the work to Meitner, who, in cooperation with her nephew Otto Frisch, was able to unravel the phenomenon they called nuclear fission. An induced nuclear fission of uranium-235 can be expressed as follows:

URANIUM PHYSICAL PROPERTIES

The specific activity, i.e. the activity, on average, of one disintegration every second (Becquerel) per milligram arising solely from the decay of the natural uranium isotopes is 25.4 Bq/mg. The International Atomic Energy Agency (IAEA) defines uranium as a low specific activity material. In nature, uranium isotopes are typically found in radioactive equilibrium with their radioactive decay products, i.e. the activity of each of the radioactive progeny is equal to the activity of the uranium parent isotope. Isotopes of natural uranium decay by emitting mainly alpha particles. The emission of beta particles and gamma radiations are low. Uranium-235 has half-life of about 704 million years and the most abundant isotope U-238, approximately 4.5 billion years. Uranium-238 has the following decay scheme (figure 3):

$$UO_2\,(N_3O_2)_2 \,.\, 2H_3\,O \xrightarrow{350\,^oC} UO_3 \xrightarrow{CO\ 350\,^oC} UO_2$$

$$U + O_2 \xrightarrow{heat} U_3O_8$$

$$UO_3 \xrightarrow{700\,^oC} U_3O_8$$

Figure 4. Interconversion of uranium oxides [7].

PHYSICOCHEMICAL AND CHEMISTRY PROPERTIES OF URANIUM

Uranium metal has atomic number of 92, melting point of 1,135 °C, boiling point of 4,130 °C. The finely divided powder is attacked by cold water. In air, the metal becomes coated with a layer of oxide. It is soluble in acids but is unaffected by alkalis.

Elemental uranium is easily oxidizable to forms ranging from +2 to +6 depending on the compound it forms or on the solution in which it is present [6]. Among the following ions U^{3+}, U^{4+}, UO_2^{+}, UO_2^{2+}, the most stable is +VI [7].

Uranium Oxides

There are several types of uranium oxides with different oxidation states giving rise to stable non-stoichiometric compounds. The forms U_3O_8 and UO_2 are the most common forms of uranium oxide, both are solid with low solubility in water and relatively stable over a wide range of environmental conditions. Triuranium octaoxide (U_3O_8) is the most stable form of uranium and most commonly found in nature. Uranium dioxide or uranium oxide (UO_2), also known as urania or uranous oxide, is a black crystalline powder that naturally occurs in the mineral uraninite.

Uranium Halides

Uranium forms halides with oxidation states of + III + IV + V, +VI. These uranium halides are hydrolyzed in water. The hexahalides dissolve in strong acids giving rise to the ion UO_2^{2+}. At higher pHs this hydrolyses to polymerize through groups OH. Uranium can form trihalides and tetrahalides with fluor, chlorine, bromine and iodine. The tetrahalides UF_4, UCl_4 , UBr_4, UI_4 are the most stable compounds, so UF_5 tends to disproportionate giving rise to UF_4 and UF_6. Some uranium +V halides are UF_5, U_2Cl_{10} and +VI uranium halides include UF_6, U_2F_9, U_4F_{14}, U_5F_{22}, UCl_6 [7].

Uranium Hydrides

Uranium reacts with hydrogen at room temperature but at a 250^0C the reaction is faster forming UH_3. The hydride is very reactive and usually more appropriate than the pure metal for the formation of other compounds [7].

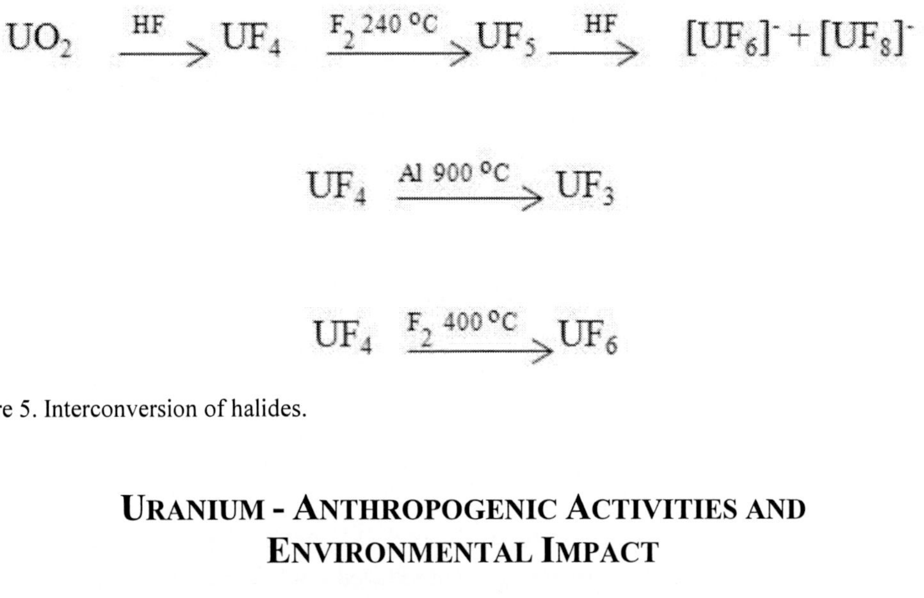

Figure 5. Interconversion of halides.

Uranium - Anthropogenic Activities and Environmental Impact

Occupational exposure to uranium can occur as a result of various human activities such as work associated with different stages of the nuclear fuel cycle, scientific research, industry, and occupations which involve the handling of materials containing enhanced concentrations of naturally occurring radionuclides.

According to United Nations Scientific Committee on the Effects of Atomic Radiation, UNSCEAR 2008 [8], the available data that correspond to occupational radiation exposures are insufficient for estimating worldwide levels of dose. So its Committee has adopted the data available for particular occupational categories, one for application to occupational exposures arising at most stages in the commercial nuclear fuel cycle and the other for general application to other occupational categories.

$$2UH_3 + 4H_2O \longrightarrow 2UO_2 + 7H_2O$$

$$2UH_3 + 4Cl_2 \longrightarrow 2UCl_4 + 3H_2$$

$$2UH_3 + 8HF \longrightarrow 2UF_4 + 7H_2$$

$$UH_3 + 3HCl \longrightarrow UCl_3 + 3H_2$$

Figure 6. Compounds formed from UF_3.

In general, phosphate ores of sedimentary origin have higher concentrations of nuclides of the uranium family. Since the phosphate rock is extensively used as a source of phosphorus for fertilizers, for making phosphoric acid and gypsum it naturally becomes a permanent source of uranium radiation exposure. In 90 % of cases, the ore is treated with sulphuric acid, the fertilizers become somewhat enriched in uranium up to 150 % relative to the ore, while 80 % of the ^{226}Ra, 30 % of the ^{232}Th and 5 % of the uranium are left in the phosphogypsum. The processing of phosphoric rocks generate gaseous and particulate emissions that contain ^{238}U and ^{226}Ra. When discharged to the environment, these nuclides lead to radiation exposure of the population. Local dump sites for phosphogypsum are usually not protected from rainfall and become hydraulically connected to surface waters and shallow aquifers [9]. Uranium presents significant hazard both from a chemical and radiological point of view. There is no data available for long-term effects of uranium ingestion on humans, information available is from intermediate-term studies on animals using safety factors to infer human inhalation levels. From intermediate duration inhalation of uranium soluble forms, the study of Rothstein [10] on dogs is initially used because a level of 0.15 mg U/m^3 in air produced the lowest observable adverse effect. A minimal risk inhalation level for humans of 0.4 µg/m^3 is derived. For the insoluble forms, the study of Rothstein [11] on dogs showed that uranium concentrations of 1.1 mg U/m^3 in air produced no observable adverse effect and minimal risk inhalation level for humans of 8.0 µg/m3 is derived. For the hazard from chronic duration inhalation of soluble uranium forms, the study of Stokinger [12] on dogs is used in which uranium concentrations of 0.05 mg U/m^3 in air produced no observable adverse effect and a minimal risk inhalation level for humans of 0.3 µg/m^3 is derived. The chemical form of the uranium compounds, the weight of its particles and its respective solubility determines how easily the compound can move through the environment, as well as how toxic it might be. Uranium in the environment is mainly found as a uranium oxide, typically as UO_2, which is an insoluble compound found in minerals and sometimes as UO_3, a moderately soluble compound found in surface waters. Several international agencies dealing on health and environmental protection have recommended maximum concentration limits for uranium in drinking water or water springs for consumption [13]. The International Commission on

Radiological Protection [14] (ICRP 1993) recommended the maximum limit in water of 0.0019 mg L^{-1} and World Health Organization [15](WHO 2004) the maximum of 0.015 mg L^{-1} used for water springs for consumption. However, according to a Brazilian environmental agency CONAMA Resolution no. 357/05, Article 15 [16], the standard uranium emissions was set at 0.002 mg L^{-1}. Soluble uranium compounds can combine with other chemical elements and compounds in the environment to form other uranium compounds. Some forms of uranium oxides are very inert and may stay in the soil for a long time without moving downward into groundwater [17]. According to Wagner and co workers [18], by products originated from uranium processing can impact human health through environmental exposure. They assessed the potential health effects among residents living near the Fernald Feed Materials Production Center, via the Fernald Medical Monitoring Program (FMMP,) using hypertension or changes in hematologic parameters as uranium exposure associated indicators, this facility have worked from 1951 to 1989 with uranium processing. The authors suggested that residents in the vicinity of the Fernald plant with elevated exposure to uranium primarily via inhalation exhibited decreases in white blood cell counts, and small, gender-specific alterations in systolic blood statistically significant. There were evidences correlating inhalation of uranium arising as a byproduct of various anthropogenic activities and blood modifications. In 1949, Dygert [19] exposed rats to ammonium diuranate dusts for 6h/day for 30days and observed decreases in red blood cell counts, lymphocytes, and white blood cell counts, and increases in neutrophils and eosinophils. Residents living within 2 miles of the plant had small increases in white blood cell counts and hemoglobin levels.

Nuclear Fuel Cycle versus Radioactive Exposition

The main source of exposure during the manufacture of nuclear fuel is uranium. This can result in external exposure by gamma emission and internal exposure through ingestion or inhalation of particles containing ^{234}U, ^{235}U and ^{238}U. Nuclear fuel is a material that can be used in nuclear fission or fusion processes to derive nuclear energy. Nuclear fuel can refer to the fuel itself, or to physical objects such as bundles composed of fuel rods. They are composed of the fuel material, mixed with structural, neutron moderating, or neutron reflecting materials. The fuel cycle comprises the following steps: uranium ore mining, milling, refining, conversion and fuel production. Uranium enrichment process gives rise to a large amount of nearly pure U-238 that is useless as fuel in nuclear reactors. For each kilogram of enriched uranium produced, 200 kg of low-enriched uranium are generated, mainly U-238, a radionuclide that emits a less harmful type of radiation but with a half-life of about 4.5 billion years. It is considered 40 % less radioactive than natural uranium, but its chemical toxicity is similar [20].

Uranium Mining and Processing

The first step of exploration is characterized by finding a natural deposit and making their economic evaluation in order to determine the location in which the extraction of ore from the ground will be performed. After the choice of geological reserve was completed, the mining procedures take place in which the ore is removed from the ground.

URANIUM MILLING

After the uranium ore is extracted from the ground it is processed by milling in which mined uranium ores normally are processed by grinding the ore materials to a uniform particle size. Then the ore is refined to extract the uranium by *in-situ* chemical leaching. The milling process commonly yields dry powder-form material consisting of natural uranium "yellow cake", which is sold on the uranium market as U_3O_8.

Uranium Conversion - Production of Nuclear Grade Uranium

Due to the high density of uranium-based compounds, the ore is first concentrated by flotation. Then it becomes calcined in air and extracted with sulfuric acid usually in the presence of an oxidizing agent such as MnO_2 to ensure that conversion to uranium with oxidation state + VI takes place. A solid compound is precipitated as sodium diuranate, $Na_2U_2O_7$ a yellow oxide. Depending on the types of ores, concentrates may be prepared by various extraction and refining methods. They may be transformed in three chemical forms useful for use as fuel, metal, UO_2, UO_3 and UF_6.

The yellow oxide is placed to react and dissolve with nitric acid to form uranyl nitrate, $UO_2(NO_3)_2.(H_2O)n$, which is further purified. The water of crystallization molecules can be 2, 3 or 6 depending on whether it was used fuming nitric acid, concentrated or diluted respectively. The uranyl nitrate is converted to UF_4 and to UF_6 gas suitable to be used in the enrichment process (figure 7) [7].

Considering nuclear reactors in which nuclear fuel is enriched, the next step is a separation process that is able to increase the content of ^{235}U. Uranium compounds may be enriched by thermal diffusion, gaseous diffusion, electromagnetic separation and centrifugal gas. The enrichment of uranium aims to increase the concentration of fissile uranium 235 which is only 0.7 % for 2 to 5 %, which enables its use as fuel. UF_6 is a gas suitable for the enrichment stage.

In world's most common reactor types, the lower water reactors (LWR), the fuel itself is the UF_6 while the pressurized heavy water reactors (PHWR) use natural uranium oxide. For PHWR reactors is possible to do the reconversion of UF_6 to UO_2 then the pellets are fabricated and fuel element is assembled.

$$UF_4 \xrightarrow{F_2} UF_{6(gas)}$$

Figure 7. Conversion to uranium hexafluoride.

Nuclear Reactors Based on Uranium *versus* Accident Risks

The use of nuclear fuels based on uranium is closely related to energy production by nuclear power plants, design and production of atomic weapons. Considering the use of nuclear fuel to generate electricity, there are different types of reactors. They are mainly classified according to some technical features. The Pressurized Water Reactors (PWR) are fueled with 3 % enriched UO_2, cooled and moderated by high pressure liquid water. They have a hermetically sealed system with a primary coolant in direct contact with the core capturing heat from nuclear fissions. The radioactive water that leaves the pressure vessel is looped through a steam generator, which in turn heats a secondary (non-radioactive) loop of water to steam that can propels turbines. In terms of large scale, their technology are considered safe and reliable because PWR reactors are very stable due to their tendency to produce less power as temperatures increase and the turbine loop is physically separated from the primary so that the water in secondary loop is not contaminated by radioactive species.

This makes the reactor easier to operate from a stability standpoint and safe in terms of radioactive contamination. On the other hand the primary coolant water must be highly pressurized to remain liquid at high temperatures requiring high pressure resistance piping and a heavy pressure vessel, something that increases the consequences of a loss-of-coolant accident (LOCA).

Another possibility of nuclear accident may occur by overheating followed by melting caused by water pumping failure since shutdown of the primary nuclear reaction requires one to three years and the fission products continue to generate considerable decay heat. Boiling Water Reactors (BWR) use fuel pellets with 2 % enriched UO_2 and operate at much lower pressures so the reactor wall should not be necessarily as strong as a PWR. BWRs operate for 18 to 24h between refueling outages [21].

Compared to PWR reactors, the BWRs has several advantages like a lower operating temperature of the nuclear fuel, a pressure vessel subject to significantly less irradiation, the possibility of having greater thermal efficiency and, depending on the design, they can be simpler and potentially more stable and safe. BWR reactors are also water cooled and moderated like a PWR. The pressure of a BWR reactor use to be about half of PWR's. Like PWRs, BWR's core damage incident is possible because they will continue to produce heat after the reactions have stopped due the presence of radioactive decay of fission products and materials that have been activated by neutron absorption. However, considering that a BWR has lower probability of a rupture causing loss of coolant compared to a PWR and less risk of nucleus damage in the case of this rupture occurs. BWRs present multiple safety systems for cooling the core after an emergency shut down. In case of emergency conditions, the BWR reactors are easy to operate because the measurement of water level in the pressure vessel is the same for both normal and emergency operations. The plant of an Advanced Boiling Water Reactor (ABWR) uses low-enriched uranium as a fuel for bringing the moderator and coolant water that boils directly at the reactor core. The resulting steam drives the turbine [22].

The Pressurized Heavy Water Reactor (PHWR), a Canadian design known as CANDU (Canadian Deuterium Uranium), is a heavy-water-cooled and moderated pressurized-water reactor. These kind of reactors are fueled with natural uranium UO_2 and are thermal neutron reactor designs that use heavy water D_2O as moderator and coolant. PHWRs may have a precise flux control in the core because they can be refueled while at full power, which makes them very efficient in their use of uranium.

Main Nuclear Power Plant Accidents - Environmental and Human Impacts

Chernobyl

The nuclear power station of Chernobyl is located 15 kilometers to the northwest of the actual Chernobyl town. Chernobyl is a small town in the Ukraine near the Belarus border, one-hundred ten kilometers north of Kiev the capital city of the Ukraine [23]. In terms of nuclear industry, the most severe accident occurred at Chernobyl nuclear power plant, on April 26, 1986, in Ukraine, about 20 km south of the border with Belarus. The Chernobyl reactor is a Pressurized Water Reactor using light water as a coolant and graphite as a moderator. The accident occurred during a low-power engineering test of the Unit 4 reactor.

Doses received by the general public came from the radionuclide releases in the damaged reactor, which led to the ground contamination of large areas. Within a few days or weeks it caused deaths of 30 power plant employees and firemen (including 28 with acute radiation syndrome) and brought about the evacuation, in 1986, of about 116,000 people from areas surrounding the reactor and the relocation, after 1986, of about 220,000 people from Belarus, the Russian Federation and Ukraine. Vast territories of those three countries, at that time republics of the Soviet Union, were contaminated, and trace deposition of released radionuclides was measurable in all countries of the northern hemisphere. During the night of the accident, approximately 600 emergency workers who were on the site of the Chernobyl power plant received the highest doses. About 16,000 Chernobyl-caused cancer deaths are projected during the lifetime of those who were exposed in Belarus, Ukraine, and western Russia and in Europe farther downwind [24].

Represented in grays (Gy), the absorbed dose measured as joules per kilogram is a measure of the energy deposited in a medium by ionizing radiation per unit mass is. Acute radiation sickness was confirmed in 134 of those emergency workers, 41 received whole-body doses from external irradiation of less than 2.1 Gy, 93 received higher doses ranging from 2.2 until 16 Gy. The skin doses from beta exposures, evaluated for eight patients with acute radiation sickness, were in the range from 400 to 500 Gy [25].

The radionuclide releases from the damaged reactor occurred mainly over a 10-day period, but with varying release rates. The amount of fuel in the lower regions of the reactor building was estimated to be 135 ± 27 t, which is about 71 % of the core load at the time of the accident (190.3 t). The remainder of the fuel was accounted for as follows: fuel in the upper levels of the reactor building (38 ± 5 t); fuel released beyond the reactor building (6.7 ± 1 t); and unaccounted for fuel (10.7 t). These radioactive particles can be classified as fuel fragments with a mixture of fission products bound to a matrix of uranium oxide, similar to the composition of the fuel in the core, but sometimes strongly depleted in caesium, iodine and ruthenium, and particles consisting of one dominant element (ruthenium or barium) but sometimes having traces of other elements [26], [27], [28], [29], [30], [31]. These monoelemental particles may have originated from its embedments in the fuel during reactor operation and released during the fragmentation of the fuel [32].

After the accident of Chernobyl, few measurements were performed in order to collect air contamination. On April 27, 1986 an air sample analysis of particles collected a few hundred meters above Chernobyl power plant presented dimensions ranging from abundant smaller sizes to larges particles with tens of micrometers [33]. In 1986, on days 14 and 16 samples were collect by aircraft above the damaged reactor and analysed by spectrometry [34]. Almost all activity of the larger particles was due to non-volatile radionuclides such as ^{95}Zr,

^{95}Nb, ^{140}La, ^{141}Ce, ^{144}Ce and transuranium radionuclides embedded in the uranium matrix of the fuel [29].

Caesium-137 was chosen as a reference radionuclide for the ground contamination due to its substantial contribution to the lifetime effective dose, substantially long radioactive half-life of 30-year, and ease of measurement. The radionuclides released in the accident deposited over most of the European territory of the former Soviet Union. The three main areas of contamination, defined as those with ^{137}Cs deposition density greater than 37 kBq m^{2} are in Belarus, the Russian Federation and Ukraine [25].

The environmental behaviour of deposited radionuclides depends on the physical and chemical characteristics of the radionuclide considered, on the type of fallout, i.e. dry or wet, and on the environment characteristics. Radioactive contamination of the ground was found to some extent in practically every country of the northern hemisphere [35]. The fuel particles deposited on the ground contained alpha-emitting transuranics, such as ^{238}Pu, ^{239}Pu, ^{240}Pu and ^{241}Am, as well as beta-emitting transuranics, such as ^{241}Pu. In an assessment of the equivalent doses to lungs for agricultural workers, it was concluded that even at sites inside the 30-km zone, lifetime committed equivalent doses to the lungs per year of work will not exceed 0.2 mSv for most individuals [36]. The unit Sievert (Sv) takes into account the biological effect on living tissue, produced by the radiation absorbed. The unit Sievert measures the equivalent dose of radiation, having the same damaging effect as an equal dose of gamma rays and is obtained from the absorbed dose multiplied by the appropriate weighting factors. Under normal conditions without accidents, the public dose limits for exposure from uranium mining or nuclear plants are usually set at 1 mSv/yr above background.

The contribution of aquatic pathways to the human dietary intake of ^{137}Cs and ^{90}Sr was usually quite small. Deposition of hot materials on seas and oceans surfaces presented low levels of radioactivity contamination because the materials were rapidly diluted into very large volumes of water. In fresh waters such as rivers and small lakes, the radioactive contamination resulted mainly from erosion of the surface layers of soil in the watershed, followed by runoff in the water bodies. In the 30-km zone, the largest surface water contaminant was found to be ^{90}Sr mostly in dissolved form, as ^{137}Cs was strongly adsorbed by clay minerals, low levels of plutonium isotopes and of ^{241}Am were also measured in the rivers of the 30-km zone [37]. In Scandinavian and Russian water lakes with long retention times, high concentrations of ^{137}Cs could be found in the muscle of fish predators [38], [39], [40]. In summer of 1996, the whole-body activity of ^{137}Cs in inhabitants of the village Kozhany located along the coast of lake Kozhany was 7.4 ± 1.2 kBq in 38 adults who did not consume lake fish but was 49 ± 8 kBq in 30 individuals who often consumed lake fish. Taking into account seasonal changes in the ^{137}Cs whole-body activity, the average annual internal doses were estimated to be 0.3 mSv and 1.8 mSv in these two groups, respectively. Also, the relative importance of the aquatic pathways, in comparison to terrestrial pathways, may be high in areas downstream of the reactor site where ground deposition was small [25]. From the radiological point of view, ^{131}I and ^{137}Cs are the most important radionuclides to consider, because they are responsible for most of the radiation exposure received by the general population. The releases of ^{131}I and ^{137}Cs are estimated to have been 1,760 x10^{15}Bq and 85x10^{15}Bq, respectively [25]. For short-lived radionuclides such as ^{131}I, the main pathway of human exposure was the transfer of the amounts deposited on leafy vegetables that were consumed within a few days, or on pasture grass that was grazed by cows or goats, giving rise to milk contamination. When the substantially long lived radionuclides such as ^{137}Cs and ^{90}Sr with half-lives of 30 and 28 years are deposited on soil, they migrate downwards and are partially absorbed by plant roots, leading in turn to upward migration into the vegetation and are determined by a number of natural phenomena, including relief features, the type of plant, the structure and makeup of the soil, hydrological conditions and weather patterns, particularly at the time that deposition occurred.

Human exposure is closely related to the presence of cesium isotopes. The gamma rays emitted by these contaminants directly raise the radiation levels and determine the long-term evacuation plans. In areas distant from the Chernobyl accident, the volatile fission products cesium isotopes accounted for 70 percent of the external dose two months after the accident and 85 percent four months after [25]. Haematological cancer registries were set up in Belarus, the Russian Federation and Ukraine shortly after the accident. However, thyroid cancer registries only started to appear during the early 1990s [41]. The Thyroid Cancer Registry has been created for the Bryansk and Tula regions including data on all cases of thyroid cancer since 1981 [42].

In Ukraine, at the Kiev Institute of Endocrinology and Metabolism has been created a clinical morphological register of thyroid cancers in patients under the age of 18 years at the time of the Chernobyl accident, as well as retrospective data since 1986 [43]. Apart from the substantial increase in thyroid cancer after childhood exposure observed in Belarus, in the Russian Federation and in Ukraine, there is no evidence of a major public health impact related to ionizing radiation 14 years after the Chernobyl accident. The risk of leukemia, one of the most sensitive indicators of radiation exposure, has not been found to be elevated even in the accident recovery operation workers or in children [25].

Fukushima

The Fukushima power station is located on the east coast of the island of Honshu, 200 km north-east of Tokyo. The coast runs north-south, facing the Pacific Ocean. Fukushima City is about 63 km north-west of Fukushima I Nuclear Power Plant, the site of the nuclear accident that followed the tsunami. On April 12, 2011, one month after a 9.0-magnitude earthquake, tsunami disabled four reactors at the Fukushima Daiichi Nuclear Power Station.

The Fukushima plant comprises six boiling water reactors designed by General Electric (GE) and maintained by the Tokyo Electric Power Company (TEPCO). At the time of the earthquake, reactors 5 and 6 were in cold shutdown for planned maintenance and reactor 4 had been de-fuelled. After earthquake, the remaining reactors shut down automatically, and emergency generators came online to control electronics and coolant systems. The tsunami broke the reactors' connection to the power grid, leading the reactors to begin to overheat. With loss of coolant, the fuel temperature rose above 1,000 °C and the hot steam reacted with zircaloy cladding in the reactor pressure vessel [44]. After few hours and days reactors 1, 2 and 3 meltdown. Even though the workers had made efforts to cool and shut down the reactant, the steam exposed zirconium from the allow removing oxygen due to water lysis and giving rise to hydrogen gas that exploded in contact with air damaging structures of the buildings. The four hydrogen explosions signaled that a loss of coolant had occurred. Thus the workers suffered radiation exposure and were temporarily evacuated at various times. Electrical power was slowly restored for some of the reactors, allowing for automated cooling.

In terms of environmental impact is possible to infer that hydrogen explosions were indicators that the fuel cladding had failed and released out of the power plant the fission products cesium-137 and iodine-131, once they have boiling points respectively of 671 °C and 184 °C and cesium-iodide 1,277 °C [24].

Three months after the disaster, Japanese government estimates cesium-137 and iodine-131 releases respectively of $0.41x10^6$ Ci and 4.3 $x10^6$ Ci into Fukushima Daiichi Nuclear Power atmosphere. To best assess these Fukushima releases one may confirm that Chernobyl's emissions of cesium-137 and iodine-131 respectively of 2,3 $x10^6$ Ci and 50 $x10^6$ Ci were larger but with comparable orders of magnitude.

One million live within 80 km radius of the Fukushima plant, in areas contaminated with cesium-137 with levels greater than 1 Ci/km^2. Considering that in Chernobyl six million people living in areas contaminated with levels of radioactive cesium-137 above 1 Ci/km^2 will cause a expected cancer death level up to 16,000, one can estimate that Fukushima's cancer deaths caused by exposure to cesium-137 may reach the order of magnitude next to thousand.

However it is not clear what the long-term health consequences of the Fukushima Daiichi accident will be. The Nuclear and Industrial Safety Agency (NISA) announced that the subsequent releases of radioactivity into the atmosphere qualified it as major accident, or a Level 7 emergency accident. Within a day of the accident at Fukushima Daiichi, nearly 130,000 people living within 20 kilometers of the power plant were ordered to evacuate. And on March 15, four days after the accident, an additional 354,000 people living between 20 and 30 kilometers of the plant were advised to stay indoors to reduce exposure to radiation. Comparing with unaffected control groups, populations that were subjected to nuclear accidents present psychological consequences such as levels of anxiety two times higher that controls and multiple unexplained physical symptoms 3 to 4 four times greater [24].

Three Mile Island

This Power plant is located at Three Mile Island in the Susquehanna River in Dauphin County, Pennsylvania, near Harrisburg, United States. The nuclear accident occurred in 1979 where the Pressurized Water Reactor was severely damaged with core meltdown. However radiation was contained and there were no adverse health or environmental consequences. The accident began at 4 a.m. on Wednesday, March 28, 1979, with failures in the non-nuclear secondary system, followed by a stuck-open pilot-operated relief valve in the primary system, which allowed large amounts of nuclear reactor coolant to escape. Radiochemical analysis of the debris indicate that it was composed of about 70 % U, 13,75 % Zr, and 13 % O. The remaining constituents are the elemental constituents of stainless steel and inconel core components that were probably melted during the relocation of the debris to the lower head of TMI-2 reactor vessel [45]. Detailed epidemiological studies made since the accident have supported the conclusion that radiation released from the accident had no perceptible effect on cancer incidence in residents near the plant, though these findings are contested by one team of researchers.

Human Exposure to Uranium Production and Disposal

Concerns regarding operational practices are usually greater at uranium mines than elsewhere, although the focus is on situations where workers are in immediate contact with the ore and processing streams [46].

Mining Wastes and Environmental Impact

The radionuclides that are released into the atmosphere by processes of mixing and dispersion will accumulate on the soil surface contaminating plants and sometimes they can enter in the food chain. These atmospheric processes are governed by weather conditions and local topography [47].

Mining activities in Central-North regions of Portugal are a good example of environmental impact since the production of 4370 ton of U_3O_8 from 1944 to 200 resulted in 13 million tons of waste rocks [48]. Chemical processes like acid drainage affected surface and groundwater quality up to 1–1.5 km downward from the mining site.

High levels of sulphate, Al, Mn, U, and low pH values (<4.5–5) made these waters unsuitable for irrigation and livestock watering. Irrigation of acid soils (pH <4.5) with contaminated waters presents risks to the crops owing to a high content of U in the available soil fraction. Consequently, maize harvested in these soils showed amounts of uranium in roots and leaves that may pose risk when it is used for animal feeding and plant residues that are used for soil fertilization [46].

Cunha Baixa mine, in the center-north of Portugal may be taken as object of contamination study presented groundwaters in the immediate surroundings of the mine with calcium or magnesium sulphate type composition that were not recommended for irrigation. The levels of some parameters such as pH, electrical conductivity, sulphate, Al, Mn, and total dissolved salts, fluorine and Co were above the recommended or admissible maximum values established by Portuguese law for irrigation water.

The uranium content also exceeds the trigger values established in Canada, Australia and New Zealand for irrigation and livestock drinking water. The use of this water for irrigation presents risks to soils in terms of acidity and salinity hazard and to crops i.e. phytotoxicity [46].

Environmental Remediation of Uranium Production Facilities

After a couple of decades extracting and milling uranium ores and producing uranium oxides the facility is closed, and a process of remediation begins. The following steps involve dismantling, decommissioning, waste management and water treatment.

Decommissioning and remediation of facilities may be successful when the rules are well established, when the companies involved in environmental remediation have enough know-how and the decommissioning process is subsided. Spain has suitable rules for licensing of these decommissioning activities with regulations requiring that the facility owner prior to finishing production activities, present a approved dismantling programme evaluated by the Spanish Nuclear Safety Council (NSC). The decommissioning task can be considered accomplished when radiological exposure in the area is not significant compared to the natural background at uranium ore sites, the open pit mine is backfilled, the procedure of dismantling and demolition of the plant is finished and a landscape integration is performed [46].

Uranium Mill Tailings versus Environmental Impact

Uranium mill tailings are the same as tailings from other types of mines or from water containment dams, a sudden failure of a large structure like this may cause death through drowning, crushing or suffocation [49].

Uranium ore mining and milling produce significant amounts of radioactive waste that comprises solid radioactive waste from low grade useless ores that is stored in dumps and another kind radioactive and chemical waste that is solid, liquid, gas from hydrometallurgical plants that produces U_3O_8. The radioactive waste contains such uranium decay products as thorium-230, radium-226, radon-222 and radon progeny elements. The chemical waste consists mainly of hazardous heavy metal such as lead and arsenic. The by-product materials that left over are stored in large reservoirs called tailings impoundments that were often arranged utilizing geomorphological depressions or by filling-in valleys. The common mode of disposal is in near-surface impoundments in the vicinity of the respective mine or mill [50]. Mining and milling activities of a single facility may give rise to residues that can exceed millions of tones and uranium is the by-product [48] and the waste originated may produce environmental contamination of the atmosphere, soil, groundwater and fresh water ecosystems.

If radioactive tailings are kept solid, the wind may suspend radioactive particle and spread gases in the air contaminating the surrounding places. The radioactive dust can be carried great distances by the wind and may induce contamination of food chain and bodies of water. Thus tailings are usually kept underwater. Since radioactivity diminishes in the course of time, waste must be typically isolated and stored for a period of time until it theoretically do not pose risk. Ponds are lined with plastic liners to meet accepted environmental protection standards for uranium mill tailings. However the stored radionuclides such as uranium-238 and thorium-230 have too high half-lives of magnitude orders respectively of billions and tens of thousands years. This means that this kind of protection will hardly ensure safety of these uranium tailing impoundments for the entire period demanded to rule out the danger for contamination.

Nowadays there are several technical resources to optimize the assembly and construction of new uranium mill tailings in order to avoid environmental impact and human exposure. According to IAEA TECDOC 1403 [49], to keep environmental emissions to a minimum over long times, the task is to find conceptual and technical solutions that will guarantee tailings more inert over prolonged time-spans with impounded materials and engineered structures stable. These deposits should have minimized the need for active maintenance and must be technically and economically feasible and acceptable to society.

In 2002 IAEA has published a comprehensive set of guidelines dealing with monitoring and surveillance of residues [51] and management of radioactive waste [52]. To set environmental protection standards, in 1983, the Environmental Protection Agency (EPA) issued two sets of standards controlling hazards from uranium mill tailings, under the authority of the Uranium Mill Tailings Radiation Control Act of 1978. According to these standards, control measures include use of fences, warning signs, the establishment of land use restrictions, the use thick earthen covers constructed protected by rock in such way to prevent seepage into the ground water [53].

However, in the past, when these uranium mill tailings were improperly built they allow environmental degradation of these tailings giving rise to instability and conflict of many levels. The environmental impact generated by the expansion of so called Soviet Union into the Central Asian republics were profoundly associated with a centralized economy of the former. As Tsarist Russia was expanding into Central Asia at that time named Turkestan, uranium natural deposits was one of the Russia's interests. Many deposits were discovered on the territory of the current name Kyrgyzstan.

The Central Asian republics Kyrgyztan, Kazakhstan, Tajikistan, Turkmenistan and Uzbekistan served primarily as exporters of raw materials and one of then was uranium. Production quotas were set by Moscow without full consideration of local factors peculiar to region [54]. Kyrgyzstan was one of the major producers of uranium in the Soviet Union during the fifties, by the end of the sixties and most of the active uranium mines were shut down and conserved [54]. Nowadays these waste products left behind now represent a challenge not just to environmental quality, public health and domestic political stability. When the territory in which the uranium tailings are located with a transboundary nature there is a need for routine monitoring with international assistance. Currently efforts are under way to conduct joint monitoring projects such as Navruz engaging Central Asian States and donor countries. Navruz project has been conducted by scientific experts from these countries facilitated by the Cooperative Monitoring Center at the United States Department of Energy's Sandia National Laboratories.

In 1958 a tailings reservoir broke and in 1994 a landslide pushed radioactive material into the river. Currently, potential landslides pose a direct threat to Mayluu-Suu. Kyrgyztani officials stated that landslide did not cause radioactive contamination, but there were explicit concerns over the threat it could have posed in terms of tailings impoundments' damages. The southwestern Kyrgyztan region is constantly subject to landslides that pose a constant risk of spreading radioactive material. But even if the landslide did not diffuse radiation there is enormous possibility of impoundment mill tailings being seriously damaged. Mayluu-Suu river in southwestern Kyrgyztan might carry away the nuclear waste accumulated in these reservoirs along its banks, which may threaten the downstream regions of Kyrgyztan and Uzbekistan.

The fundamental reason to remediate tailings piles is to reduce the level of hazards and thus lessen the risk of significant impacts upon human health and the environment, or to reduce ongoing levels of impact where effects are already apparent.

The IAEA technical document 1403 pay attention to different chronic failure of uranium mill tailings such as:

- Dispersal of radioactive dust caused by wind-blown distribution of dust in tailings surfaces allowed to dry out, e.g. Durango in Colorado, Wismut operations - Germany, West Mecsek - Hungary, Bukhovo - Bulgaria, Kyrgyzstan, Rum Jungle - Australia, Ulba, Kazakhstan;
- Erosion of Tailings from outer surfaces of the containment, e.g. Rum Jungle - Australia, Wismut operations - Germany;
- Seepage through the floor and/or walls of the containment, e.g. Mecsek in Hungary, Schneckenstein in Germany, Kyrgyzstan, some of the Far West Rand mines in South Africa, Anhua in northern China;

- Effluent discharge as decant allowed to drain or overflow from the containment directly or indirectly into natural waterways, e.g. Jaduguda in India, several Wismut operations in Germany, Elliot Lake in Canada.

CONCLUSION

Unsuspected before, anthropogenic activities with uranium have been increasing environmental radioactivity levels, being, nowadays, a potential source of exposure. Therefore, papers have been published on this topic, looking for the development of technologies for remediation and disposal treatment since ore extraction and uranium processing releases radionuclides into the atmosphere by processes of mixing and dispersion. Radionuclides mixed on drinkable water, deposited on soil surface and plant that enter in the food chain may cause chemical toxicity and radionuclides delivered in the air may provoke radiological toxicity. On the other hand, nuclear accidents have been put employees and thousands, millions of citizens in risk that can extend for decades. Nuclear accidents may give rise to psychological consequences such as higher levels of anxiety and higher multiple unexplained physical symptoms. Large-scale nuclear accident causes a turning point in the global development of nuclear power, and the number of reactors under construction tend to decline after each accident. With so many evidence we should seriously reconsider other alternative energy sources because several sequels of nuclear accidents comprise social and economic impacts of entire populations and taint permanently the cities involved.

REFERENCES

[1] International Commission on Radiological Protection. Uranium Chemical and Radiological Risk Assessment for Freshwater Ecosystems Receiving Ore Mining Releases: Principles, Equations and Parameters. *The Natural Radiation. 8th International Symposium* (NRE VIII). 2008 1034, 333-336.

[2] International Atomic Energy Agency. Uranium 2007: *Resources, Production and Demand.* In: The '2007 Red Book'. Paris,Fr: OECD/NEA, 2008.

[3] International Atomic Energy Agency. World Distribution of Uranium Deposits (UDEPO) with Uranium Deposit Classification. *IAEA-TECDOC*-1629, Vienna, Au: IAEA, 2009.

[4] Dahlkamp, FJ. *Uranium Ore Deposits*. Berlin,Al: Springer, 1993.

[5] Hahn, O. *"From the natural transmutations of uranium to its artificial fission.* Nobel Lecture, December 13, 1946. Nobel Foundation. http://www.nobelprize.org/nobel_prizes/chemistry/laureates/1944/hahn-lecture.pdf#search=%27meitner%27.

[6] Wang, KX; Chen, JS. Extended structures and physicochemical properties of uranyl organic compounds. *Acc. Chem. Res.*, 2011 44(7), 531–540.

[7] Lee, J.D. *Concise Inorganic Chemistry*, 5th ed., 1072pp, New Jersey, USA: Wiley-Blackwell;1999.

[8] United Nations Scientific Committee on the Effects of Atomic Radiation (2008). Report: Volume I, Sources and Effects of Ionizing Radiation, Annex B: Exposures of

the Public and Workers from Various Sources of Radiation. *United Nations Scientific Committee on the Effects of Atomic Radiation*. New York: UNSCEAR; 2010; 273-439.

[9] Vandenhove, H. European sites contaminated by residues from the ore extracting and processing industries. In: *International Symposium on Restoration of Environments with Radioactive Residues*, Arlington, Virginia, USA. Proceedings Series. STI/ PUB/1092. Vienna, Au: IAEA; 2000; 61-89.

[10] Rothstein, A. Uranyl Fluoride. In: Voegtlin C, Hodge HC (eds.). *Pharmacology and toxicology of uranium compounds I-IV*. New York; 548-560.

[11] Rothstein, A. Uranium Dioxide. In: Voegtlin C, Hodge HC (eds.). *Pharmacology and toxicology of uranium compounds I-IV*. New York; 614-621.

[12] Stokinger, H.E.; Baxter R.C.; Dygert H.P. et al. Chapter 21. Uranium Tetrachloride. Toxicity following inhalation for 1 and 2 years. In: *Pharmacology and Toxicology of Uranium Compounds* I-IV, New York: Voegtlin, C.; Hodge, H. C. eds.;1953; 1522-1553.

[13] Gerenutti, M; Gonçalves, MM; Rissato, SR; Oliveria-Jr, JM; Reigota, MAS; Galhiane M Environ Monit Assess Assessment of liquid disposal originated by uranium enrichment at Aramar Experimental Center São Paulo—Brazil. *Environ. Monit. Assess,* 2011.

[14] International Commission on Radiological Protection (1993). Protection against Radon-222 at Home and at Work. ICRP Publication 65. *Ann. ICRP* 23 (2). Oxford: Pergamon.

[15] WHO. (2004). Guidelines for drinking-water quality, vol.1, recommendations, 3rd ed. Geneva, Ch: *World Health Organization.*

[16] CONAMA (2005). Resolution no.357, March 17. Brasília: *Diário Oficial da União.*

[17] IAEA. Features: *Depleted Uranium. Questions and Answers*. 2011. http://www.iaea.org/newscenter/features/du/du_qaa.shtml#q10.

[18] Wagner, SE; Burch, JB; Bottai, M; Pinney, SM; Puett, R; Porter, D et al. Hypertension and hematologic parameters in a community near a uranium processing facility. *Environ. Res.*, 2010 110, 786–797.

[19] Dygert, HP. Pharmacology and Toxicology of Uranium Compounds. New York,: McGraw-Hill Book Inc,1949 apud Pinney, S.M., et al., Health effects in community residents near a uranium plant at Fernald, Ohio, USA. *Int. J. Occup. Med. Environ. Health*, 2003(16); 139–153.

[20] Valdović, V.; Bošković, R. *Radioactivity in the Environment*. Zagreb: Elsevier. 2000.

[21] Boiling water reactor (BWR) basics. *Union of Concerned Scientists*. 2008. http://www.ucsusa.org/assets/documents/nuclear_power/20081000-bwr-basics.pdf.

[22] Toshiba. Advanced Boiling Water Reactors. 2011 http://www.toshiba.co.jp/nuclearenergy/english/business/reactor/abwr01_1.htm.

[23] Marples, D.R. *The Social Impact of the Chernobyl Disaster*. New York,Us: St. Martin's, 1988, 3.

[24] von Hippel, F.N. The radiological and psychological consequences of the Fukushima Daiichi accident Frank. *Bull. At.Sci.*, 2011 67; 27-36.

[25] United Nations Scientific Committee on the Effects of Atomic Radiation (2008).Report: Volume I, Sources and Effects of Ionizing Radiation, Annex B: Exposures of the Public and Workers from Various Sources of Radiation. *United Nations Scientific Committee on the Effects of Atomic Radiation.* New York: UNSCEAR; 2010; 273-439.

[26] United Nations Scientific Committee on the Effects of Atomic Radiation (2000). *Sources and Effects of Ionizing Radiation*. Vol. II, Annex J: Exposures and effects of the Chernobyl accident. www.unscear.org/unscear/publications/2000_2.html.
[27] Devell, L. Nuclide composition of Chernobyl hot particles. In: Von Philipsborn, H; Steinhäusler, F. *Hot Particles from the Chernobyl Fallout. Proceedings of an International Workshop held in Theuern 28/29 October 1987*. Schriftenreihe des Bergbau- und Industriemuseums, Ostbayern Theuern, 1988; 23-34.
[28] aracz, P., E. Piasecki, S. Mirowski et al. Analysis of gamma-radioactivity of "Hot Particles" released after the Chernobyl accident. I. Calculation of fission products in hot particles (A detective approach). *J. Radioanal. Nucl. Chem.*, 1990, 141(2), 221-242.
[29] Jaracz, P. ; Piasecki, E. and Mirowski, S. Analysis of gamma-radioactivity of "Hot Particles" released after the Chernobyl accident. II. An interpretation. *J. Radioanal. Nucl. Chem.*, 1990, 141(2), 243-259.
[30] Kutkov, VA; Arefieva, ZS; Muraviev, YuB; Komaritskaya, O.I., Unique form of airborne radioactivity: nuclear fuel "hot particles" released during the Chernobyl accident. p. 625-630. In: Environmental Impact of Radioactive Releases. *Proceedings of a Symposium*. STI/PUB/971. Vienna, Au: IAEA; 1995; 625-630.
[31] Kutkov, V.A. Application of human respiratory tract models for reconstruction of the size of aerosol particles through the investigation of radionuclides behaviour in the human body. *Radiat. Prot. Dosim.*, 1998 79, 265-268.
[32] Steinhäusler, F. Summary of the present understanding of the significance of hot particles in the Chernobyl fallout. *Proc. Int. Workshop on Hot Particles from the Chernobyl Fallout*. Ostbayern Theuern: v. Philipsborn, H. and Steinhäusler, eds. 1988, band 16, 143-144.
[33] Devell, L. Composition and properties of plume and fallout materials from the Chernobyl accident. In: The Chernobyl Fallout in Sweden - *Results from a Research Programme on Environmental Radiology*. Swedish Radiation Protection Institute, Sweden: L. Moberg, ed. ; 1991; 29-46.
[34] Izrael, Yu.A.; Vakulovskii, S.M.; Vetrov, V.A.; et al. Chernobyl: *Radioactive Contamination of the Environment*. Gidrometeoizdat, Leningrad, 1990.
[35] Borisov, N.B.; Verbov, V.V.; Kaurov, G.A. et al. Contents and concentrations of gaseous and aerosol radionuclides above Unit-4 of the Chernobyl NPP and in the remote zone in May 1986. In: *Questions of Ecology and Environmental Monitoring*, Moscow,1992 1, 11-17.
[36] United Nations (1988). Sources, Effects and Risks of Ionizing Radiation. *United Nations Scientific Committee on the Effects of Atomic Radiation*. New York: United Nations.
[37] Jacob, P.; Roth, P.; Golikov, V. et al. Exposures from external radiation and from inhalation of resuspended material. In: The Radiological Consequences of the Chernobyl Accident. *Proceedings of the First International Conference*, Minsk, Belarus: Karaoglou, A.; Desmet, G.; Kelly, G.N. et al. eds; 1996; 251-260.
[38] Amano, H.; Matsanuga, T.; S. Nagao et al. The transfer capability of long-lived Chernobyl radionuclides from surface soil to river water in dissolved forms. *Org. Geochem.*,1999 30, 437-442.
[39] Hakanson, L. Radioactive caesium in fish in Swedish lakes after Chernobyl – Geographical distributions, trends, models, and remedial measures. In: The Chernobyl

Fallout in Sweden - Results from a *Research Programme on Environmental Radiology. Swedish Radiation Protection Institute*. Sweden: L. Moberg, ed.; 1991; 239-281.

[40] Konoplev, A.V.; Bulgakov, A.A.; Zhirnov, V.G. et al. Study of the behavior of Cs-137 and Sr-90 in Svyatoe and Kozhanovskoe lakes in the Bryansk region. *Meteorol. Gidrol.*, 1998, nº 11, 78–87.

[41] Ryabov, I.; Belova, N.; Pelgunova, L. et al. Radiological phenomena of the Kojanovskoe Lake. In: The Radiological Consequences of the Chernobyl Accident. *Proceedings of the First International Conference*, Minsk, Belarus: Karaoglou, A.; Desmet, G.; Kelly, G.N. et al. eds; 1996; 213-216.

[42] van Hoff, J.; Averkin, Y.I. ; Hilchenko, E.I. et al. Epidemiology of childhood cancer in Belarus: review of data 1978-1994, and discussion of the new Belarusian Childhood Cancer Registry. *Stem Cells*, 1997 15, Suppl. 2, 231- 241.

[43] Remennik, L.V. ; Starinsky, V.V.; Mokina, V.D. et al. Malignant neoplasms on the territories of Russia damaged owing to the Chernobyl accident. In: The Radiological Consequences of the Chernobyl Accident. *Proceedings of the First International Conference*. EUR 16544 Minsk, Belarus: Karaoglou, A.; Desmet, G.; Kelly, G.N. et al. eds; 1996; 825-828.

[44] Tronko, N.D.; Bogdanova, T.; Komissarenko, I. et al. Thyroid cancer in children and adolescents in Ukraine after the Chernobyl accident (1986-1995). In: The Radiological Consequences of the Chernobyl Accident. *Proceedings of the First International Conference*, Minsk, Belarus: Karaoglou, A.; Desmet, G.; Kelly, G.N. et al. eds; 1996; 683-690.

[45] Alvarez, R. ; Beyea, J. ; Janberg, K. ; Kang, J. ; Lyman, E. ; Macfarlane, A. ; Thompson, G. and von Hippel, F. Reducing the hazards from stored spent power-reactor fuel in the United States. *Science and Global Security*, 2003, vol.11, nº.1, 1-51.

[46] Akers, D.W.; Jensen, S.M. ; Schuetz, B.K (1994). *Exanination of relocated fuel debris adjacent to the lower head of TMI-2 reactor vessel*. http://www.eoearth.org/article/Uranium?topic=49557.

[47] International Atomic Energy Agency (2004). Environmental Contamination from Uranium Production Facilities and their *Remediation Proceedings of an International Workshop Lisbon*, López Romero, M. Criado Martín Enusa Industrias Avanzadas SA, Madrid, Spain.

[48] Cardoso, G.V. (2004). Distribuição geoquímica e transferência solo-planta de radionuclídeos naturais em solos sob a influência de uma unidade de mineração e atividade de urânio. Seropédica, Universidade Federal Rural do Rio de Janeiro 67 pp., Rio de Janeiro, Brazil. Master's Thesis.

[49] Silveira, B.C. Impacte radiológico da exploração de urânio em Portugal. *Geonovas, Rev. Assoc. Portg. Geólogos,*2001, nº 15, 71-86.

[50] International Atomic Energy Agency (2004). The long term stabilization of uranium mill tailings.Final report of a co-ordinated research project 2000–2004. *IAEA TECDOC* 1403. Vienna, Au: IAEA.

[51] International Atomic Energy Agency (1992) Current practices for the management of uranium mill tailings. IAEA *Technical Report Series* TR-335. Vienna, Au: IAEA.

[52] International Atomic Energy Agency (2002). Monitoring and Surveillance of Residues from the Mining and Milling of Uranium and Thorium. *Safety Report Series* No. 27. Vienna, Au: IAEA.

[53] International Atomic Energy Agency (2002). *Management of Radioactive Waste from the Mining and Milling of Ores, Safety Standards Series* No. WS-G- 1.2. Vienna, Au: IAEA. EHSO (2001). Radioactive Waste from Uranium Mill Tailings http://www.ehso.com/NuclearuraniumMills.htm.

[54] Sevcik, M. Uranium tailings in Kyrsgyzstan: Catalyst for cooperation and confidence building? *The Nonproliferation Review*, 2003, p.147-154.

In: Uranium: Characteristics, Occurrence and Human Exposure ISBN: 978-1-62081-207-5
Editors: Alik Ya. Vasiliev and Mikhail Sidorov

Chapter 7

URANIUM HYDROTHERMAL DEPOSITS

Miloš René
Institute of Rock Structure and Mechanics,
Academy of Sciences of the Czech Republic, Czech Republic

ABSTRACT

Hydrothermal ore deposits represent a significant type of industrial important U mineralization occurring in highly variable rock environments. These deposits constitute 30% of global U reserves and about 30% of world U production. Hydrothermal U deposits can be grouped in the four main categories based on their geologic setting, namely in vein-like deposits, metasomatic deposits, breccia complex deposits and volcanic–associated deposits. The first three groups usually occur in metasedimentary and in granitic, partly structurally deformed, rocks series. The volcanic–associated deposits occur together with felsic volcanic rocks, mainly in continental extensional settings. The typical ore grades of vein-like and volcanic associated deposits are between 0.1 and 2% U, whereas the ore grade of metasomatic and breccia complex deposits is distinctly lower (0.03–0.2% U). The metasomatic and breccia complex deposits are restricted to the Proterozoic era (Ukraine, Australia), whereas the vein-like deposits are restricted from the Upper Paleozoic to Mesozoic era (Western- and Central Europe, China). The hydrothermal U deposits are commonly related to late phases of orogenic cycles. The ore-forming fluids can be of various types including juvenile, post-magmatic, connate, diagenetic and meteoritic waters. The deposition of uranium minerals may be caused by changes in Eh, pH, pressure, and temperature of the ore-forming fluids. A significant feature of hydrothermal alterations occurring in the area of hydrothermal U deposits is the occurrence of alkali-rich rocks described usually as episyenites. The episyenites often form irregular pipe- or lenses-shaped bodies and are formed by albite, chlorite and hematite in high variable concentrations. The hydrothermal U deposits are thought to have formed by the remobilization of uranium from a variety of pre-existing uranium-rich rocks as black shales and uranium–enriched granitic rocks. The main mineral phases of the hydrothermal U deposits are pitchblende, uraninite and coffinite. The majorities of hydrothermal U deposits are monometallic, but also occur complex-mineralized deposits with economically interested concentrations of Cu, Au, Ag, Pb and Zn.

INTRODUCTION

Hydrothermal ore deposits represent a significant type of industrial sources of uranium. These U deposits have been the source of uranium since its discovery by German chemist M. H. Klaproth in 1789. During his chemical experiments, Klaproth was able to precipitate sodium diuranate by dissolving pitchblende in nitric acid, which mistakenly assumed for oxide of a yet-undiscovered element. The element was extracted by Klaproth from pitchblende, collected from vein deposits that were mined for silver in the Ore Mountains (Erzgebirge), part of the Hercynian orogenic belt in Central Europe. However, the first use of uranium compounds was as coloring agents in glassmaking and porcelain industry [1]. The new history of uranium is coupled with the discovery of its radioactivity by H. Becquerel (1896) [2], but the new atomic era was started by a curious observation made by M. Curie, which found that pure U actually had a lower radioactivity than natural U compounds, inclusive pitchblende from hydrothermal uranium deposit in Jáchymov (Ore Mountains, Czech Republic) [3]. After the discovery of radium (Curie and Curie 1898) [4], the uranium ores were used to extract the radium for medical purposes and the making of luminescence colors.

Uranium hydrothermal ore deposits occur in variable rock environments and can be grouped in the four main categories based on their geotectonic setting, namely in vein-like deposits, metasomatic deposits, breccia complex deposits and volcanic–associated deposits.

VEIN-LIKE URANIUM DEPOSITS

Vein-like uranium deposits are epigenetic concentrations of U minerals, typically uraninite, pitchblende and coffinite; in open spaces (faults and fissures) and/or in relatively closed shear zones accompanied by irregular areas of huge hydrothermal alterations. In veins are uranium minerals commonly associated with carbonate and quartz. For U-mineralization evolved in shear zones is the common association of uranium minerals, predominantly coffinite and pitchblende with chlorite, albite and other products of intensive hydrothermal altered surroundings rocks series. Hydrothermally altered rocks enriched in albite, but co-existed with removal of quartz from surrounding rocks are generally described as episyenites or eysites. The term episyenite is frequently used by French geologists who have described these rocks firstly from hydrothermal uranium deposits in the Western European Hercynian granite belt, particularly from the Massif Central and the Armorican Massif in the France. The term was used for the first time by Lacroix (1920) [5] to describe the alteration, which brings about the removal of the original magmatic quartz during hydrothermal processes while the rock assumes an almost syenitic character. The term eysite, used usually in the Russian papers about uranium vein-like and metasomatic deposits, was suggested by Omelyanenko et al. (1974) [6] for the granites affected by a series of hydrothermal alterations (hematitization, albitization and chloritization), which accompany certain hydrothermal uranium deposits. They used a phonetic transcription for the name of Canadian uranium deposit Fay–Ace–Verna. Episyenites connected with uranium mineralization was firstly described from intragranitic deposits in France [7, 8], but this hydrothermal alteration also occurs in U deposits coupled with shear zones without association with granites (e.g., Rožná–Olší ore district, Bohemian Massif, Czech Republic, [9]).

The vein-like U deposits are classified using some different criteria. These deposits are often distinguished according to their metallic mineral composition: (1) monometallic, where uranium is the principal ore constituent; and (2) polymetallic, where U is associated by other metals, such as Ni, Co, As, Bi, Cu, Ag and Au. Some other classification is based on the geological setting of these deposits, respectively their relation to granitic bodies, which are often suggested as potential sources of uranium. Vein-like deposits hosted by granitic rocks are intragranitic deposits; whereas vein deposits developed in metasedimentary or metavolcanic rocks at their contacts with granitic plutons are perigranitic deposits.

Intragranitic vein deposits are typically evolved in the Western European Hercynian belt (*La Crouzille ore district*, Le Bernardan, Massif Central, Vendée ore district, Armorican massif, France, Figure 1) and/or in the South China (*Xiazhuang ore district,* Figure 2), whereas the perigranitic U-deposits occur in the Central European part of the European Hercynian belt (*Schlema–Alberoda ore district,* Germany, *Příbram ore district*, Czech Republic, Bohemian Massif).

The U ore deposits connected with shear zones are typically evolved on margin of granitic plutons (Zadní Chodov, Okrouhlá Radouň, Bohemian Massif, Czech Republic) and/or in areas without some direct connection to granitic plutons (*Rožná–Olší ore district,* Czech Republic, Bohemian Massif). All these uranium deposits are part of the European Hercynian belt.

Intragranitic vein deposits that occur in the Massif Central, particularly in the *La Crouzille ore district* and/or in the Vendée ore district of the Armorican Massif, are distinctly associated with Hercynian granites of the both geotectonic units. The uranium ore veins occur predominantly in two-mica leucogranites and are associated with zones of albitization and K-alteration.

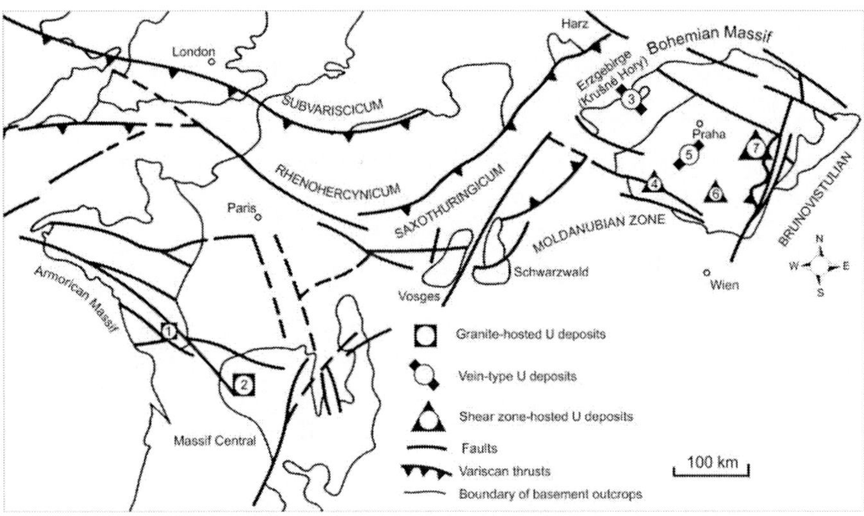

Figure 1. Hydrothermal granite-hosted, vein-type and shear zone-hosted uranium deposits in the West- and Central-European Hercynian belt.

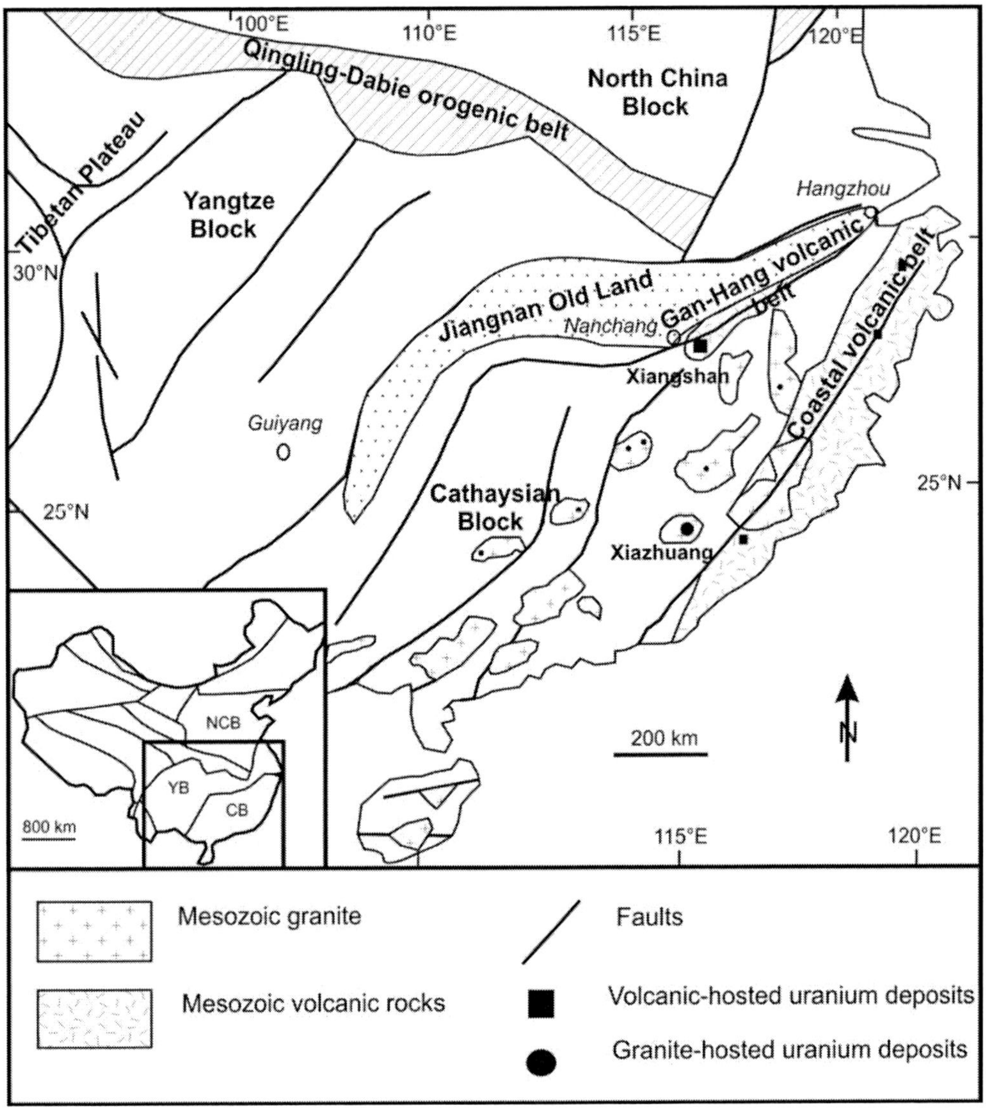

Figure 2. A sketch geological map of the south-eastern China (modified from Hu et al. 200 [27]).

La Crouzille Ore District

The uranium deposits of the La Crouzille uranium district are located in the Saint-Sylvestre plutonic complex, approximately 20 km N of Limoges (Haute-Vienne, France, Figure 3). This plutonic complex is composed of biotite-sillimanite Brame granite in the western part and the two-mica Saint-Sylvestre granite in the eastern part. These granites were emplaced in metamorphic formations (micaschists, gneisses and leptinites) of the Massif Central around 336 Ma ago (U–Pb method). Two fault systems have been recognized in the Saint-Sylvestre granite. The older, E–W and the NNE–SSW oriented fault system controls muscovitization and desilicification of the two-mica granites and emplacement of lamprophyric and microgranitic dykes. The uranium-bearing veins are related to a second

NW–SE faulting system [10]. Uranium mineralization immediately followed the emplacement of the lamprophyres and the micaeous episyenitization of the granite with a minimum age of 276 Ma (U–Pb). However, in some samples, a remobilization stage with alteration of primary pitchblende at 170–175 Ma (U–Pb) was found [11]. Two textural types of U mineralization are distinguished, vein and disseminated mineralization. Vein-like mineralization occurs as brecciaed veins with pitchblende, coffinite pyrite, hematite, quartz, cemented with muscovite and clay minerals. Sulfide minerals are represented by marcasite, galena and sphalerite. Disseminated-type mineralization represents nearly 40 vol.% of uranium mineralization at the Margnac mine and 30 vol% at the Fanay mine, the two most important U deposits of this ore district. Pitchblende is the primary U mineral, closely accompanied by pyrite. Quartz is more rare than in the veins and appears together with marcasite and coffinite. Coffinite represents an important part of U mineralization and is followed by fluorite, barite and calcite. The microthermometry based on the study of fluid inclusion shows that episyenites originated between 210–310 °C. Uraniferous fluids were carbonatic and their temperature was about 340–350 °C. During crystallization of coffinite and fluorite, temperature fluids decreased slowly from 140 °C to below 100 °C [12]. In the La Crouzille area, the U ores were exploited at Margnac, Peny, Fanay, le Fraise, les Gorces, Bellezane and Montulat deposits (Ruzicka 1993). The total mined amount of uranium from this district was about 23,600 t U.

Schlema-Alberoda Ore District

The Schlema-Alberoda uranium ore district is located in the Western Ore Mountains (Erzgebirge), in Germany near the state boundary to the Czech Republic. This ore district is evolved in the intersection of the SW–NE striking Loessnitz-Zwoenitz syncline with NW–SE trending Gera-Jáchymov fault zone (Figure 4). The Loessnitz-Zwoenitz syncline is one from sectional tectonic structures, which are ingredients of the Erzgebirge–Fichtelgebirge anticlinorium in the fold framework of the Saxothuringian Zone (Bohemian Massif). The most important and central tectonic element of the Gera-Jáchymov fault zone is the vein structure Red Ridge (Roter Kamm), also defining the border between the Schlema-Alberoda ore district in NE and the Schneeberg uranium deposit in SW. In the Loesnitz-Zwoenitz syncline, predominantly Upper Ordovician-Silurian-Middle Devonian "productive" rocks are folded into lower Ordovician schists of the northern edge zone of the Erzebirge-Fichtelgebirge anticlinorium. The rocks series of the "productive unit" are phyllites with intercalations of metamorphosed black shales and metacarbonates. The uranium-bearing veins occur in the contact metamorphic zone of the syncline beneath the late-Variscan Aue granite pluton. This granite body, located within the Gera-Jáchymov fault zone, intruded early-Variscan metasediments, especially low-grade garnet phyllites and medium-grade mica schists formed during an amphibolite-facies, Barrow-type metamorphism. The Aue granite body comprises various biotite granite of the Aue granite suite (324 Ma, U–Th–total Pb age on uraninite). The Aue granite suite should have served as a major source for U accumulated in post-granitic deposits of Schneeberg and Schlema-Alberoda ore districts [14, 15, 16].

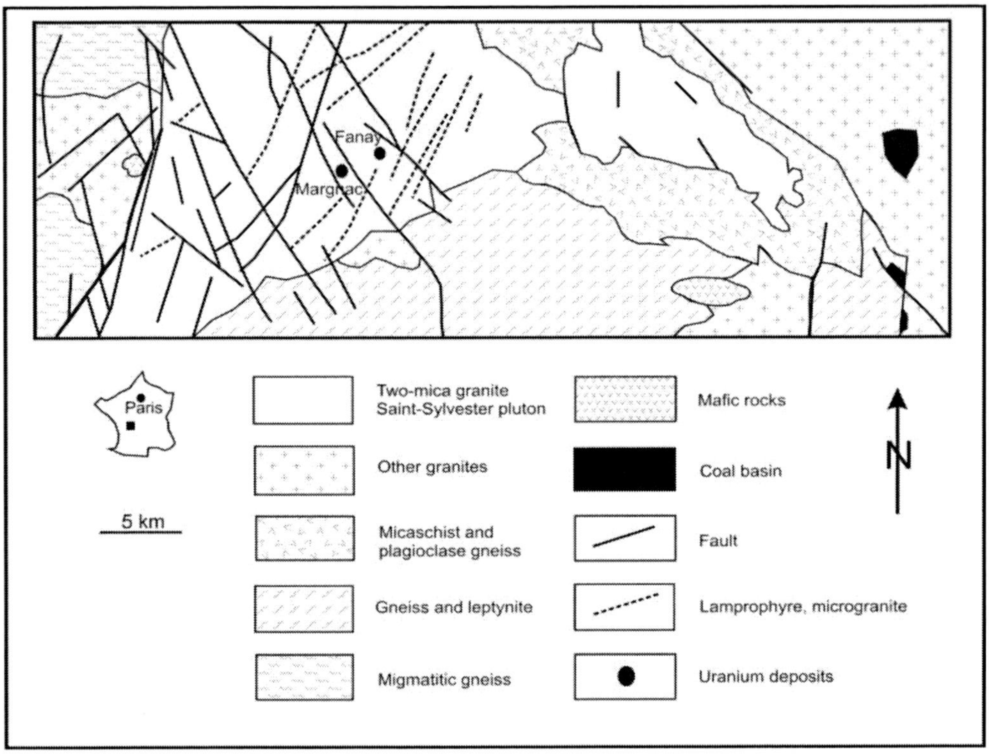

Figure 3. Schematic geological map of the La Crouzille uranium ore district (modified from Leroy 1978 [10]).

An extraordinary number of faults, fissures and veins of different directions intersect the rocks series of the ore district area. These fissures are formed by tectonic processes such as contraction and relaxation before, during and after granite intrusion. During repeated activation by further fault-block movements, the fissures served as paths of migration for circulating hydrothermal solutions and for deposition of their mineral content. There are faults oriented approximately in the direction of strata striking, but most of them are trending in WNW–ESE, NW–SE and NNW–SSE directions. The post-collisional lamprophyre dykes are controlled by intersections of deep-rooted fault zones and lineaments similar to the uranium deposits (Figure 5). The uranium ore veins have a common thickness from 0.1 to 0.3 m with a maximum of 1 m. However, some ore veins show a massive pitchblende mineralization with a thickness up to 2 m, which were mined down to a depth of about 2,000 m. The hydrothermal mineralization is usually divided into three main stages [17, 18]. The most important, first pitchblende-quartz-calcite-fluorite-sulphide stage has age 300–250 Ma (pitchblende, U–Pb). The temperature of hydrothermal solutions deducted from the fluid inclusions study was 143–156 °C and salinities of fluid inclusions were relatively low (0–1.2 wt% NaCl eq.). The second, post-Variscan dolomite-selenide-pitchblende stage with U–Pb pitchblende age of 200–140 Ma displays temperature of fluids 117–137 °C and some higher salinity of fluids (6.9–7.7 wt% NaCl eq.). The third, Bi–Co–Ni stage with a predominance of arsenides and sulphides Co and Ni, native Bi and late-stage Ag mineralization displays pitchblende age 170–90 Ma, temperature of fluids from 48 °C to 90 °C and high salinities of fluids (16–23 wt% NaCl eq.). The latter two mineral stages often caused cataclastic alteration and remobilization of older pitchblende generations [19]. The total amount of uranium, mined in the Schlema-Alberoda ore district from 1946 to 1990 was 88,270 tons of U. This ore district was the most productive uranium district in the whole European Hercynian belt. The main mined ores were uranium ores and accompanying ores (Co, Ni, Ag, Bi, Se and/or Pb, Zn and Cu from older quartz-sulfide veins) have been extracted only temporarily and in small quantities [17, 20].

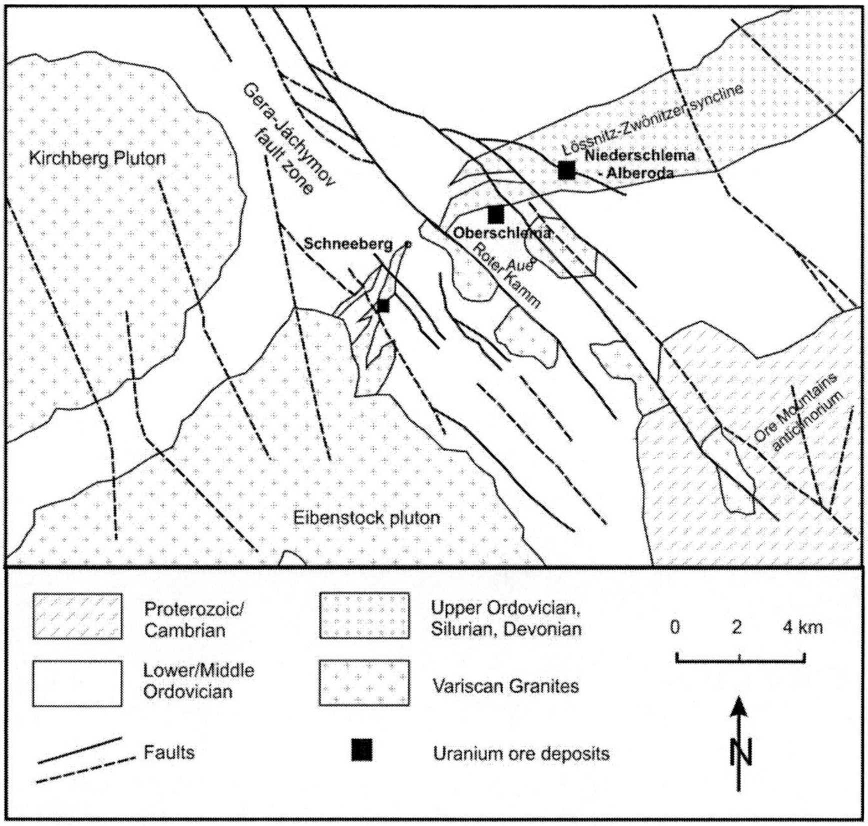

Figure 4. Schematic geological map of the Schlema–Alberoda uranium ore district (modified from Hiller and Schuppan 2008 [20]).

Příbram Ore District

The Příbram uranium ore district extends along the northwestern contact of the Central Bohemian Plutonic complex (CBPC) with Neoproterozoic and Cambrian rocks of the Teplá-Barrandian Zone (Figure 6). The 3,200 km^2 complex crops out between two contrasting crustal units, in the Teplá–Barrandian Zone to the NW and Moldanubian Zone to the SE, in the central part of the Bohemian Massif. The plutonic complex is made up of multiple individual plutons and smaller magmatic bodies that vary in age, petrographic and geochemical characteristics, shape, size, internal fabrics and relationships to the host rock structures, from older calc-alkaline to potassium-rich calc-alkaline and ultrapotassic magmas. In the NW margin of the CBPC, in area of the Příbram ore district, crop out biotite and biotite-amphibole granodiorites of the Blatná suite (346 Ma, U–Pb), together with the Marginal, high-K calc-alkaline biotite granites [21]. The Neoproterozoic, a slightly metamorphosed flysh sequence, up to 2,000 m thick, are overlain by a Lower Cambrian sediments, containing thin layers of quartz-pebble conglomerate at its base and slates at a higher stratigraphic position. A volcano-sedimentary complex underlies the Neoproterozoic flysh sequence, comprising intercalated claystones, sillstones and conglomerates. Both the Lower Cambrian and Neoproterozoic rocks are contact metamorphosed by the CBPC within an aureole that extends 1,000 to 1,200 m from the intrusive contact. This aureole is also cut by aplite and lamprophyre dykes. The Neoproterozoic host rocks form a simple fold, the Příbram anticline, with NE-trending axis, roughly parallel to the CBPC contact. Brittle structures in the Příbram ore district may be classified relative to the axial plane of the Příbram anticline as: (i) The most prominent, longitudinal NE-striking faults, i.e. subparallel to boundary of Neoproterozoic metasediments with the CBPC, (ii) transverse NW-striking faults, and (iii) oblique, E- or N-striking faults. Ore veins in the ore district strike NW (44 % of veins), N (43 % veins), and NE (13 % veins). The ore veins are from a few meters to several kilometers long, from a few centimeters to more than 10 meters wide and comprises three mineral assemblages (from older to younger): (1) sulphidic with Pb, Zn and Cu-Fe sulphides, (2) uraninite bearing, and (3) sulphide-selenide-carbonate. The main U-bearing minerals are uraninite and U-anthraxolite, coffinite being far less abundant. Uranium minerals occurred as veinlets, coatings and pods in calcite gangue [22]. The most precise dating of uraninite gives age 275–278 Ma [23]. A fluid inclusion study of carbonate-uraninite stage gives temperatures 80–140 °C and low salinities up to 5 wt% NaCl eq. [24]. The deposit has been mined to a depth exceeding 1500 m and total mined amount of U was about 51,000 tons.

Rožná-Olší Ore District

The Rožná-Olší uranium district lies in the NE part of the Moldanubian Zone of the Bohemian Massif (Figure 7). The high-grade metamorphic rocks of this zone were overthrust on its NE boundary by the Svratka Crystalline unit and on the easterly located Cadomian Brunovistulian foreland. The high-grade paragneisses of the Moldanubian Zone are subdivided into Monotonous, Varied and Gföhl Unit. The Rožná-Olší ore district is located in the uppermost Gföhl unit. The host rocks of the Rožná U deposit consist mainly of biotite paragneisses and amphibolites with small bodies of calc-silicate rock, marble, serpentinite and pyroxenite. The subsequent exhumation of these rocks series to middle crustal levels was associated with kilometer-scale isoclinal folding. Longitudinal N–S to NNW–SSE striking ductile shear zones (Rožná and Olší shear zones) dip WSW at an angle of 70–90° and strike parallel to the tectonic contact between the Gföhl unit and the Svratka Crystalline Unit. The main longitudinal faults of the Rožná shear zone are designated as Rožná 1 (R1) and Rožná 4 (R4) and host the main part of the disseminated uranium mineralization. The less strongly mineralized Rožná 2 (R2) and Rožná 3 (R3) fault zones host numerous separate pinnate carbonate veins (Figure 8). Longitudinal fault structures are crosscut and segmented by steep, ductile to brittle NW–SE and SW–NE-striking fault zones that host post-uranium carbonate-quartz-sulfide mineralization.

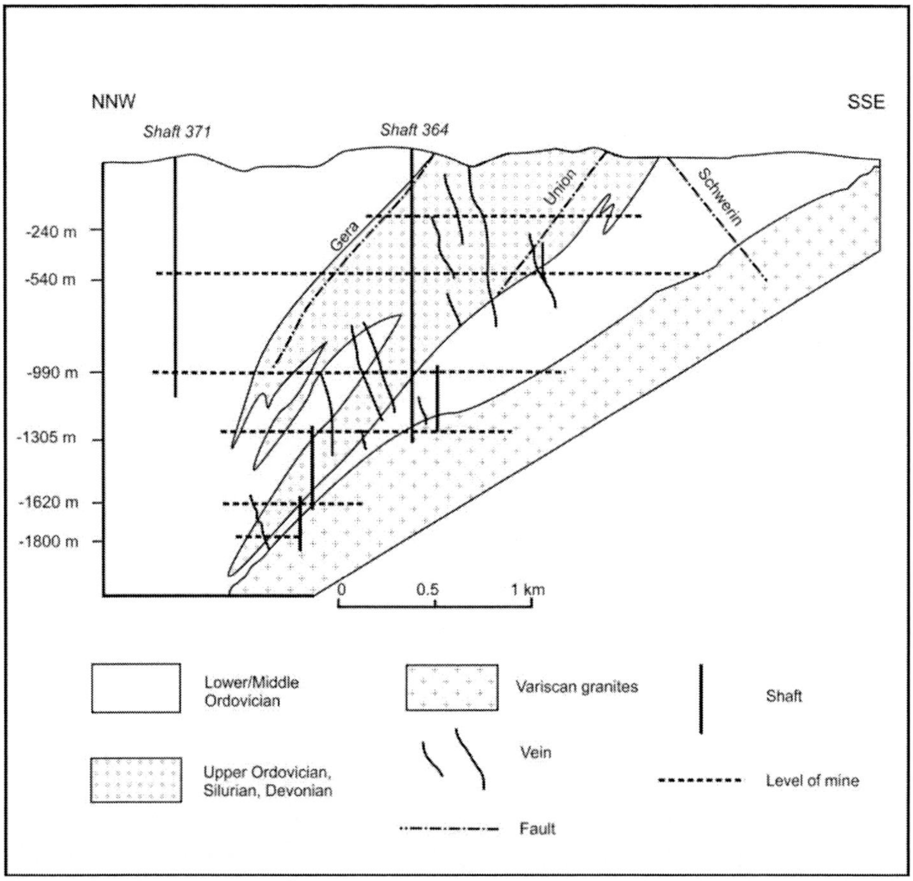

Figure 5. Schematic cross section of the Schlema–Alberoda uranium ore deposit (modified from Hiller and Schuppan 2008 [20]).

Uranium mineralization forms (i) disseminated coffinite>uraninite>U-Zr-silicate ore in chloritized, pyritized, carbonatized, and graphite-enriched cataclastites of longitudinal faults, (ii) uraninite>coffinite ore in carbonate veins, (iii) disseminated coffinite>uraninite in desilicified, albitized and hematitized gneiss (episyenite) adjacent to longitudinal faults and (iiii) mostly coffinite ore bound to the intersection of the longitudinal structures. Ore lenses of disseminated ore in fault zones R1 and R4 are 3.5 thick on average, ore grade is around 0.5% U, up to 10% U locally. Ore bodies in ore zones R2 and R3 host a large number of carbonate veins up to 2 m thick with U-mineralization of the average grade 0.6% U. Ore bodies in episyenites with predominance of coffinite on uraninite and U–Zr silicate have U-mineralization of a grade 0.1–0.15% U, exceptionally 0.3% U. Disseminated U-mineralization bound to oblique fault zones is usually hosted by quartz-carbonate-sulfide breccias at intersections with diagonal and longitudinal structures. Compared to other types of mineralization, the ore bodies are small but contain relatively high-grade ore of average grade 0.8% U and up to 20% U in some ore shoots. Total mine production of the Rožná-Olší ore district was 23,000 tons U with average grade of 0.24% U [9]. The Rožná uranium deposits is the last recently mined uranium deposit in Central Europe with an annual production about 300 t U.

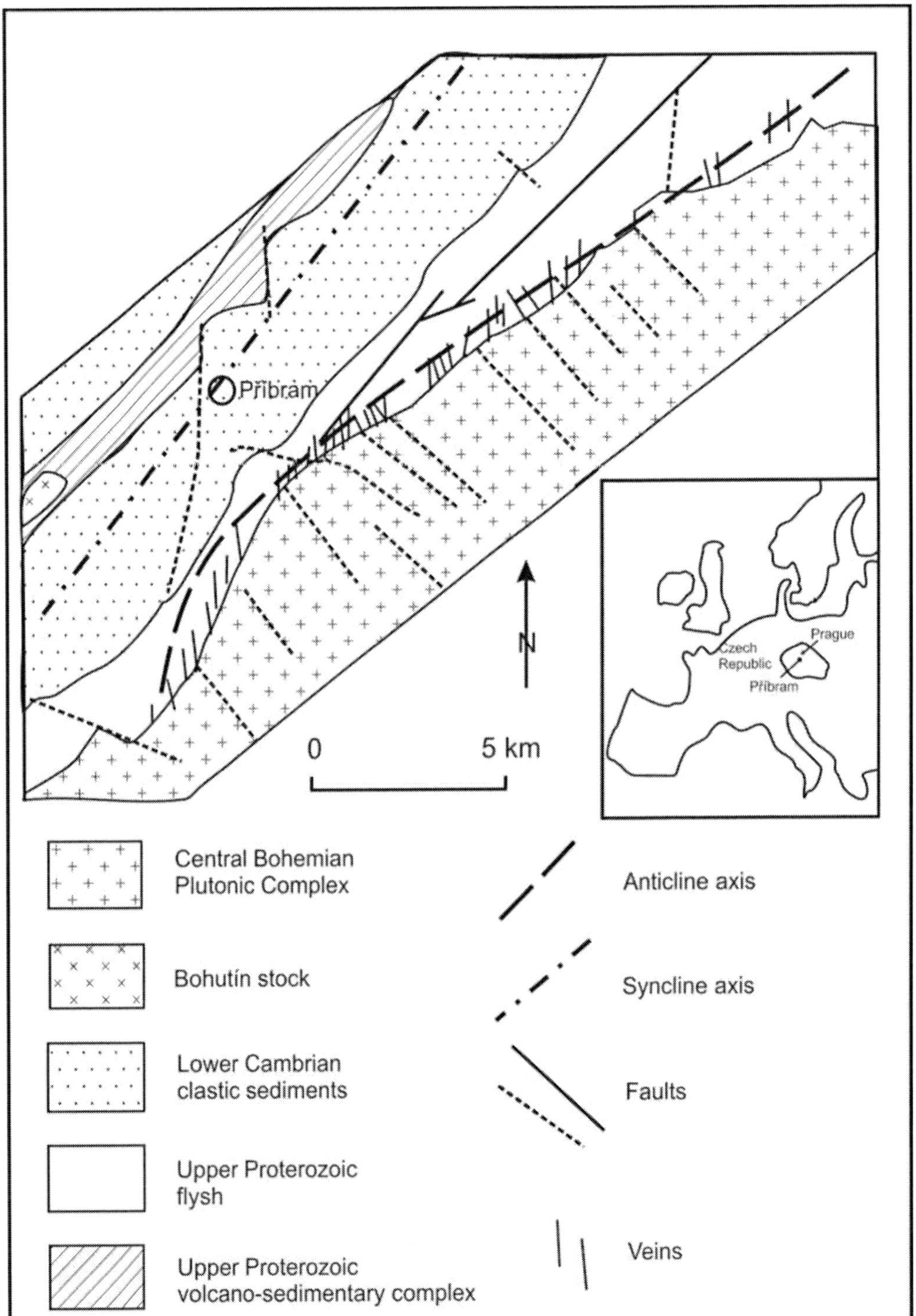

Figure 6. A sketch geological map of the Příbram uranium ore district (modified from Křibek et al. 1999 [22]).

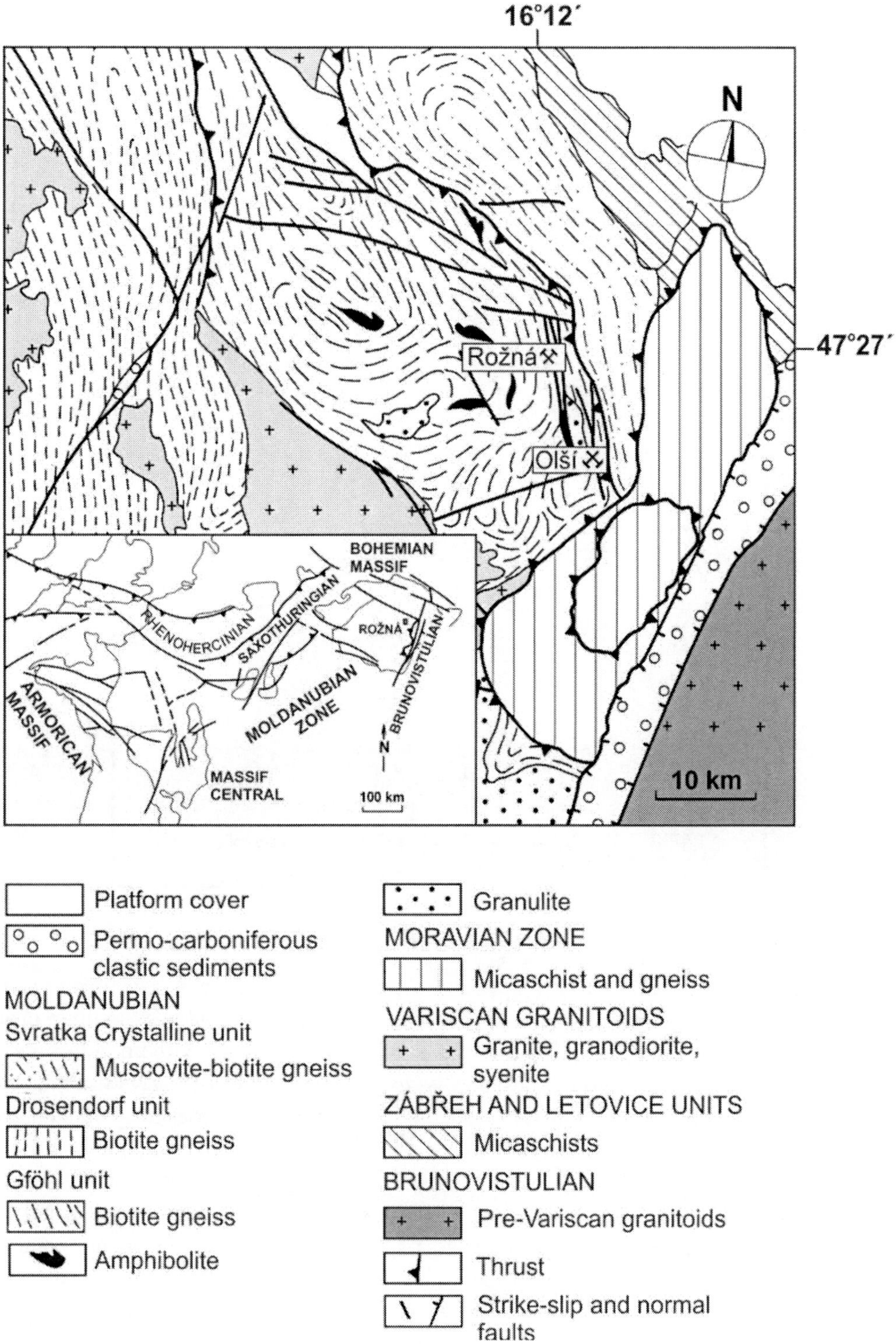

Figure 7. Schematic geological map of the Rožná–Olší uranium ore district (modified from Křibek et al. 2009 [9]).

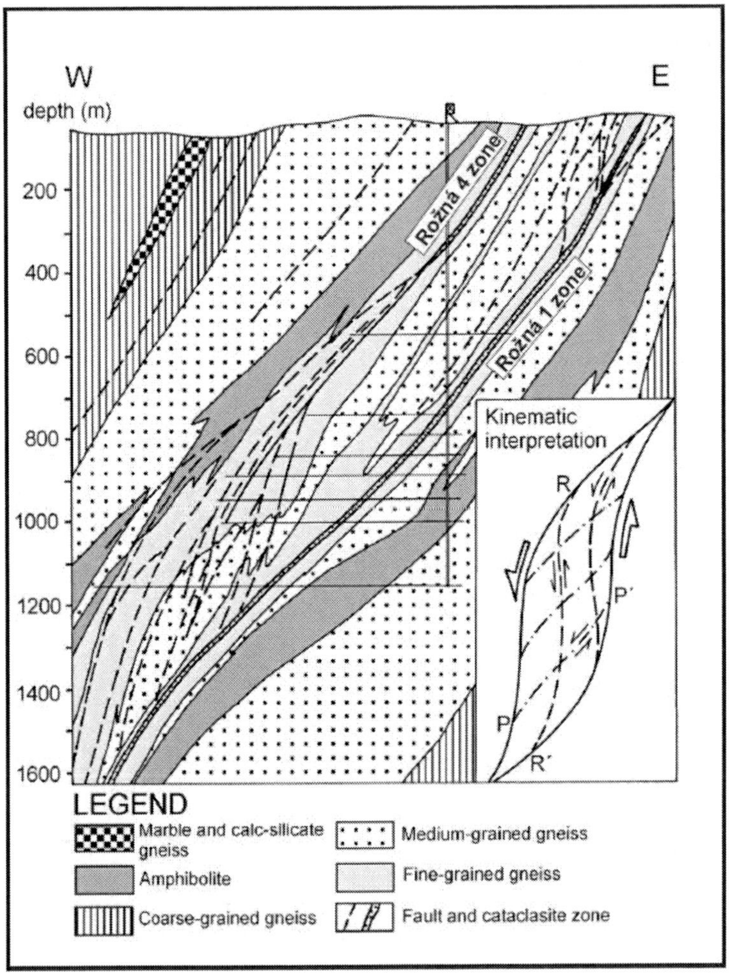

Figure 8. Schematic cross section of the Rožná uranium deposit.

Uranium mineralization at Rožná ore deposit comprises pre-uranium stage quartz-sulfide and carbonate-sulfide mineralization, uranium stage mineralization and post-uranium quartz-carbonate-sulfide mineralization. Uranium stage can be subdivided into the pre-ore, ore and post-ore substages. The pre-ore substage is characterized by widespread albitization, hematitization, chloritization, and carbonatization of host rocks series. Uranium (and selenide) mineralization formed during the ore substage and comprises uraninite, coffinite, (U–Zr)-silicates, complex U–Ti–Si–Zr phases, and less important brannerite. Post-ore substage is formed by carbonates, chlorite, sulfides, and subordinate quartz. Sulfide minerals are represented by pyrite, marcasite, galena, sphalerite and some other rare sulfides. Homogenization temperatures for fluid inclusions from ore substage occurred in carbonates are 152–174 °C and their salinities are highly variable (3.1–23 wt% NaCl eq.). The age of the main ore stage is according to U–Pb dating of uraninite 260–280 Ma [25]. The origin of the uranium mineralization is associated with the infiltration of oxidized basinal brines of the Upper Stephanian and Lower Permian basins into the crystalline basement along deep brittle structures that opened during the late Hercynian transcurrent crustal block reorganization. Infiltration of basinal brines effectively leached U from U-bearing accessories and silicates in the crystalline rocks. Apart from the vein-type deposits in the Ore Mountains (Schlema-Alberoda U district) and the Příbram U district, which are characterized by proximity of granitic plutons and by very weak hydrothermal alteration, the Rožná deposit shows extensive hydrothermal alterations of the country rocks and a mostly disseminated style of mineralization. Reactions of hydrothermal fluids with the host rocks can also be held responsible for the much wider variability in the chemical composition of the fluids compared to vein-type deposits [9].

Xiazhuang Ore District

The Xiazhuang ore district occurs in Guandong Province, southern China (Figure 2). The ore deposits of this district (e.g., Zhushanxia, Shijiaowei and Xiwang) are associated with granitic rocks of the Yanshanian structural unit. The host Guidong granites were emplaced 185–135 Ma ago, whereas the uranium mineralization was deposited later, 87–71 Ma ago [26, 27]. The orogenic igneous activity of the Triassic–Jurassic age is interpreted as the result of westernward subduction of the Paleo-Pacific plate. The various tectonic elements of South China were amalgamated into a single terrane in the Early Triassic and thereafter had a similar geologic evolution. The uranium ore district occurs in two-mica, relatively Ca-poor, S-type granites in the eastern part of the Guidong batholith, where granites are crosscut by numerous Cretaceous (91 Ma) mafic dykes (lamprophyres, basalts). These dykes usually intruded along NE–SW faults and were generally accepted to be products of Cretaceous–Tertiary crustal extension. The extension phase corresponds broadly to the major episodes of uranium mineralization. Uranium mineralization is monometallic. Mineral assemblage includes pitchblende, pyrite, hematite, quartz, fluorite and calcite. Hydrothermal alteration of host granites includes silicification, hematitization, sericitization, carbonatization and argillization. The main ore stage is usually connected with silicification and hematitization, the other hydrothermal alterations are later. From a study of fluid inclusions on quartz temperature of ore-fluids was estimated at 150–250 °C. Inclusions have low to moderate salinity (1.3–8.9 wt% NaCl eq.). They can be generally classified into H_2O rich and H_2O–CO_2 inclusions. Hydrogen and oxygen isotopes support the suggestion that the water in ore-forming fluids was dominantly meteoritic. Uranium was considered to have been derived from U-rich granites in which uranium exists dominantly in uraninite. Reserves of particular ore deposits in this district are from 3,000 to 10,000 tons in grades 0.1–0.3% U [27].

Metasomatic Uranium Deposits

Metasomatic deposits are hosted in structurally deformed rocks series that have been affected by hydrothermal aided alteration with significant participation of Na-enriched solutions. During alteration original rocks series were enriched in alkali-rich minerals as well as albite, K-feldspar, alkali pyroxenes and alkali amphiboles. This alkali metasomatism is a widespread geological process and is most commonly associated not only with enrichment on these alkali-rich minerals but also with quartz dissolution, oxidation of Mg- and Fe-enriched minerals. During this alkali-enriched alteration discontinuous occurrences of metasomatic zones over tens of kilometers were formed. However, the uranium–enriched metasomatic zones in these areas are usually several meters wide and several hundred meters to several thousands of meters long. The alkali metasomatism is sometimes connected with carbonatization. All of these alterations are mostly concentrated on regional evolved shear zones. Ore grades of metasomatic deposits are typically low at less than 0.2 wt% U_3O_8, but may range up to 3 wt% U_3O_8.

Two subtypes of metasomatite deposits are defined on the basis of host rocks:

Metasomatized metasedimentary rocks series
Metasomatized granitic rocks

Altered metasediments are mostly found in middle to high rank, partly dynamothermally, metamorphosed rocks series adjacent to granite intrusions in mobile belt terranes. For these alterations is significant feldspars occurrence together with alkali pyroxenes and amphiboles (typically aegirine and arfvedsonite). Transition zones between metasomatized zones and parent rocks series are commonly narrow. Alteration is reflected by the enrichment of Na_2O, depletion in SiO_2 and K_2O. Locally highly intense hematitization and/or carbonatization are evolved. Principal uranium minerals are Th-enriched pitchblende, uranothorianite, uranothorite, thorite, brannerite and Th-enriched, highly metamictized zircon (malacon). Uranium minerals are disseminated in altered rocks and partly concentrated in thin veinlets.

Uranium mineralization associated with metasomatized metasediments occurs predominantly during two discrete periods of earth history. The major one corresponds to the 1.4–1.9 Ga highly mineralized the Central Ukrajine ore district (in the older Russian paper called the Krivoy Rog ore district), the Valhalla ore deposit in Australia and the Liangshanguan ore deposit in northeastern China. A less widespread uranium metasomatic ore event is associated with the Pan-African-Brazilian ore deposit of Espinharas, northeastern Brazil (cca 500 Ma) and at the Kitongo ore deposit (Cameroon).

The most significant representation of uranium metasomatic deposits originated by alteration of metasedimentary rocks series is *Central Ukrainian Uranium Province (CUUP)* which occurred in the central part of Ukrainian Shield (Figure 9) [28, 29]. Two important uranium ore districts (Kryvy Rig and Kirovograd) are hosted in the banded iron ore formation of Kryvy Rig-Kremenchug shear zone and in gneisses, migmatites and granites of Novoukrainka pluton along the Kirovograd shear zone. Alkali metasomatism and uranium mineralization were controlled by tectonic activity along N-S evolved shear zones with ductile to brittle regimes of deformation. Altered rocks series are made by dequartzified rocks (episyenites), aegirine-albitites (Na-metasomatites), andradite–epidote–Ca-amphibole–calcite-rich rocks (Ca-Mg-metasomatites) and biotite-enriched albitites (K-metasomatites). Uranium ores, predominantly consisting of brannerite and uraninite, with lesser amounts of U-ferropseudobrookite, are associated with Ca-Mg and K-metasomatized zones. The most reliable age for uranium mineralization origin is provided by U-Pb isotopic age determination on uraninite and pitchblende (approximately 1.8 Ga). For Ca-Mg-K, metasomatism is a significant enrichment of altered rocks in Ca, Mg, K, U, Zr, Fe and Hf. The hydrothermal fluids were oxidized, moderately saline and highly alkaline. The high alkalinity of the fluids explains the dissolution of quartz, deposition of albite, the high Zr/Hf ratios (up to 224) and the high solubility of U, Th and Zr. Na-Ca-Mg metasomatism together with uranium mineralization is controlled by shear zones. The metasomatic mineralization can be related to four and/or five stages of hydrothermal activity [29, 30]. The first stage corresponds to quartz dissolution and limited albitization (episyenitization) of original metasediments. Mafic minerals are replaced by chlorite I and locally by epidote I. The second, Na-metasomatic stage produced massive albitization of host rocks and formation of albitite (episyenite) bodies. Aegirine, riebeckite, crossite or arfvedsonite crystallized from the Fe liberated by the alteration of original mafic minerals. During this stage, hematite, titanite and rutile originated. Third, Ca-Mg metasomatic stage corresponds to the replacement of aegirine and sodic amphiboles by andradite-titanite-epidote-diopside and amphibole-calcite mineral associations. During this stage main part of uranium mineralization formed by U-pseudoferrobrookite and brannerite originated, together with metamict hydrothermal Th- enriched zircon (malacon) and apatite. Brannerite is the most abundant U mineral in these ore deposits. The fourth, K-metasomatic stage is connected with the origin of pitchblende (uraninite), coffinite and boltwoodite. Pitchblende and/or uraninite usually replace brannerite and U-pseudoferrobrookite and in these U-oxides can be later altered to coffinite and uranophane or boltwoodite. The last, late chlorite-epidote stage corresponds with the origin of new crystallized chlorite II and microcrystalline epidote-calcite aggregates.

Total resources of the CUUP exceed 300,000 t U. However, uranium mineralization is commonly of low grade, generally 0.01% U, only occasionally can reach up to 3% U. The uranium ore district of the Kryvy Rig-Kremenchug shear zone could be represented by the Zhovta Richka ore deposit (18,900 t U) (in older Russian papers called the Zheltorechenskoye deposit and/or the Zheltye Vody ore district). The distinctly more important Kirovograd ore district could be represented by the Novokostantynivka ore deposit (93,600 t U) and the Michurinske deposit (26,800 t U).

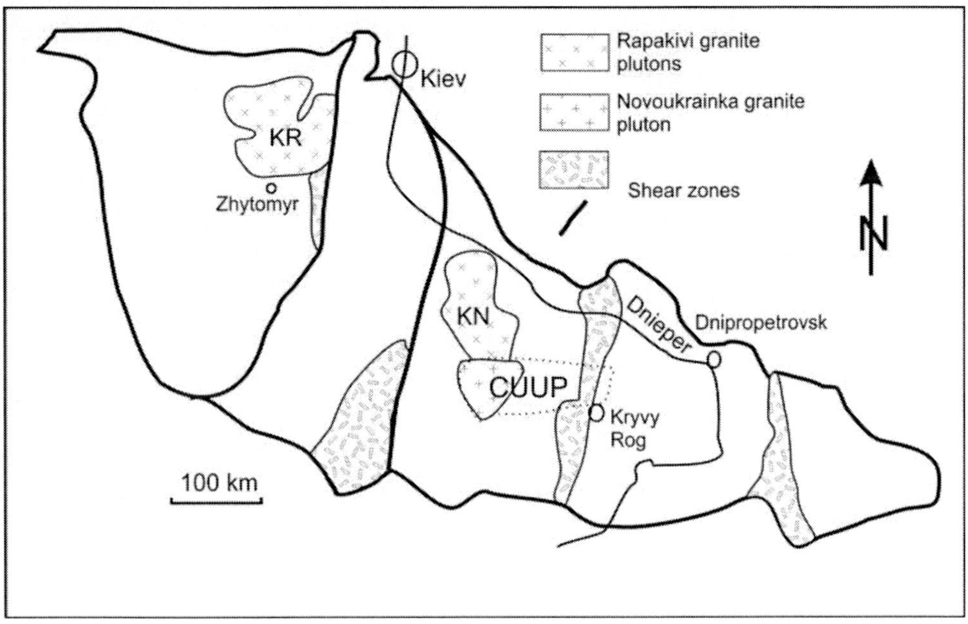

Figure 9. A sketch geological map of the Ukraine shield (modified from Cuney et al. 2011 [29]).

Zhovta Richka Ore Deposit

The Zhovta Richka ore deposit covers an area of 1.5 × 0.7 km of the central part of the Zhovta Richka syncline on the southern flank of the Kryvy Rig-Kremenchug synclinorium zone (Figure 10). Brecciated shear zones are preferentially evolved conformably with the foliation of the host rocks series. The shear zones control the origin of metasomatic ore bodies, which are usually tens of kilometers long and are several tens of meters thick. The mineralized zones are commonly evolved as highly variable lens- and pipe-like bodies. Uranium mineralization mostly occurs in the magnetite-albite-riebeckite albitites, but aegirine- and riebeckite-enriched rocks may also be mineralized. The original host rocks series of the Zhovka Richka deposit are formed by magnetite quartzites (banded Fe-ore formation), quartz-biotite-amphibole schists, dolomitic limestones and granitized gneisses. The richest uranium ore zones occur on the contact between albitites and riebeckite-enriched rocks. Uranium minerals (uraninite, brannerite, sooty pitchblende, nenadkevite) form spots, bands and veinlets in brecciaed host rocks. In the Ca-bearing metasomatites, apatite-zircon–enriched mineralization also occurs with higher amounts of titanite. Dolomite-pitchblende veinlets are commonly associated with antraxolite and coffinite.

Novokostantynivka Ore Deposit

The Novokostantynivka uranium deposit covers an area of 1.5 × 1.5 km in the northern part of the Novoukrainska granite pluton, near of its contact with the Korsun-Novomyrgorod rapakivi granite pluton which was emplaced in the Archean metasediment approximately before 1.7 Ga in an anorogenic geodynamic setting. Granite of Novoukrainska granite pluton represents a potential uranium source with 2 to 19 ppm U and Th/U ratios of 2 to 21, as already proposed by Belevtsev et al. (1995) [31]. Distribution of ore mineralization is controlled by Novokostantynivka N-S shear zone. Within the shear zone, breccias extend overall of several tens to hundreds meters and are up to 10–15 m thick. Lens- and stockwork-type ore bodies are concentrated into three ore zones. The uranium mineralization is predominantly localized in biotite, aegirine, actinolite-riebeckite and chlorite-epidote albitites, which are surrounded by episyenites. The most abundant uranium mineral is brannerite, whereas uraninite occurs in subordinate amounts. The fluid inclusions in episyenites homogenize between 110 and 210 °C with salinities of 4.5 to 10.5 wt% NaCl eq. The H and O isotopes of inclusion fluids indicate a meteoric/formation water source of the albitizing fluids [29].

Michurinske Ore Deposit

The Michurinske uranium deposit occurs along the Kirovograd shear zone and is adjacent to the Kirovograd granite pluton. It covers an area of 0.5 × 1.2 km and is evolved in migmatized biotite gneisses and biotite granites. Na-metasomatic zones comprise aegirine-riebeckite and chlorite albitites passing to dequartzified chloritized gneisses, migmatites and granites. Ore bearing leticular bodies occurs in brecciaed shear zones within albitites. Uranium mineralization consist brannerite and uraninite is rather scarce.

Another example of the Achaean metasomatic uranium mineralization in metasediments may be the *Lianshanguan ore deposit* in northeastern China [32] and the *Valhalla ore deposit* in Australia.

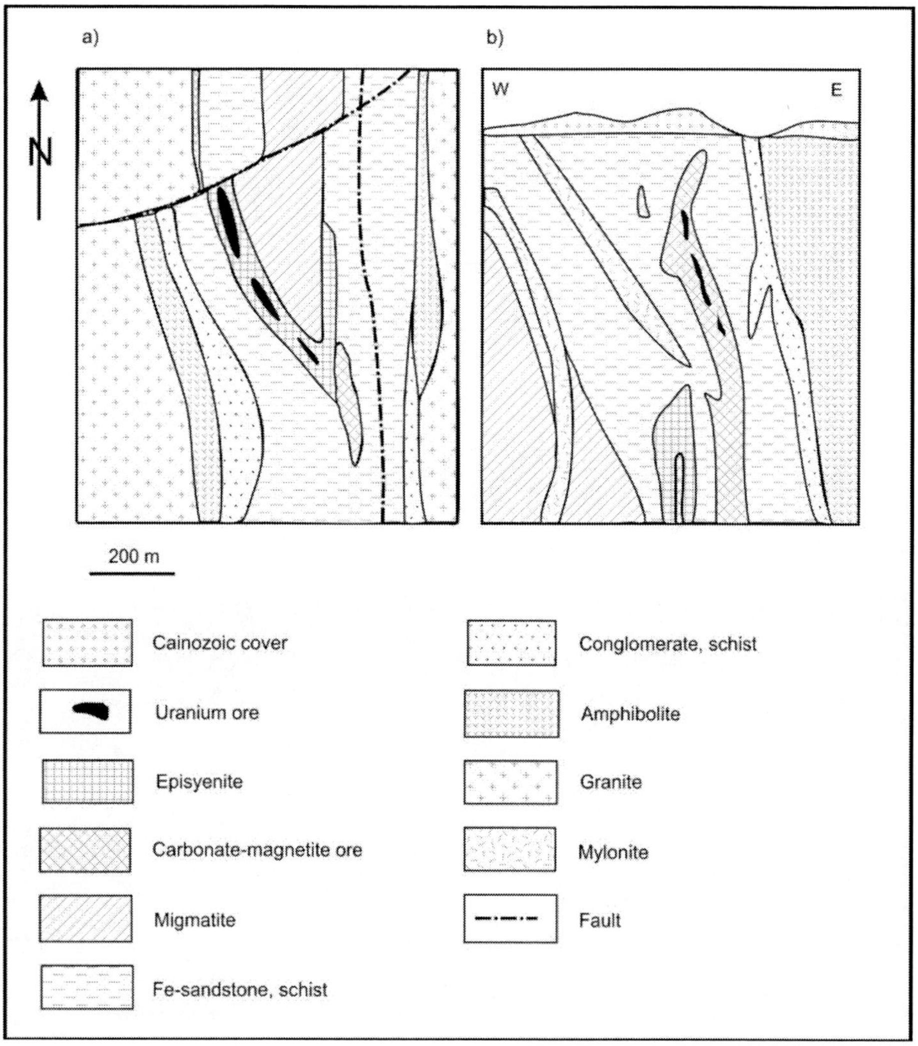

Figure 10. Schematic geological map (a) and cross section (b) of the Zhovta Richka uranium ore deposit (modified from Belevtsev et al. 1995 [31]).

Lianshanguan Ore Deposit

The Lianshanguan uranium deposit is located in the Liaoning Province of northeastern China. The deposit occurs in the transitional zone between an Archaean craton of the Northern China platform and a Lower Proterozoic geosynclinale sequence of the Liaohe group, which occurs on the southern edge of the Lianshanguan anticline. Three tectonostratigraphic units are presented in this area, namely the Archaean basement, a Lower Proterozoic metasedimentary sequence and an Upper Proterozoic-Phanerozoic platform cover. The Archaean basement is composed of K-rich monzogranite. The Lower Proterozoic metasedimentary unit comprises sandstone, conglomerate, graphite-bearing schist and marble intercalated with acid volcanics. The Upper Proterozoic-Phanerozoic cover comprises sandstone, marl, shale and limestone. The uranium mineralization occurs in quartzites and schists at the base of Lower Proterozoic and in metasomatic zones developed in granite within the Langzishan shear zone (Figure 11). The ore-bearing horizons belong to the Liahone group, the total thickness of which is 300–600 m. However, the economic uranium mineralization is distributed in NW–SE trending shear zone between granite complex and the Lower Proterozoic metasediments. The uranium ore bodies are located in the brecciaed zone within metasomatic rocks (Figure 12).

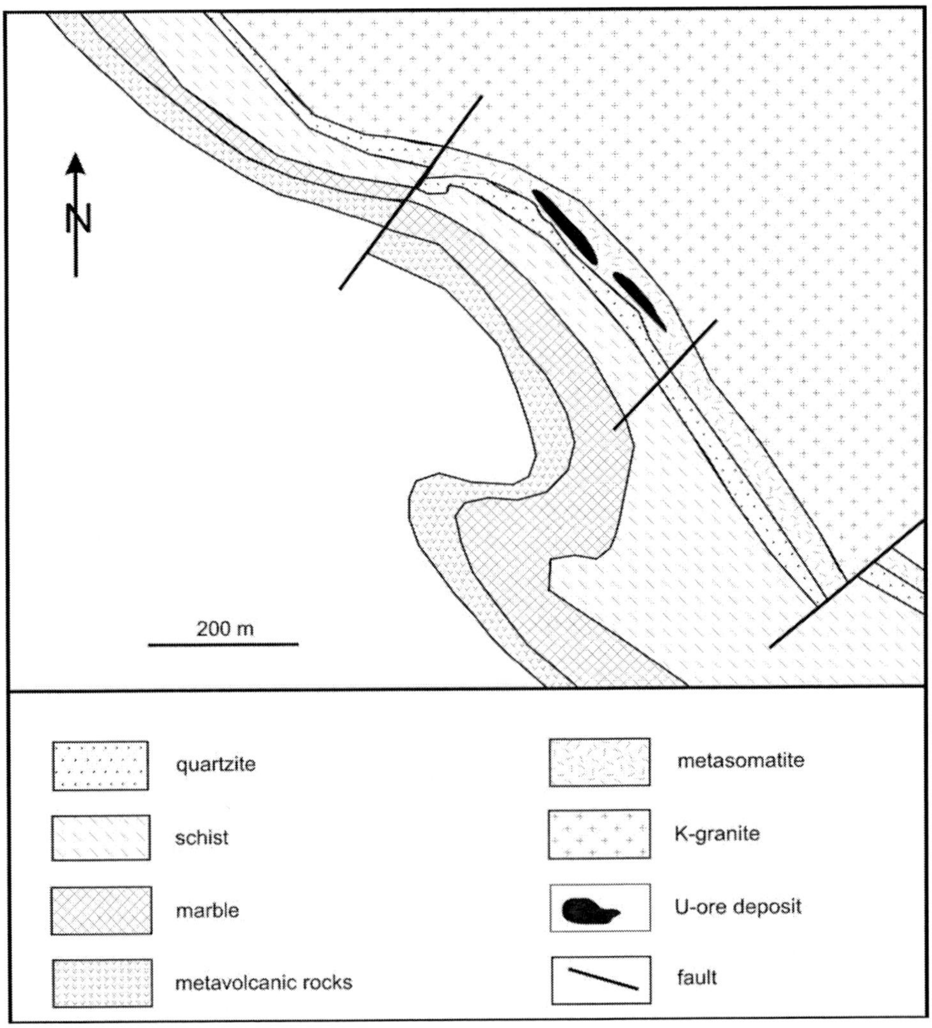

Figure 11. Schematic geological map of the Lianshanguan uranium ore deposit (modified from Zhong and Guo 1988 [32]).

The ore bodies evolved in metasediments of the Liahone group are stratiform with thickness 1–3 and ore grade is less than 0.1% U. Metasomatic rocks in the Langzishan shear zone originated by hydrothermal alteration of quartzite and granite with origin of metasomatic albite and K-feldspar and an average Na_2O/K_2O ratio 1.7. The main economic U mineral is pitchblende occurring usually in small crystals and/or having oolitic texture. The ore grade is relatively rich, locally enriched 5% U. The hydrothermal mineral assemblage is formed of pyrite, galena, sphalerite, chalkopyrite, calcite, fluorite, quartz and muscovite. The age of this mineralization dated by U–Pb method is 1,891 Ma. For sources of uranium are suggested the Archaean granite (6–10 ppm U) and relatively U–enriched quartzites (8–15 ppm). The formation temperature of metasomatic U mineralization suggested from a fluid inclusions study of fluorite, calcite and quartz was 280–350 °C. The formation temperatures of some younger hydrothermal U mineralization originated by hydrothermal fluids filling within fractures of altered granite are 160–230 °C. The main mineral assemblages of this mineralization are pitchblende, calcite, fluorite and quartz. The U–Pb age of the hydrothermal filling stage is 1,829 Ma [32].

Valhalla Ore Deposit

The Valhalla uranium deposit, located 40 km north of Mount Isa, Queensland, Australia, is a metsomatic Mesoproterozoic ore deposit similar to metasomatic U-deposits that occurred in the Ukrajine and Brazil. It is one of approximately one hundred known U occurrences that have been found in the Lower Proterozoic metasediments and metabasalts belonging to cca 1,800 Ma Eastern Creek volcanics in the western part of the Mount Isa basin. Uranium mineralization is hosted in a thick package of sandstones, arkoses and siltstones and is controlled by two NNW–SSE striking shear zones. This mineralization can be divided into three stages. Laminated and intensely altered rocks comprising albite, riebeckite, calcite, titanomagnetite and brannerite represent the early stage. The main mineralization stage is dominated by brecciaed and altered rocks, which are cemented by brannerite, apatite, U-enriched zircon, uraninite, anatase, albite, riebeckite, calcite and hematite. The late stage comprises uraninite, coffinite, hematite, dolomite calcite, chlorite and sulphides (pyrite, galena, chalkopyrite, bornite and sphalerite). Brannerite U–Pb age of the main stage is about 1.5 Ga. The isotopic O study gives origin temperatures of calcite and riebeckite between 340–380 °C. Natrium and Ca-dominated alteration fluids according of O and H isotopic composition have basinal and magmatic fluid characteristics. Mixing between two fluids during peak metamorphism appears to have occurred. The richest part of the ore deposit has about 0.6% U. The indicated reserves are about 14,000 t U at 0.07% U [28, 33].

A less widespread uranium metasomatic mineralization is associated with the Pan-African-Brazilian event (approximately 500 Ma). During this Na-metasomatic episode, uranium deposits in Brazil (Itataia, Espinharas) and in Cameroon (Kitongo) originated [28].

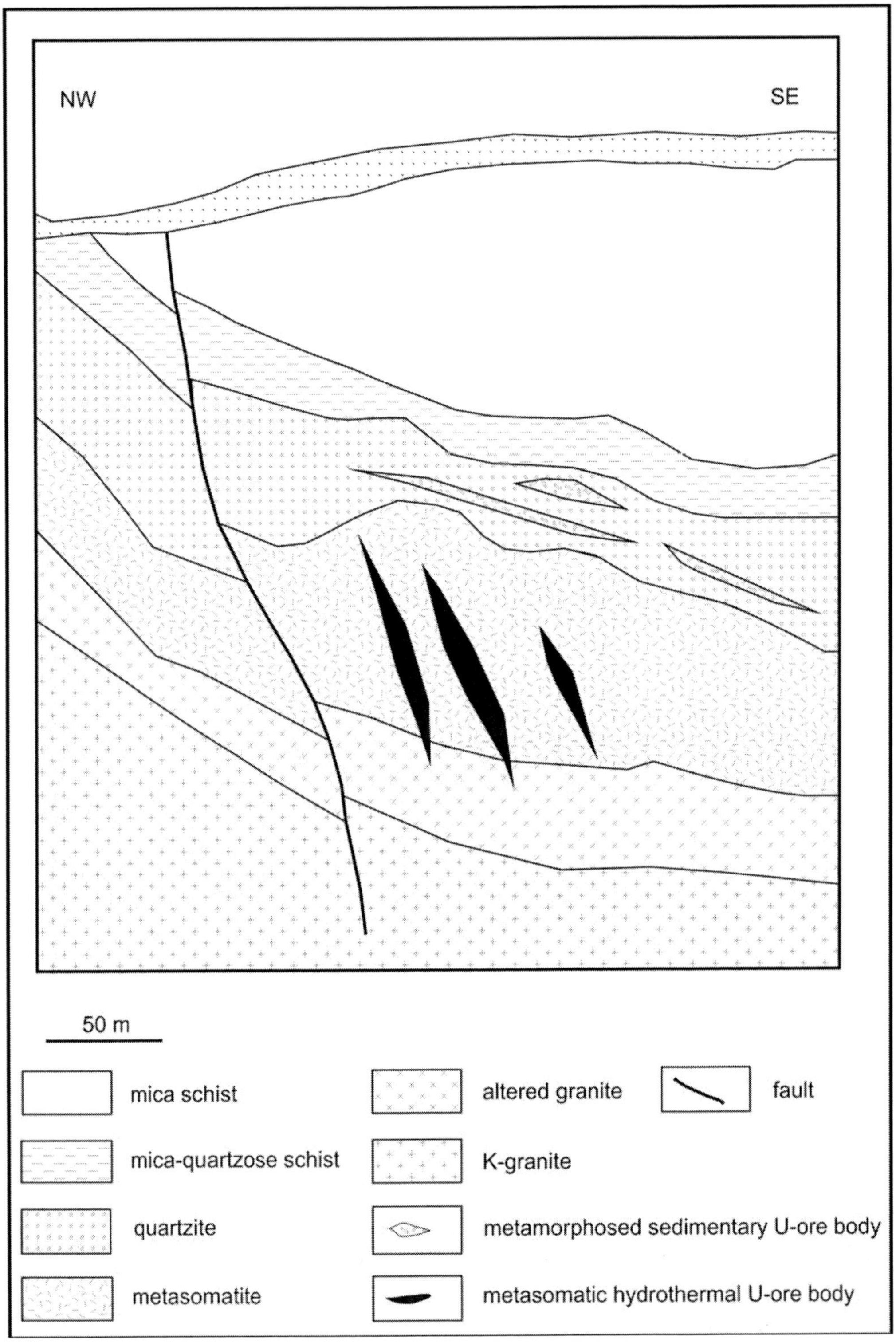

Figure 12. Schematic cross section of the Lianshanguan uranium ore deposit (modified from Zhong and Guo 1988 [32]).

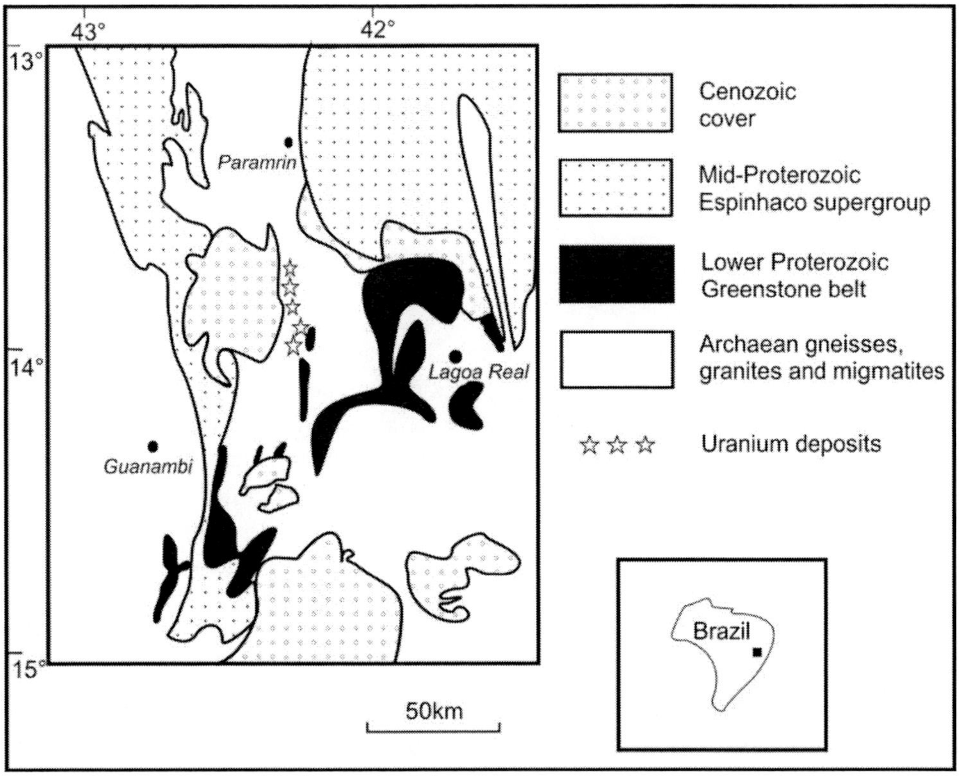

Figure 13. Simplified geological map of the Lagoa Real uranium ore district (modified from Lobato et al. 1983 [48]).

Espinharas Ore Deposit

Uranium deposit Espinharas in the northeastern part of the Brazilian shield is related to episyenitization of medium- to high-grade polymetamorphic gneisses and microgranites. Undeformed microgranite dykes intruded Proterozoic granitic othogneisses and biotite-amphibole paragneisses of the Caico group. All of these rock types were albitized in different degrees along NE–SW shear zone outcropping over more than 1.5 km. The extent, dimension and form of metasomatic rocks together with the occurrence of high-grade uranium mineralization depended on the primary rock composition and pre-existing regional foliations. Thus, the episyenitization was best developed in microgranites with low grades of U mineralization of a few hundred ppm U. Metasomatism of biotite-amphibole gneisses sometimes-yielded higher U contents up to 1,500 ppm. Alteration of leucocratic granitic gneisses yielded the highest U content where up to 40,000 ppm was detected [34]. The metasomatic alteration generally resulted in the total dissolution of quartz, alteration of Fe-Mg-Ti minerals, albitization of K feldspars and plagioclases and finely disseminated hematitization followed by pyritization and the precipitation of new minerals in the porosities caused by the quartz dissolution. Mineralization includes two generations of coffinite, albite, chlorite, apatite and calcite. Two mineralization stages were established with two generations of albite, coffinite, and calcite. Uranium mineralization is associated with the enrichment of heavy REE, which are concentrated in coffinite. Coffinites of the second hydrothermal stage are distinctly Th–enriched, indicating an evolution from initially more U-rich to a final Th–enriched pulse of mineralizing solutions.

Metasomatite deposits bounded on altered granitic rocks are distinctly rare and could be represented by the *Lagoa Real ore district* in Brazil and the *Ross Adams ore deposit* in Bokan Mountain, Alaska.

Lagoa Real Ore District

Uranium deposits of the Lagoa Real district in Brazil occur within ductile shear zones transecting the Proterozoic Lagoa Real granitic complex, enclosed in Archean and Proterozoic basement gneisses of the Sao Francisco craton. The Sao Francisco craton lies in the South-Central part of Bahia State, and comprises three main units: (1) Archean and lower Proterozoic (3.1 to 1.8 Ga) including granulitic complex, migmatites, granitic-gneissic and granite-greenstone belts; (2) mid-Proterozoic (ca. 1.7 to 1.0 Ga) Espinhaco fold belt and the Chapada Diamantina metasedimentary belt; (3) late Proterozoic and Phanerozoic covers (Figure 13).

The Lagoa Real granitic complex is bound to the east by the Archean–Lower Proterozoic migmatites and to the west by the Espinhaco fold belt and the Urandi granite-greenstone belt. The Lagoa Real complex consists predominantly of homogeneous metagranites, resulting from the transformation of the Fe-rich subalkaline granites. This complex has been thrust over metasediments of the Espinhaco and the Urandi belts.

Metagranites of the Lagoa Real complex consist of K-feldspar, prevailing over plagioclase (An_{5-18}), scarce biotite and in some facies amphibole (ferropargasite to hastingsite). Magnetite, ilmenite, titanite, fluorapatite zircon, Nb–Ti–REE–U oxides, monazite, uranothorite, fluorocarbonates and fluorite represent accessory minerals. Amphibole and ilmenite are partly altered to biotite, titanite, and magnetite ± calcite assemblage. According U–Pb dating of zircon the original Lagoa Real granite pluton was emplaced during late Paleoproterozoic at 1.7 Ga [35].

The south-central part of the Lagoa Real complex contains a very large number of discontinuous lenticular bodies of albitized and dequartzified metagranites developed about 50 km along approximately meridional shear zones [36]. Each albitite body may be up to 3 km long and up to 30 m wide. The albitite bodies may contain several uranium mineralizations or be barren. The altered zones present zonation from metagranites to albitized and dequartzified metagranites with pure albite and/or with albite, pyroxene (diopside-hedenbergite), garnet (andradite), uraninite, titanite and magnetite assemblage. The richest uraniferous albitites are characterized by pyroxene–andradite assemblage. The uranium was deposited at ca. 1.4 Ga. The later extensive recrystallization of original albitites during the deformation event related to the Brazilian orogeny (ca. 480 Ma) [35]. The sequential metamorphic-metasomatic process in this area shows a continuous oxidation trend, with uraninite precipitation being controlled by the reduction of fluid phase, via progressive oxidation of mafic minerals. Optimum precipitation conditions are connected to the maximum oxidation state of the rocks series at the magnetite-hematite buffer. The temperature of hydrothermal alterations according to isotopic thermometers was 500–550 °C. Fluid inclusion data on albite and late-stage calcite from mineralized and unmineralized albitites reveals fluids of variable salinities (0–20 wt% NaCl eq.) [36]. Of the 35 major uranium-bearing anomalies discovered by the Nuclebras Company in the Lagoa Real complex, 11 are considered as uranium deposits. The presently mined Cachoeira deposit has measured geological reserves of ca. 20,000 tons U at 0.35% U [28].

Ross-Adams Uranium Deposit

The Ross-Adams uranium deposit occurs on the southeast flank of the Bokan Moutain in southeastern Alaska. Uranium mineralization is evolved within the peralkaline granite along cataclastic N-S striking shear zone. The host rock is peralkaline granite that was altered in episyenites with distinctly enrichment of sodium with the occurrence of alkali pyroxenes and amphiboles (aegirine, arfvedsonite) and locally accumulated hematite. During the later stage feldspars were argitilized. Uranium mineralization is formed by Th-enriched uraninite, uranothorianite, uranothorite, brannerite and coffinite. Allanite, bastnaesite, monazite, xenotime, galena, pyrite, magnetite, zircon, apatite, fluorite and calcite are accessories. Mineralization occurs in numerous ore-bearing veinlets between 0.1 and 0.8 mm thick. The ore body in the Ross-Adams ore deposit is ca. 100 m long, 8–22 m wide and contain 850 t U_3O_8 at average grades ranging from 0.1 to 1% U_3O_8, occasionally to 3% U_3O_8. The content of Th is up to 6%. This ore deposit was mined very shortly in 1957 and resulted in the production of approximately 15,000 tons of uranium ore containing more than 0.8% U_3O_8 [37].

Breccia Complex Deposits

The breccia complex deposits occur in coarse breccias comprised largely of polymict granite- and hematite-dominated fragments in a hematite-chlorite-sericite matrix. Mineralization consists of iron oxide-copper-gold (IOCG) mineralization in rare cases enriched also in U, Ag, REE, Mo, Zn, Co, W and Bi. This class of ore deposits was first recognized by discovery of the super giant *Olympic Dam copper-gold-uranium deposit* in South Australia. However, the Olympic Dam is mainly represented by uranium mineralization evolved in this ore deposit type, whereas other IOCG deposits founded in South America have only valuable concentrations of Cu and Au (La Candelaria, Mantoverde, Mantos Blancos, Chile; Cristalino, Sossego, Solabo, Brazil; Marcona, Pampa de Pongo, Peru). IOCG ore deposits range in age from Late Archean to the Pliocene and are found in a number different settings (e.g., rift, subduction zone, basin collapse). They display a variety of morphologies but generally show a spatial relationship to major, crustal-scale fault zones. The metal content of the deposits is highly variable. However, the Fe oxides are the dominant mineral components. IOCG deposits occur in mafic to felsic igneous, metamorphic, or sedimentary rocks. They are generally spatially and temporally associated with batholithic complexes of intermediate to acid composition and they share a characteristic suite of hydrothermal alterations. Magnetite-bearing, albite enriched alteration in more felsic host rocks and albite-actinolite-diopside assemblage in more mafic rocks, is the dominant alteration type in the most IOCG ore deposits. Potassic alteration generally post-dates albite-enriched alteration. A number of IOCG deposits contain late and structurally high-level zones of hydrolytic alteration characterized by replacement martite (hematite) after magnetite and evolution of muscovite, carbonates and quartz. Ore components may be precipitated during any of the alteration stages, but significant enrichment occurs during potassic alteration, although the Olympic Dam deposit is associated with hydrolytic alteration [38].

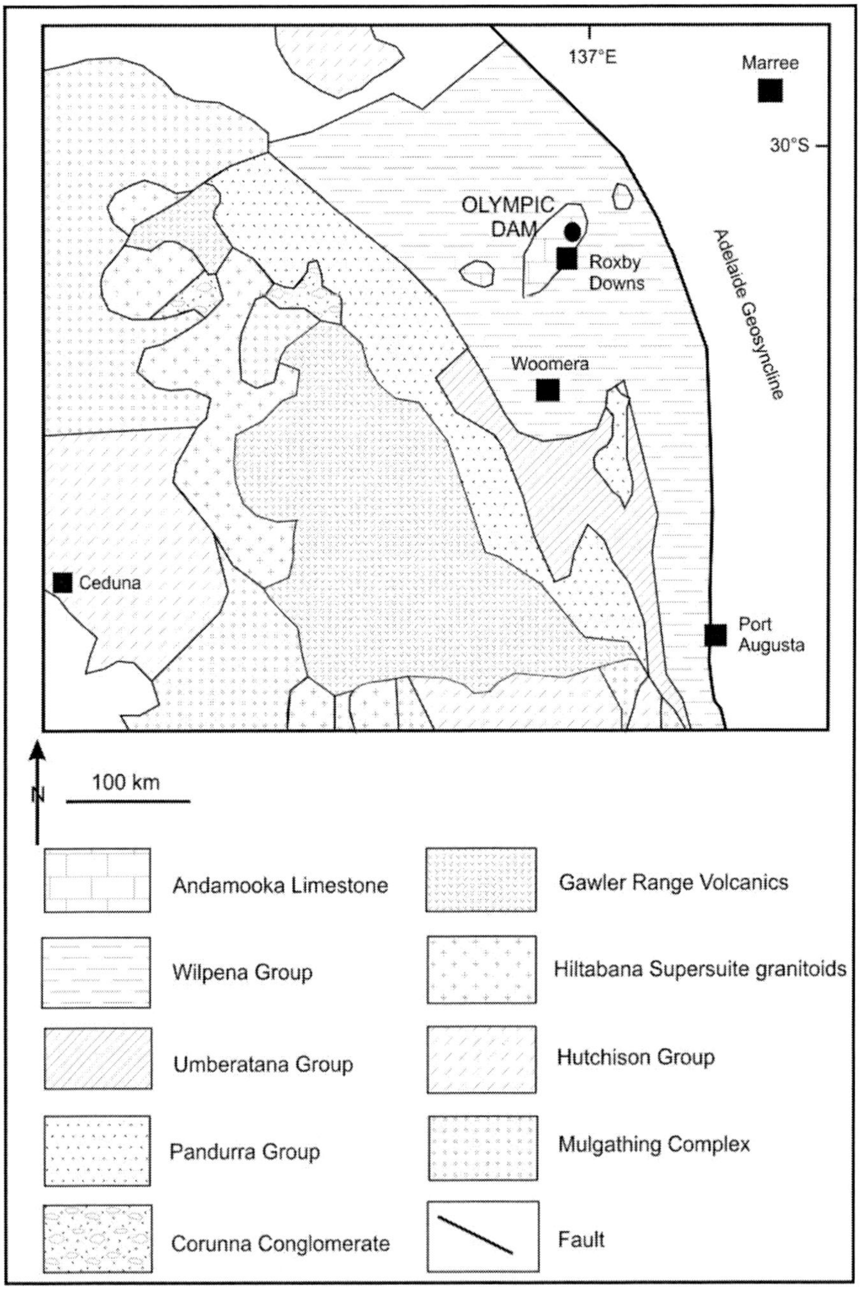

Figure 14. Schematic geological map of central South Australia and location of the Olympic Dam uranium ore deposit (modified form Haynes et al. 1995 [41]).

Most IOCG ore deposits, including the Olympic Dam, have extremely low U grades, in comparison to other types of uranium deposits. They have a range of U contents from negligible (Chile) to a maximum grade of about 0.1% of U_3O_8. Uranium in IOCG deposits generally occurs as uraninite, although brannerite and coffinite have also been recognized in some deposits. Uranium minerals are generally associated with Cu-Fe sulfides and occur in Cu-enriched zones, commonly with potassically-altered zones.

Olympic Dam Deposit

The Olympic Dam deposit is located on the Gawler Craton, South Australia. A thick Mesoproterozoic to Cambrian sedimentary sequence of the Stuart shelf covers the Olympic Dam breccia complex. The breccia complex is hosted in the Roxby Downs granite of Hiltaba granite suite, dated at approximately 1.59 Ga (U–Pb method). This complex extends over 5 to 7 km along a NW-SE trend. It is zoned and presents a central barren quartz-hematite breccia associated with volcanic rocks, which are relict from an eroded volcanic suite equivalent to the Gawler Range volcanics, rimmed by a variably mineralized hematite-granite breccia. Several types of breccia result from variable processes including hydraulic fragmentations, tectonic faulting, hydrothermal alteration and gravity collapse. Mafic to felsic dykes intruded the breccia complex and represent the final stage of magmatic activity associated with hydrothermal alteration.

The Roxby Downs granite represents the most fractionated member of the large Mesoproterozoic Burgoyne batholith, which belongs to the Hiltaba suite, a voluminous unit across the Gawler craton. Granite to quartz monzonite is enriched in Th, U, REE, Nb, Y and Zr and represents A-type, high-K calc-alkaline granite. The Gawler Range volcanic complex underlies a great part of the Gawler Craton and consists mainly of felsic volcanics with lesser amounts of mafic to intermediate volcanic rocks [39] (Figure 14).

Hydrothermal activity connected with Cu–Au–U mineralization is presumed to be related to the magmatic episode related to the emplacement of the Roxby Downs granite and the extrusion of the Gawler Range volcanic complex. Alterations are characterized by sericite and hematite with less abundant magnetite, chlorite, quartz and siderite. The earliest phase of alteration was associated with magnetite. Later, fluids overprinted the primary magnetite mineralization and Cu–Au–U mineralization is connected with this late hematite alteration. Although the ore zones account for only a small part of the whole volume of the breccia complex, Cu, U, Ag and Au minerals are widespread within the deposit (Figure 15). The main Cu-bearing phases are chalcopyrite, bornite and chalcolite. Uranium mineralization is predominantly present as fine grains of uraninite dispersed within the mineralized breccia together with pitchblende, coffinite, brannerite and davidite. The distribution of uraninite are correlated with Cu sulphide and locally enriched in discontinuous zones 10–20 m wide within the bornite–chalcolite ore. Gangue minerals are hematite, barite, fluorite, quartz and magnetite. Dating indicates that magmatic activity, brecciaing, and Cu-Au mineralization were synchronous [40]. Very little information is published concerning the origin of the U mineralization. According to Hitzman and Valenta (2005) [38], the leaching of U from the host rocks by the hydrothermal fluids would produce the enrichment in the IOCG ore deposits. Uranium enrichment in the ore is 10 to 40 times larger than in unaltered host rocks.

Uranium ore deposition resulted from the mixing of at least two fluids: a high-temperature, saline magmatic or deeply circulated meteoric fluid, which may have been boiling prior to mixing, and a lower temperature, moderately saline, more oxidized water of meteoric origin. Mixing occurred repetitively in a high-level subvolcanic setting. The U-bearing mineralization, according to fluid inclusion and O isotope study precipitated in temperatures between 200 and 300 °C from solutions with salinities ranged mostly between 7 and 23 wt% NaCl eq. Many minerals of the younger stage (hematite, Cu-sulfides, and Cu-Fe-sulfides precipitated at temperatures less than 300 but greater than 150 °C, from no boiling waters [41].

The calculations suggest the Olympic Dam deposit contains measured ore resources of 650 million tons of 1.5 wt% Cu, 0.05 wt% U_3O_8 and 0.5 g/t Au [38]. Furthermore a variable REE mineralization (bastnaesite, florencite, monazite and xenotime) is disseminated across the deposit. Since 1988 about 40,000 t U have been already produced. In the year 2007 the Olympic Dam mining complex produces around 9 million tons of ore and recovers approximately 200,000 tons of refined Cu, 3,500 tons of U_3O_8, 820,000 ounces of Ag and 90,000 ounces of Au.

Volcanic Associated Deposits

The oldest U ore deposits truly related to hydrothermal activity connected with volcanic complexes are probably the Devonian Ben Lomond of Queensland, Australia and those in Late Silurian Chuili-Kandyktas uplift of northern Kazachstan [42]. A few U ore deposits are also related to Carboniferous and Permian volcanism (e.g., Permian deposits in European Alps and Carpathian) [42]. All these deposits are relatively small size, with total reserves on the order of 10,000 t U. The main hydrothermal-volcanic originated during Late Jurassic (*Streltsovka*, Transbaikalia, Russia), Cretaceous (*Xiangshan*, China) and Miocene (*McDermitt*, Virgin Valley, Nevada, Lakeview, Oregon, Marysvale, Utah, USA). The most mineralized ore districts (Streltsovka, Mc Dermitt) developed in volcanic caldera, forming large depressions, resulting from the collapse of the magmatic chamber roof, during intracratonic basin and range-scale extensional events. The depressions are filled with altered volcanics, and subordinate amounts of siliciclastic sediments, but acidic volcanics generally represent the largest proportion of the extruded magmas. The pyroclastic rocks represent particularly favorable source rocks because they are relatively permeable and a high proportion of the U is hosted by the glassy matrix of which it can be easily mobilized during devitrification. Among volcanic rocks, peralkaline magmas represent the most efficient uranium source. The origin of large ore deposits requires also the existence of a relatively shallow magma chamber, lasting overall several million years. The U ore bodies are predominantly trapped along basement faults, which are reactivated after major part of the caldera filling has been completed. They may also occur along the ring structure of caldera and/or at the margin of protruded domes [42].

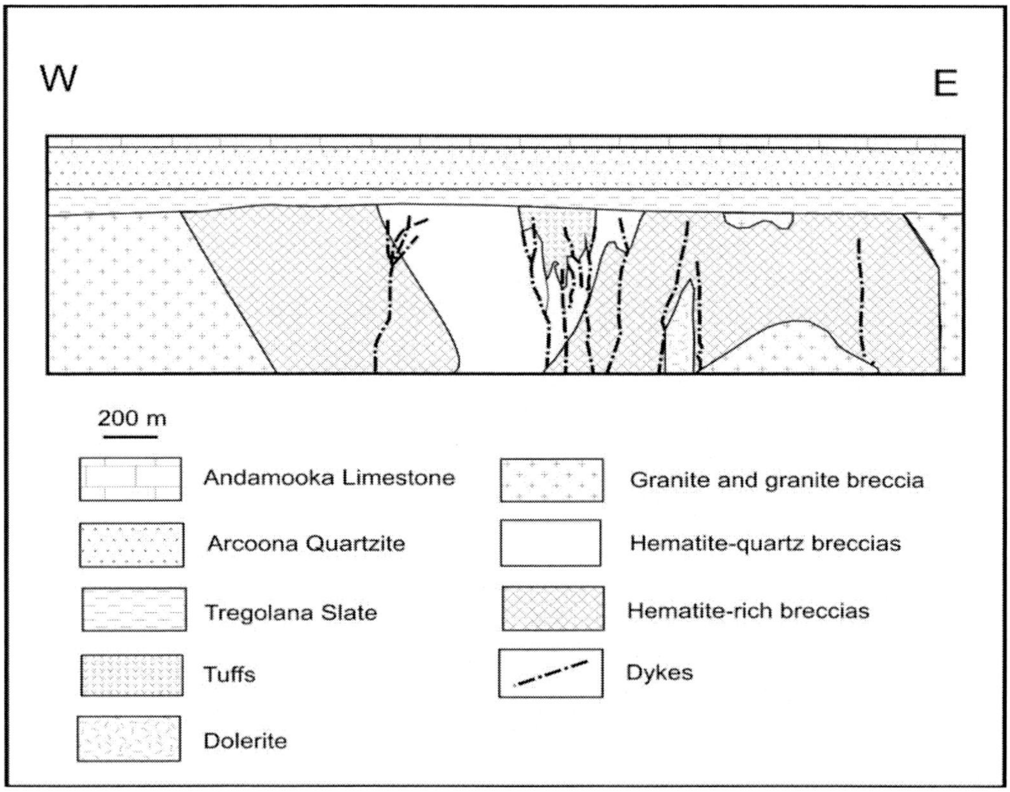

Figure 15. Schematic cross section of the Olympic Dam uranium ore deposit (modified from Haynes et al. 1995 [41]).

Streltsovka Ore District

The uranium deposits of the Streltsovka caldera represent a unique example of a giant uranium ore district associated with volcanic rocks series. The Streltsovka district is located in Transbailkalia (Russia) near the Chinese-Mongolian border. The caldera has a diameter of about 20 km and was filled during Late Jurassic with a pile of volcanic rocks (basalt, andesite, trachydacite, rhyolite) and interlayered sedimentary horizons. The present thickness of the pile locally reaches more than 1 km. Rhyolites which correspond to the last erupted volcanics were dated at 142 Ma by K–Ar method. They may have represented about 30 % of the volume of the volcanic pile. The basement is mainly composed of Hercynian granites (Figure 16).

Uranium mineralization may be located in any of the different volcano-sedimentary horizons. This mineralization generally made subvertical veins or stockworks. Age determinations (U–Pb of pitchblende, 133 Ma) indicate that the uranium mineralization was formed immediately after the end of the volcanic activity. Small uranium deposits were also discovered in thee surroundings of the caldera. Uranium contents generally amount to about 0.2%, up to 0.6% in large stockworks, and up to 1.0% in veins. The ore resources estimate grading above 0.2% U reach more than 280,000 t U_3O_8.

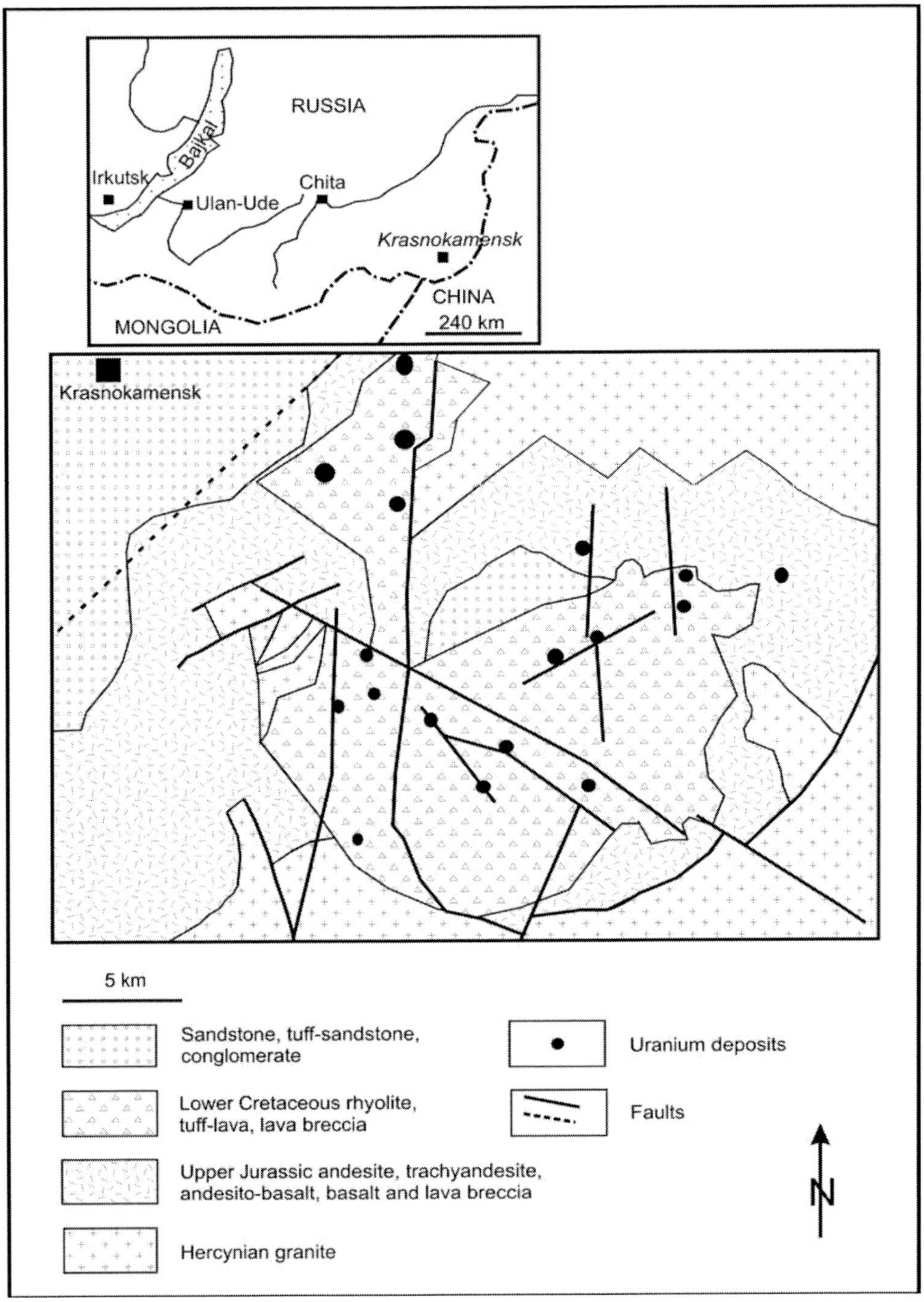

Figure 16. A sketch geological map of the Streltsovka caldera (Transbaikalia) (modified from Chabiron et al. 2003 [44]).

The mineralization can be divided in to five stages. First stage corresponds with local albitization in the deep part of the ore deposit. The most intensive, hydrothermal alteration of the second stage corresponds to the origin of clay minerals (illite, phengite and/or chlorite). The chlorite geothermometer [43] gives temperatures between 290 and 390 °C for all chlorite types. After this alteration stage quartz veins with variable amounts of carbonate, pyrite and hydromica were formed during third stage. Main uranium deposition occurred during fourth stage together with deposition of quartz and molybdenite. Pitchblende, which is the main U mineral, occurred as isolated spherolites or as disseminated grains in the cement of breccias. Anhedral coffinite occurs at the margin of pitchblende spherolites, associated with pitchblende in the breccia cement, and with pyrite. Brannerite occurs as small, euhedral grains deposited earlier than coffinite or pitchblende. Pyrite, galena and molybdenite may be associated with the U minerals. Brannerite occurs at depths of more than 1,300 m in the basement, mixed pitchblende-coffinite. Brannerite ores occur between 800 and 1,300 m, and molybdenite-pichblende ores occur at higher levels. During last, fifth stage fluorite and calcite originated. The main source of uranium was rhyolites, which display in unaltered samples average content 19 ppm U. Some other source rocks represent subalkaline granites in the basement. Thorite, allanite, monazite, zircon and apatite in these granites were the major hosts for U, whereas thorite, allanite and monazite were the major hosts for Th [44].

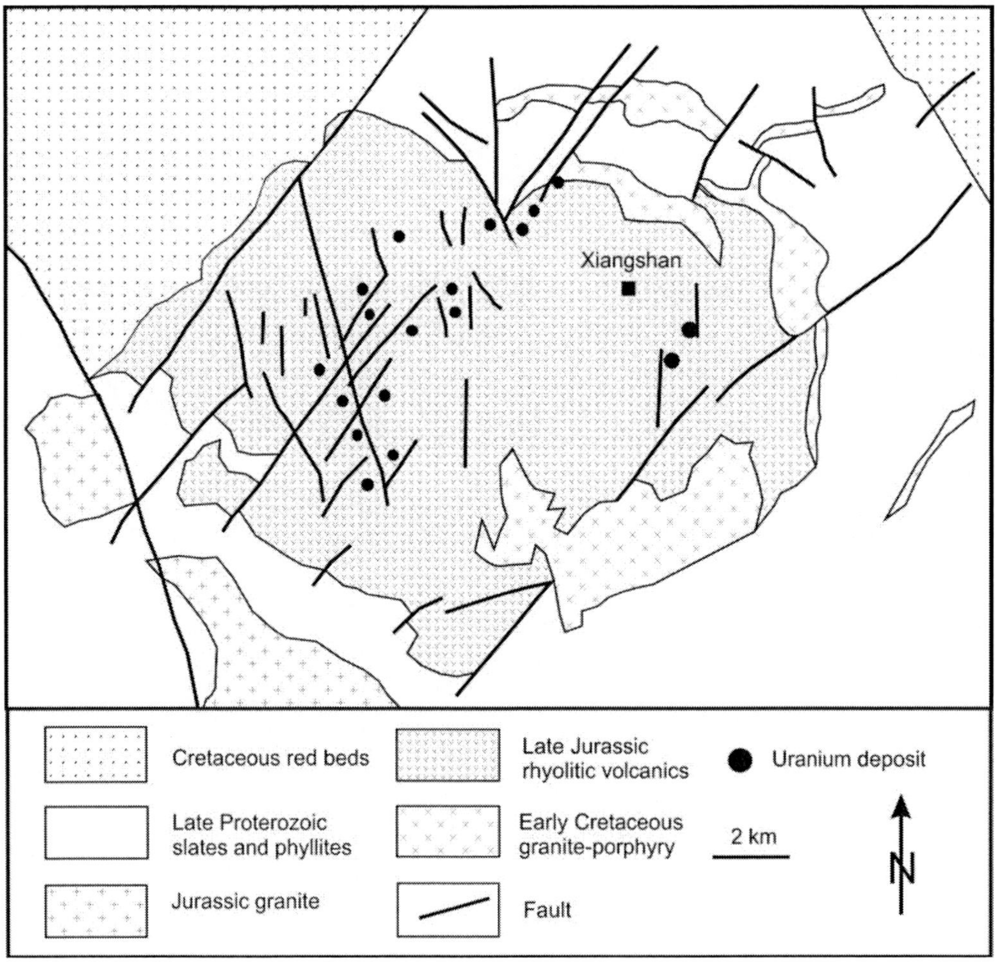

Figure 17. Schematic geological map of the Xiangshan uranium ore district (modified from Hu et al. 2009 [45]).

Xiangshan Ore District

The Xiangshan ore district, the largest hydrothermal uranium deposit in China, is hosted in late Jurassic felsic volcanic rocks. This ore district is hosted within the Mesozoic Gan-Hang volcanic belt, which extends from Zhejiang Province in the east to Jiangxi Province in the west, measuring about 600 km in length and 50 km in width. Both the Jiangshan–Shaoxing and Wuzhou–Yongfeng deep faults structurally control the volcanic belt. The Gan-Hang volcanic belt consists a series of volcanic basins filled predominantly with Late Jurassic felsic volcanics thought to be related to the subduction of the Pacific plate beneath the Eurasian plate during the Late Jurassic.

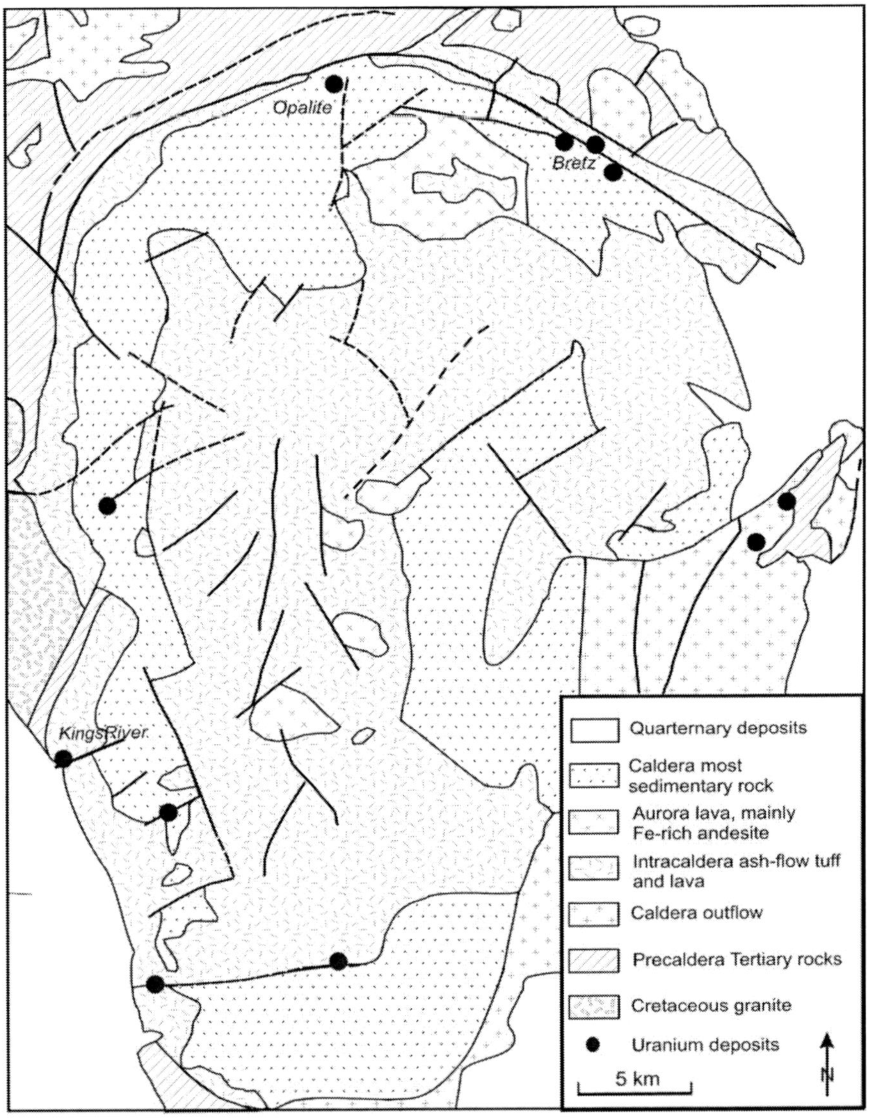

Figure 18. Generalized geological map of the McDermitt caldera (modified from Castor and Henry 2000 [46]).

The Xiangshan volcanic basin, one of the components of the Gan-Hang volcanic belt, is composed of and largely dominated by rhyolitic rocks and sub-volcanic rocks (granite porphyry), with thickness more than 2 km. The basement of volcanic basin is composed of Precambrian slates and phyllites [45] (Figure 17).

The uranium mineralization is hosted within ignimbrites, rhyolites and granite porphyries in the Xiangshan volcanic basin. Hydrothermal alteration of host rocks includes albitization, silicification, sericitization and carbonatization, which extends away from faults and fractures along more brecciaed rocks in zones several tens of centimeters wide. The mineral assemblage of U mineralization is relatively simple, formed by quartz, calcite, pyrite and pitchblende. This mineral assemblage usually fills second-order faults and shear zones near major faults in the host rocks to form epigenetic networks and irregular vein-like bodies. The hydrothermal assemblages textures indicate three stage of mineralization. During the first stage, albitization, together with origin of quartz veins, took place, whereas silicification is part of ore-forming stage. During this stage, pitchblende, pyrite, hematite, quartz, fluorite and calcite originated. Brannerite, broggerite and molybdenite occur as relatively rare minerals. The main stage yields U–Pb age of 119 to 99 Ma on pitchblende. This age coincides with the age of mantle-derived mafic dykes formed during crustal extension with the Xiangshan basin (K–Ar ages of 100 to 120 Ma). Fluid inclusions in quartz deposited together with U mineralization display low to moderate salinity (2.5–8.0 wt% NaCl eq.) and homogenization temperatures 150–250 °C. The water in the ore-forming fluids is mainly of meteoric origin, as evidenced by H and O isotopic composition of these fluids. On the other hand, isotopic composition of He, Ar of ore fluids together with C isotope data of calcites associated with U mineralization evidenced mantle-derived gaseous components in ore-forming fluids. Thus, the ore-forming fluids resulted from a mixture CO_2- and ^{3}He-rich mantle derived fluid and CO_2-depleted meteoric fluid. The Cretaceous crustal extension promoted uranium mineralization by supplying mantle-derived CO_2, which triggered remobilization of U from U-rich felsic rocks to form the Xiangshan U ore district. Sericitization, argillitization (illite, kaolinite) and carbonatization occur in the third, post-ore stage [45]. This ore district contains about 30,000 tons of U at an average grade about 0.1–0.3% U and the Xiangshan district has provided uranium for the country for over three decades.

McDermitt Ore District

The largest uranium deposits in the Tertiary volcanic rocks are in the USA, in northwestern Nevada and southeastern Oregon. These deposits occur in various geotectonic settings, some of them are in or directly adjacent to calderas, others are in outflow ash-flow tuff distant from source calderas, and still others are associated with rhyolite intrusions or flow unrelated to calderas. All of these deposits are associated with U-rich rhyolitic volcanic rocks.

Uranium deposits in the McDermitt caldera are present along the southwestern (Kings River) and northern (Bretz–Opalite) margins of the 30 × 40 km McDermitt caldera complex. Possibly four overlapping calderas and three satellitic calderas erupted regional; peralkaline to metaluminous ash-flow sheets about 16 Ma ago (Ar–Ar method) to form the 20,000 km^2 McDermitt volcanic complex. The ash-flow sheets overlie mafic volcanic rocks of the

Orevada View volcanic series. The main caldera complex was subsequently filled with a series of tuffs, lava (mostly rhyolite but including Fe-rich andesite), and rhyolite intrusions. Overlying volcanoclastic sedimentary rocks are exposed in a nearly continuous peripheral moat [46] (Figure 18).

In the Kings River area, uranium mineralization is focused along NW to NE striking faults that make up the western margin of the McDermitt caldera complex. This mineralization is formed by a few primary (U-enriched zircon, coffinite, pitchblende) and many secondary U minerals (metazeunerite, autunite, torbernite, uranyl silicate). The U-rich zircon occurs as a dominant U mineral. The amount of U in ore is distinctly variable from 0.006 to 0.13% U_3O_8. The U-rich rocks are also markedly enriched in Zr, As and Mo, as well as in Ag, Au, Sb and Te. Notable is a wide range of sulfides, mainly arsenopyrite and pyrite. Gangue minerals are dominantly quartz, K-feldspar and fluorite. Fluid inclusions in quartz indicate deposition of ore mineralization at ca. 330 °C [47].

For the Bretz–Opalite uranium deposit is a significant occurrence of Hg mineralization. The Hg and U mineralization are located along a series of subparallel faults that make up the northern ring fracture zone of the McDermitt caldera complex and in rocks that accumulated in a caldera moat adjacent to these faults. Hydrothermal alteration in this area consists of silicification and strong argilitization in and adjacent to the U and Hg mineralization. The U and Hg contents range up to 0.1% U_3O_8 and 1.6% Hg. Associated trace elements are As, Mo, Sb, W and Pb. Uraninite, coffinite, uranyl phosphates and silicates make this mineralization and is associated with pyrite and Ti oxides. Alteration minerals include sulfides, fluorite, K-feldspar, chalcedony and opal. Total reserves of the McDermitt ore district are estimated in 9,500 t U_3O_8 in grade 0.13% U_3O_8 [46].

CONCLUSION

Hydrothermal ore deposits represent an important type of uranium mineralization. These deposits can be grouped into four main categories, namely in vein-like deposits, metasomatic deposits, breccia complex deposits and volcanic–associated ore deposits. Vein-like deposits are evolved as vein deposits filling open spaces (faults and fissures) and/or relatively closed shear zones. According to their relation to hosted rocks series are vein-like deposits classified into intragranitic and perigranitic U deposits. Metasomatic U deposits are hosted in structurally deformed rocks series that have been affected by hydrothermal aided alteration with significant participation of Na-enriched solutions. During alteration, original rocks series are enriched in alkali-rich minerals and depleted in quartz. These hydrothermal altered rocks are usually termed as episyenites. The breccia complex deposits recently form the most important group of hydrothermal uranium deposits, accordingly to the giant Olympic Dam deposit in South Australia. The volcanic U deposits are concentrated chiefly in Mesozoic and Tertiary volcanic calderas formed by acidic and volcanic rocks (rhyolite, trachyte).

REFERENCES

[1] Kirchheimer, F. *Das Uran und seine Geschichte;* E. Schweizerbarsche Verlag, Stuttgart, 1963; pp 1–372.

[2] Becquerel, A. H. *Compt. Rend. Seanc. Sci.* 1896, 122, 420–421.

[3] Curie, M. *Compt. Rend. Seanc. Sci.* 1898, 126, 1101–1103.

[4] Curie, P.; Curie, M. *Compt. Rend. Seanc. Sci.* 1898, 127, 175–178.

[5] Lacroix, M.A. *Acad. Sci., Seance Lundi* 1920, 685–690.

[6] Omeljanenko, B.I.; Lisicina, G.A.; Naumov, S.S. In *Metasomatism and ore deposits*; Nauka: Moscow, 1974; pp 160–161.

[7] Leroy, J. *Miner. Deposita* 1984, 19, 26–35.

[8] Cathelineau, M. *J. Petrol.* 1986, 27, 945–965.

[9] Křibek, B.; Žák, K.; Dobeš, P.; Leichmann, J.; Pudilová, M.; René, M.; Scharm, B.; Scharmová, M.; Hájek, A.; Holeczy, D.; Hein, U. F.; Lehmann, B. *Miner. Deposita* 2009, 44, 99–128.

[10] Leroy, J. *Econ. Geol.* 1978, 73, 1611–1634.

[11] Leroy, J.; Holliger P. *Chem. Geol.* 1984, 45, 121–134.

[12] Poty, B.; Leroy, J.; Cathelineau, M.; Cuney, M.; Friedrich, M.; Lespinasse, M.; Turpin, L. In *Vein type uranium deposits;* Fuchs, H.; Ed.; IAEA: Vienna, 1986; IAEA-TECDOC, 316, pp 215–246.

[13] Ruzicka, V. *Ore Geol. Rev.* 1993, 8, 247–276.

[14] Barsukov, V.L.; Sokolova, N.T.; Ivanitskii, O.M. *Geochem. Int.* 1996, 34, 1041–1056.

[15] Barsukov, V.L. *Geochem. Int.* 2006, 44, 591–612.

[16] Förster, H.-J.; Romer, R.L.; Gottesmann, B.; Tischendorf, G.; Rhede, D. *Neu. Jb. Miner. Abh.* 2009, 186, 163–184.

[17] Schuppan, W.; Büder, W.; Lange, G. In *Mineral deposits of the Erzgebirge/Krušné Hory* (Germany/Czech Republic); Von Gehlen, K.; Klemm, D.D.; Ed.; Gebr. Borntraeger: Berlin, Stuttgart, 1994; pp 191–207.

[18] Hein, U.F.; Lehmann, B.; Křibek, B.; René, M. In *Uranium deposits: from their genesis to their environmental aspects;* Křibek, B.; Zeman, J.; Ed.; Czech Geological Survey: Prague, 2002; pp 61–64.

[19] Seifert, T. In the 33th International Geological Congress. *Abstract* 1344759. Oslo, 2008.

[20] Hiller, A.; Schuppan, W. Geologie und Uranbergbau im Revier Schlema-Alberoda;*Bergbau in Sachsen* 14; Sächsisches Landesamt für Umwelt un Geologie: Dresden, 2008; pp 1–200.

[21] Holub, F.V.; Machart, J.; Manová, M. *Sbor. geol. Věd, LG* 1997, 31, 27–50.

[22] Křibek, B.; Žák, K.; Spangenberg, J.; Jehlička, J.; Prokeš, S.; Komínek, *J. Econ. Geol.* 1999, 94, 1093–1114.

[23] Anderson, E. B. *Czechoslovak uranium industry report* 1987, 1862–87, CSUP: Příbram, pp 1–32.

[24] Žák, K.; Dobeš, P. *Rozpr. ČSAV, Ř. mat. přír. Věd*, 1991, 1–109.

[25] Anderson, E.B.; Ivanov, P.A.; Komínek, *J. Geol. Hydrometal. Uranium* 1988, 12, 70–88.

[26] Li, T.; Huang, Z. In *Vein–type uranium deposits*; Fuchs, H.; Ed.; IAEA: Vienna, 1986; IAEA-TECDOC, 316, pp 293–305.

[27] Hu, R.Z.; Bi, X.W.; Peng, J.T.; Su, W.Ch.; Liu, S.; Qi, H.W. *Econ. Geol.* 2008, 103, 583–598.

[28] Cuney, M.; Kyser, K. In *Recent and not-so-recent developments in uranium deposits and implication for exploration;* Raeside, R.; Ed.; Short Course Ser., Mineralog. Assoc. Canada; Quebec City, 2008; Vol. 39, pp 1–258.

[29] Cuney, M.; Emetz, A.; Mercadier, J.; Mykchaylov, V.; Shunko, V.; Yuslenko, A. *Ore Geol. Rev.* 2011, doi:10.1016/j.oregeorev.2011.09.007.

[30] Zhukova, V.I. In *Uranium ore deposits: zonation and mineral associations;* Atomizdat: Moscow, 1970; pp 5–34.

[31] Belevtsev, Y.N.; Koval, V.V.; Bakarjiyev, A.H.; Anisimov, V.A.; Batashov, V.G.; Valter, A.A. In *Genetic types and regularities in placement of uranium deposits of Ukrajine;* Belevtsev, Y.N.; Koval V.V.; Ed.; Naukova Dumka Press: Kiev, 1995; pp 4–17.

[32] Zhong, J.; Guo, Z. *Precam. Res.* 1988, 39, 51–64.

[33] Polito, P.A.; Kyser, K.T.; Stanley, C. *Miner. Deposita* 2009, 44, 11–40.

[34] Porto da Silveira, C.L.; Schorscher, H.D.; Miekeley, N. *J. Geoch. Explor.* 1991, 40, 329–347.

[35] Turpin, L.; Maruejol, P.; Cuney, M. *Contrib. Mineral. Petrol.* 1988, 98, 139–147.

[36] Lobato, L.M.; Fyfe, W. S. *Econ. Geol.* 1990, 85, 968–989.

[37] MacKevett, E.A. U.S. *Geol. Survey Bull.* 1963, 1154, 1–125.

[38] Hitzman, M.W.; Valenta, R.K. *Econ. Geol.* 2005, 100, 1657–1661.

[39] Roberts, D.E.; Hudson, G.R.T. *Econ. Geol.* 1983, 78, 799–822.

[40] Johnson, J.P.; Cross, K.C. *Econ. Geol.* 1995, 90, 1046–1063.

[41] Haynes, D.W.; Cross, K.C.; Bills, R.T.; Reed, M.H. *Econ. Geol.* 1995, 90, 281–307.

[42] Cuney, M. *Econ. Geol.* 2010, 105, 553–569.

[43] Cathelineau, M. *Clay Miner.* 1988, 23, 471–485.

[44] Chabiron, A.; Cuney, M.; Poty, B. *Miner. Deposita* 2003, 38, 127–140.

[45] Hu, R.Z.; Burnard, P.G.; Bi, X.W.; Zhou, M.F.; Peng, J.T.; Su, W.Ch.; Zhao, J.H. *Chem. Geol.* 2009, 266, 86–95.

[46] Castor, S. B.; Henry, Ch. D. *Ore. Geol. Rev. 2000*, 16, 1–40.

[47] Rytuba, J.J.; Glanzman, R.K. *Bur. Mines. Geol., Rep.* 1979, 33, 109–117.

[48] Lobato, L.M.; Forman, J.M.A.; Fuzikawa, K.; Fyfe, W.S.; Kerrich, R. *Canad. Mineral.* 1983, 21, 647–654.

In: Uranium: Characteristics, Occurrence and Human Exposure ISBN: 978-1-62081-207-5
Editors: Alik Ya. Vasiliev and Mikhail Sidorov

Chapter 8

THE LEGACY OF URANIUM MINING ACTIVITY IN CUNHA BAIXA: A PORTUGUESE STUDY CASE

M. O. Neves[1,*], M. M. Abreu[2,†] and M. J. Matias[1,†]

[1]Centro de Petrologia e Geoquímica, Instituto Superior Técnico, Universidade Técnica de Lisboa (TULisbon), Portugal

[2]Unidade de Investigação de Química Ambiental, Instituto Superior de Agronomia, Universidade Técnica de Lisboa (TULisbon), Portugal

ABSTRACT

In the 20th century Portugal had a significant history of radioactive mining industry. Unfortunately, some of the 60 mines that have been exploited and abandoned represent pollution sources, which are associated with some environmental changes that may place public health or safety at risk. The purpose of this chapter is to review the previous studies done at Cunha Baixa mine-site, one of the most important uranium mines exploited (1970-1993) in the centre of Portugal, in order to assess the environmental impact from the past mining activity on the region, which was nowadays classified by the authorities as a prior area requiring reclamation measures. The proximity of the population to the mine site (100 m SW) called for the diagnosis of the level and extension of the environmental problems as well as an assessment of the related chemical and radiological risks. Among other chemical elements, the uranium and radioactive daughters concentration analyzed in rocks, mine-wastes, surface and groundwater, sediments, soils and plant samples (foodstuff) was the support to evaluate potential risks to the surrounding exposure population and to recommend site-specific strategies for rehabilitation actions.

* orquidia.neves@ist.utl.pt.
† manuelaabreu@isa.ist.utl.pt.

INTRODUCTION

At all places where uranium mining has occurred more or less extensive environmental impacts resulted, being this exploitation the most problematic of all the mining activities since radiological impacts occur jointly with chemical ones (Vandenhove et al., 2007). With the decline of uranium prices on the world market in the last century decades, mining became too expensive for most deposits; many mines have been shut down since then in most European countries. At many old mines, the exploitation companies often left sites with improper disposal of wastes and without any clean up after production ceased. Contamination of soils, groundwater, sediments, surface water and air with metals (As, Pb, Cu, Zn...) and uranium decay radionuclides (e.g. ^{226}Ra and ^{222}Rn) were the major problems in such abandoned mining sites. Human and other living species from uranium mining regions may be exposed to: (i) radiation doses emitted by wastes; (ii) radioactive dust and (iii) contaminated water and foodstuffs.

In 2007 was celebrate 100 years of the first uranium-radium ore deposit discovered in Portugal. The first license for mining (the Rosmaneira, Bendada) was issued in 1909. Before this exclusive interest for uranium in 1944, the exploitation was only for radium.

The uranium deposits in Portugal are mainly situated in central-north regions of the country. They are located within Hercynian granitic batholiths or in the metasediments of exocontacts and occupy a structurally high position, which can be deduced from the presence of numerous pegmatites and metasedimentary enclaves (Matos Dias and Soares Andrade, 1970). The granites, occurring in the form of circum-scribed massifs, belonging to a calc-alkaline series, generally post-tectonic and are heavily fractured, particularly the NNE-SSW to ENE-WSW and NNW-SSE to NW-SE systems, pointing to an N-S maximum compression direction (Goínhas, 1987). The Portuguese uranium metallogenetic province demonstrated reserves in the known deposits in the Beiras and Alto Alentejo regions around 8200 Mg of uranium metal.

Until to 2000, there were produced 4370 Mg of uranium oxides, in addition to the production of radium salts, mainly in 60 mines in the districts of Guarda, Viseu and Coimbra (Centre-North Portugal), resulting about 13 Tg of mining residues (Nero et al., 2003). Today no uranium mines are working, but recently several international companies showed interest in the Portuguese uranium potential, especially in the Nisa deposit, where 2.5 Tg with 0.11% U_3O_8 (cut off 0.05%) of geological resources are estimated (Martins, 2010) which possible exploitation in the future has been causing strong controversy among the public opinion.

The legacy of these historic radioactive mining sites raised concern, as they are often located near villages where the agriculture and livestock pastures frequently take place. Also, most of those mining activities were performed before the development of environmental legislation and radiation protection standards. So, the knowledge of the environmental impacts that have already occurred and its potential future hazards are main issues, in order to minimize or to adopt the best decisions about reclamation actions to be implemented in the affected areas.

THE CASE OF CUNHA BAIXA MINE

Mining Activities

The Cunha Baixa was one of the most important Portuguese uranium mines as its production (1000 Mg of U_3O_8) represented a quarter of the total exploitation in the country. The mine (40° 34' N latitude and 7° 45' W longitudes; Figure 1) is located 20 km southeast of Viseu (Central Portugal). The uraniferous structures (brecciate quartz veins and dissemination pockets) cut calc-alkaline Hercynian granite (medium to coarse grained, porphyritic two-mica granite) and enclosing metasedimentary rocks from the schist-greywacke complex of Upper Precambrian age. The main ore minerals were secondary uranium phosphates: autunite ($Ca(UO_2)_2(PO_4)_2 \cdot 10\text{-}12H_2O$) and torbernite ($Cu^{2+}(UO_2)_2(PO_4)_2 \cdot 8\text{-}12H_2O$) but manganese oxides, pyrite and arsenopyrite were also present in the mineral association (Matos Dias, 1972). The exploitation began as an underground mine (1970), but later as open pit (1984-1991). Heap leaching with sulphuric acid solutions (pH 1-2) was used until 1993 to recover uranium from low-grade ore (500 000 Mg) stored in the open pit area that was not rendered impermeable. The uranium bearing liquors were pumped to the surface where they flowed through adsorption resin columns. The high-grade ore ($U_3O_8 > 0.05\%$) and the loaded resins were transported to Urgeiriça mill, 25 km away from Cunha Baixa mine, to produce the yellow cake.

When mining activity ceased one Tg of solid waste materials were deposited in several dumps surrounding the mine area, which were not restored and consequently were exposed to weathering processes. As in other mines there is the potential for acid mine drainage during or after uranium mining. In addition to the threat of radionuclides contamination, acid mine drainage would present additional environmental problems including the releasing of toxic metals into the environment. Additionally, mine tailings and waste rock that remain at the mine site will continue to release radioactive elements that are more easily dispersed into water and air from mine wastes than from the original underground ore. Mine tailings also continually release radioactive radon gas into the air.

Despite the fact that in the past the concerns about mining impacts were focused on those affected productivity only on the mine's life time, the last administrator of Cunha Baixa mine (ex ENU – National Uranium Enterprise) has done some works to minimize environmental damage such as pumping mine groundwater to a treatment plant and the cultivation of cropping pines (*Pinus pinaster* Aiton) on the dumps and heap leaching area (Figure 2).

Research in Mine Site

The treatment of mine groundwater pumped out through the main shaft of the mine began in 1990 in order to neutralize pH and precipitate radium. The mine water after mixed with $CaCO_3$, $Ca(OH)_2$ and $BaCl_2$ solution was discharged into a settling pond without impermeable basement. This operation continued after closure of the mine and also to prevent open pit flooding during the rainy season.

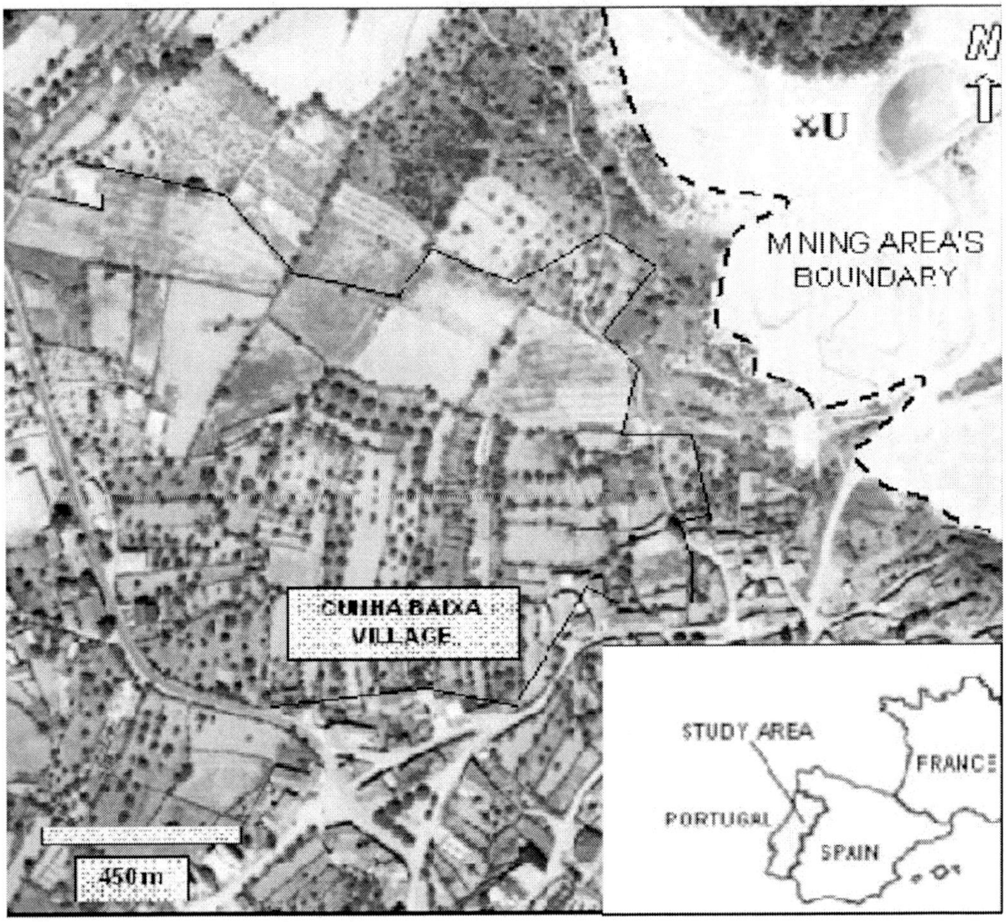

Figure 1. Location of Cunha Baixa village, uranium mining area's boundary and limits of the agricultural land.

The analysis of the discharged water (Neves and Matias, 2008; Pereira et al., 2005) shown that the treatment procedures and settling/precipitation process were efficient in the removal of ^{226}Ra and other chemical species (> 80%) but lower for the removal of uranium and manganese (< 50%). According to Neves and Matias (2008), the low uranium efficiency was probably related with the water pH (> 8) where uranium remains mainly as aqueous carbonate complexes. Until 1999, this effluent was released into the waste-rock dump and afterwards into the open pit area. It was a Ca–SO_4 type water with high electric conductivity (EC), SO_4, F, Ca, U, total hardness (TH) concentration and ^{226}Ra activity (Table 1). The mine water treatment during exploitation and after mine closure was indicated as responsible for the high total hardness of groundwater wells, an uncommon characteristic in alkaline granitic regions. Based on saturation indices calculated for the effluent composition Neves and Matias (2004) suggested an increase of the water retention time in the settling pond to improve the chemical effluent quality by solid phases precipitation (e.g. aluminium and calcium sulphates, manganese and calcium carbonates, aluminium hydroxides and calcium and fluorine halides).

Table 1. Chemical concentrations and radiological activity in the effluent released from the settling pond of Cunha Baixa mine

EC µS/cm	SO_4 mg/L	F mg/L	Ca mg/L	U µg/L	^{226}Ra Bq/L	TH mg eqCaCO3/L	Reference
2900 - 1863	2017 - 1143	7.5 - 3.0	645 - 375	2900 - 222	0.05 - 1.71	2226 - 1100	Neves (2002)* Neves and Matias (2008)*
–	–	–	–	1478 - 186	0.10 - 0.53	–	Pereira et al. (2005)**

EC: Electrical Conductivity; TDS: Total Dissolved Solids; *: water samples collected between September 1996 and April 1998; **: water samples collected in June and November 2000, respectively.

The sludge from the settling pond with a concentration of 3850 mg U/kg and composed mainly of a mixture of gypsum, calcite, and clay minerals (Neves and Matias, 2004) was removed during the cleaning settling pond operations and transported to the open pit area. The results of gamma spectrometry reported by Carvalho et al. (2007a) indicate that this material also contained high concentrations of uranium series radionuclides (^{238}U: 7.4±0.4, ^{235}U: 0.31±0.04,^{234}U: 7.3±0.4, ^{230}Th: 17.8±1.8, ^{226}Ra: 5.1±1.6, ^{210}Po: 5.3±0.4 kBq/kg) and low ^{232}Th concentration (0.30±0.04 kBq/kg). Therefore, exposure of these solid phases to rainwater can originate the remobilization of some elements and radionuclides and its transfer to the surrounding environment, contributing as a potential source of contamination.

Antunes et al. (2007a) used elutriates from the sludge-settling pond (1:4 w/v) as a tool to simulate the remobilization of sediment-bound contaminants to the water column and in the assessment of their potential toxicity to planktonic organisms from local freshwater resources, in case of mechanical disturbance (e.g. rainfall). The elutriates presented, in spring and winter sampling periods, high conductivity (900 – 750 µS/cm) and high average concentrations of chemical elements – 3200 and 1033 µg U/L, 101 and 135 mg Ca/L, and 40 and 34 mg Mg/L, respectively. However, the results using two daphniids (*Daphnia magna*; standard tests-organism and *Daphnia longispina*; native species) according to standard assay protocols, indicate that sludge from settling pond do not pose environmental concerns in terms of toxicity for the aquatic system (Antunes et al., 2007a).

To improve toxicity information about contaminated soils in order to better assess the ecological risk of this mine site, Lourenço et al. (2011) exposed adult earthworms (*Eisenia andrei*) for 56 days, to a moderately alkaline (pH: 7.79) and uranium rich soil (soil B: 215.72±8.50 mg/kg) collected in the area where the sludge from the sedimentation pond are cyclically dumped and to a natural reference soil (LUFA 2.2). The results have shown bioaccumulation of metals (e.g. U: 47 – 74 µg/g dry weight) and radionuclides (^{238}U: 110 – 145, ^{235}U: 4.7 – 6.6, ^{234}U: 110 – 153, ^{230}Th: 99 – 30.2, ^{226}Ra: 29.8 – 19.7, ^{210}Pb: 25.7 – 13.6; ^{210}Po: 108.2 – 128; ^{232}Th: $<$ 0.2 Bq/kg), as well as growth reduction, DNA damages and cytotoxicity in earthworms exposed to soil B. The results also show that comet assay and flow cytometry are useful techniques to evaluate the toxicity of contaminants such as metals and radionuclides in earthworms.

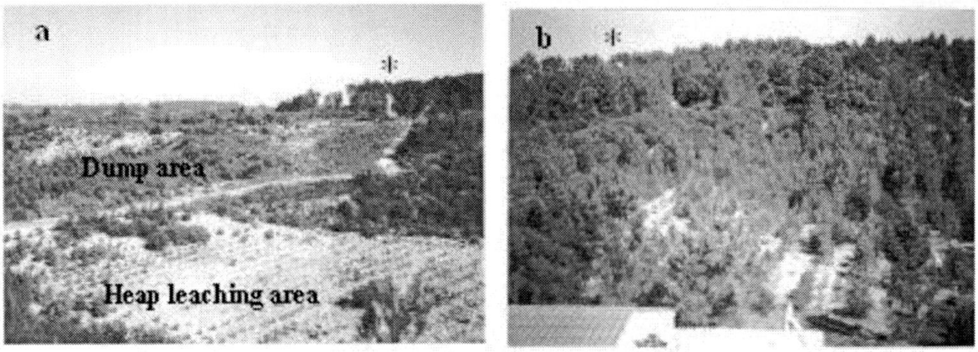

Figure 2. View of landscape rehabilitation of the Cunha Baixa mining area: (a) in September 1996 and (b) in September 2006 (* indicates the same reference point) (Adapted from Neves and Abreu, 2010).

During the rainy season the open pit leaching acid area left a temporary pond that is subjected to variable water volume depending on the fluctuations of the underground water level. In summer and in years of severe drought the pond could disappear.

Santos Oliveira and Ávila (2001) reported that trace elements concentrations of the solid wastes from this area did not match with those of the geological basement (granitic and shale rocks) except for total uranium concentrations (average of 318 mg/kg). The radionuclide ^{226}Ra was also analyzed and presented an average of 1950 Bq/kg.

The three surveys conducted in the heap leaching waste area, in 1997 by the former ENU, at the request of the National Laboratory for Energy and Geology – LNEG (formerly Geological Survey – IGM), provided averages of 83, 201 and 164 mg/kg for the total uranium concentration (Neves, 2002) in drill core samples that reached the bedrock, respectively, at 35, 16.5 and 21 m depth. The wastes from this area (sediments and drill core samples) that were submitted to static and kinetic "humidity cell tests", following MEND (1991) procedures, showed low acid neutralizing capacity and low capacity to generate acid leachates (pH< 4.5) but potentialities to release sulphate and metals as Al, Fe, Mn, Ni, Zn and U (Neves et al., 2005).

Analysis of soil samples from the heap leaching area, carried out in September 2006, show that the topsoil layer (0–25 cm) was acidic (pH: 4.08), non-saline (electric conductivity: 1500 μS/cm) and presented 36.4% of the total uranium concentration (118 mg/kg) in the available forms (Neves and Abreu, 2010). These results suggest that the leaching performed for long time by rainwater, combined with the environment acidity has not been enough to remove significantly the uranium from the wastes, although the large proportion of the element in mobilisable forms (soil solution and exchangeable complex).

Measurements of ambient radiation doses and determination of radionuclides concentrations in mining wastes and soils (0-15 cm) were performed by Carvalho et al. (2007a) in the 60 former radioactive mine areas, mostly located in Viseu and Guarda regions, as the Cunha Baixa mine. Radiation dose rates measured on the top of some waste dumps were in excess of 10 μSv/h, reaching occasionally 20 μSv/h, while the radiation background in reference areas of the region was 0.1 – 0.2 μSv/h. The observation of the ambient radiation dose rates measured in the field, as well as radionuclides concentrations determined in sample materials at the laboratory, indicate that many of the old mining sites do not have high radioactive materials left on site. The results of gamma spectrometry analyses performed on samples from mining wastes heaps showed that ^{238}U concentrations were clearly higher than ^{226}Ra concentrations (Carvalho et al., 2007a).

The first measures to reduce the visual landscape mining impact adopted by ENU, in the waste dump and heap leaching areas were not very satisfactory since the chosen plant species (pine trees) presented difficulties of implantation (Figure 2a). However, nowadays the pine trees seem to be well adapted to the physic-chemical properties of the soils developed on the waste materials without apparently manifesting visible symptoms of toxicity (Figure 2b). In addition, the area is being colonized by wild species, as is the case of the genus Cytisus (Neves and Abreu, 2010). The studies of Petrova (2006) related with the development of pines (*Pinus* sylvestris L. and Pinus nigra Arn.) in uranium mine waste dumps (Bulgaria) concluded that in the early years the pines growth was significantly suppressed, but later this type of vegetation was adapted to the unfavorable environmental conditions and survive with an apparently normal vegetative growth.

The uranium concentration (dry weight) in the shoots of Pinus and Cytisus species grown on the soils from the Cunha Baixa heap leached area ranged in the following order: pine (P. pinaster): 13.9 mg/kg> white Spanish broom (C. *multiflorus* (L'Hér.) Sweet): 6.26 mg/kg > yellow Portuguese broom (C. *striatus* (Hill) Rothm): 0.96 mg/kg (Neves and Abreu, 2010). The levels of uranium in the analyzed species are similar to those determined by Araújo et al. (2001) in pine needles and brooms (<10 mg/kg) that grew up in old heap wastes (with more than 40 years) from the Urgeiriça mine. The authors reported that these species had excluder behavior in relation to uranium, making it impossible to reach toxic levels despite the high total element concentrations in sampled soils (69 to 356 mg U/kg).

In the mining wastes dumps, none of the studied plant species were uranium accumulator (Bioaccumulator Coefficient = [plant element]/[total soil] < 1 or Bioconcentration Coefficient = [plant element]/[element bioavailable soil fraction] < 1). The species that colonized waste materials were well adapted to the substratum despite the high total and available uranium concentration (118 mg/kg and 43 mg/kg, respectively). As a consequence, and also due to the effective vegetation cover, these plants may provide effective protection to wind and water erosion and contribute to the stabilization of the mining wastes dumps (Neves and Abreu, 2010).

In the open pit pond, the physic-chemical and radiological analyses of the collected water showed an acid environment with high electrical conductivity values and high concentrations of several ionic species (Table 2). The water quality was worst in summer due to the effect of evaporation and present similar composition to the mine water that was pumping from the main shaft (Table 3). According to Neves et al. (2005) water percolating through these mine wastes resulting in acidic drainage water features. Therefore all this area was a potential source of surface and groundwater contamination of the shallowest aquifer. In order to assess the ecotoxicological effects of the Iberian green frogs (*Rana perezi* Seoane) and to perceive how this amphibian inhabited in this extreme environment, Marques et al. (2008) exposed its eggs and laboratory hatching larvae to different dilutions of the water from the open pit pond. The egg masses, for the acute toxicity tests (96 h) were collected in an uncontaminated river located few kilometers from Mangualde (Vouga River), where the presence of the amphibians has been recorded. The author's research revealed that effects on frogs mainly occurred at higher water concentrations and after the hatching of the egg masses. Reduced growth, small tail deformities, darker pigmentation and reduced movement in response to a stimulus were some of the verified abnormalities. The results also seem consistent with the absence of amphibian larvae in the open pit pond, which is used by adults mainly as a shelter.

The geochemical signature of Cunha Baixa groundwater that was pumped from the main shaft of the mine (Table 3) was the result, not only of the residual products of the acid mine waste leaching process, but also of sulphide minerals oxidation as pyrite (FeS_2). This mineral (< 0.5%) was observed in the schist rocks exposed in the open pit mine walls (Neves, 2002). In the groundwater mine was also detected *Thiobacillus ferrooxidans* (Neves, 2002), a bacterium associated with acid mine drainage, which has a powerful action on direct oxidation of sulphides (Banks et al., 1997).

High radionuclides concentrations (Bq/L$\pm1\sigma$) for Cunha Baixa' mine groundwater (^{238}U: 2.2±0.2, ^{235}U: 0.10±0.01, ^{234}U: 1.9±0.2, ^{226}Ra: 0.084 ±0.005, ^{210}Pb: 0.13±0.006, ^{210}Po: 0.247±0.013) were also reported by Carvalho et al. (2009a).

Between September 1999 and September 2004, Neves and Matias (2008) observed in the mine groundwater a decrease on the concentrations of the following analyzed parameters: up to 30% for Al, 30 to 40% for SO_4, F, Ca, Mg and 40 to 60% for Mn and U.

Research in the Vicinity of the Mine Site

Surface and Shallow Groundwater

Since shallow groundwater is the major source of irrigation water in Cunha Baixa village, the water quality and contamination are essential issues for the region as it was frequently used for agricultural practices during the irrigation season (May to September) or occasionally to watering farm animals. The first studies carried out by Neves (2002) for the waters sampled in shallow private wells (< 10 m depth), located in the surroundings of Cunha Baixa mine area, allowed to identify the contaminated well waters. The private wells are most contaminated if located near the mine and within the convergent ESE-WNE groundwater flow path direction. This contamination was identified up to 1 km down gradient from the mine site (Neves, 2002). The contaminated waters belong to the Ca/Mg–SO_4 hydrochemical facies and showed variations in water quality, which is worst in summer season (Table 4). For some water quality parameters (pH, EC, TDS, SO_4, Al, Mn, Co, Ni, Zn) the maximum recommend value (MRV) or even the maximum allowable value (MAV), established by the Portuguese regulations for water irrigation (DL 236/98) was exceeded (Table 4). Concerning the uranium concentration, these waters cannot also be recommended for irrigation. In fact, in some waters, uranium values reached concentrations 20 to 36-fold higher than the trigger value of 100 μg/L established in Australia and New Zealand (ANZECC, 2000) for short-term water irrigation (20 years). These waters are also not recommended for watering animals, a common practice in the region, because they also exceed the trigger value allowed in those countries for livestock drinking water (200 μg/L). When compared with the average values for the uncontaminated waters, the degree of contamination was especially high with respect to sulphate (a factor of 40), aluminium and manganese (a factor of 60 and 70, respectively) and uranium (a factor of 120). The measured ^{226}Ra activity concentration were always lower than the trigger value admitted by ANZECC (2000), except for one water sample collect in September 1996 (5.97 Bq/L).

Table 2. Physico-chemical and radiological data for the water pit pond from the Cunha Baixa mine

pH	EC mS/cm	TDS mg/L	SO_4 mg/L	Fe mg/L	Al mg/L	Mn mg/L	Zn mg/L	Ni mg/L	U mg/L	^{226}Ra Bq/L	Reference
2.9 - 3.7	2.90- 1.86	3.34 - 2.12	2.3 - 1.2	36.6 - 7.2	69.3 - 37.2	34.5 - 17.8	2.5 - 2.9	0.7 - 0.1	6.3- 1.2	5.8 - 12.2	Neves (2002)* Neves and Matias (2008)*
4.2	1005	_	_	3.26	5.07	3.88	_	_	1.404	_	Antunes et al. (2007b)

EC: Electrical Conductivity.
TDS: Total Dissolved Solids.
*: Water samples collected between September 1996 to April 1998.

Table 3. Physico-chemical and radiological data for the Cunha Baixa mine groundwater

pH	EC mS/cm	Ca mg/L	SO_4 mg/L	Fe mg/L	Al mg/L	Mn mg/L	Zn mg/L	Ni mg/L	U mg/L	^{226}Ra Bq/L	Reference
3.1 - 4.3	2.82 - 1.35	307 - 138	2.0 - 1.1	49 - 12	57.2 - 17.0	31.9 - 10.2	1.9 - 0.2	0.4 - 0.3	5.7 - 0.6	7.3 - 1.7	Neves and Matias (2008)*
3.9	_	_	1.4	_	_	_	_	_	_	0.08	Carvalho et al. (2009a)**

EC: Electrical Conductivity.
*: Water samples collected between September 1996 to April 1998.
**: Water samples collected in May 2002.

Table 4. Physico-chemical and radiological values of the water quality parameters in the contaminated water wells from Cunha Baixa mine area

Parameter		Summer		Winter		MRV	MAV
		Median (Maximum – Minimum)					
pH		4.4	(5.5 – 3.7)	4.5	(5.7– 3.9)	6.5 – 8.4	4.5 – 9.0
EC	µS/cm	1728	(2640 – 537)	905	(2350 – 107)	1000	nf
TDS	mg/L	1647	(2630 – 862)	752	(2252 – 98)	640	nf
SO_4	mg/L	1213	(1849 – 510)	426	(1563 – 20)	575	nf
F	mg/L	0.86	(1.58 – 0.43)	0.36	(1.32 – 0.12)	1	nf
Al	mg/L	7.9	(49.0 – 0.13)	4.0	(21.64 – 0.08)	5	20
Co	µg/L	69	(414 – < 10)	< 10	(244 – < 10)	50	10000
Mn	mg/L	6.0	(20.5 – 0.10)	2.1	(14.8 – 0.049	0.2	10
Ni	µg/L	156	(626 – 29)	56	(390 – < 10)	500	2000
U	µg/L	1200	(3700 – 52)	330	(2000 – 12)	100*	nf
Zn	µg/L	880	(3288 – 134)	313	(1991 – 11)	2000	10000
^{226}Ra	Bq/L	0.60	(5.97 – 0.06)	0.6	(4.67 – 0.15)	5*	nf

MRV and MAV: Maximum Recommend Value and Maximum Allowable value established in Portuguese regulations (DL 236/98) for water irrigation; EC: Electrical Conductivity; TDS: Total Dissolved Solids; nf : not fixed; *: trigger value established in Australia and New Zealand (ANZECC, 2000) for short-term water irrigation.

In Cunha Baixa, Carvalho et al. (2005) also reported moderate concentration (mBq/L) of the uranium series radionuclides in water (^{238}U: 11 – 256; ^{235}U: 0.5 – 10.8; ^{234}U: 10.5 – 243; ^{226}Ra: 12 – 104; ^{210}Pb: 10 - 32; ^{210}Po: 4.4 – 72) from private supplies (superficial wells). A well water in the village of Aguieira, located 7 km W of Cunha Baixa, in an area not impacted by uranium mine activities, presented the following radionuclides concentrations (mBq/L, Carvalho et al., 2009b): (^{238}U: 80.4±2.5; ^{235}U: 3.9±0.3, ^{234}U: 73.3±2.3; ^{232}Th: 73.3±2.3, ^{226}Ra: 97.6±11.3, ^{210}Pb: 61.6±1.8, ^{210}Po: 6.6±0.2, ^{232}Th: 0.02±0.009) that were not very different from those measured in Cunha Baixa. However, these concentrations were higher than the concentrations measured in the rivers (Alva, Seia, Mondego, and Távora) of the centre-north regions of Portugal (^{238}U: 2.5 – 8.0; ^{235}U: 0.1 – 0.44; ^{234}U: 2.6 – 8.8; ^{226}Ra: 4 – 21; ^{210}Po: 6 – 15; ^{232}Th: 0.13 – 0.28 mBq/L) suggesting an enhancement of the natural radionuclides concentration (Carvalho et al., 2005). The more acidic waters contained the highest concentrations of radionuclides belonging to uranium series and Carvalho et al. (2009a) reported that a special attention must be given to the ^{226}Ra. According to Pereira et al. (2005) these ^{226}Ra concentrations had its origin in the water mine waste as well as in radium leaching from wastes piles disposed near the mine.

The results obtained for the 1995–2008 monitoring period, showed that the quality of the contaminated shallow groundwater sampled in the wells from the area affected by the mining activities has improved (Figure 3). This behavior could be caused by attenuation in the natural environment and/or by some actions developed in the mining area such as the decrease of the mine water pumping and the improvement of the quality of the effluent water discharged from the settling pond into the open pit area or into the natural environment. However, the improvement of the contaminated water by natural attenuation may require tens, or even hundreds of years (IAEA, 2001) if other remediation efforts were not applied.

In the period 1995–1998, surface water contamination was detected only in the main watercourse (Castelo creek) that drains the mine area, to approximately 1–1.5 km downstream. Values for parameters such as pH (3.7 – 5.9), SO_4 (135 – 1892 mg/L), Zn (14 – 782 µg/L) and Ni (18 – 176 µg/L) frequently exceeded the Portuguese regulatory limits for surface water (DL 236/98). Uranium (60 – 3200 µg/L) and ^{226}Ra activity (0.4 – 2.0 Bq/L) in this surface water were higher than in other surface waters sampled in the area (Neves, 2002; Neves et al., 2005).

In May 2002, Carvalho et al., (2009a) determined in the same watercourse (Castelo creek), that presented an average pH value of 5.1 and 16 to 12 mg/L of SO_4, the following radionuclides concentrations (mBq/L): ^{238}U: 256 – 75; ^{235}U: 108 – 3.7; ^{234}U: 243 – 75; ^{226}Ra: 45 – 47; ^{210}Pb: 32–92; ^{210}Po: 21 – 32. These radionuclides concentrations decreased downstream uranium mine site. The referred authors also reported that radionuclides in Castelo Stream water increased with the discharge of wastewater originated in Cunha Baixa and Quinta do Bispo (located near 2 km W of Cunha Baixa) mines. Radionuclides concentration in surface water samples collected during May–July 2007 at Castelo Stream, 2 to 2.5 km downstream of both uranium mines, showed no significant enhancement of radionuclides concentrations (^{238}U: 56.5±1.7; ^{235}U: 3±0.2; ^{234}U: 58.4±1.8; ^{230}Th: 2.4±0.2; ^{226}Ra: 26.7±5.7; ^{210}Po: 8.8±0.4; ^{232}Th: 0.27±0.05 mBq/L) (Carvalho et al., 2008). However, these samples were collected during a period of no mine water discharge (treated or untreated) into the watercourses and thus the results may contrast with situations of high radionuclides concentrations observed before.

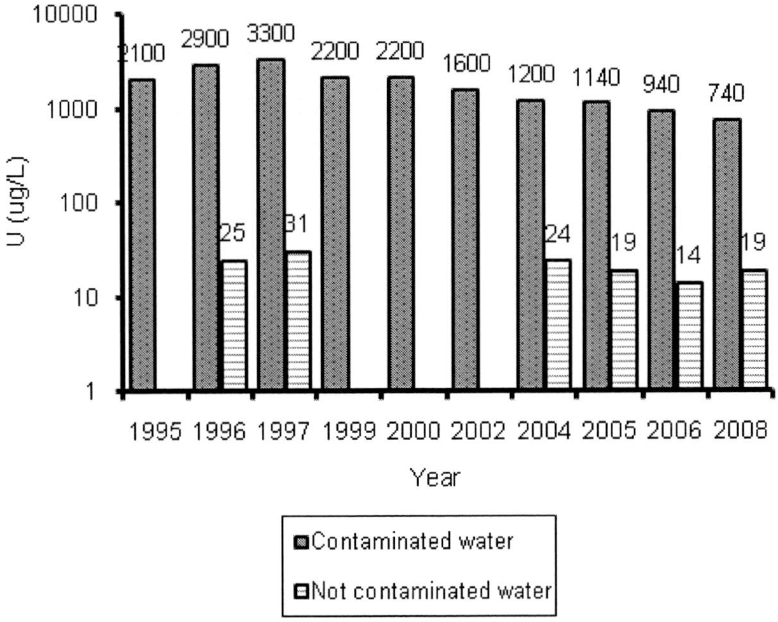

Figure 3. Variation of uranium concentration in contaminated water (P15) and not contaminated water (P24) wells used for irrigation purpose in the vicinity of the uranium Cunha Baixa mine during 1995 – 2008 monitoring period. (Wells P15 and P24 in Neves (2002); Neves and Matias (2008)).

Agricultural Soils

As previous reported the surface and shallow groundwater quality of Cunha Baixa village was affected by the mining activities and its abandon. The proximity of agricultural soils to the mine site and the villagers' dependency on the shallow water pumped from wells for farmer purposes, as soil irrigation, called for the diagnosis of the effects on the soil characteristics as well as in the quality and security of the cultivated food crops. In the summer season this purpose may create soil quality problems and even the consequent accumulation of toxic elements in the crops because of the greater amounts of water used.

The soils of the area, classified as Cambisols or Alluvisols (Carta de Solos, 1978), with a sandy-loam or loam texture, are used for cultivation of vegetables foodstuff for the villager's own consumption.

In 1996, Neves (2002) observed that agricultural soils that were irrigated with contaminated uranium water (2 to 2.5 mg/L) were very acidic (pH < 4.5) and presented very low to moderately salinity (2 < CE < 12 mS/cm). Also, the total (225 to 362 mg/kg) and soil available uranium concentration (158 to 128 mg/kg) were greater than those found in literature for normal concentration range (0.3 to 11.7 mg/kg; Bleise et al., 2003) or exceeded the uranium value of 12 mg/kg estimated as the average geochemical background content for Alluvial soils (Santos Oliveira et al., 2005), developed on sediments originated from the regional granitic rocks, as well as the value of 14 mg/kg for soils outside the influence of the uranium Urgeiriça mine area (Araújo et al., 2001). A soil sampled at 1.5 km SW from Cunha Baixa mine area, used for agricultural practices and irrigated with non contaminated water, which can be considered as a regional reference level, presented uranium concentration of 9.8 mg/kg (Neves, 2002).

After a period of ten years there was an improvement in the soil quality with regard to characteristics such as pH (4.6 < pH < 5.5) and salinity (2 < CE < 8 mS/cm) or the available fraction uranium concentration (< 13.3 mg/kg) which did not exceed 17% of the total uranium concentration (68 to 259 mg/kg), although it is, on average, 10 times higher than the above cited values for the regional reference level (Neves and Abreu, 2010). It was suggested that this situation could be related to the fact that uranium is preferentially bound to the non-crystalline iron oxides fraction present in these soils (Neves et al., 2009). Also, the low cation exchange capacity (CEC) of these soils, as a consequence of the low clay fraction content (< 16%), combined with its clay mineralogy (kaolinite > illite), as well as the low total organic carbon content (< 20 g/kg) (Neves, 2002; Neves et al., 2008, 2009), may all be responsible for the small percentage of the uranium available fraction.

In Portugal and in European Union, there is no legislation to evaluate environmental impacts on soils, so it is recommended by the Portuguese Environmental Agency to estimate its quality using the Canadians norms or even Dutch norms. In the sampled Cunha Baixa agricultural soils the uranium concentrations often exceeded 23 mg/kg, which is the preliminary health soil quality guideline value for uranium ($PSQG_{HH}$) for agricultural land use in Canada (CCME, 2007). As the concentration of this element in the analyzed soils was still very high, they cannot be recommend for that purpose. However, according to the predicted no-effect concentration (PNEC) of 250 mg U/kg dry soil reported by Sheppard et al. (2005), for non-human biota, plants grown in the studied Cunha Baixa soils would not be affected by uranium toxicity.

In soils located nearby the Cunha Baixa mine Pereira et al. (2004, 2005) measured in topsoil samples (0–15 cm) ^{226}Ra activity between 800 and 2100 Bq kg^{-1}, that exceeded, in some cases, six to seven times the maximum permissible limit proposed by USNRC (2000); a reference value of 185 Bq/kg above the background (which was assumed at time by the authors as 115 Bq/kg).

In two different soils which had similar characteristics but different uranium content (45±9 and 128±22 mg/kg) and used in a field experiment in 2006, Carvalho et al. (2009c) observed that concentrations of uranium isotopes in both soils were very similar, although ^{226}Ra (1203±131 and 793±62 Bq/kg, respectively) and ^{230}Th (132±8.8 and 507±24 Bq/kg, respectively) displayed some differences between soil samples. According to Carvalho et al. (2007a) forest and agriculture soils collected in the regions of the old mines display both ^{238}U and ^{226}Ra concentrations in general one to two orders of magnitude lower than the mining wastes.

The neighborhood of the Cunha Baixa uranium mining zone and its influences, as well as the extensive use of uranium rich waters for irrigation, could explain the enhancement of uranium and radionuclides concentration in soils, as was reported by Amrhein et al. (1993) for agricultural soils in the San Joaquim Valley (USA). As a consequence, villagers and cattle could be exposed to the intake of metals and radionuclides concentrations higher than the background values.

Crops

Different crops are cultivated during the year in Cunha Baixa agricultural soils, but very little was known about its chemical and radiological quality. The transfer of elements and radionuclides from soil to plants is the first step in the pathway of its incorporation to humans and hence for the evaluation of health risks.

In plants, uranium has not shown to be an essential or beneficial element, even though many species have the capacity to absorb and concentrate chemical and radioactive species into their tissues (Gulati et al., 1980; Sheppard et al., 1989; Sarić et al., 1997; Anke et al., 2009).

The first results on Cunha Baixa crops were reported by the authors (Neves, 2002; Neves et al. 2003) on maize (*Zea mays* L.) grown in acid soils (pH < 4.5), with high uranium content in soil available forms ($U_{available}$: 128 – 158 mg/kg) and irrigated with uranium contaminated water (2.1– 3.7 mg U/L). The data showed that the plants accumulate more uranium in their root system (20 – 129 mg/kg dry weight) than in leaves (0.6 – 1.2 mg/kg dry weight), stems (121 – 166 μg/kg dry weight) or grain (< 26 μg/kg dry weight). Maize grown in a soil located 1 km far from the mine site, with low uranium concentration (U_{total}: 9.8 mg/kg, $U_{available}$: 1.04 mg/kg) and irrigated with uncontaminated water (< 3μg/L) shown low uranium concentrations (roots: 1.1 mg/kg, leaves: 43 μg/kg, shoots: 7 μg/kg). The authors also observed unusual twinned maize ears (*Zea mays* L.), which occurred with a high frequency in plants irrigated with contaminated water (> 100 μg/L). Such abnormality may reflect the impairment of biological functions subsequent to plant exposure to uranium in contaminated irrigation water. Although the addition of low concentrations of uranium salts to fertilizers can have a positive effect plant growth (i.e., hormesis), subsequent increases in uranium concentrations can negatively impact plants development (Lamas, 2005). The ability of this chemical and radiological element to cause hormesis is further supported by the results of Meyer et al. (1998), where hormesis was observed for grass plants cultivated in soil pots contaminated with depleted uranium.

In Cunha Baixa field experiments carried out by the authors, in 2005, on lettuce (*Lactuca sativa* L.) in two different soils (with low and high total uranium concentrations: 38 and 106 mg/kg, respectively) watered with non contaminated and uranium contaminated (998 to 1140 μg/L, respectively) irrigation waters, showed that although the plants accumulate uranium in their roots (up to 6 mg/kg dry weight) they only have the capacity to transfer 30% of the uptake uranium to the leaves (up to 2.6 mg/kg dry weight) whatever the soil or irrigation water quality (Neves et al., 2008).

These same field experiments were repeated in 2006, between April and October, with crops as potatoes (*Solanum tuberosum* L.), beans (*Phaseolus vulgaris* L.) and lettuce (*Lactuca sativa* L.) The average uranium concentration for these crops after harvest are presented in Table 5 as well as the ones determined in carrot (*Daucus carota* L.), cabbage (*Brassica oleracea* L.), apple (Malus domestica Borkh) and maize (*Zea mays* L.) and sampled by the authors in October 2008 (Neves et al., 2011). This work was undertaken as part of the uranium related health risk assessment for this area in order to address concerns expressed by the local farmers.

Among the analyzed vegetables foodstuff the highest uranium concentrations were found in lettuce, and the lowest in apple (with peel). The data reported by Neves et al. (2011) and the relationships obtained between them suggested that the uranium content in the studied foodstuff was dependent on the uranium concentrations in irrigation water or in the soil available fraction, as well as the plant uranium uptake capacity and its distribution within the plant tissues. However, the foodstuffs grown in the Cunha Baixa agricultural study area have concentrated, in the edible parts, higher uranium levels than the reported data in the available wider literature. Also, Anke et al. (2009) found that wild and cultivated plants from the immediate vicinity of uranium mining waste dumps (Ronneburg, Germany) stored 8-fold more uranium than control plants.

The results obtained for potatoes shown that the tubers peel was a sink for uranium, as it concentrates 90 to 96% of the total uranium found in the whole tubers (69 to 589 mg/kg dry weight). The tuber peel removal before being ingested (a common practice in this region) is important to minimize human exposure through this route (Neves et al., 2011, 2012). The use of tubers peel to feed livestock (e.g. *Sus shofa domesticus*, L.), which is common among farmers in the region, is also not recommended as a measure to reduce uranium animal exposure.

Table 5. Average uranium concentration (µg/kg dry weight) in the edible tissues of crops grown in agricultural soils of Cunha Baixa mine area

Crop	U_{total} Soil: 68 – 108 mg/kg U Water irrigation: 218 – 1035 µg/L	U_{total} Soil: 45 – 94 mg/kg U Water irrigation: 8 – 19 µg/L
Lettuce (leaves)	1241 – 5373	189 – 758
Potato (tubers peeled)	7.7 – 19.9	6.5 – 17.8
Green bean (pods)	141 – 194	239 – 94
Carrot (peeled)	67 – 187	6
Cabbage (leaves)	149 – 255	14
Apple (with peel)	11	ns
Maize (grain)	22	ns

ns: not sampled.

The cultures harvested in soils irrigated with uranium contaminated water showed high uranium concentrations, which could present some risks for soils if the plant residues (non edible parts) were used for soil fertilization (e.g. stems and leaves of maize, potatoes, and green beans), an agricultural practice common in this area, or when were used for animal feeding.

Other research conducted by Ferreira (2007), on mushrooms (*Macrolepista procera*, Scop ex. Fr), an appreciated foodstuff consumed by the Cunha Baixa villagers, also showed high uranium concentrations (dry weight) in the stalks and in the caps (152±106 mg/kg and 5.44±1.34 mg/kg, respectively).

Investigation of the human radiotoxic exposure arising from the transfer of radionuclides to soils and crops from the use of uranium contaminated irrigation waters have recently begun. In some preliminary studies Carvalho et al. (2007b, 2009c) checked the radioactivity on lettuce and potatoes grown on soils with enhanced levels of uranium isotopes and ^{238}U

series alpha-emitting radionuclides (U_{total}: 47 mg U/kg (^{238}U: 455 – 483; ^{235}U: 23.2 – 23.5; ^{234}U: 453 – 489 Bq/kg), ^{230}Th: 132 – 507; ^{226}Ra: 793 – 1203; ^{210}Po: 563 – 628 Bq/kg, respectively) and irrigated with water containing lower (U_{total}: 0.3 Bq/L (^{238}U: 289; ^{235}U: 12.7; ^{234}U: 282 mBq/L); ^{230}Th: 0.6; ^{226}Ra: 2698; ^{210}Po: 330 mBq/L) and higher uranium activity (U_{total}: 13.7 Bq/L (^{238}U: 13738; ^{235}U: 650; ^{234}U: 12844 mBq/L); ^{230}Th: 22; ^{226}Ra: 1189; ^{210}Po: 76.4 mBq/L). The results suggested that irrigation with water containing low or high uranium activity is unlike to increases the concentration of uranium, thorium and polonium in the studied vegetables, even growing in uranium rich soils. Nevertheless, irrigation with water containing high radium concentrations may further contribute to enhance the uptake of this radionuclide by those foodstuffs. Lettuce accumulates more uranium, radium and polonium (e.g. ^{238}U: 28.3; ^{234}U: 26.6; ^{226}Ra: 118; ^{210}Po: 15.9 Bq/kg dry weight) than the whole potato tubers (e.g. ^{238}U: 7.0; ^{234}U: 6.6; ^{226}Ra: 15; ^{210}Po: 0.88 Bq/kg dry weight) grown in the same soil and irrigated with the same water quality (Carvalho et al., 2007b). The ^{226}Ra concentration in potato tubers was related to the ^{226}Ra concentration in the irrigation water but peeling will remove about 50% of the potatoes' radium content (Carvalho et al., 2009c). In other study with lettuce, grown in different kitchen gardens near Cunha Baixa mine and irrigated with water containing enhanced levels of radioactivity, Carvalho et al. (2008) reported higher radionuclide concentrations (^{238}U: 0.54 – 154; ^{234}U: 0.56 – 152; ^{230}Th: 0.39 – 86; ^{226}Ra: 3 – 100; ^{210}Po: 4.7 – 101, ^{232}Th: 0.28 – 2 Bq/kg dry weight) in comparison with reference levels (baseline) for the same species (^{238}U: 0.52; ^{234}U: 0.53; ^{230}Th: 0.49; ^{226}Ra: 1.7; ^{210}Po: 2.1, ^{232}Th: 0.092 Bq/kg dry weight).

Risks to Human Health

For communities related to mine operations or to non-occupationally exposed populations, mining activities can impact health at various levels. Adverse health effects might result from environmental exposures to noise, air, water, soils and foodstuff contamination. Internationally there are relatively few studies of community health impacts due to mining operations and the majority of studies come from Europe and North America. Effects of uranium mining on human health are not immediate and it may take several years before any adverse consequence is recognized. Historically most epidemiology on uranium mining has focused on mine workers and radon exposure. The chemical toxicity of uranium is the primary environmental health hazard, being the radioactivity of uranium a secondary concern. Uranium has been identified as a nephrotoxine, which effects are more likely due to its chemical properties rather than to its radioactivity, although ingested uranium may have a radiological effect on other deposition tissues (Zamora et al., 1998).

In Cunha Baixa the villagers may be exposed to uranium by inhalation (dust) or through the diet (water or foodstuff). In the region, water ingestion was not considered a significant exposure route of concern as the villagers consume water from the public network. This tap water has a total uranium concentration below 2 μg/L (Neves et al., 2008), which is lower than the provisional guideline value (15 μg/L) considered by the World Health Organization, since 2005, for this propose. Also, the analysis of public water supplies (tap water), in some towns and cities in uranium centre-north regions of Portugal, for total alpha and total beta radioactivity showed concentrations generally below the maximum recommended limits applicable to water for human consumption (0.1 and 1 Bq/L, respectively) as set in EUCD 98/83/EC (Carvalho et al., 2005). The analysis of tap water from Mangualde (2 km far from

Cunha Baixa) also showed low concentration of alpha radionuclides: ^{238}U: 20.2; ^{235}U: 1.0 ^{234}U: 18.4; ^{226}Ra: 12.8; ^{210}Pb: 48; ^{210}Po: 3.6; ^{232}Th: 0.14 mBq/L (Carvalho et al., 2005).

The chemical health risk posed to Cunha Baixa' villagers by uranium exposure through the ingestion of the previous studied foodstuffs (lettuce, potatoes with and without peel, green bean pods, carrot without peel, cabbage, apple with peel and grain maize (used to make bread) was assessed based on the hazard quotient (HQ = EDing/RfD). The uranium ingestion exposure dose (EDing) for an adult (lifetime 50 years; 70 kg weight) and a child (5 – 11 years old; 32 kg weight) was estimated considering a Cunha Baixa site-specific scenario, namely food ingestion rate (g/person day) and exposure frequency (days/year) typical of the local consumption habits. The Reference Dose (RfD) for chronic oral uranium exposure was established by the American Environmental Protection Agency at 0.003 mg/kg day (IRIS, 2010).

The risk analysis calculated for non-cancer health effects and related only to the exposure through the consumption of this basic foodstuffs, revealed safety (hazard quotient < 1) for Cunha Baixa villagers (adults and children) even when crops was grown on uranium enriched soils and irrigated with contaminated water (Neves et al., 2011).

Food chain transfer of uranium descendent radionuclides to man is other concern in uranium mining areas because of the potential harmful biological effects of alpha radiation. Among the radionuclides measured in Cunha Baixa foodstuff, ^{226}Ra might be the critical one for radiation exposure of the local population. Although a proper risk assessment would require the measurement of radionuclides in the complete diet of the local population, Carvalho et al. (2009c) evaluated the health risk through only the ingestion of potatoes. For the calculations, it was assumed, on a provisional basis, a potatoes consumption rate of 73 kg per year per capita, a radium concentration of 5 Bq/kg in peeled potatoes, and a gut transfer factor of 0.20, with the committed effective dose per unit intake via ingestion for adult members of the public of 2.8x10-7 Sv/Bq (IAEA, 1995), the annual effective dose rate was estimated at 0.1 mSv. This value was 10% of the annual dose limit allowed for members of the public (1 mSv/y, EUCD 96/29/Euratom). Although, in Cunha Baixa potatoes are one of the main vegetable foodstuffs consumed during the year (it is consumed in at least one meal per day), future radiological risk analysis may include others foodstuff products. However, the radionuclides concentrations detected in horticulture and farm products around old uranium mining sites near Mangualde (Cunha Baixa, Quinta do Bispo, Espinho) suggested low soil to plant transfer of radionuclides and low food chain transfer of them to man (Carvalho et al., 2007c).

Environmental Remediation Project

In Portugal, the DL 198-A/2001 assigned the concession and the responsibility of the activity of environmental reclamation of abandoned mine sites to EDM S.A. (Mining Development Company). Priority has been given to the radioactive and sulphide mines.

Environmental reclamation of the Cunha Baixa mining area presents itself as overall very positive and necessary, as it will solve environmental problems and will contribute to the improvement of the environmental village and residents' quality.

The interpretation of all data collected in literature and in the field, allowed to establish a diagnosis of the main problems and to present a project to solve or minimize the geotechnical, the chemical and the radiological risks taking into account the village proximity (EDM, 2009).

The mine site is a point source of pollution either for soil and water environment as well as an area with uncontrolled deposits of sludge and wastes that in some cases, present a high degree of contamination. The release of acid effluents and the presence of large open pits determine risks to people and animals. In addition, open pits provoke a strong visual impact on the landscape. The proximity of the mining area to the village, also rising concerns on public health given the potential issue of increased flows of radon.

The EDM (2009) project proposes the following actions:

- creation of a confinement cell in the main open pit area for the deposition of confined sludge, enriched wastes, materials from the areas to clean up and wastes from the demolition of the old structures within the boundaries of the mining site;
- naturalization of the water line located immediately downstream the open pit which crosses fields, preventing both the riverbanks erosion and overflowing, and increase of water quality in order to promote fields irrigation;
- modeling and sealing, or just sealing, the open pits using natural and synthetic materials. This action will prevent surface water to feed the volume of acidic groundwater;
- construction of a water river control and passive treatment on a old well to control and, when necessary, to treat in time the sub-surface flows;
- construction of a new mine water treatment plant, to working temporarily until the water mine reaches standards quality;
- modeling the wastes dam to ensure slope stability and implementation of landscape quality and installation of fencing and appropriate signage;
- redirection of the existing drainage throughout the area of intervention;
- provide conditions for the creation of an environmentally pleasant area, with possible use for recreational purposes as construction of a picnic area with water mirrors nearby, by an adequate landscape project providing all the security conditions, and
- recovery of the agricultural land located downstream mining area.

ACNOWLEDGMENTS

The authors thank the financial support given by Praxis XXI Project (2/2.1/CTA/235/94), IAEA research contract n° 9117/RO and Project POCI/ECM/59118/2004 supported by Program POCI 2010 and Feder European Communitarian Funds. The authors also thank to the Centres of Petrology and Geochemistry and Pedology of Technical University of Lisbon (Portugal). Special thanks to Dr. A. K. Mathur for editing the English.

Conclusion

The environmental reclamation of old uranium mining sites is a priority in Portugal as well as in many European countries. The environmental degradation associated with uranium mining can impact not only the immediate ecosystems, but also human health. For protection of the communities that leaves around these mine sites a complete surveillance of mining wastes, waters, soils and agricultural products must be done as well as human and animals risk health assessment in order to evaluate all the potential risks and to recommend site-specific strategies for rehabilitation actions. The studies performed in uranium Cunha Baixa mine area and its surroundings were an attempt to achieve the best characterization of the environmental impacts created by the mining activities and its abandonment. This knowledge is crucial to undertake the correct measures for the area recovery.

References

Amrhein, C; Mosher, P; A, Brown, A D. (1993). The effects of redox on Mo, U, B, V, and As solubility in evaporation ponds. *Soil Science*, 155(4), 249-255.

Anke, M; Seeber, O; Müller, R; Schäfer, U; Zerull, J. (2009). Uranium transfer in the food chain from soil to plants, animals and man. *Chemie der Erde-Geochemistry*, 69, 75-90.

Antunes, SC; Pereira, R; Gonçalves, F.(2007a). Evaluation of the Potential Toxicity (acute and chronic) of Sediments from Abandoned Uranium Mine Ponds. *Journal of Soil Sediments*, 7, (6), 368-376.

Antunes, SC; Pereira, R; Gonçalves, F. (2007b). Acute and Chronic Toxicity of Effluent water from an Abandoned Uranium Mine. *Archives Environmental Contamination and Toxicology*, 53, 207-213.

ANZECC, Australian and New Zealand Environment and Conservation Council (2000). *Water Quality Guidelines*. The Guidelines Vol.1. Available at http://www.mfe.govt.nz/issues/water/ANZECC/about.html.

Araújo, MF; Barbosa, T; Madruga, MJ; Faria, I. (2001). Dispersão de contaminantes e sua transferência no sistema solo-planta nas escombreiras da mina de urânio da Urgeiriça. *Actas Congresso Internacional Sobre Património Geológico e Mineiro*, (567-573). Beja, Portugal.

Banks, D; Younger, PL; Arnesen, R; Iversen, ER. Banks, SB. (1997) Mine-water chemistry: the good, the bad and the ugly. *Environmental Geology*, 32 (3),157-174.

Bleise, A; Danesi, PR; Burkart, W. (2003). Properties, use and health effects of depleted uranium (DU): a general overview. *Journal of Environmental Radioactivity*, 64, 93-112.

Carta de Solos (1978). *Unidades Pedológicas segundo o esquema da FAO para a Carta dos Solos da Europa (escala 1:1000 000).* Atlas de Ambiente, Comissão Nacional do Ambiente, Portugal.

Carvalho, FP; Oliveira, JM; Libânio, A; Lopes, I; Ferrador, G; Madruga, MJ. (2005). Radioactivity in public water supplies in the uranium mining regions of Portugal. *IAEA Proceeding Series, International Workshop on Environmental Contamination from Uranium Production Facilities and their Remediation (Lisboa, Portugal)*, paper 3, 41-51.

Carvalho, FP; Madruga, MJ; Reis, MC; Alves, JG; Oliveira, JM; Gouveia, J; Silva, L. (2007a) Radioactivity in the environment around past radium and uranium mining sites of Portugal. *Journal of Environmental Radioactivity*. 96, 39-46.

Carvalho, FP; Oliveira, JM; Neves, O; Abreu, MM; Vicente, EM. (2007b). Agriculture in an area impacted by past uranium mining activities. *Proceedings 11th International Conference on the Environmental Remediation and Radioactive Waste Management,* (977 - 980). Bruges, Belgium.

Carvalho, FP; Oliveira, JM; Torres, LM. (2007c). Environmental radioactivity assessment around old uranium mining sites near Mangualde (Viseu), Portugal. *Proceedings 11th International Conference on the environmental remediation and radioactive waste management,* (1325 - 1331). Bruges, Belgium.

Carvalho, FP; Oliveira, JM; Malta, M. (2008). Agriculture and soil-to-plant radionuclide transfer in the surrounding of an old uranium mine. Available at: http://radioecology.info/Bergen2008/ proceedings/76.%20Carvalho%20uranium%20OP.pdf.

Carvalho, FP; Oliveira, JM; Faria, I. (2009a). Alpha Emitting Radionuclides in Drainage from Quinta do Bispo and Cunha Baixa Uranium Mines (Portugal) and associated Radiotoxicological Risk. *Bulletin of Environmental Contamination and Toxicology*, 83(5), 668-673.

Carvalho, FP; Oliveita, JM; Malta, M. (2009b). Analysis of radionuclides in soil, water and agriculture products near the Urgeiriça uranium mine in Portugal. *Journal of Radionalytical and Nuclear Chemistry*. 281, 479-484.

Carvalho, FP; Oliveira, JM; Neves, O; Abreu, MM; Vicente, EM. (2009c). Soil to plant (Solanum tuberosum L.) radionuclide transfer in the vicinity of an old uranium mine. *Geochemistry: Exploration, Environment and Analysis,* 9 (3), 275-278.

CCME, Canadian Council of Ministers of the Environment. (2007). *Canadian Soil Quality Guidelines for Uranium: environmental and human health*. Scientific supporting document. CCME, Winnipeg. Available at: http://www.ccme.ca/assets/pdf/ uranium_ssd_soil_1.2.pdf.

DL 198-A/2001. *Principles and objectives that should rule the environmental remediation of degraded and abandoned mining areas in Portugal*. DR 155/2001-Série I-A 1° Supl (6 July 2001).

DL 236/98, *Portuguese Legislation on Water Quality*. DR 176/98, annexes XVI and XXI (1 August 1998).

EDM, Empresa de Desenvolvimento Mineiro. (2009). *Remediação ambiental da área mineira da Cunha Baixa*. Nota justificativa. Candidatura QREN-POVT, Novembro 2009.

EUCD 96/29/EURATOM, European Union Council Directive 96/29/EURATOM of 13 May 1996. Basic safety standards for the protection of the health of workers and the general public against the dangers arising from ionizing radiation. *Official Journal of European Communities,* L 159, 29.06.96.

EUCD 98/83/EC, European Union Council Directive 98/83/EC of 3 November 1998. On the Quality of Water Intended for Human Consumption. *Official Journal of European Communities*, L 330, 05.12.1998.

Ferreira, MJS. (2007). *Toxicidade de solos uraníferos em cogumelos e plantas comestíveis.* Dissertação de Mestrado (MSc Dissertation), University of Aveiro, Portugal.

Goinhas, J. (1986). Cadre géologique et metallogénique des ressources minières du Portugal. *Chronique de la Recherche Minère*, 489, 25-42.

Gulati, KL; Oswall, MC; Nagpaul, KK. (1980). Assimilation of uranium by wheat and tomato plants. *Plant and Soil*, 55 (1), 55-59.

IAEA, International Atomic Energy Agency. (1995). Planning and management of uranium mine and mill closures. *IAEA-TECDOC*, 824, Vienna.

IAEA, International Atomic Energy Agency. (2001). Manual of Acid in situ Leach Uranium Mining Technology. *IAEA-TECDOC*, 1239, Vienna.

IRIS, Integrated Risk Information System. (2010). *Uranium, soluble salts*. Available at: http://www.epa.gov/iris/subst/o421.htm.

Lamas, MC. (2005). Factors affecting the availability of uranium in soils. *Braunschweig, Germany: FAL Agricultural Research*, Landbauforsch Volk SH 2005; 278 (Special Issue).

Lourenço, JI; Pereira, OR; Silva, AC; Morgado JM; Carvalho, FP; Oliveira, JM; Malta, MP; Paiva, AA; Mendo, SA; Gonçalves, FJ. (2011). Genotoxic endpoints in the earthworms sub-lethal assay to evaluate natural soils contaminated by metals and radionuclides. *Journal of Hazardous Materials*, 186, 788-795.

Marques, SM; Gonçalves, F; Pereira, R. (2008). Effects of a uranium mine effluent in the early-life stages of *Rana perezi* Seoane. *Science of Total Environment*, 402, 29-35.

Martins, LP. (2010). *Mineral Resources of Portugal*. On line version. Available at: http://www.scribd.com/doc/46430033/Mineral-Resources-of-Portugal.

Matos Dias, J. (1972). A região uranífera da Cunha Baixa-Quinta do Bispo. *Memórias e Notícias*, 73, 26-27.

Matos Dias, J; Soares Andrade, AA. (1970). Portuguese uranium depositd. *Memórias e Notícias*, 70, 1-22.

MEND, Mine Environment Neutral Drainage. (1991). *Acid Rock Drainage Prediction Manual*. MEND Project 1.16.1b, Coastech Research Inc.

Meyer, MC; McLendon, T; Price, D. (1998). Evidence of depleted uranium-induced hormesis and differential plant response in three grasses. Journal of Plant Nutrition, 21, 2475-2484

Neves, O. (2002). *Minas desactivadas e impactos geoquímicos ambientais. O caso da mina de urânio da Cunha Baixa (Viseu)*. Dissertação de Doutoramento (Ph.D. Dissertation), Universidade Técnica de Lisboa, Instituto Superior Técnico, Portugal.

Neves, O; Matias, MJ. (2004). Pollution sources in Cunha Baixa uranium mining area (Viseu, Portugal). *Cadernos Laboratorio Xeolóxico de Laxe*, 29, 187-202.

Neves, O; Matias, MJ. (2008). Assessment of groundwater quality and contamination problems ascribed to an abandoned uranium mine (Cunha Baixa region, Central Portugal) *Environmental Geoogy*, 53 (8), 1799-1810.

Neves, O; Abreu, MM. (2010). Uranium concentration in plants developed on agricultural soils and waste dumps in Cunha Baixa mine site (Mangualde). *Revista de Ciências Agrárias*, 33 (1), 217-225.

Neves, O; Abreu, MM; Matias, MJ. (2003). Comportamento do urânio, alumínio e manganês no milho cultivado em solos na área mineira da mina de urânio da Cunha Baixa. *Memórias e Notícias*, 2 (Nova Série), 265-78.

Neves, MO; Matias, MJ; Abreu, MM; Magalhães, MCF; Basto, MJ. (2005) Abandoned mine site characterization for remediation: the case of the Cunha Baixa uranium mine (Viseu, Portugal). *IAEA Proceeding Series, International Workshop on Environmental Contamination from Uranium Production Facilities and their Remediation (Lisboa,*

Portugal), paper 14, 159-169. Available at http://www-pub.iaea.org/MTCD/publications/PDF/Pub1228_web.pdf.

Neves O, Abreu, MM; Vicente, EM. (2008). Uptake of uranium by lettuce (*Lactuca sativa* L.) in natural uranium contaminated soils in order to assess chemical risk for consumers. *Water Air and Soil Pollution,* 195, 73-84.

Neves, O; Abreu, MM; Matias, MJ. (2009). Uranium distribution in the solid phases of soils from Cunha Baixa mining site (Portugal). *Revista de Ciências Agrárias*, 31 (1), 195-204.

Neves, MO;. Abreu, MM; Figueiredo, VR. (2011). Uranium in vegetable foodstuffs: should residents near the Cunha Baixa uranium mine site (Central Northern, Portugal) be concerned? *Environmental Geochemistry and Health*, DOI: 10.1007/s10653-011-9428-9.

Neves, MO; Figueiredo, VR; Abreu MM. (2012). Transfer of U, Al and Mn in the water-soil-plant (Solanum tuberosum L.) system near a former uranium mining area (Cunha Baixa, Portugal) and implications to human health. *Science of the Total Environment*, 416, 156-163.

Nero, JMG; Dias, JMM; Pereira, AJSC; Godinho, MM; Neves, LJPF. (2003).Metodologia integrada para a caracterização do cenário ambiental em minas de urânio desactivadas. *III Seminário sobre Recursos Geológicos, Ambiente e Ordenamento do Território*, (91-100). Vila Real, Portugal.

Pereira, AJSC; Neves, LJPF; Dias, JMM; Barbosa, SVT. (2004). Evaluation of radionuclides contamination in the vicinity of Cunha Baixa and Quinta do Bispo old uranium mines (Central Portugal). *Radioprotecção,* 2, 4/5, 103-117.

Pereira, AJSC; Neves, LJPF; Dias, JMM; Barbosa, SVT. (2005). Environmental hazards at the former uranium mines Urgeiriça, Cunha Baixa and Quinta do Bispo. *IAEA Proceeding Series, International Workshop on Environmental Contamination from Uranium Production Facilities and their Remediation (Lisboa, Portugal)*, paper 16, 181-191. Available at: http://www-pub.iaea.org/MTCD/publications/PDF/Pub1228_web.pdf.

Petrova, R. (2006). Accumulation of natural radionuclides in wooden and grass vegetation from abandoned uranium mines. Opportunities for phytoremediation. *In* Merkel, BJ; Hasche-Berger, A. (Eds) *Uranium in the Environment. Mining Impact and Consequences (*507-518). Berlin Heidelberg: Springer.

Santos Oliveira, JM; Ávila, PF. (2001). Geoquímica na área envolvente da mina da Cunha Baixa (Mangualde, no centro de Portugal). *Estudos, Notas e Trabalhos,* 43, 25-47.

Santos Oliveira, JM; Canto, MJ; Pedrosa, MY; Ávila, P; Machado Leite, MR. (2005). Geochemical evaluation for the site characterization of Cunha Baixa Uranium Mine Central Portugal. *IAEA Proceeding Series, International Workshop on Environmental Contamination from Uranium Production Facilities and their Remediation (Lisboa, Portugal)*, paper 21, 233-243. Available at: http://www-pub.iaea.org/MTCD /publications/ PDF/Pub1228_web.pdf.

Sarić, MR; Čonkić, LJ; Bikit, I. (1997). Concentration of uranium in root-crops, bulbous and tuberous plants. *Acta Horticulturae*, 462, 43-547.

Sheppard, SC; Evenden, WG; Pollock, RJ. (1989). Uptake of natural radionuclides by field and garden crops. *Canadian Journal of Soil Science*, 69, 751-767.

Sheppard, SC; Sheppard MI; Gallerand M; Sanipelli, B. (2005) Derivation of ecotoxicity thresholds for uranium. *Journal of Enviromental Radioactivity*, 79, 55-83.

USNRC, United States Nuclear Regulatory Commission (2000). Draft working Document As of 02/01/2000. Available at: http://pbadupws.nrc.gov/docs/ML0608/ML060870642.pdf

Vandenhove, H; Van Hees, M; Wannijn, J; Wouters, K; Wang, L. (2007). Can we predict uranium bioavailability based on soil parameters? Part 2: soil solution uranium concentration is not a good bioavailability index. *Environmental Pollution*, 145, (2), 577-586.

Zamora, ML; Tracy BL; Zielinski JM,;Meyerhof, DP; Moss, MA. (1998). Chronic ingestion of uranium in drinking water. A study of kidney bioeffects in humans. *Journal of Toxicological Sciences*, 43, 68-77.

In: Uranium: Characteristics, Occurrence and Human Exposure ISBN: 978-1-62081-207-5
Editors: Alik Ya. Vasiliev and Mikhail Sidorov

Chapter 9

GEOPHYSICAL STUDIES TO EVALUATE GEOLOGICAL AND HYDROGEOLOGICAL FEATURES IN OLD URANIUM MINES

E. C. Ramalho[1], J. Carvalho[1], S. Barbosa[2] and F. A. Monteiro Santos[3]

[1]Laboratório Nacional de Energia e Geologia, Estrada da Portela, Bairro do Zambujal – Alfragide, Amadora, Portugal
[2]Empresa de Desenvolvimento Mineiro, Portugal
[3]Universidade de Lisboa, IDL–Laboratório Associado, Campo Grande, Lisboa, Portugal

ABSTRACT

Portugal is a country with a large tradition in uranium exploitation in the XXth century with 61 registered old uranium abandoned mines. Those uranium exploitations, mostly located in the central part of the country, created open pits that in most cases were simply abandoned, causing environmental problems that sometimes lasted for decades. More recently, following the establishment on 2001 of the "Concession Act for Environmental Remediation of Abandoned Mines", a public service addressed to "EDM", an owned Portuguese State company, a major old abandoned mines recovery plan for Mainland Portugal was developed comprising the possible use of old open pits from abandoned uranium mining sites as "Waste Containment Deposits", where mining waste from other adjacent mines would be stored. The old mining sites major plan and classification were carried out by "EDM" accordingly to its location, accessibility, geological and hydrogeological conditions and three mining areas were previously selected. Considering the purpose of the future use for the old open pits of uranium mining sites, detailed information about the development in depth of the shallow geological features was needed. Mining waste deposition in the open pits may however cause additional environmental problems related with geological and hydrogeological features that must be predicted and prevented before a particular site is chosen. The identification of faults and conductive zones that may promote groundwater circulation and the spread of contaminated waters are of great importance, since sometimes the surrounding areas are populated or used in agriculture. The ability of geophysical methods in identifying either faults or conductive zones was used to understand geological and hydrogeological critical zones near previously selected abandoned uranium mine sites in Portugal. The possible negative environmental impacts of the presence of such potential waste disposal sites were therefore assessed using geological, hydrogeological and geophysical methods.

Deeper studies where developed in two open pit mines, the abandoned Quinta do Bispo and Prado Velho uranium mines, initially considered favourable sites to be used in the near future as a "Waste Containment Deposit" due to their location and regional geological and hydrogeological characteristics. But the need for the local characterization to prevent any possible negative environmental impact required different baseline studies, such as the use of geophysical methods to complement geological and hydrogeological information. These methods comprised the joint use of electromagnetic, electrical and both refraction and reflection seismic. They allowed the development of 2D schematic models that highlight the most important characteristics that may jeopardize the adequate geotechnical solutions to adopt. These 2D schematic models have been constructed showing alteration and fault zones in depth that control groundwater circulation and therefore, future water circulation problems with negative environmental impact.

Mining waste deposition in open pits of old uranium mines may cause environmental problems that can be predicted using the joint use of geological, hydrogeological and geophysical methods, preventing future problems with adequate solutions. This chapter shows some of the environmental assessment works that have been conducted in Portugal with this purpose.

1. Introduction

EDM S.A. is a holding company representing the Portuguese State interests in the Mining Sector and a Concessionaire of the Portuguese State for "Environmental Remediation of Abandoned Mines". This Concession was established, in 2002, between both Environmental and Economy Portuguese Ministries.

In 2002, EXMIN S.A. – Companhia de Indústria e Serviços Mineiros e Ambientais, S.A. (thereafter designated EXMIN), an enterprise of EDM Group Company with its activity focused on "Environmental Remediation of Abandoned Mines" Concession, started the development of the environmental characterization studies and remediation projects for the prioritary old mines, including the old Uranium mines like Quinta do Bispo and Prado Velho. These studies aim at establishing the geo-environmental feasibility of each site. These old uranium mining sites are located in Central Portugal, where uranium was extensively explored during several decades in the 20th century. Since 2002, different types of site characterization studies, comprising detailed geology and hydrogeology studies are being conducted by Empresa de Desenvolvimento Mineiro S.A. (thereafter designated EDM), the former EXMIN.

Among the 61 old uranium abandoned mines in Mainland Portugal that were characterized, two of those mining areas were selected for the construction of a "Waste Containment Deposit" (collecting site of mining waste from other adjacent mines), according to "The Waste Management Criteria and Strategy" developed by EXMIN (Exmin, 2002-2003).

Since a "Waste Containment Deposit" is defined as a collection site of mining waste from other adjacent mines, it must be carefully chosen according to its geological characteristics and environmental risks. One of the objectives of site investigations developed was to get a better understanding of the local and natural drainage conditions and to predict the potential of water interaction between the "Containment Deposit" and the contaminants spreading (radionuclides and metallic elements) in fractured aquifer systems.

For a successful achievement of these purposes, a good understanding of the groundwater drainage system is required. Detailed studies were conducted by EXMIN-EDM, but some technical doubts arose, namely regarding the hydrogeological conditions in relation with local fracturing. These characteristics are important to define, since a choice based on insufficient information may lead to unexpected environmental problems, as the process reaches an advanced stage. Other types of information are therefore requested and geophysical methods were considered as a possible solution to overcome this situation. Therefore, geophysical surveys involving several methods in the geological and hydrogeological critical areas were conducted with the aim of characterizing the open pit limits and confirm mapped faults and fractures, which may increase the risk of interactions between the waste deposit and groundwater circulation. The results for Quinta do Bispo and Prado Velho old uranium mines are presented in this study.

2. Geophysical Methods – Data Acquisition and Processing

A set of geophysical methods was applied to understand the features related with the hydrogeological conditions in relation with local fracturing in the area of the old uranium mining areas. These methods were chosen according to the physical properties that each one responds to. Radiometric, electric, electromagnetic and seismic methods were applied in the surrounding areas of Quinta do Bispo and Prado Velho old mining sites.

2.1. Radiometry

Radiometry is an expedite geophysical method, widely used to measure natural gamma radiation emitted by geological formations, in the energy bands of ^{238}U, ^{232}Th and ^{40}K. The radiometric information was acquired in total counts per second. Quinta do Bispo and Prado Velho were both covered with surface radiometric prospecting using a SPP2 scintillometer. This is a device that has been used since the first radioactivity surveys conducted by the Portuguese authorities in the 1950´s (Grasty et al., 1995). The surveys were conducted with a mesh of 5x5 in the dumps and its vicinities and were developed in 2002 at Quinta do Bispo and in 2005 at Prado Velho. The results are represented in Figs. 1 and 2, respectively.

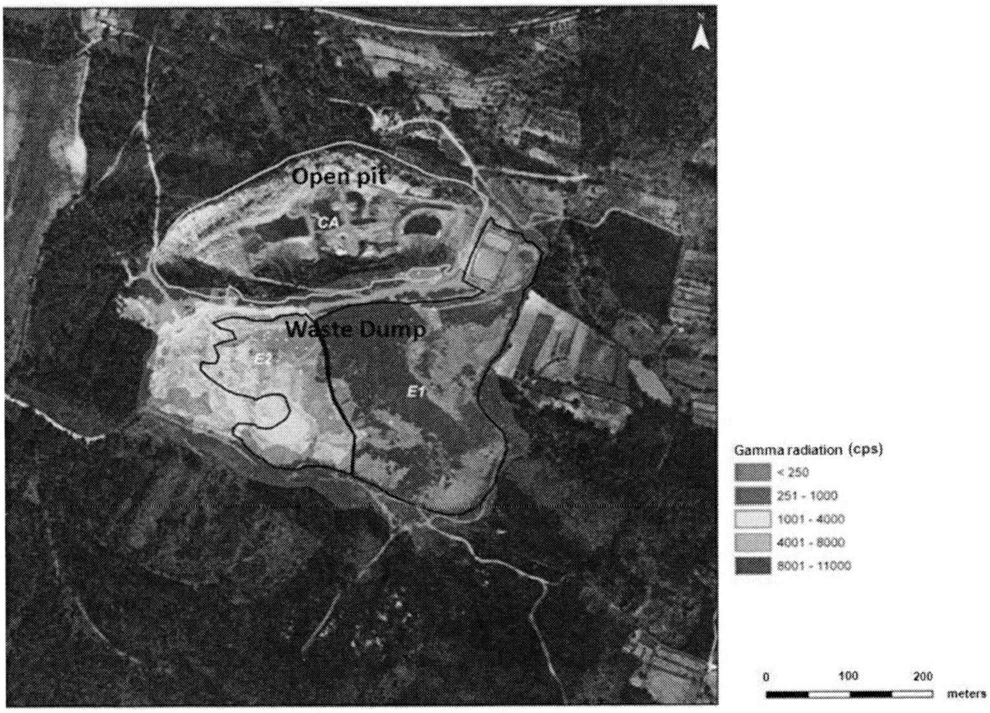

Figure 1. Superficial gamma radiation map elaborated using a scintillometer at Quinta do Bispo (units in cps). Average value of measurements: 1291 cps (n=466).

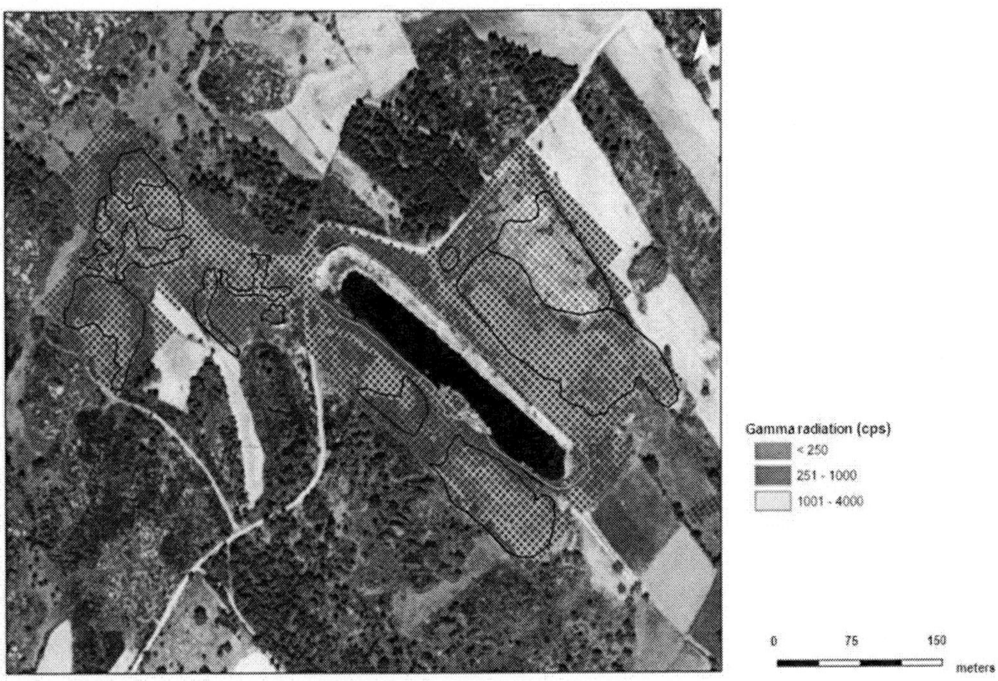

Figure 2. Superficial gamma radiation map elaborated using a scintillometer at Prado Velho (units in cps). Average value of measurements: 691 cps (n=1043).

2.2. Electromagnetics EM34

Frequency domain electromagnetic (FDEM) surveys using the Geonics EM34-3 device were carried out over the study areas of Quinta do Bispo and Prado Velho old mining sites. Lines to survey were chosen according with critical directions and available space. The utility of the Geonics EM34 system in hydrogeological and environmental studies is well-known. As described in Geonics (1990), the FDEM Geonics EM34 is composed by a transmitter coil Tx and a receiver coil Rx located at predefined distances from Tx of 10, 20 or 40 m. The equipment measures the apparent soil electrical conductivity (mS/m). Both possible configurations, vertical and horizontal coil modes, create horizontal or vertical dipoles, respectively, reaching different investigation depths to the same inter-coil distance. Cable lengths are 10m, 20m and 40m, corresponding respectively to 6.4 kHz, 1.6 kHz and 0.4 kHz (Geonics, 1990), reaching apparent depths ranging from 7.5 to 60 meters.

FDEM Geonics EM34-3 was conducted every 10 meter in lines surveyed in both sites, using 10, 20 and 40 m length cables. Data inversion was accomplished with software EM34-2D developed by Monteiro Santos (2004), using a 1D laterally constrained method that has proved to be effective to delineate the main geological structures using the Geonics EM34 device (Monteiro Santos, 2004). The inverse problem is solved using a smooth inversion algorithm, where each 1D conductivity model, obtained beneath each measurement site, is constrained by its neighbors (Monteiro Santos, 2004). The final model is a rough representation of a two-dimensional resistivity or conductivity model and it has allowed profile modeling into three different resistivity layers. The Rms errors of the models presented in this study are lower than 20%.

2.3. Vertical Electrical Soundings

Vertical electrical soundings (VES) were also acquired over the FDEM lines in Quinta do Bispo and Prado Velho using the Schlumberger array (Telford et al., 1995) to calibrate results with those obtained from the EM34 survey and, whenever possible, integrate them. The maximum distance for current electrodes depended on the topography and available space. The AB/2 distance ranges from 1 to 125-200 m. The VES were carried out in every line, at coincident sites with EM34 measuring points and/or boreholes, and without the probable presence of faults. At each profile, the VESs were oriented according with the profile direction. In some cases, however, due to topography, accessibility and equipment operation abilities, the VESs did not reach the unaltered bedrock, which is reflected in data quality and final adjustment of the model. Nevertheless, each VESs model ended with growing resistivity values. The data have been inverted assuming 1D model and using the ridge-regression method (Johansen, 1977). Rms values ranged from 2 to 6%. Where available, borehole data were used to perform interpretation adjustments, namely confirming depths and refining the geological alteration intervals.

2.4. Seismic Refraction and Reflection

Seismic refraction data was acquired along almost the full length of the lines in both mines, to provide information on the degree of alteration from the velocity models. A layout of 24 or 48, 50 Hz vertical geophones with a 2 m spacing was employed. Several spreads were therefore used for each line. End-off shots were fired with 1 tm and 20 m (where possible) offsets to the nearest receiver. To constrain the uppermost layers, shots were fired inside the receivers spread every 10 m. Reflection data was also acquired in confirmed or probable fault zones. Although the seismic reflection method is mostly adequate for sedimentary environments, some success has been obtained in crystalline or hard rocks settings (e.g. Kim et al., 1994; Stanley and Pavlis, 1999; Eaton et al., 2003). A shot spacing of 2 m was therefore used in the above mentioned areas inside the fixed receiver layout, with a source to nearest receiver offset of 1 m. In both methods, data were acquired using an accelerated weight-drop seismic source (Gisco AWD-550T) and a 24 bit digital data acquisition system (Seistronix RAS-24). A 0.5 ms sampling rate with a total recording length of 500 ms was used.

The reflection data was processed with the Seismic Processing Software-SPW™ using the following flow: trace editing, gains, frequency filtering, first arrival mute, deconvolution, frequency filtering and amplitude equalization, velocity analysis, NMO correction and stack, automatic residual statics (surface consistent), DMO integral correction (Yilmaz, 2001), FK migration with stretch (Stolt, 1978) and post-stack filtering. Depth conversion of the profiles was performed using interval velocities calculated from the stacking velocities.

Seismic refraction data was acquired along almost full length of the lines, in order to provide information about the alteration degree from the velocity models. The refraction interpretation of the data was performed using commercial software based on the delay-times method and ray tracing (Haeni et al., 1987). The delay-times method is used to obtain a first 2.5D velocity model which is refined with a three step iterative process of observed and calculated arrival time residuals minimization. Velocities are determined by a weighted (by receiver number) average of a simple linear regression of the time-distance curves and the generalized reciprocal method (Palmer, 1980) where reciprocal times are available.

3. Quinta Do Bispo

Quinta do Bispo abandoned uranium mine (Figure 3) has a central geographic location (Figure 4), in the Viseu district, near one of main roads that serve Central Portugal. Special care had to be taken in this site evaluation, since it is located in the suburbs of the city of Mangualde, in an area with a high population density that is also highly industrialized.

The Quinta do Bispo ore was discovered in 1957. The open pit exploitation began in 1979 and ended in 1987; mine annexes and dump occupy a total area of about 158,000 m^2 (Exmin, 2000-2001). The open pit reached a depth of 75 m below ground level and more than 460,000 tons of ore was produced during that time, with an average concentration of about 910 ppm (Exmin, 2000-2001). From 1987 to 1992, waters in the open pit were monitored, and due to the large amount of uranium, an ionic permute unit was installed. After 1992, “poor” ore from dumps was re-processed with concentrations between 300 and 500 ppm and deposited in the open pit for an in situ leaching process (Exmin, 2000-2001). In the Quinta do Bispo mining site, the open pit has dimensions of about 300×90×80 m, which corresponds to an excavation volume of about 2,000,000 m^3. There is a "poor" ore dump with about 100,000 tons and a sterile dump partially reforested with about 1,500,000 tons.

Figure 3. View of the Quinta do Bispo mine towards SW, located NNE from the mining area.

3.1. Geological and Hydrogeological Framework

According to Exmin (2000-2001), the main geological unit that outcrops in the region is the medium to coarse grainned, two micas granite, sometimes kaolinitised with many ferruginous impregnations and/or haematization (Figure 2). Meta-sediments of the Schist Greywake Complex also crop out on several places, with strong contact metamorphism, sometimes with the occurrence of migmatites. The ore is located in a contact of granite with embedded meta-sediments, composed by brownish, ferruginous, clayey-mica altered schists. The main fault system strikes N48°W, with clayey filling. A N-S fracturing system has also expression in the area (Exmin, 2000-2001).

Mineralization is composed by autunite and torbernite, and black uranium minerals are spread in the contact zone, either in the meta-sediments or in the granites. In the meta-sediments, disseminated sulphides occur.

Regional groundwater flows towards SE and the shallow aquifer is contaminated in the vicinities of the old mine with infiltrated water with acidic leachates from the dumps and residual water treatment muds, due to possible outflows from the decantation basins, mainly in the rainy season. Table 1 shows the average piezometry, water conductivity and water temperature at the time of the year the geophysical surveys were conducted (May). These parameters were measured in the boreholes located nearby the selected lines (see Figure 4).

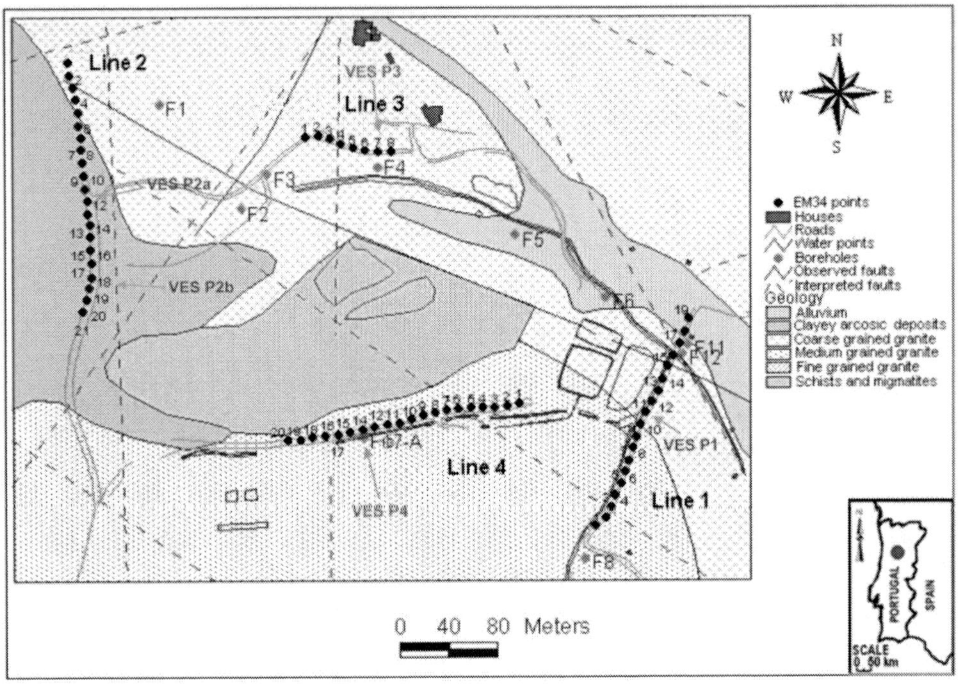

Figure 4. Geological map of the old Quinta do Bispo Mine, with location of the four lines where geophysical profiles were acquired using different methods, vertical electrical soundings (green) and location of the drilled boreholes (F#). Numbers in the four lines correspond to EM locations, used for common referential of all geophysical surveys (Ramalho et al., 2009).

Borehole water conductivity ranges from 145 μS/cm in F4 to 1600 μS/cm in F12 and 3120 mS/m in F8 (Ramalho et al., 2009), which is considerably high (Table 1). The soil contamination potential with heavy metals and radioactive elements from the dumps is high. The acidic characteristics of the embedded ore rock and the ore leached inside the open pit have contributed to the high conductivities of the surface water, with pH ranging from 3 to 3.5 and water conductivity ranging from 400 to 600 mS/m, with rock permeability varying between 1.2×10^{-7} and 9×10^{-8} m/s (Exmin, 2000-2001). Surface water has been the object of a neutralization process and chemical precipitation of radionuclides (^{226}Ra) (Exmin, 2000-2001).

Table 1. Average piezometry, water conductivity, water temperature and pH at the time of the year the geophysical survey was conducted (May) measured in the boreholes located nearby the selected lines (Ramalho et al., 2009)

Piezometer	Depth (m)	Average piezometry in May (m)	Water conductivity (μS/cm)	Temperature (°C)
F4	92	9.3	145	15.1
F8	30	1.8	3120	15.7
F11	37	1.8	571	16.5
F12	66	1.5	1600	15.2
F17-A	42	9.0	688	14.8

3.2. Geophysical Surveys – Data Interpretation

Detailed information concerning the geophysical surveys conducted in Quinta do Bispo old uranium mining site may be found in Ramalho et al. (2009). During the electromagnetic surveys, some problems occurred during data acquisition due to the electromagnetic cultural noise caused by industries and the "Beira Alta" railroad. The high level of noise was observed mainly in the northern part of Line 2. Overload was also detected in different degrees in all surveyed lines, but both problems were overcome according to the method suggested by McNeill (1980).

Description of data interpretation of the lines studied and showed in Figure 4 is as follows.

Line 1

According to the surveys results, this line shows a clear disturbance between points 14 and 17, in agreement with a mapped fault (Figure 4). The seismic reflection survey was acquired only approximately between points EM11 and 17 and its interpretation shows a similar structure to the one determined from the refraction profile (Fig. 5). Seismic horizons dip to depths greater than 20 m towards the NE and the presence of the above mentioned important deep faults is seen between points 15 and 17. The fault zone is composed of two separate faults around EM points 14-15 and 16-17. Although there is no indication about screen location in boreholes F11 (total depth of 37 m with water electrical conductivity of 57.1 mS/m) and F12 (total depth 66 m with 160 mS/m), located 10 m away from each other, a hydraulic connection between the shallow alteration zone and the probable fault identified between points 16 and 17, might exist. This fault is mapped in the NNW-SSE direction in Figure 4 and its alteration zone becomes narrower with depth, until about 60 m (see Figure 5).

Regarding the observed outcropping fault with WNW-ESE direction, crossing the line perpendicularly at point 15, its depth effects seem to mix with the other fault. The high shallow conductivities registered between points 15 and 16 are the result of the streamline that crosses Line 1 at that point. Water inflows in borehole F12 are, therefore, consistent with water table depth inferred from EM resistivity data interpretation (see Figure 5).

As observed in the refraction profile (Figure 5), both faults are clearly marked and have the strongest expression down to about 40 m depth. Borehole F12 shows a lithological sequence similar to the refraction profile; below the shallowest 3 m comprising vegetable coverage, altered "grey granite with quartz fragments" was found until 7 m depth, where "grey granite" was crossed, but not detected in the refraction profile. The 3rd seismic horizon is detected in the borehole log at an 11 m depth while the 4th seismic horizon is detected at a 24 m depth. The reason why the "grey granite" layer without water was not detected with the refraction interpretation is related with the fact that this layer has a relatively reduced thickness in the borehole lithological log description (5 m). According to this interpretation (which admits that the 2nd seismic horizon corresponds to the integration of the "grey granite with quartz fragments" and the "grey granite" without water layers), this layer will be even thinner in other areas of the profile. Since P-velocity waves in the "grey granite" layer with and without water will not have a significant variation, this is a typical "hidden layer" problem, from which results the non-detection of the upper thin layer (Palmer, 1980; Telford et al., 1995).

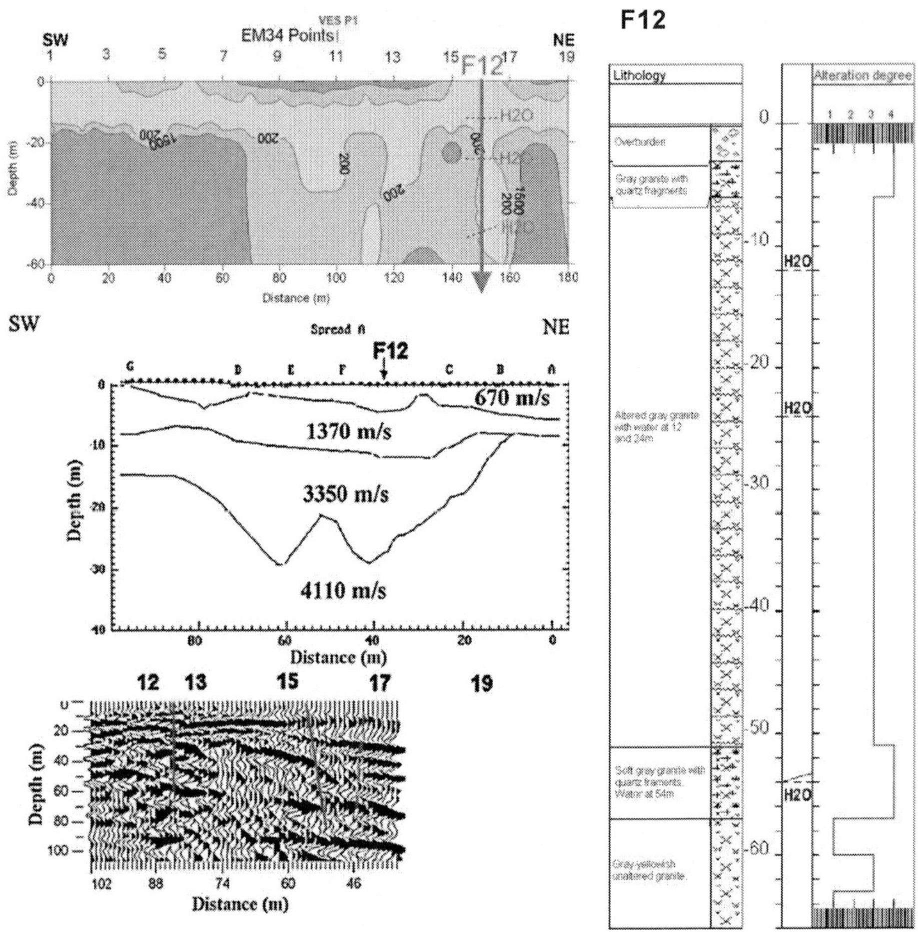

Figure 5. Line 1 – (Above left): Resistivity section of the inverse EM34 model showing F12 borehole and VES P1 locations. Units in ohm.m. (Right): Borehole F12 log. Alteration degree: 1-Weakly altered; 2-Low alteration; 3-Altered; 4-Highly altered. (Middle left): Seismic refraction interpretation of Line 1, with location of borehole F12 indicated. Average velocities of P-waves for each spread (receiver layout) are shown. A single spread was used for Line 1. Letters represent the position of shotpoints used in the refraction interpretation. (Bottom left): Miggrated, depth-converted stacked seismic reflection section from points EM11 to 17, with fault interpretation overlaid. Red lines correspond to interpreted faults. Images selected from Ramalho et al. (2009).

The 4th seismic horizon coincides again with water inflow into the borehole, and, although the borehole does not show changes in the alteration degree, the increase in seismic velocity reflects a change in the "grey granite" properties that was not detected in borehole F12.

Since the true velocity of this 4th seismic horizon is quite lower than the apparent velocity measured (4110 m/s) due to large dips of its interface, the change of properties from the 3rd to the 4th seismic horizons is quite less pronounced than as suggested by the refraction model. This might explain why this change of properties was not detected in the borehole. As mentioned before, the interface between the 3rd and 4th model layers shows very high irregularities, due to the presence of the two important faults between points EM14-15 and EM16-17. On the other hand, there is no evidence in depth about the N-S interpreted fault crossing Line 1 between points 2 and 3. The water table seems to be at about 8 m depth by the time of the FDEM surveys were acquired in most length of the profile, reaching the surface in the zone of borehole F12. Shallow resistivity values, lower than 200 ohm.m, registered from EM points 1-8, are clearly related with the streamline along the line and with conductive fluid from the old open pit. Between points 12-13 a deep fault was also identified in the reflection profile, in which, considering the low electrical resistivity values encountered at the site, may also have fluid circulation. The latter is clearly less important than the other features mentioned above, since it is not mapped in Figure 4, and therefore, has no surface expression. In the refraction profile, there is no sign of its presence, probably because it was absorbed by the major accident of points 14-15, which is much more important.

Line 2

The main characteristic of the refraction profile of Line 2 (Fig. 6) is the large depression with high conductivity and low P-wave velocity located at the southern part of the profile, beyond point EM12. Since the EM profile continues further south than the seismic refraction profile, the former shows that the depression ends approximately at point EM17. Therefore, the seismic profile has not its deeper part conveniently constrained, as expected. The depth reached by this zone is about 50-60 m either in the EM or the refraction profile. However, geological mapping stretches the deposits located N and S from these points; the fact that this low resistivity/low seismic velocity zone is located close to a N-S oriented probable fault (see Figure 6), may indicate that although the profile doesn't cross this fault, it is still strongly affected at depth by its presence. Besides the existence of very low electrical resistivity values, the low seismic velocities observed in this zone suggest the presence of a much altered granite zone, since such a fast change in its properties and geometry does not make an increase of the clayey-arkosic sediments thickness very credible. These low seismic velocities include the appearance in the southern part of the profile of a superficial layer with a very low velocity of 460 m/s, the decrease in velocity of the second layer (from 1410 m/s at the N to 900 m/s at the south) and the deepening of the third interface of the model, which represents a lateral change of velocity at about a 30-50 m depth from 4350 m/s to 2270 m/s. Keeping in mind these considerations, it is assumed that this N-S fault may have significant dimensions.

The NW-SE fault identified near point 5 in the stacked reflection section (Figure 6) does not seem to affect the shallow surface (the first 15-20 m). Its presence is also suggested by the deepening of the deepest interface of the refraction interpretation (Figure 6) and it is clearly seen in the EM interpretation near point 4. This structure may correspond to the mapped fault near point 3 in Figure 4; the low resistivity zone, with values between 200 and 2000 ohm.m, is particularly visible at point 4, below approximately 20-30 m depth, still showing signs of its existence at 60 m depth. Further south, the fault interpreted in outcrop near point 8 (see Figure 6) is not visible neither in the refraction profile nor the electrical resistivity profile. In the reflection stacked section, however, it seems to reach the near surface continuing down to a depth of about 60 m. This discrepancy can be explained by resolution issues of the different methods and that the fault does not reach the surface. The reflection profile was only conducted over refraction spread A.

It shows a deepening towards the N of the deepest horizons (below about 15 m depth) in the N part of the profile (points 4-5), a high at the center of the spread and a new deepening in its southern part (in this case, also in the shallow horizons). It shows a good coincidence with the interpretation of the refraction profile.

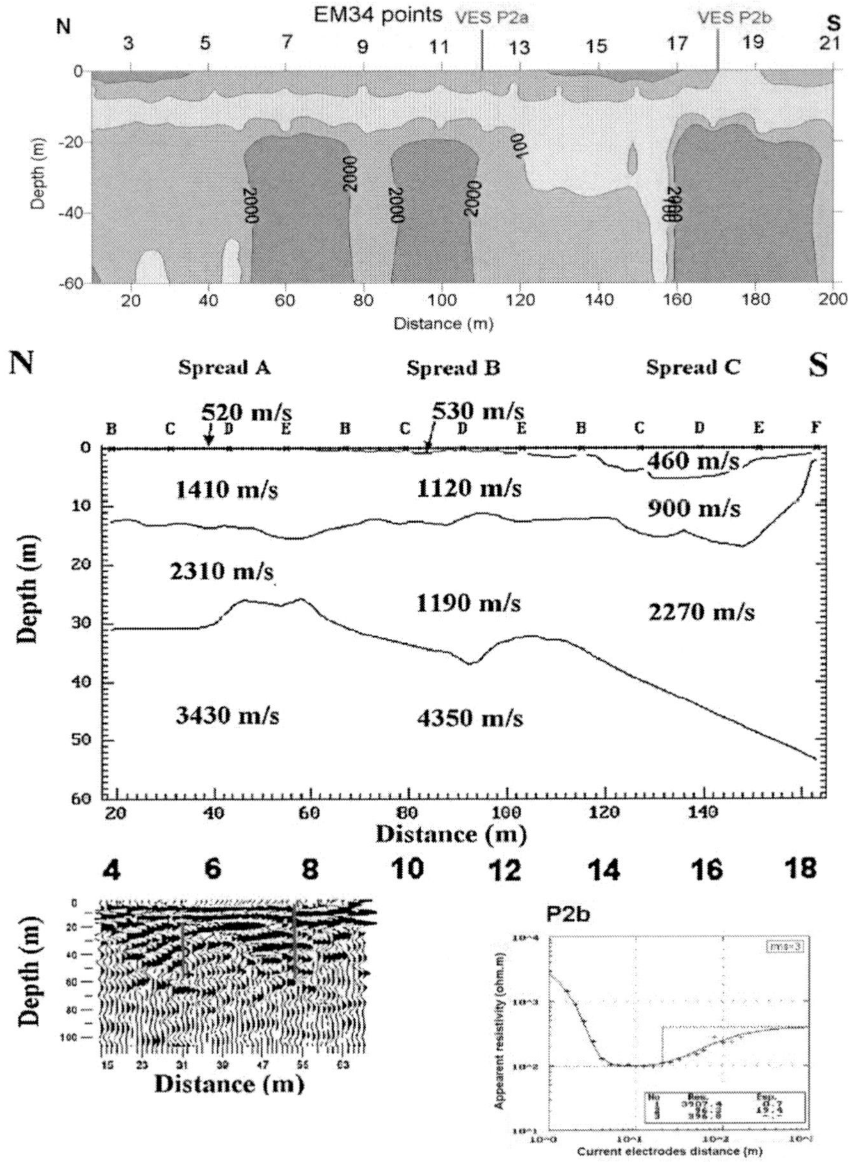

Figure 6. Line 2 – (Above): Resistivity section of the inverse EM34 model (units ohm.m). Units in mS/m. Location of the VES acquired over this profile are indicated. (Middle): Seismic refraction interpretation of Line 2. Average velocities of P-waves for each spread (receiver layout) are shown. Letters represent the position of shotpoints used in the refraction interpretation. (Bottom left): Miggrated seismic reflection profile, with fault interpretation overlaid. Depth conversion was performed with stacking velocities. (Bottom right): P2b inverse models of soundings located over Line 2. The dashed line represents the true resistivity versus depth model, with the uppermost layer omitted. The dotted lines are displayed to help visualization. Current electrode distance corresponds to AB/2 (m). Images selected from Ramalho et al. (2009).

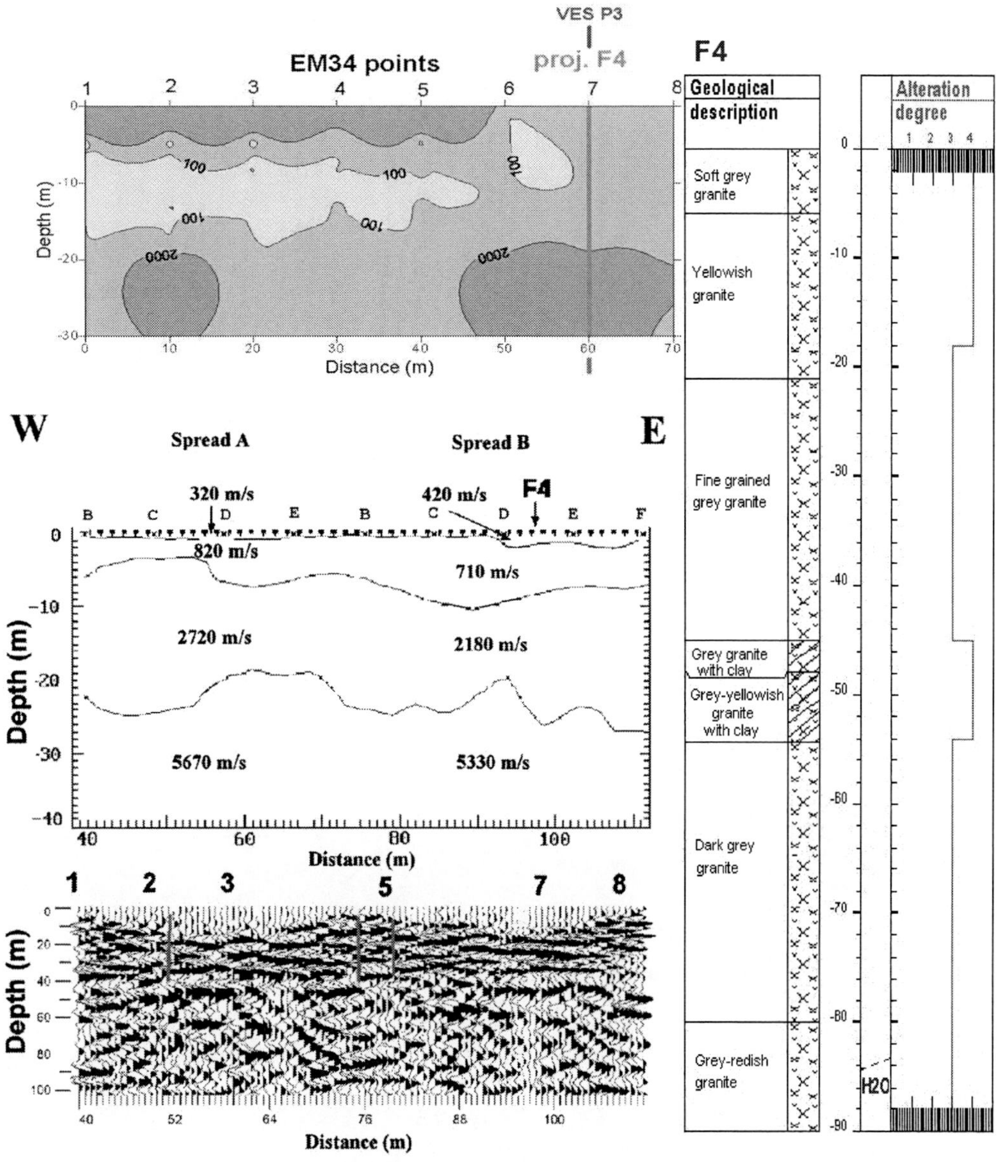

Figure 7. Line 3 – (Above left): Section of the inverse EM34 model of Line 3. Units in mS/m. Location of VES acquired over this line is indicated. Projection of borehole F4 in the line is also shown. (Right): Borehole F4 log. Alteration degree: 1-Weakly altered; 2-Low alteration; 3-Altered; 4-Highly altered. (Middle left): Seismic refraction interpretation of Line 3. Average velocities of P-waves for each spread (receiver layout) are shown. Letters represent the position of shotpoints used in the refraction interpretation. Location of borehole F4 is indicated. (Bottom left): Depth-converted migrated seismic reflection section from points EM1 to 8. Images selected from Ramalho et al. (2009).

Line 3

The most important aspect to highlight in Line 3 regards the confirmation that the probable fault location mapped in outcrop between points 3 and 5 prolongs in depth, but however, according to the EM profile (Figure 7) and the conductivity section, it does not seem to reach the near surface. The inexistence of fractures with water circulation in the shallowest meters of borehole F4 (Figure 7) and the low electrical conductivity of its water

(14.5 mS/m) also support this conclusion. The interfaces of all geophysical sections (Figure 7) are relatively smooth, except the deepest interface, which presents some important irregularities, possibly due to some important fractures. They are located at points EM2-3, 5 and 7, and apparently they don't reach the shallowest 5-6 m. The two faults around point EM5 are coincident with fractures interpreted with the refraction, reflection and EM methods, and extend until a depth of a few tens of meters.

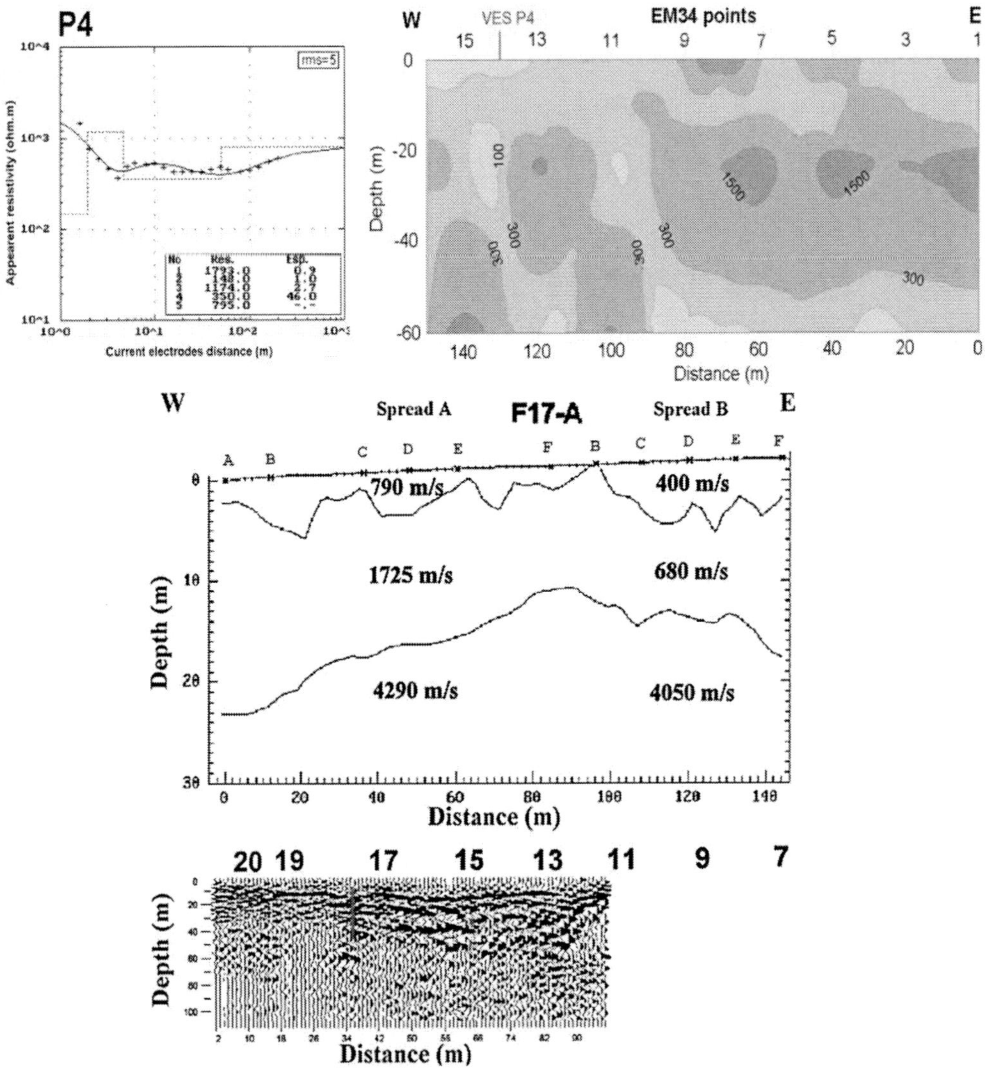

Figure 8. Line 4 – (Above left): Iinverse model of the VES P4, located over Line 4. The dashed line represents the true resistivity versus depth model, with the uppermost layer omitted. The dotted lines are displayed to help visualization. Current electrode distance corresponds to AB/2 (m). (Above right): Resistivity section of the inverse EM34 model. Units in ohm.m. Location of VES acquired over this Line and borehole F17-A are indicated. (Middle): Seismic refraction interpretation of Line 4. Average P-wave velocities for each spread (receiver layout) and layer are shown. Letters represent the position of shot points used in the refraction interpretation. Location of borehole F17-A is indicated. Bottom: Miggrated seismic reflection section from EM points 12 to 20. Images selected from Ramalho et al. (2009).

The two interpreted faults in the seismic reflection section around point EM5 were apparently integrated in the same fault zone by both refraction and EM methods, whose sections show two fault zones centered in points EM2-3 and 5 extending in depth. A possible fault in point EM7 detected with the seismic refraction interpretation does not show up neither in the reflection profile nor in the EM profile and there are not also evidences of its existence in borehole F4.

Line 4

The probable fault mapped in outcrop and shown in Figs. 1 and 2, near point EM6 of Line 4 has a deep rooting with a consequent decrease of the electrical resistivity values (Figure 8).

However, it does not seem to be as important as the ones between points EM11 and 14, which are not visible at the surface. Nevertheless, except for the high conductivity zones attributed to faults, there is a slight trend for a conductivity "dipping" towards east, which is more visible after points 5-6. On the other hand, a shallow alteration zone about 15 m deep is especially visible in its eastern part. The refraction model (Figure 8) shows the third refractor surface is deepening towards the west (higher EM points), but its surface is relatively smooth. A layer plunge can be seen at the eastern edge of the profile, near point EM7. This fact, together with velocities decrease observed in spread B in every layer, suggests the presence of a deep fracture beyond point EM7, in accordance with the interpreted fault in outcrop near point EM6 and the conductivity section. The EM profile also shows a higher conductivity zone near point EM6. The reflection profile (Figure 8) does not cover this area. The upper interface of the refraction model is rather irregular, indicating an important fracturing degree in its two shallowest layers or a very irregular alteration surface. The most pronounced irregularities show up near points 18–19, W of the EM profile and about points EM10-11. They are accompanied by a decrease of the seismic P-waves velocity and therefore, they were interpreted as fractures, while the other irregularities have been attributed to granite alteration and small fractures. The fracture located about points EM10-11 is accompanied by an irregularity in the layer below, so it can be extended in depth. The EM resistivity profile supports this possibility.

The plunging zone of the low altered/fractured granite (points higher than EM13), includes the shallow fault interpreted in the seismic reflection section near point 18 (Figure 8). The corresponding EM profile also shows a zone with lower resistivities in its western end (points EM12-15), just like a low resistivity area is seen in the correspondent section to the west of point EM11. Therefore, the area between approximately points EM12 and 20 is probably an important fault zone, with a large deformation zone. From the observation of Figure 4, it can be seen that the prolongation of two faults mapped in outcrop to the N and to the S of Line 4 would pass around points EM16-17, which suggests that both outcropping faults are in fact the same structure.

3.3. Conclusions

An integrated interpretation of all the results from the different geophysical surveys and borehole data confirmed the previously mapped faults based on geological outcrop studies and their influence area. Previously unknown faults were also detected. This study also allowed assessing their local importance regarding the groundwater circulation. Figure 9 shows a composition of the schematic conceptual models for the four studied lines till a depth of about 60 m corresponding to the integrated interpretation of all used geophysical methods (Ramalho et al., 2009). The model of Line 3 reaches only a depth of 30 m, due to electrical constraints. Each schematic model is a compromise among each method used, since these reflect distinct geophysical properties of the subsurface and that are in good agreement with the mapped faults in outcrop shown in the geological map of Figure 4. In these schematic models, three major layers were defined for each profile, integrating information from electrical, electromagnetic, seismic refraction, seismic reflection, borehole and lithological data.

Average values of physical properties attributed to the materials composing the layers in the schematic models of Lines 1, 2, 3 and 4 were estimated and presented in Table 2. These layers comprise a wide range of alteration degree, ranging from 1 (highly altered formations or clayey-arkosic deposits) to 3 (low altered and unfractured granites) (see Table 2 and Figure 9). The interpreted faults are also included in Figure 9, and their relationship with previously mapped in outcrop faults is also highlighted.

The EM, electrical and seismic refraction data showed a good agreement either in the determination of the physical properties of the area, the presence of faults, or the water conductivities distribution related with groundwater flow and its characteristics. Seismic reflection data, though better suited for sedimentary environments (such as those in Profile 2), was useful to complement the other geophysical information. These results have shown that the joint use of these geophysical methods can be successfully employed to provide a better understanding of how the geological features develop with depth. With these synthetic models and surface geological studies, a detailed program regarding waterproofing of the critical areas can be created so that drainage problems can be minimized and overcome, allowing the installation of a safe "Waste Containment Deposit" in the old Quinta do Bispo Mine open pit.

4. Prado Velho

The Prado Velho open pit (Figure 10) is an old uranium mining site, also located in the Central Region of Portugal (Figure 11) in a flat area at 780-790 meters high, about 4 km away from one of the main roads serving the central part of the country. Prado Velho was exploited from 1977 to 1982, producing an average annual mean of 1.1 tons of uranium ore. Initially, it was exploited as an underground mine that, for technical and economic reasons, was progressively developed into an open pit. It is not located near any significant urban center, but its fields support a locally important agricultural area located downwards surrounded by pine trees.

Another open pit in the NW part of the ore was exploited and closed afterwards. The major open pit remains, occupying a partially flooded area of about 18,000 m^3. About 35,211 tons of ore were exploited with an average concentration of 0.16% of U_3O_8, producing 59,655 kg of U_3O_8. Since there was no leachate process in this mine, 25,000 ton of ore was transported to another mine to conduct that process.

Table 2. Physical properties attributed to the materials in the schematic models of Lines 1, 2, 3 e 4. Average values are presented. Vp: Seismic velocity of longitudinal waves (Ramalho et al., 2009)

Material	Characteristics	Resistivity (Ω.m)	Vp (m/s)
1	Highly altered/fractured granites or clayey-arkosic deposits	<150	400-1500
2	Altered granites	150-1500	1500-4000
3	Low altered/unfractured granites	>1500	>4000

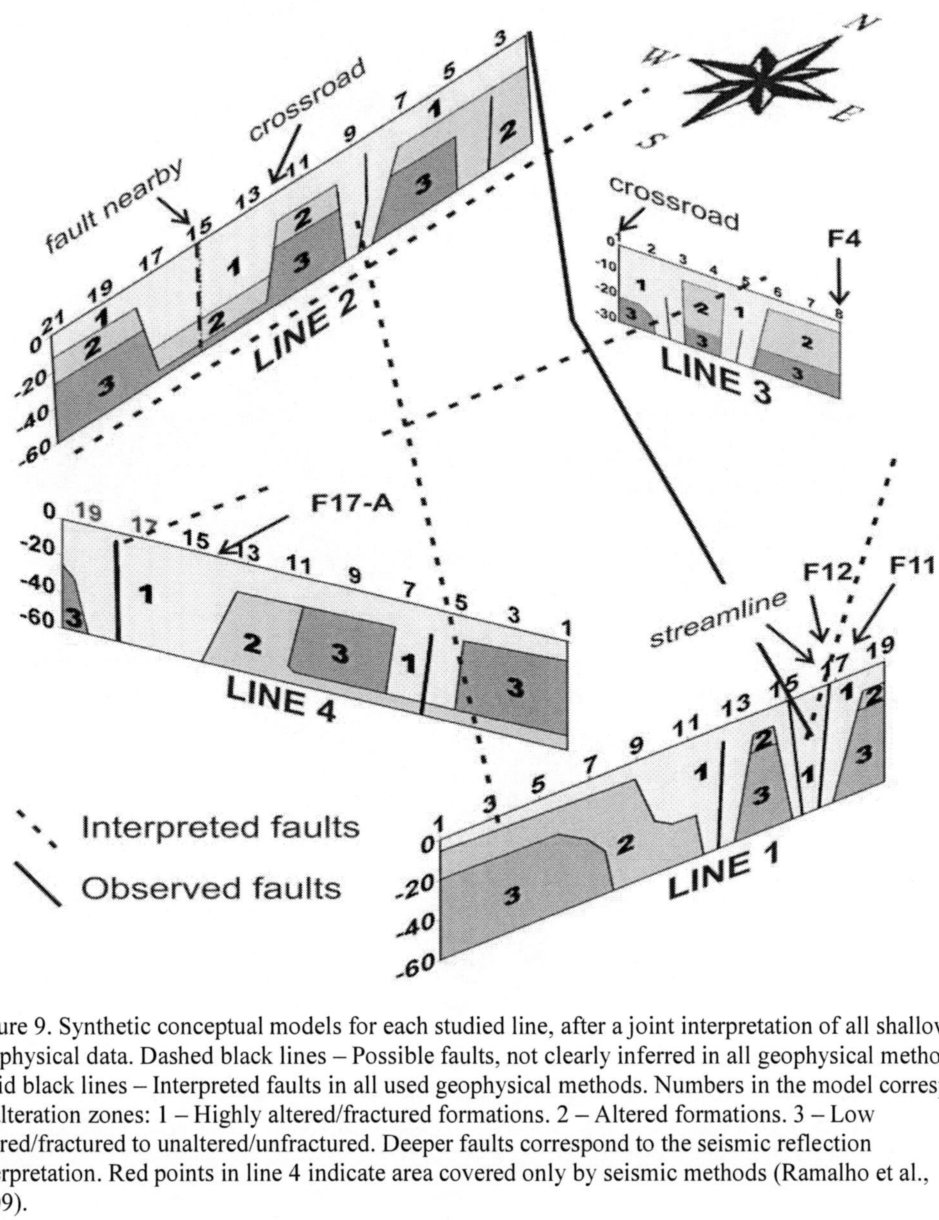

Figure 9. Synthetic conceptual models for each studied line, after a joint interpretation of all shallow geophysical data. Dashed black lines – Possible faults, not clearly inferred in all geophysical methods. Solid black lines – Interpreted faults in all used geophysical methods. Numbers in the model correspond to alteration zones: 1 – Highly altered/fractured formations. 2 – Altered formations. 3 – Low altered/fractured to unaltered/unfractured. Deeper faults correspond to the seismic reflection interpretation. Red points in line 4 indicate area covered only by seismic methods (Ramalho et al., 2009).

Figure 10. Partial view of the Prado Velho open pit.

Detailed information concerning the geophysical surveys conducted in Prado Velho old uranium mining site may also be found in Carvalho et al. (2005).

4.1. Geological Framework

The regional lithology is a two-mica porphyritic coarse-to-medium-grained granite (Figure 11) (Exmin, 2003), where transmissivity is usually extended along short distances, through more permeable fractures.

The mineralized structure is a dyke of basic rock and ferruginous quartz, sometimes compact and more or less homogeneous other times breccia-type, silicifying parts of the basic rock. The predominant direction of the basic rock is N55°W, sub-vertical (Figure 11). Exmin (2003) also mapped a granitic breccia filled with clay and individualized different systems defined by the following directions: N10°W to N10°E, N30-35°E, N60-65°E, N70°E to N70°W, N50-55°W, N40-45°W. In the mining area and immediate vicinities, the N40-45°W and N50-55°W directions are dominant. Some of these fractures are filled with dykes of basic rock, such as the one occurring in the open pit that was exploited. According with Exmin (2003) the mining area belongs to the hydrographic basin of Telhal River (flowing towards NNE); small tributary stream lines confine the ore and drain towards E, with some storage wells especially in the alluvium zone by the River. Compiled data and in situ observations (Exmin, 2003) showed that pH in the open pit water is 7.10, water conductivity 93 μS/cm and the waters in the drainage channel towards the Telhal River have pH of 6.42 and an electrical conductivity of 30 μS/cm.

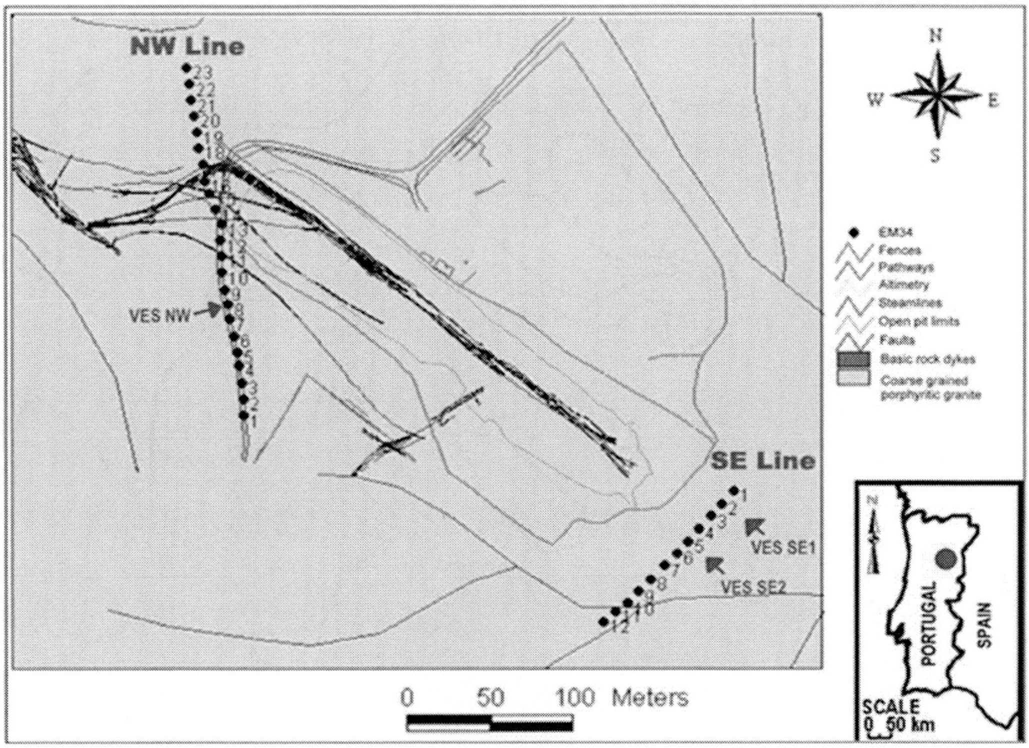

Figure 11. Geological map of the old Prado Velho Mine, superimposed with location of geophysical profiles surveyed with different methods (Carvalho et al., 2005).

Later, a hydrogeological inventory in the area comprising 132 points was conducted and *in situ* parameters were determined in water samples: pH, electrical conductivity and water level depths. The pH and conductivity values are inside the expectable limits with average pH almost neutral and the very low electrical conductivity values reflect the low water mineralization (EDM, 2006).

From the hydrogeological point of view, although higher transmissivity values can be expected in the immediate vicinities of the open pit due to the high fracturing status of the slopes, this same transmissivity may lose its spacial expression and it may become null with the increasing distance from the open pit (Exmin, 2003). Its hydrogeological behavior is directly dependent on the local fracturing conditions; although the Prado Velho open pit is crossed by a large sub-vertical fracture, this one is filled by basic rock, which will contribute to its waterproofing due to the existence of clayey materials resulting from its alteration (Exmin, 2003).

In hydrochemical terms, from 1997 to 2000, too anomalous situations never were detected in the open pit waters (Exmin, 2003). The values of total U are always below the limit defined by the standard US EPA CFR 40 "Protection of Environment – Part 141", as well as those obtained by the radionuclide Ra-226, and water samples did not show any hydrochemical contamination (EDM, 2006).

4.2. Geophysical Surveys – Data Acquisition, Processing and Interpretation

Radiometric information from scintillometer SPP2 was obtained in 3489 points. The radiometric values obtained have a high variability, ranging from 150 to 2600 cps.

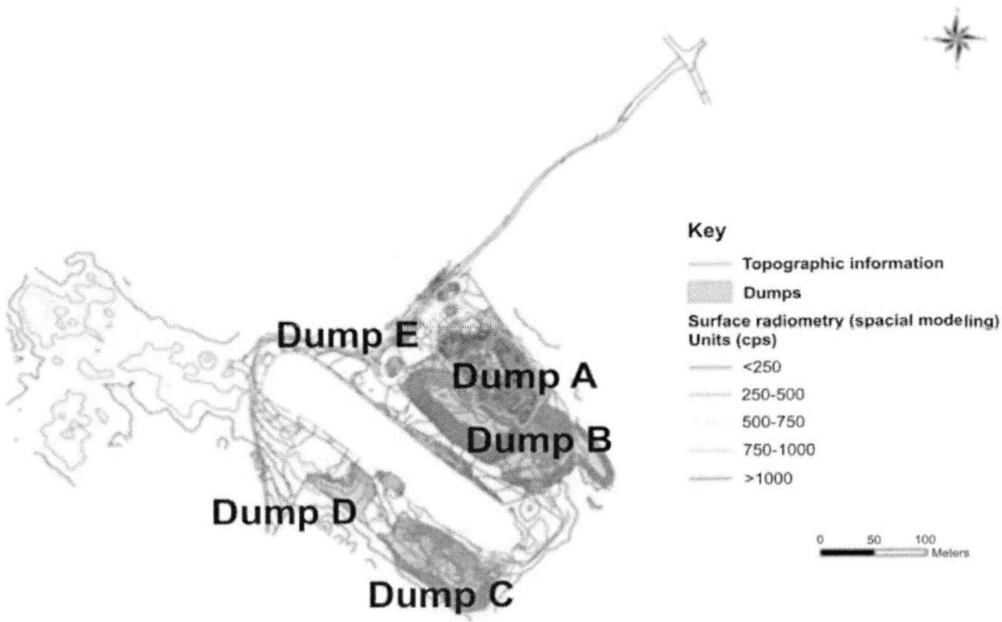

Figure 12. Spacial modeling of surface radiometry (Exmin, 2006).

According to EDM (2006), the lowest radiometric value in dumps is in dump B, with an average value of 453 cps; on the other hand, the highest gamma radiation values also with the highest radiometric heterogeneity are located in dump A (max. 1253 cps). Dumps C, D and E show average radiometric values of intermediate intensity (516, 550 and 731 cps, respectively).

Among the works carried out, radon flows measurements were conducted and obtained values vary between 0.3 a 2.3 $Bq.m^{-2}.s^{-1}$ (Exmin, 2006). The lowest values were obtained in geological bedrock composed by granitic rock. In the alloctonous materials, the flows are higher and the highest values were recorded in the "poor" ore dumps (A and B).

Keeping these results in mind, two lines were chosen to acquire FDEM Geonics EM34, electrical, seismic reflection and refraction surveys. These lines were located NW and SE of the old mine open pit, accordingly to the direction of the main geological features identified (Figure 11).

NW Line

Each result obtained for this line shows a high heterogeneity, due to the geological complexity of the area (Figure 13).

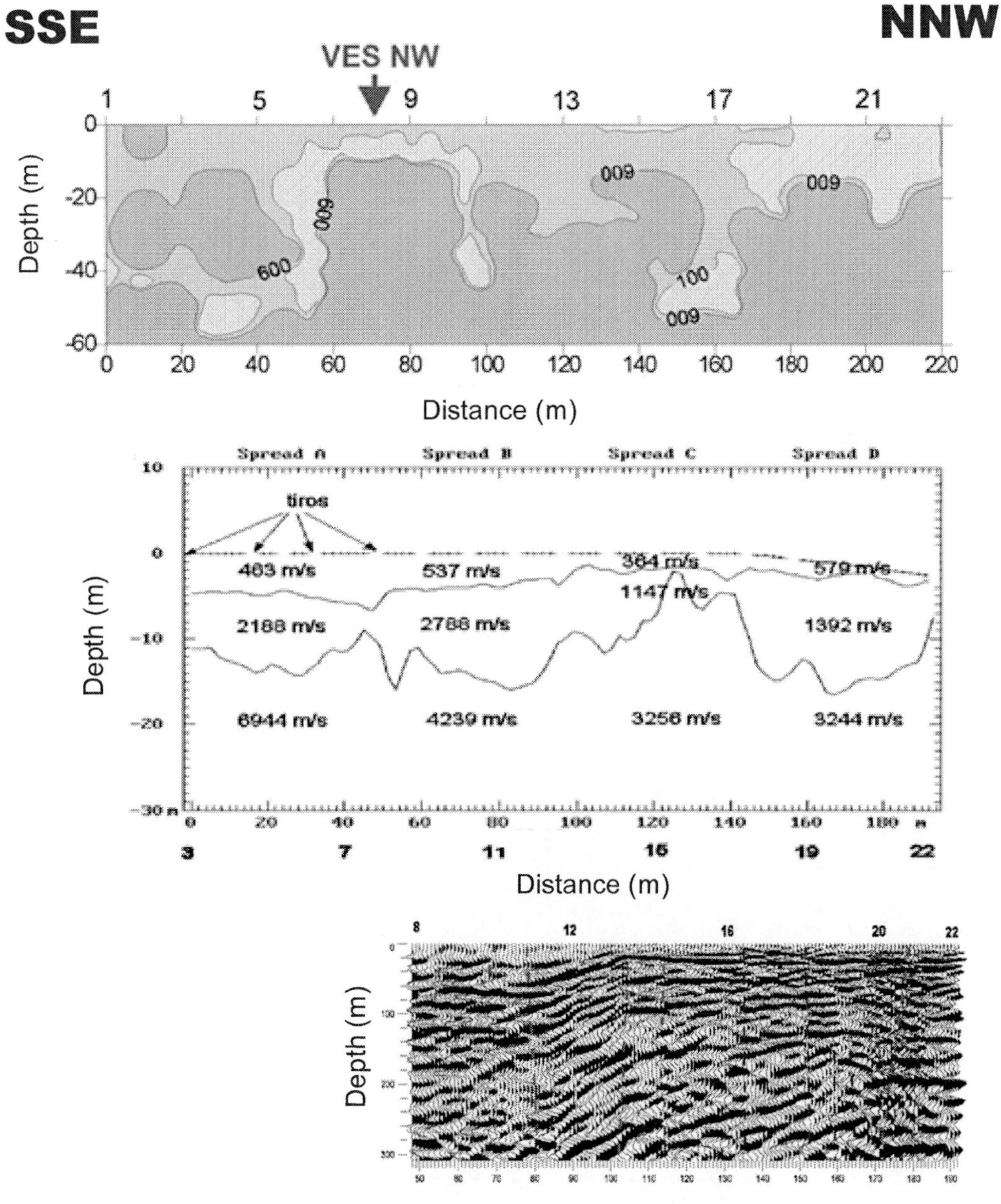

Figure 13. NW Line – (Above): Resistivity section of the inverse EM34 model (units, ohm.m). (Middle): Seismic refraction interpretation. Average velocities of P-waves for each layer and spread are indicated. (Bellow): Miggrated reflection profile: depth conversion from stacking velocities. Red dashed lines correspond to probable faults. Blue correspond to observed and interpreted dykes. Images selected from Carvalho et al. (2005).

SSE part of the profile appears to have lower shallow alteration, with resistivity values between points EM1 and EM5-6 that generally correspond to altered granite. P-wave velocities are higher than in the NNW part of the profile supporting this conclusion. In point EM7, there is a fracture that can go deeper, as seen in Figure 13. At depth, results seem to indicate the existence of a fractured zone with a complex geometry; near point EM10, several fractures almost reaching surface are seen, causing discontinuities in the resistivity values and in the refractors depth, down to about 40 m. The most important feature in this profile is from point EM17 towards NNW, an area highly altered and fractured, maybe filled with the mapped basic rock dykes. This is inferred from the low resistivity values and low P-wave velocities. This zone, located in the northern prolonging of the N40-45°W direction fault system mapped inside the open pit, has characteristics that correspond to higher altered deeper zones.

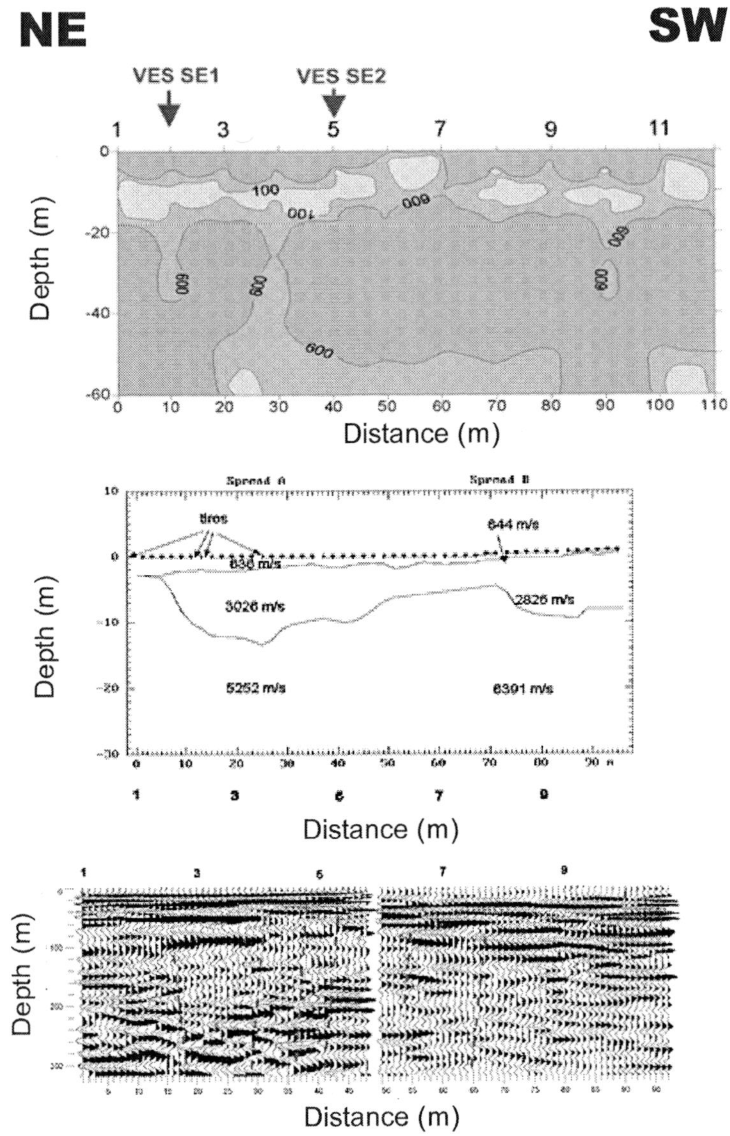

Figure 14. SE Line – (Above): Resistivity section of the inverse EM34 model (units, ohm.m). (Middle): Seismic refraction interpretation. Average velocities of P-waves for each layer and spread are shown (Bellow): Miggrated reflection profile: depth conversion from stacking velocities. Red dashed lines correspond to probable faults. Images selected from Carvalho et al. (2005).

SE Line

In this line (Figure 14), granite alteration is lower and laterally more uniform in the near surface (about 20 m) compared to the NW line.

The zone showing lowest resistivities above 12 m depth is related with the water table location, dividing saturated from unsaturated zones. This homogeneity is broken by the existence of deep faults, from the fault system with N40-45°W, crossing longitudinally the open pit (points EM2, EM4-5) and, probably, the fault system with N50-55°W trend, near point EM10, showing a shallow fault that does not seem to reach the surface. Between points EM3 and EM9, the resistivity section shows an altered granite zone at about 50 m depth, which according to the interpretation of the seismic reflection profile may be the result of deeper fractures, not reaching the ground surface. Resistivity decrease in points EM11 and EM12 at about 50 m depth may also correspond to a deeper fault.

4.3. Conclusions

The bedrock nature and the fault filling, mostly with basic rock are factors that contribute to the confinement of the negative environmental impacts.

Table 3. Physical properties attributed to the materials of the synthetic conceptual models for NW and SE lines of Prado Velho old uranium mining area (average values). V_P: P-wave seismic velocity (Carvalho et al., 2005)

Material	Characteristics	Resistivity (Ω.m)	Vp (m/s)	Porosity (%)
1	Highly altered/ fractured granites	<100	<1100	>1.92
2	Altered granites	100-600	1100-3500	0.62-1.92
3	Low altered/fractured to unaltered/unfractured granites	>600	>3500	<0.62

The existence of a shallow fault that does not reach ground surface may correspond to the continuation in depth of the fault system trending N50-55°W, seldom seen in outcrop. The existence of this fault is seen in both profiles (between points EM9 to EM10 of SE line and 5 to 8 of the NW line. The fault system trending N40-45°W and crossing the open pit in its major axis (line SE, point EM2 and from point EM17 forward) appears to continue in depth, although major alteration fracturing levels occur in the NW extreme of the open pit down to about 50 meters deep.

Average values of physical properties attributed to the materials of the synthetic conceptual model for the NW and SE lines are shown in Table 3. In both profiles, porosity was estimated considering only average resistivity values of each material, using Archie's Law with parameters defined by Shon (1996).

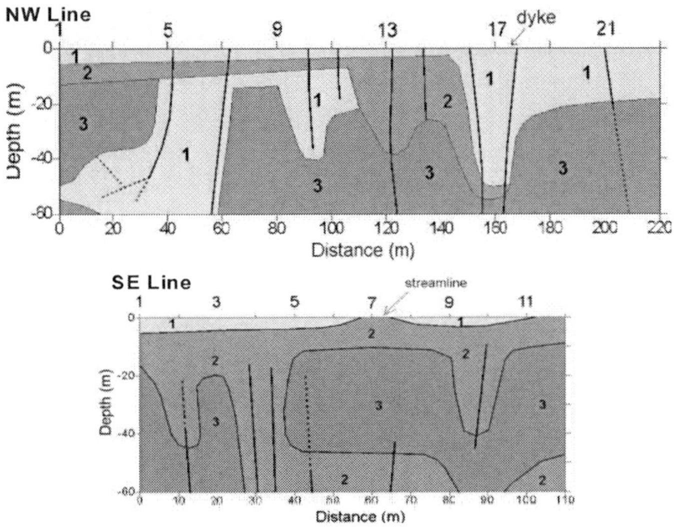

Figure 15. Synthetic conceptual models for NW (above) and SE (bellow) profiles, after a joint interpretation for all shallow geophysical data. Numbers in the models correspond to physical properties of Table 1. Dashed black lines: possible faults, not clearly inferred in every geophysical method. Solid black lines: interpreted faults in all applied geophysical methods. Deeper faults correspond to the reflection seismics interpretation in Figs. 13 and 14 (Carvalho et al., 2005).

CONCLUSION

Joint interpretation of all used geophysical methods confirmed mapped faults and detected previously unknown faults in Quinta do Bispo and Prado Velho old uranium mines. It also allowed assessing their local importance regarding groundwater circulation and therefore contamination vulnerability. The synthetic conceptual models showed in figures 9 and 15 correspond to the interpretation of all used geophysical methods till a depth of about 60 m in the studied profiles. These models are a compromise among every method used, since they reflect different material properties. EM, electrical and seismic refraction data showed a good agreement in the determination of the physical properties of both areas. Seismic reflection data, though better suited for sedimentary environments, was useful to complement the other geophysical information. These results have shown that these methods can be successfully used to understand how geological features develop in depth. With these synthetic models, a detailed program regarding critical areas waterproofing can be created so that drainage problems can be minimized and overcome, allowing the installation of a safe "Waste Containment Deposit" in the old Quinta do Bispo and Prado Velho old mining areas.

REFERENCES

Carvalho, J., Ramalho, E. C. and Barbosa, S., 2005. Using geophysical methods to identify alteration zones in the Prado Velho abandoned uranium mining site, Portugal. *67th European Association of Geoscientists and Engineers (EAGE) Conference and Exhibition*, Madrid - Extended Abstracts, pp. 3227-3230.

Eaton, D.W., Milkerreit, B. and Salisbury, M.H., 2003. Hard Rock Seismic Exploration, (Eds.), *Geophysical Developments* No. 10, SEG, 270.

EDM, 2006. Área mineira do Prado Velho. Relatório final. 84p+annexes.

Exmin, 2002-2003. *Estudo Director de Áreas de Minérios Radioactivos - 2.ª Fase*, Internal Report.

Exmin, 2003. *Estudo Director de Áreas de Minérios Radioactivos – 2ª Fase*. Volume 3. Excerto do Texto Base - Mina do Prado Velho. 17p.

Exmin, 2000-2001. Estudo director de áreas de minérios radioactivos. 1ª fase. *Exmin Internal Report.*

Geonics Ltd., 1990. *EM34-3 and EM34-3XL Operating Instructions.*

Grasty, R., Tauchid, M. and Torres, L., 1995. Standardization of old gamma ray survey data. Application of uranium exploration data and techniques in environmental studies. *Proceedings of a Technical Committee meeting held in Vienna,* 9-12 November 1993, pp.35-46.

Haeni, F. P., Deborah, G. and Ellefsen, K., 1987. *Micro-computer based version of SIPT – A program for the interpretation of seismic refraction data, USGS Open-File Reports*: 87-103-A, U.S. Geological Survey.

Johansen, H. K., 1977. A man/computer interpretation system for resistivity soundings over horizontally stratified earth. *Geophysical Prospecting*, 25, pp.677-691.

Kim, J. S., Wooil, M. M., Lodha, G., Serzu, M. and Soonavvala, N., 1994. Imaging of reflection seismic energy for mapping shallow fracture zones in crystalline rocks, *Geophysics* Vol. 59, (5), 753-765.

McNeill, J.D., 1980. EM34-3 survey interpretation techniques. *Technical Note* TN-8. 6 pp.

Palmer, D.,1980. The generalized reciprocal method of seismic refraction interpretation, K. B. S. Burke (Ed.), *Soc. Explor. Geophys.*

Ramalho, E., Carvalho, J., Barbosa, S. and Monteiro Santos, F. A., 2009. Using geophysical methods to assess geoenvironmental impacts in the Quinta do Bispo abandoned uranium mining site, Portugal. *Journal of Applied Geophysics*, nº67, pp.14-33.

Monteiro Santos, F., 2004. 1-D laterally constrained inversion of EM34 profiling data. *Journal of Applied Geophysics,* v.56, pp.123-134.

Schon, J. H., 1996. Physical Properties of Rocks – Fundamentals and Principles of Geophysics. In: *Handbook of Geophysical Exploration – Seismic Exploration* – Klaus Helbig and Sven Treitel (Editors). Vol. 18. Pergamon. 583p.

Stanley, J.R. and Pavlis, G.L., 1999. High-frequency reflections in granite? Delineation of the weathering front in granodiorite at Piñon Flat, California. *Geophysics* 64, pp.1828-1835.

Stolt, R. H., 1978. Migration by Fourier transform, *Geophysics,* 43, 23-48.

Telford,W.M., Geldart, L.P., Sheriff, R.E., 1995. *Applied Geophysics.* Cambridge University Press, New York, 770p.

Ylmaz, Ö., 2001. *Seismic Data Analysis*, Investigations in Geophysics, 10, S. M. Doherty (Ed.), *Soc. Explor. Geophys,* 2027p.

In: Uranium: Characteristics, Occurrence and Human Exposure ISBN: 978-1-62081-207-5
Editors: Alik Ya. Vasiliev and Mikhail Sidorov

Chapter 10

URANIUM: FROM OCCURRENCE, MINING, HUMAN EXPOSURE TO REMEDIATION

O. Saïdou*

Nuclear Physics Laboratory, Faculty of Science, University of Yaoundé
Nuclear Technology Section, Institute for Geological and Mining Research, Yaoundé, Cameroon

ABSTRACT

Energy is very important for the socio-economic development of the world requiring sufficient access to reliable energy resources. Renewable, water and fossil energy cannot totally cover the increasing needs in power supply. Nuclear energy is considered as a resort to overcome this limitation. Before the nuclear accident of Fukushima, many countries already considered nuclear power in their energetic grid. Operating nuclear power plants requires the availability of nuclear fuel.

Uranium exploration, re-assessment and mining are observed nowadays around the world to supply power plants in nuclear fuel. Moreover, for most human activities involving minerals and raw materials, the levels of radiological exposure are not significantly greater than normal background levels. Such exposures, while having been the subject of much research, are not of concern for radiation protection. However, certain activities can give rise to significantly enhanced exposures that may need to be controlled by regulation. It is for instance the case of the activities related to uranium mining.

The objective of this paper is to deal with the natural radioactivity, the uranium occurrence, mining and related radiological exposure. Finally, regulatory aspects of uranium mining and milling, the radiological protection of the public and workers, the site remediation will be taken into account. The case of the uranium deposit of Kitongo in Cameroon will be tackled.

* E-mail: saidous2002@yahoo.fr, Tel.: 00237 74 17 44 73.

INTRODUCTION

Natural uranium consisting of 99.3% ^{238}U and 0.7 % ^{235}U is the basic raw material for nuclear fuel. The present generation of nuclear power reactors derives energy from the fission of ^{235}U, the only fissile isotope in nature. These reactors also transmute the more abundant ^{238}U to man–made fissile isotope ^{239}Pu, which could be subjected to multiple recycling, as fuel, in fast reactor for efficient utilization of natural uranium resources and to ensure long term sustainability of nuclear energy.

In recent years, rising expectation for nuclear power has led to increase in uranium exploration, mining and ore processing activities all over the world. Several new countries, with a limited experience, have embarked on uranium exploration, mining and production. Uranium and its daughter products are radioactive and health hazardous. Radiological safety is a major challenge in uranium production cycle and in uranium mine and mill remediation and reclamation [1].

In 2008 the International Energy Agency (IEA) projected that, with current government policies, total primary energy and electricity demand will increase respectively by more than 50% and 90% by 2030, and that the great majority of this increase will be met by fossil fuel sources, even though global climate change and security of energy supply remain key issues.

Nuclear energy is an established technology offering a reliable option to address these issues. The size of its contribution will depend as much on the capabilities of governments and the nuclear industry to address society's concerns about safety, waste disposal and proliferation concerns. In 2008, installed nuclear generating capacity of around 372 GWe and annual production of some 2 700 TWh supplied about 16% of the world's electricity [2].

There are many uranium deposits already exploited or whose mining is under way around the world, creating disequilibrium in the environment [3]. World identified uranium resources accessible thanks to the UDEPA website (World Distribution of Uranium Deposits) of the International Atomic Energy Agency (IAEA), uranium intake, the radiological protection of the public and workers, waste management and site remediation will be tackled. The case of the Kitongo deposit in Cameroon will be dealt with [4-5].

URANIUM OCCURRENCE

The Uranium occurrence is defined as a naturally occurring, anomalous concentration of uranium. Naturally occurring radionuclides of terrestrial origin are present in various degrees in all media in the environment. Geological and hydrogeological conditions can sometimes lead to their enrichment in the environment, creating uranium or thorium deposits over a geological time scale.

Many uranium deposits are observed around the world. Most of the main deposits are already exploited or ongoing for nuclear applications to generate electricity, nuclear weapons and radio-isotope production. Since 2010, Kazakhstan is ranked the first producer of uranium in the world followed by Canada, Australia, USA, South Africa, Namibia, Brazil, Niger, Russian Federation, etc.

To have an overview about the worldwide distribution of uranium, the International Atomic Energy Agency created a database called World Distribution of Uranium Deposits

(UDEPA) [3]. The database and its web site are being updated continuously. Uranium resources can be assigned on the basis of their geological setting to the following categories of uranium ore deposit types (arranged according to their approximate economic significance): unconformity related deposits, sandstone deposits, hematite breccia complex deposits, Quartzpebble conglomerate deposits, Vein deposits (granite related deposits), Intrusive deposits, Volcanic and caldera related deposits, Metasomatite deposits, Surficial deposits, Collapse breccia pipe deposits, Phosphorite deposits, etc. Each type of deposit is well described in the reference [3].

As uranium is mostly present in oxide form, it is commonly reported as either uranium or its oxide U_3O_8.

For instance many studies concluded that the metasomatic uranium deposit of Kitongo in northern Cameroon could contain a historic resource of 10,000 t U_3O_8 at a grade of 0.1% [6]. This is based on the work completed, the geological setting and the mineralized occurrences observed outside the area of the initial drilling. The geological similarity of this deposit compared with those of other metasomatic uranium deposit sizes further confirms this assessment [5]. Additional ongoing re-evaluation could lead to a resource well greater than 13,000 t U_3O_8. UDEPA software gives a value ranging between 10,000-25,000 t U_3O_8.

To identify uranium deposits through anomalies, a helicopter equipped with the NaI(Tl) detector is frequently used to make aerial measurements [7]. The spectral analysis of gamma ray surveys using NaI(Tl) detectors normally gives three bands of information. ^{40}K is obtained from the peak at 1.46 MeV. Bands eU (uranium equivalent) and eTh (thorium equivalent) represent the main peaks used for measurement of U- and Th-series activity respectively.

Although often referred to as the uranium and thorium peaks, these are essentially associated with γ -rays which arise in the decay of ^{214}Bi (1.73 and 1.76 MeV) and ^{208}Tl (2.62 MeV). The common use of the terms uranium and thorium for the two peaks at 1.73–1.76 MeV and 2.62 MeV implies an assumption that the U and Th series can be considered in secular equilibrium for the purposes of interpretation of field γ spectrometry data.

World identified uranium resources (5.45 million tU) are adequate to meet projected future high case nuclear power requirements until 2050, providing that mine production is increased significantly.

Supplying uranium requirements beyond this date will require the identification of additional resources. Considerable effort will be required to bring about the substantial increase in mine production required to meet future Nuclear Energy Agency (NEA) demand projections. In addition to these challenges, uranium producers must continue developing and implementing best practices globally in order to improve public perception of uranium mining [2].

UDEPO database provides general, technical and geological information, including references, about the worldwide uranium deposits. UDEPO has been published on the internet which allows the users to register freely and to work with datasets (http://www-nfcis.iaea.org).

The web site provides filtering and navigation to the data from the database. It has also a statistical tool which provides summary information on number of deposits and uranium resources by type and status, and by country and status.

Uranium Mining

Average concentrations of uranium in typical soils and rocks are about 3 mg/kg U_3O_8. This background uranium is partly responsible for natural background radiation. In order to mine uranium economically using existing technology, this concentration has to reach at least 0.03% U_3O_8, with most uranium mines historically ranging between 0.1 to 0.5% U_3O_8. The most common types of mineable economic uranium ores are listed above.

Uranium is mined using traditional techniques such as open cut or underground mining, but sandstone deposits can also be mined by in situ leaching [8]. Once the ore is mined it is finely ground and the uranium is chemically extracted through conventional processes involving leaching with acid or alkali, concentration and then purification to uranium oxide. For in situ leaching, the acid or alkali is injected directly into the ore zone and pumped back to the surface. After leaching from the ore, the uranium is further concentrated using solvent extraction or ion exchange, followed by chemical precipitation to an impure oxide using ammonia called yellowcake. Finally, the yellowcake is heated at high temperature to remove the ammonia and leave relatively pure uranium oxide (>97% U_3O_8).

Before any mining and milling processes it is important to make a baseline study for a better assessment of their impact on the environment and the members of the public. It is for example the case of the uranium deposit of Kitongo in northern Cameroon [4] where a baseline study was undertaken before a probable mining.

Waste Management

The radioactive waste generated in mining and milling activities, especially those involving uranium and thorium (U, Th) ores, differs from that generated at nuclear power plants and most other industrial operations and medical facilities. Waste from mining and milling activities contains only low concentrations of radioactive material but it is generated in large volumes in comparison with waste from other facilities. The management methods to be employed are therefore different and will usually involve waste disposition on or near the surface, in the vicinity of the mine and/or mill sites. Furthermore, the waste will contain long lived radionuclides, and this has important implications for its management because of the long time periods for which control will be necessary [9].

An ore grade of 0.3% U_3O_8 means that 99.7% of the ore is left as solid waste, known as tailings (the minor loss of uranium is easily made up by the amount of chemicals added during leaching). Uranium mill tailings retain about 85% of the original radioactivity of the ore, and must be managed so as to minimize releases of radioactive decay products such as radium and radon as well as heavy metals [8].

Radioactive waste arises from all stages of mining and milling processes and includes, in addition to mill tailings, waste rocks, mineralized waste rocks and process water, including leaching solutions. Rainfall and snowmelt runoff and seepage from stockpiles and areas of uranium process plants should also be managed [10].

The hazards to humans or to the environment posed by mining and milling waste arise not only from its radioactivity but also from the presence of toxic chemicals and other materials in the waste. Achieving a consistent regulatory approach to protect against these different hazards is a challenge for national regulators.

As described in the reference [1], the volume of tailings piles varies from several hundred to several tens of millions of cubic meters, depending on the size, nature and duration of the operation. The activity of the tailings depends on the grade of ore mined and varies from less than 1 Bq/g to more than 100 Bq/g. The chemistry of tailings depends mainly on the leaching process (acid or alkaline) and the mineralogy of the ore. Wastes from the mining of sulphidic ores tend to produce acid leachate, which can mobilize radionuclides and other hazardous components of the tailings. Trace element concentrations in the ore and process chemicals may have a major influence on the tailings chemistry and its total environmental impact.

The coarse, sandy nature of some mill tailings makes them an attractive resource for re-use as a building material, mainly as sand for concrete making. A number of design steps and procedures are required to assure the effectiveness of isolation of tailings. These can be divided into containment preparation, tailings preparation, tailings discharge and deposition, tailings dewatering, tailings surface treatment, decant water treatment, seepage control, and capping. They usually require ancillary infrastructure in addition to the containment structure itself, and are best managed as an integrated operation [1].

Human Exposure

The exposure of persons to radiation arising from mining and milling waste is influenced by the physical, chemical and radioactive properties of the waste materials. The radionuclides in undisturbed and unprocessed ores exist in conditions close to secular equilibrium, but following mining and processing of this material to extract uranium, the state of secular equilibrium will be altered. Of more importance, however, is the changed physical and chemical state in which the radionuclides exist after the extraction operations are completed. This changed state may enhance their ability to migrate into the human environment and it may increase their accessibility for entering into the bodies of humans.

As described in the reference [11], uranium mining and milling operations involve handling of materials containing several radionuclides. However, some are of more concern than others with respect to potential radiation exposure. These include ^{230}Th, ^{226}Ra, ^{222}Rn and its short-lived progeny, longer lived radon progeny. Wherever it is found, ^{230}Th will continue to generate ^{226}Ra and thus will be a source of future loadings of radium. In the soil, radium and thorium migrate slowly, but can move via groundwater from the tailings management areas into sediments or surface waters. The ^{226}Ra will produce radon gas, ^{222}Rn that can be released into the atmosphere at a rate influenced by the condition of the tailings and any cover system employed. ^{226}Ra and ^{230}Th can also be dispersed through the air pathway on dust particles, but in open systems this results in a dilution of their radiogenic effects [11].

Radon emanating from mines and mill tailings (and/or residues) is dispersed easily in the air as a radioactive noble gas with the possibility of contributing exposure to humans through inhalation and takes up of the radionuclide. Nevertheless, the main pathway for human exposure comes from the inhalation of short-lived progeny attached to the aerosol particles in air very quickly after their sequential arising from radon [11]. The dose from inhalation of radon itself is less important since its deposition in the lungs is much less than for particulate matter.

Similarly, the radon decay products attached to fine particulate matter can be dispersed down wind, eventually settling out on the earth's surface. These longer lived daughter products (^{210}Pb and ^{210}Po) are of potential concern since they may give rise to exposure to humans through a variety of pathways [4]. We developed some α- and γ-spectrometry methods to measure ^{210}Pb and ^{210}Po in environmental solid samples [12].

Environmental radiation surveys and dose assessment programs require rapid determination of radionuclides because of the sampling size. To facilitate the environmental survey during and after uranium mining and milling processes, we proved by some statistical tests that γ-spectrometry is a valid alternative to time-consuming α-spectrometry for the determination of natural uranium and thorium activity in soil [13].

The baseline study to evidence probable impact during mining and milling processes is carried out in the uranium-bearing region of Poli where is located the Kitongo deposit. This study required sampling soil, water and foodstuffs representative of the radioactivity exposure and food consumption patterns of the population of this area. After sampling and radioactivity measurements were taken, our results indicated that the activities of natural series in soil and water samples are low. However, high levels of ^{210}Po and ^{210}Pb in foodstuffs (vegetables) were discovered and elevated activities of ^{40}K were observed in some soil samples. Most of the total dose is attributable to the intake of radon and high levels of ^{210}Po and ^{210}Pb contained in vegetables, food items which constitute an important part of the diet in Northern Cameroon. Consequently, bringing uranium ore from underground to the surface might lead to an increased dose for the population of Poli through a higher deposition of ^{222}Rn decay products on leafy vegetables [4]. An extended campaign of radon measurements in dwellings of the uraniferous regions of Cameroon where are located the Kitongo and Lolodorf deposits is ongoing for a better estimation of the inhalation dose.

URANIUM INTAKE

Uranium occurs mainly in oxidation states IV and VI. Workers may be exposed to uranium in a variety of chemical and physical forms. Some forms, like the metal and oxides may be encountered as depleted uranium (~ 0.2% ^{235}U), natural (0.7% ^{235}U) or enriched (>0.7% ^{235}U) uranium. The chemical behavior of any given uranium compound will be similar irrespective of whether the metal is present in natural, depleted or enriched form. It should be noted that intakes of the more rapidly absorbed uranium compounds are limited by considerations of chemical toxicity, rather than radiation dose [14].

Workers and public can incorporate uranium by ingestion and inhalation due to mining and milling processes. The ingestion/inhalation dose corresponding to the uranium requires incorporated radionuclide concentration and ingestion/inhalation dose coefficient. The ingestion/inhalation dose is calculated from the dosimetric and biokinetic models of the radionuclides.

The biokinetic model of uranium is well described in the reference [14]. For instance, the ingestion dose coefficient of α-, β-, αγ-, and βγ-emitters is given by the following equation:

$$e_{ing} = \frac{1}{4} k \sum_i w_{R,i} p_i E_i \sum_{T,S} w_T U_S \left[\upsilon \left(\frac{1}{M_T^c} + \frac{1}{M_{T'}^c} \right) + 2 \left(\frac{AF(T \leftarrow S)}{M_T} + \frac{AF(T' \leftarrow S)}{M_{T'}} \right) \right]$$

where p_i is the emission probability at the energy E_i of the uranium-234/235/238, $w_{R,i}$ is the radiation weighting factor. U_S is the number of disintegrations in each organ and tissue, $AF_i(T \leftarrow S)$ is the fraction of energy emitted in the source region S that is absorbed within target region T. M_T is the mass of target region T, and is the mass of the walled organ. $AF_i(T \leftarrow S) = 1$ for α-particles due to the self-absorption within the source. The Absorbed Fraction is multiplied by υ=0.01 to account for the reduced α-dose to radiosensitive cells in the wall.

The parameter values used are representative of ranges because individuals exhibit metabolic variation, which can influence the intake, distribution and retention of radionuclides, and they also vary in size and shape, which influences dosimetric quantities. In addition, there is uncertainty associated with each of the parameter values used according to the quality and reliability of the data on which they are based.

The relative uncertainty associated to the above ingestion dose coefficient corresponding to α-and β- emitters is given from the following equation:

$$\left(\frac{u_{e_{ing}}}{e_{ing}}\right)^2 = \left(\frac{u_{w_{R,i}}}{w_{R,i}}\right)^2 + \sum_i \left(\frac{u_{p_i}}{p_i}\right)^2 + \sum_R \left(\frac{u_{E_i}}{E_i}\right)^2 + \sum_T \left(\frac{u_{w_T}}{w_T}\right)^2 + \sum_T \left(\frac{u_{U_T}}{U_T}\right)^2 + \sum_{T,T'} \left[\upsilon \left(\frac{1}{M_T^c} + \frac{1}{M_{T'}^c} \right) + 2 \left(\frac{AF(T \leftarrow S)}{M_T} + \frac{AF(T' \leftarrow S)}{M_{T'}} \right) \right]^{-2} \times$$
$$\times \left[\begin{array}{l} \upsilon^2 \left(\frac{1}{M_T} + \frac{1}{M_{T'}} \right)^2 \left(\frac{u_\upsilon}{\upsilon} \right)^2 + 4 \left(\frac{AF(T \leftarrow S)}{M_T} \right)^2 \left(\frac{u_{M_T}}{M_T} \right)^2 + 4 \left(\frac{AF(T' \leftarrow S)}{M_{T'}} \right)^2 \left(\frac{u_{M_{T'}}}{M_{T'}} \right)^2 + \left(\frac{\upsilon}{M_T^c} \right)^2 \left(\frac{u_{M_T^c}}{M_T^c} \right)^2 + \left(\frac{\upsilon}{M_{T'}^c} \right)^2 \left(\frac{u_{M_{T'}^c}}{M_{T'}^c} \right)^2 + \\ +4 \left[\left(\frac{AF(T \leftarrow S)}{M_T} \right)^2 + \left(\frac{AF(T' \leftarrow S)}{M_{T'}} \right)^2 \right] \left(\frac{u_{AF(T \leftarrow S)}}{AF(T \leftarrow S)} \right)^2 \end{array} \right]$$

where u_i represents the standard uncertainty of the variable *i*. For a given source, correlations between *SAF*s of corresponding targets are neglected. This is justified by the total absorption of α- and β-particles within the source. All the details on the ingestion dose coefficient derivation and uncertainty assessment for α-, β-, αγ-, and βγ-emitters are given in the reference [15].

Radiological Protection of Workers, Members of the Public

Workers at mines or mills may receive radiation doses from ores, concentrates, the product of the milling process, associated airborne dust, radon and thoron daughter products, and radioactive waste. The protection of workers from the radiological hazards of mining and milling waste should not be considered in isolation without considering these other sources of radiation exposure. The operators of the mine and mill should have in place a comprehensive radiological protection program [9].

It is required that the dose due to occupational exposure of workers at mines or mills, all sources of exposure including radioactive waste being considered, not exceed: an effective dose of 20 mSv per year averaged over five consecutive years; an effective dose of 50 mSv in any one year; an equivalent dose to the lens of the eye of 150 mSv in a year; and an equivalent dose to the extremities (hands and feet) or the skin of 500 mSv in a year.

Releases of radionuclides from radioactive waste to the environment during mining and milling activities and subsequent waste management activities may result in the radiation exposure of members of the public.

Such releases are subject to the criteria that are applicable to releases from any practice in which radioactive material is being handled and, as with occupational protection, national requirements for radiological protection should be consistent with the Basis Safety Standards [16]. However, since mine and mill tailings will continue to present a potential hazard to human health after closure, additional analyses and measures may be needed to provide for the protection of future generations. Such measures should not be left until closure but should be considered and implemented throughout the design, construction and operation of the mining and milling facilities.

The protection of the public, from the beginning of operations to post-closure, should be considered in its entirety from the beginning of the design of the facilities. Although mining and milling waste contains only naturally occurring radionuclides, these radionuclides cannot be considered to be in their original states or concentrations, since their physical and chemical forms may have been altered substantially, and exposures may be influenced by the operation of the waste management facilities.

Exposures attributable to such waste should not be regarded as exposure to natural background radiation and exposures of the public attributable to all mining and milling waste should be included in the system of radiation protection for practices as required in the BSS [16].

REMEDIATION

The remediation of sites contaminated by radioactive materials after mining and milling of uranium deposits is challenging and become a priority for the International Atomic Energy Agency and Member States. Periodically scientific events are organized to update knowledge from learned lessons for a better radiological protection of workers, members of the public and the environment. For instance in order to plan and implement the measures needed to close out mining and milling sites, it is important to characterize their local situation including pre-operational conditions, and to carry out an assessment of the short term and long term impacts on human health and the environment. The current practice in many countries is generally restricted to estimating actual and potential radiation doses to persons resulting from migration of radionuclides from remediated sites or from human intrusion. In addition, the hazards associated with non-radioactive substances from uranium production may be significant and need to be assessed accordingly. The fundamental reason to remediate tailings piles is to reduce the level of hazards and thus lessen the risk of significant impacts upon human health and the environment, or to reduce ongoing levels of impact where effects are already apparent [1].

Remediation of uranium mill tailings piles has been undertaken and/or is in progress in virtually every country, where uranium has been or is being produced. The typical illustration stems from the Central Asia countries like Kazakhstan and Kirghistan with their legacy sites from Soviet Union period. As knowledge on environmental sensitivity grows and regulations and standards evolve to reflect this knowledge in response to public concern, the design standards for remediation will become more demanding.

Much is being learned from the remediation programs as they are undertaken and as monitoring data are collected after completion. A distinctive message that comes out of the work done to date is that there is no single best approach. Each site must be studied closely to determine the optimal approach.

The amount and type of work, and the associated costs, will differ markedly from site to site. Methods of construction and design that are harmonious with and emulate nature are preferable to engineered designs that are built to resist, rather than to harness natural processes for long term stability. Various types of techniques of remediation are well described in the reference [1].

CONCLUSION

The tendency of many countries to embark on nuclear program to generate electricity increased uranium exploration, mining and production over the world. Operating nuclear power plants goes with the availability of nuclear fuel like uranium. The entire process from uranium mining, production and use requires the application of radiological safety rules to protect workers and members of the public against ionizing radiation effects. Many uranium deposits are already mined over the world and the site remediation is challenging for governments. Some lessons were learned from the legacy sites met around the world. The proper evaluation of the impact of uranium mining and milling on the environment, workers and members of the public needs a baseline study. It is for instance the case of the uranium-bearing region of Poli in Cameroon where is located the Kitongo deposit. The IAEA organizes frequently scientific events to update knowledge related to uranium exploration, mining, milling, production and use over the world. All decision-makers, operators, regulators and users should consult related publications for application.

REFERENCES

[1] International Atomic Energy Agency, IAEA. The long term stabilization of uranium mill tailings. Final report of a co-ordinated research project 2000-2004, *IAEA TECDOC* 1403, Vienna, 2004.

[2] International Atomic Energy Agency, IAEA URAM 2009. International symposium on uranium raw material for the nuclear fuel cycle. Exploration, mining, production, supply and demand, economics and environmental issues. URAM 2009. *Books of abstracts*, Vienna, 2009.

[3] International Atomic Energy Agency, IAEA. World distribution of uranium deposits (UDEPA) with uranium deposit classification. *TECDOC* 1689, Vienna, 2009.

[4] Saïdou, Bochud F, Baechler S, Kwato Njock MG, Ngachin M, Froidevaux P. Natural radioactivity measurements and dose to the public calculations: case of the uranium-bearing region of Poli in Cameroon, *Radiation Measurements* 46 (2011) 254-260.

[5] Meadon HM. Nuclear Energy Corporation, Geological review of the Kitongo uranium deposit, Poli area, Northern Cameroon, 2006.

[6] Thoste V. Uranium exploration in North Cameroon, region of Poli. Project definition, execution and results, Jan. 1982- Dec. 1984. *Unpublished report*, 1985, Hannover.

[7] Winkelmann I, Thomas M, Vogl K. Aerial measurements on uranium ore mining, milling and processing areas in Germany. *J Environ Radioactivity* 53 (2001) 301-311.

[8] Gavin MM. Uranium mining: Australia and globally, 2006. Website http://www.energyscience.org.au.

[9] International Atomic Energy Agency, IAEA. Occupational Radiation Protection in the Mining and Processing of Raw Materials. IAEA Safety standard series. *Safety guide* No. RS-G-1.6., Vienna, 2004.

[10] International Atomic Energy Agency, IAEA. Planning and management of uranium mine and mill closures. *Proceedings of a Technical Committee meeting,* Liberec, Czech Republic. Vienna, 1994.

[11] International Atomic Energy Agency, IAEA. Closeout of uranium mines and mills: A review of current practices. *TECDOC* 939. Vienna, 1997.

[12] Saïdou, Bochud F, Laedermann J-P, Buchillier T, Kwato Njock M, Froidevaux P. Calibration of an HPGe detector and self-attenuation correction for ^{210}Pb: Verification by alpha spectrometry of ^{210}Po in environmental samples. *Nucl. Instr. and Meth. A* 2007; 578: 515-522.

[13] Saïdou, Bochud F, Laedermann J-P, Kwato Njock MG, Froidevaux P. A comparison of alpha and gamma spectrometry for environmental radioactivity surveys. *Appl. Radiat. and Isot.* 2008; 66: 215-222.

[14] International Commission of Radiological Protection, ICRP. Individual Monitoring for Internal Exposure of Workers: Replacement of ICRP Publication 54, ICRP Publication 78. Ann. *ICRP,* 27 (3/4), 124-130.

[15] Saïdou, Bouba O, Bingana MSA, Dallou GB. Ingestion dose coefficient update of ^{210}Po, ^{210}Pb, ^{238}U and uncertainty assessment in biokinetic and dosimetric models of α-, β-, αγ-, and βγ-emitters, submitted, *Radiation Protection Dosimetry.*

[16] Food and Agriculture Organization of the United Nations, International Atomic Energy Agency, International Labour Organization, OECD Nuclear Energy Agency, Pan American Health Organization, World Health Organization, International Basic Safety Standards for Protection against Ionizing Radiation and for the Safety of Radiation Sources, *Safety Series* No. 115, Vienna, 1996.

In: Uranium: Characteristics, Occurrence and Human Exposure ISBN: 978-1-62081-207-5
Editors: Alik Ya. Vasiliev and Mikhail Sidorov

Chapter 11

REMOVAL OF URANIUM FROM AQUEOUS SOLUTION, URANIUM EFFLUENT AND SEAWATER USING VARIOUS MICROBES ISOLATED FROM THE URANIUM DEPOSITS

Takehiko Tsuruta[1]
Department of Biotechnology and Environmental Engineering,
Hachinohe Institute of Technology, Myoh, Hachinohe, Aomori, Japan

ABSTRACT

Attempts were made to recover uranium (U) occurring in nuclear fuel effluents and mine tailings using microorganisms isolated from U deposits in Canada, the United States, Australia, and Japan. To establish which microorganisms absorb the most U, hundreds of strains of microorganisms were screened. Extremely high U absorbing ability was detected in some bacteria isolated from North American U deposits. Arthrobacter and Bacillus sp. absorbed approx. 2500 μmol U per gram dry wt. of microbial cells within one hour. Cells removed U from refining waste water with high efficiency. Cells also absorbed thorium with high efficiency. Lactobacillus cells isolated from Japanese U deposits removed more U from seawater than the other bacteria that had superior U removal capacity from non-saline U solutions. Cells immobilized with polyacrylamide gel had excellent handling characteristics and can be used repeatedly in U adsorption-desorption cycles. These bacteria from U deposits can be used as an adsorbing agent for the removal of the nuclear fuel elements, which may be present in nuclear effluents, mine tailings, seawater, and other waste sources.

Keywords: Uranium absorption, microorganisms, *Arthrobacter, Lactobacillus, Bacillus*, immobilized microorganisms, thorium

---

[1] tsuruta@hi-tech.ac.jp.

INTRODUCTION

The recoveries of nuclear fuel elements, such as uranium (U) and thorium (Th), from aqueous systems have become a focus of interest for exploitation of undeveloped energy resources. The removal of radioactive elements and toxic heavy metals from contaminated sources is also a worthwhile priority for environmental protection initiatives. In this regard, efforts have concentrated on studying absorption of U by microorganisms, including bacteria (Andres et al. 1993; Byerley et al. 1987; Friiss and Myers-Keith 1986; Gorab et al. 1991; Hu et al. 1996; Marques et al. 1991; Strandberg et al. 1987), fungi (Byerley et al. 1987; Galun et al. 1983a; Galun et al. 1983b; Tsezos and Volesky 1981; White and Gadds 1990) and yeasts (Strandberg et al. 1981; Shumate et al. 1978).

We have investigated U absorption from aqueous systems using bacteria isolated from U mines, among which some strains were found to possess extremely high U absorbing ability (Sakaguchi et al. 1996). Microbial biomass may thus be considered for use as a removal agent for the recovery of U from metallurgical effluents, mine tailings, seawater and other waste sources.

In U deposits, it can be presumed that some microorganisms having a high absorbing ability for U and different species having an ability to leach U from ore may exist in mine soil and aqueous systems. It would, therefore, be beneficial to isolate microorganisms having an enhanced ability to absorb U from mines.

Recently, we screened hundreds of types of microorganisms existing in U deposits located in North America, Australia, and Japan for their ability to absorb significant quantities of U, and identified new strains which absorbed large quantities of U, such as *Bacillus subtilis* in Australia, *Arthrobacter* and *Bacillus* sp. in North America, and *Lactobacillus* and *Bacillus* sp. in Japan (Sakaguchi 1998). In this chapter, new strains of bacteria identified in North American, Australian, and Japanese U deposits, especially *Arthrobacter, Bacillus* isolated in U. S. and *Lactobacter,* and *Bacillus* sp. isolated in Japan, are discussed for their potential for removal of nuclear fuel elements such as U from U refining wastewater and seawater.

SCREENING OF MICROORGANISMS ISOLATED FROM U DEPOSITS FOR THEIR U REMOVING ABILITY

To determine the ability of microorganisms isolated from U deposits in Canada, the United States, Australia, and Japan to absorb U, hundreds of strains of microorganisms were screened.

The medium for growing microorganisms contained 3 g/L beef extract, 5 g/L peptone and 5 g/L NaCl in deionized water. The microorganisms were maintained on agar slants and grown in 300 mL of the medium in a 500 mL flask with continuous shaking (120 rpm) for 72 h at 30°C. Cells were collected by centrifugation, washed thoroughly with deionized water and then used in the following absorption experiments.

U was supplied as $UO_2(NO_3)_2$. The pH of the solution was adjusted to 5.8 with 0.1 M HNO_3. Resting microorganisms (15 mg dry wt.) were suspended in 100 mL solution (pH 5.8) containing 84 μM U and the suspension was shaken for 1 h at 25 °C. Cells were collected by filtration through a membrane filter (pore size 0.2 μm). The amounts of U absorbed by the cells were determined by measuring U in the filtrate using an inductively coupled plasma quantometer (ICPS8000, Shimadzu Corporation, Kyoto, Japan). The microbial strains were identified by our co-worker (Sakaguchi, 1998).

The amounts of U absorbed by the cells ranged from a minimum of 12.6 % to a maximum of 98.3 % (Table 1). Of special interest to this discussion is the wide range of effectiveness with which different species of microorganisms absorb U.

Table 1. Absorption of uranium using microorganisms isolated from soils or water at uranium deposits

Strain number	Uranium absorbed (%)	Strain number	Uranium absorbed (%)	Strain number	Uranium absorbed (%)	Strain number	Uranium absorbed (%)
US-1	14.0	CAN-1	12.6	AUS-1	19.4	JPN-1	19.9
US-2	17.1	CAN-2	23.7	AUS-2	25.4	JPN-2	21.2
US-3	23.9	CAN-3	33.9	AUS-3	36.6	JPN-3	32.5
US-4	38.7	CAN-4	49.8	AUS-4	46.2	JPN-4	44.7
US-5	43.9	CAN-5	55.7	AUS-5	53.3	JPN-5	54.9
US-6	58.9	CAN-6	67.9	AUS-6	66.8	JPN-6	65.2
US-7	72.0	CAN-7	74.2	AUS-7	74.7	JPN-7	76.8
US-8	84.4	CAN-8	85.7	AUS-8	85.1	JPN-8	86.2
Bacillus sp. US-9	95.9	CAN-9	95.4	AUS-9	96.9	*Bacillus* sp. JPN-9	97.6
Arthro-bacter sp. US-10	96.4	CAN-10	96.9	*Bacillus* sp.AUS-10	98.3	*Lactobacillus sp.*JPN-10	97.8

Of these microorganisms tested, extremely high U absorbing ability was found in *Arthrobacter* (96.4 %) and *Bacillus* sp. (95.9 %) found in US, *Lactobacillus* (97.8 %) and *Bacillus* sp. (97.6 %)found in Japan, and *Bacillus* sp. (98.3 %) found in Australia (Sakaguchi,1998), which removed large amounts of U per gram dry wt. of microbial cells within 5 min. The features of the absorption of U using these strains were reported in details.

EFFECT OF PH ON U ABSORPTION

Effect of pH on U absorption was examined as mentioned below. The pH of a test solution was adjusted to the desired value with 0.1 M HNO_3 or 0.1 M NaOH. Resting *Arthrobacter* and *Lactobacillus* cells (15 mg dry wt.) were suspended in 100 mL solution containing 84 μM U for 1 h at 25 ^{o}C. Zeta potential was measured by the electrolysis method using 30-50 bacteria (bacterial concentration was 2.5 x 10^8 cells/mL) at pH 2-9 in 0.01 M $NaNO_3$ solution (ZC-2000; Microtec-Nichion, Chiba, Japan).

The effects of pH on U removal using *Arthrocacter* sp. US-10 (closed circles) and *Lactobacillus* sp. JPN-10 (opened circles) are shown in figure 1. The amounts of remained U were lowest at approx. pH 5-8 and increased with increasing acidity below pH 4 using *Arthrobacter* cells. On the other hand, the amounts of remained U using *Lactobacillus* were lowest at pH 6, and rapidly increased below pH 5 and above pH 7. Low pH is the result, of course, of higher proton concentrations, which can compete for binding with U, and *Arthrobacter* cells clearly remained less affected than *Lactobacillus* by the decreasing pH until about pH 4. Conversely, the high pH is, of course, the result of reduced proton concentration and increased hydroxide ions, which compete with microbial cells. Again, *Arthrobacter* exhibited a higher pH range for binding capacity than *Lactobacillus*, achieving good absorb at pH values as high as 8, while *Lactobacillus* performed well only at pH 6.

The effects of pH on Zeta potential of the cell surface are shown in Table 2. The amounts of U remained are lowest at approx. pH 5 and decrease below pH 4 using *Arthrobacter* cells. Thus, the absorption of U using *Arthrobacter* is markedly affected by solution pH. Zeta-potential of the cell surface increased with increasing acidity of the solution.

The main U species at pH 6 includes some cations such as $(UO_2)_3(OH)_5^+$ and UO_2OH^+ (Tsuruta 2006). On the other hand, Zeta- potential of the cell surface was -23.18 mV at the same pH. Therefore, cationic uranyl species readily bond with negatively- charged cell surfaces of *Arthrobacter* sp.

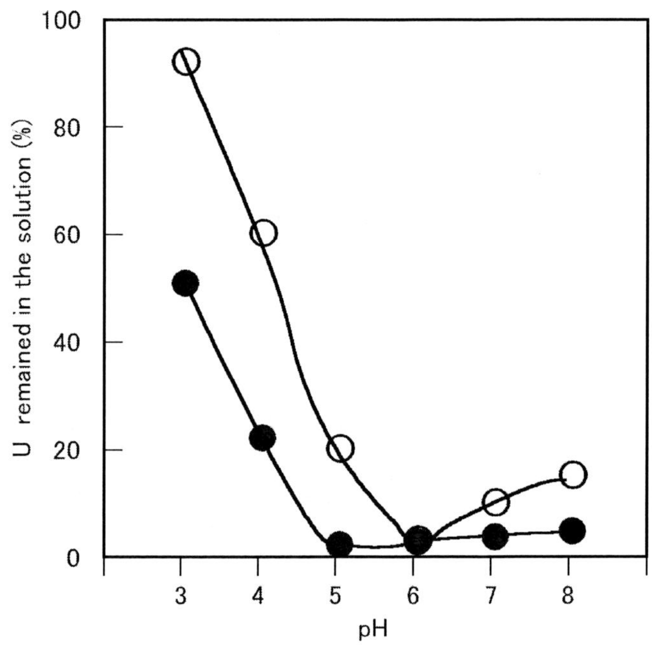

Figure 1. Effects of pH on the amounts of uranium absorbed using Arthrobacter US-10 and Lactobacillus JPN-10 cells.
Symbols: close circles, uranium(VI) absorbed (mmol/g dry wt. cells) using Arthrobacter cells; open circles, uranium(VI) absorbed (mmol/g dry wt. cells) using Lactobacillus cells.

Table 2. Effect of pH on Zeta-potential of *Arthrobacter* cell

pH	2.07	3.00	4.04	5.09	6.07	6.97	8.06
Zeta-potential(mV)	3.69	-1.50	-7.33	-21.01	-23.18	-22.11	-23.06

Effect of U Concentration on U Absorption

Resting *Arthrobacter* and *Lactobacillus* cells (15 mg dry wt. basis) were suspended in 200 mL of solution (pH 5.8) containing a specified quantity of U for 1 h at 25 °C.

Quantities of absorbed U (mmol U/g dry wt. cells) using *Arthrobacter* (closed circles) and *Lactobacillus* (opened circles) sp. increased as U concentration increased, whereas total absorbed U (%) using *Arthrobacter* (closed squares) and *Lactobacillus* (opened squares) sp. decreased (figure 2-a). *Arthrobacter* and *Lactobacillus* cells absorbed 2480 and 2120 mmol U per gram dry wt. of cells, respectively.

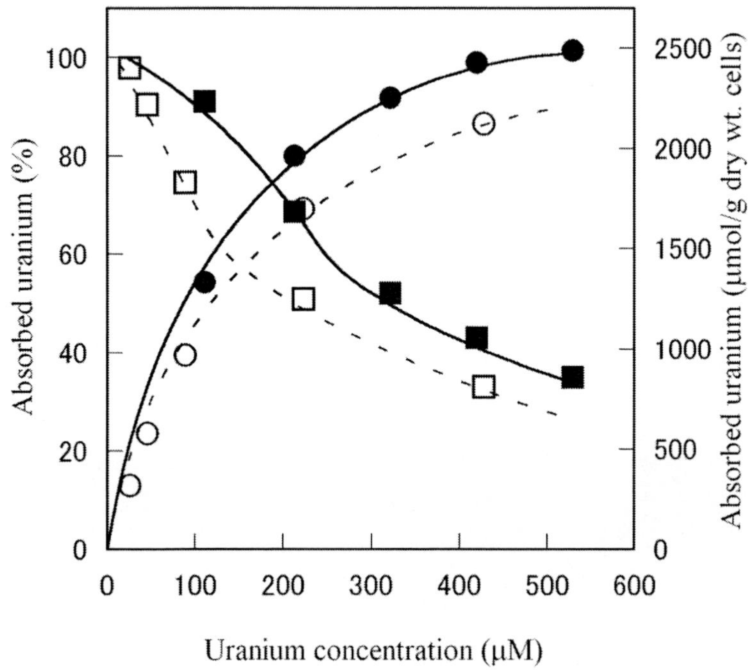

Figure 2-a. Effect of uranium concentration on uranium absorption using Arthrobacter US-10 and Lactobacillus, JPN-10 cells.
Symbols: circles, absorbed uranium (μmol/g dry wt. cells); squares, absorbed uranium (%), close, absorbed uranium using Arthrobacter sp., open, absorbed uranium using Lactobacillus sp.

Additionally, as show in figure 2-b, U absorbed using these cells, especially *Lactobacillus* sp (opened circles), does not obey the *Langmuir* isotherm over the entire U concentration range tested. It appears that the experimental data shows a dual pattern. The dotted line was calculated using less than a 21.8 mM residual U concentration (as the initial uranium concentration was 85.0 mM) using *Lactobacillus* cells. The solid line was separately calculated using residual concentrations below and above 21.8 mM. When the initial U concentration increased beyond 85.0 mM (with a 21.8 mM residual U concentration), the absorbed U amounts increased to values far greater than those calculated based on the relationship for the low concentration range. On the other hand, the dotted line in figure 1.2.3b for *Arthrobacter* cells (closed circles) was calculated using a residual U concentration below 153 mM (and an initial U concentration of 318 mM). The solid line was separately calculated for *Arthrobacter* cells using residual concentrations below and above 153 mM.

When the initial U concentration increased beyond 318 mM (with a residual U concentration of 153 mM), the absorbed amounts increased to values a little greater than those calculated based on the relationship in the low concentration range. A similar result using *Lactobacillus* cells was obtained from the absorption of potassium using barley roots (Epstein, 1966). The estimated m, n, and maximum accumulated U capacity $Q(U)_{max}$ (=1/m) are summarized in Table 3. A high $Q(U)_{max}$ value of 2580 and 2370 mmol U/g dry wt. cells is estimated from the high U concentration region using *Arthrobacter* and *Lactobacillus* cells, respectively.

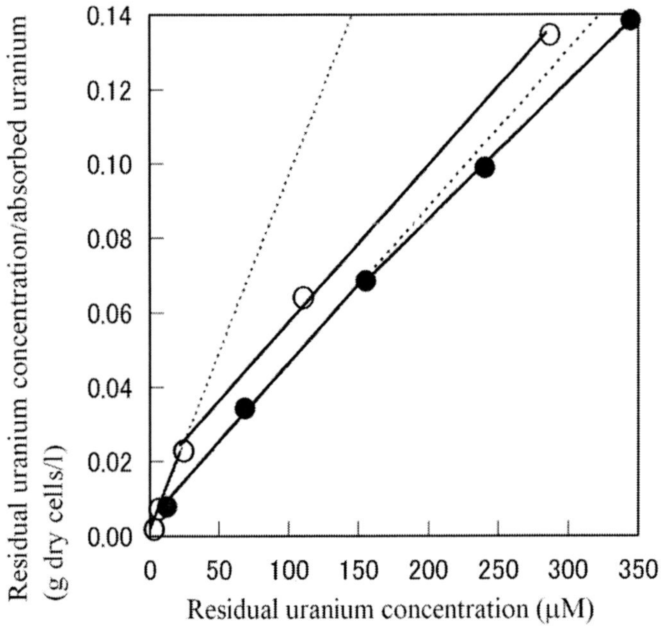

Figure 2-b. Equilibrium isotherms of uranium absorption using microbial cells.
Symbols: close, absorbed uranium using Arthrobacter sp., open, absorbed uranium using Lactobacillus sp. Ce(U)/QU = mCe(U) + n, where QU indicates the amount of absorbed uranium (mmol uranium/g dry wt. cells), Ce is the residual uranium in the solution (mM), and m and n are constants.

Table 3. Estimated *Langmuir* constants and Q(U)max from the *Langmuir* isotherm

Strains	Initial U Conc.(mM)	Residual U Conc (mM)	M	n	Q(U)max (mmol/g dry wt. cells)
Arthrobacter sp.	107-318	10.0-153	4.22 x 10-4	4.32 x 10-3	2371
	318-325	153-343	3.69 x 10-4	1.13 x 10-2	2710
	107-525	10.0-343	3.88 x 10-4	6.45 x 10-3	2580
Lactbacillus sp.	21.0-85.0	0.483-21.8	9.51 x 10-4	2.05 x 10-3	1050
	85.0-423	2.8-284	4.22 x 10-4	1.53 x 10-2	2750
	21.0-423	0.438-284	4.56 x 10-4	7.69x 10-3	2200

TIME COURSE OF U REMOVAL

Resting *Arthrobacter* (closed circles) and *Lactobacillus* (opened circles) cells (15 mg dry wt.) were suspended in 100 mL solution (pH 5.8) containing 84 μM U at 25 °C (figure 3).

Amounts of U absorbed by *Arthrobacter* and *Lactobacillus* cells increased rapidly during the first 5 minutes following application of U.

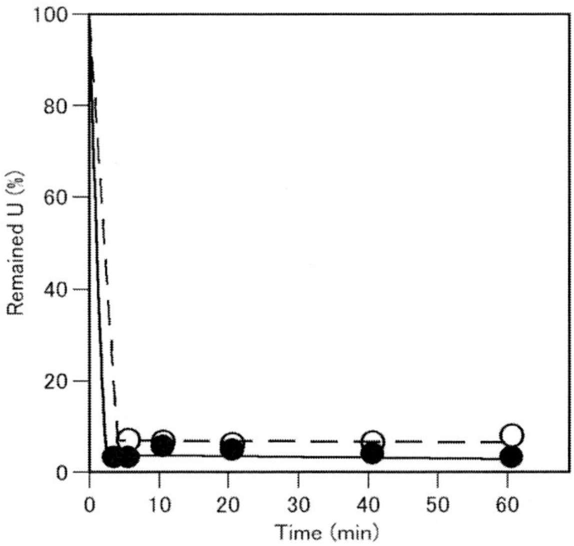

Figure 3. Time course of uranium absorption using *Arthrobacter* and *Lactobacillus* cells.
Symbols: close, absorbed uranium using Arthrobacter sp., open, absorbed uranium using *Lactobacillus* sp.

RELEASE OF U FROM CELLS BY WASHING WITH EDTA

Distribution of U in Microbial Cells

The present experiments were undertaken to determine which parts of the cells had absorbed U in *Arthrobacter*, US-10 and *Lactobacillus,* JPN-10 cells.

Resting cells of *Arthrobacter* and *Lactobacillus* (640 mg fresh weight) were suspended in 1000 mL solution (pH 5.8) containing 500 μM of U for 1 h at 25 °C. The cells were fractionated as described in figure 5-a. The freeze-dried *Arthrobacter* and *Lactobacillus* cells, cell wall and intracellular particle fractions were digested in the mixed solution of same volume of concentrated HNO_3/H_2SO_4.

As shown in figure 5-b, the amounts of absorbed U in each fraction were recorded in the following order

cell wall > intracellular particle > intracellular soluble

Abundant U was determined in the cell wall fraction and small amounts occurred in the intracellular fractions. In the case of *Arthrobacter* sp., this result coincides with the results noted above for EDTA washing. On the other hand, the results using *Lactobacillus* sp. suggested that most absorbed U was coupled with the cell wall fraction; half of the absorbed U was not released by washing with EDTA. Therefore, the bond of U with cell wall of *Lactobacillus* cells appears to be strong.

On the basis of these findings, it seems reasonable to postulate that the absorbed U using the cells, especially *Lactobacillus* sp. is mostly dependent on the physical- chemical binding relationships with cell wall components.

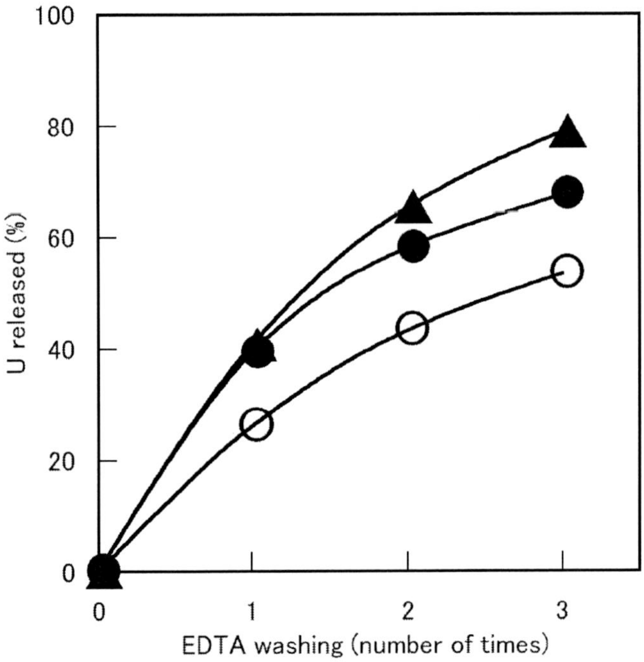

Figure 4. Release of uranium from Lactobacillus sp., JSP-10, Arthrobacter sp. US-10 or Bacillus sp. US-9 cells by washing with EDTA.
Symbols: triangles, absorbed uranium using Bacillus sp, closed circles, absorbed uranium using *Arthrobacter* sp., opened circles, absorbed uranium using *Lactobacillus* sp.

SELECTIVE ABSORPTION OF U USING *ARTHROBACTER*., US-10 CELLS

To determine which heavy metal ions are most readily absorbed by bacterial cells, the selective absorption of ions using *Arthrobacter*, US-10 cells from a solution containing six metal cations and UO_2^{2+} was examined. Resting cells (15 mg dry wt. basis) were suspended in 100 mL of a solution (pH 5.0) containing 4×10^{-5} M of Mn^{2+}, Co^{2+}, Ni^{2+}, Cu^{2+}, Zn^{2+},Cd^{2+} and $UO_2^{2+}$for 1 h at 25 °C.

The relative degree of heavy metal ions absorbed using *Arthrobacter* US-10 cells was (figure 6): $UO_2^{2+} > Cu^{2+} >$ others.

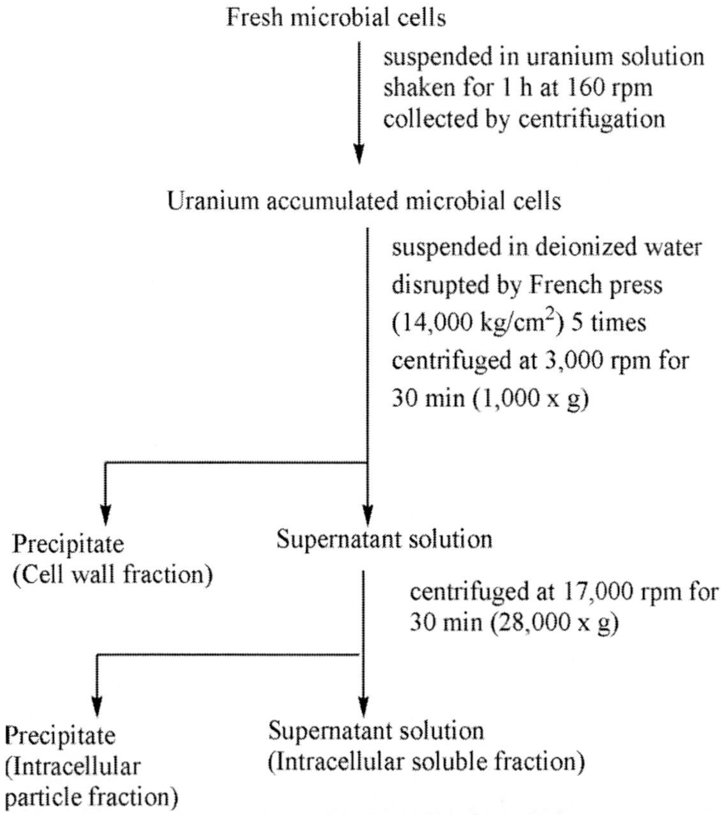

Figure 5-a. Fractionation of Lactobacillus sp., JPN-10 and Arthrobacter sp. US-10 absorbed uranium.

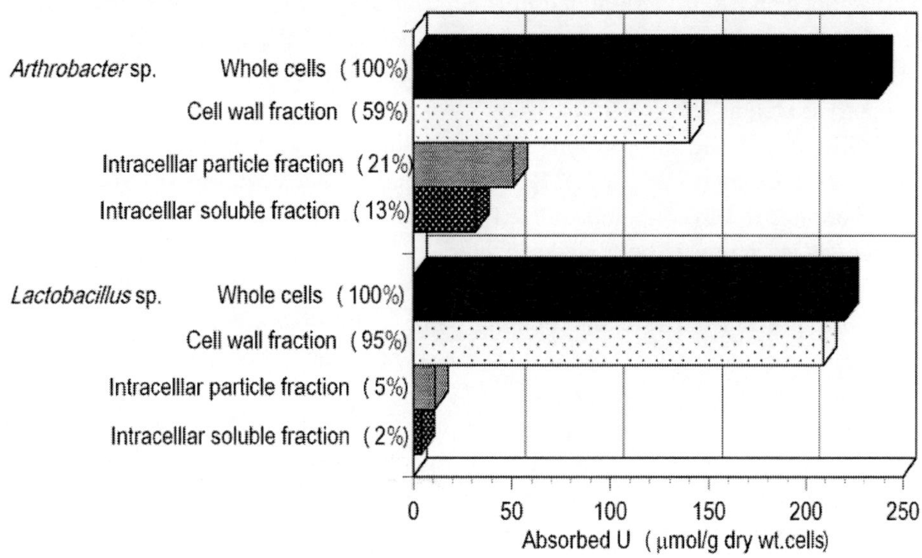

Figure 5-b. Distribution of uranium in microbial cells.

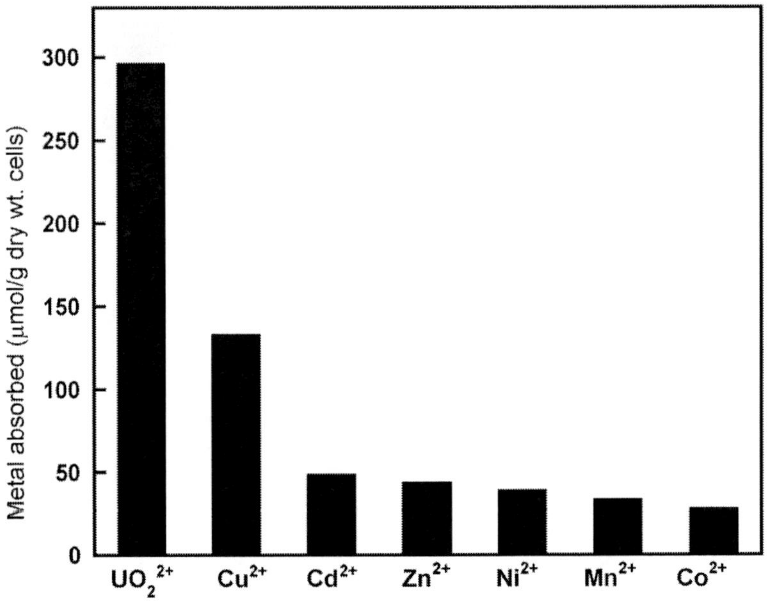

Figure 6. Selective absorption of heavy metals using Arthrobacter US-10 cells.

Absorption of Th and Selective Absorption of Th and U by Bacteria

As described above, bacteria found in U deposits absorbed U with high efficiency. In this course of our study, the question was raised as to whether these strains had the ability to absorb Th, another common waste from nuclear processing (and an environmental contaminant) as well as U.

As thorium hydroxide is precipitated in a solution containing thorium at pH 4.0, the absorption of Th examined at pH 3.5. Resting cells (15 mg dry wt. basis) were suspended in 100 mL solution (pH 3.5) containing 50 μM Th (as $Th(NO_3)_4$) and/or U for 1 h at 25 °C.

As shown in figure 7, both *Arthrobacter* sp. US-10 and *Bacillus* sp. US-9 can also absorbed Th with high efficiency. However, the amounts of U and Th absorbed by *Lactobacillus* sp. from the solution containing one metal only were lower than those using *Arthrobacter* and *Bacillus* sp. These results appear reasonable, because solution pH strongly affects the absorption of both elements using *Lactobacillus* cells. The amounts of absorbed U from the solution containing both elements at pH3.5 by *Arthrobacter* was half that from the solution containing U only; however, that absorbed by *Bacillus* from the solution containing both elements was far lower than that from the U-only solution. Accordingly, the effect of Th on U absorbed by *Bacillus* sp. is greater than that by *Arthrobacter* sp. Thus, the *Arthrobacter* sp. appears to be the most efficient choice for removing both elements from a mixed solution of Th and U.

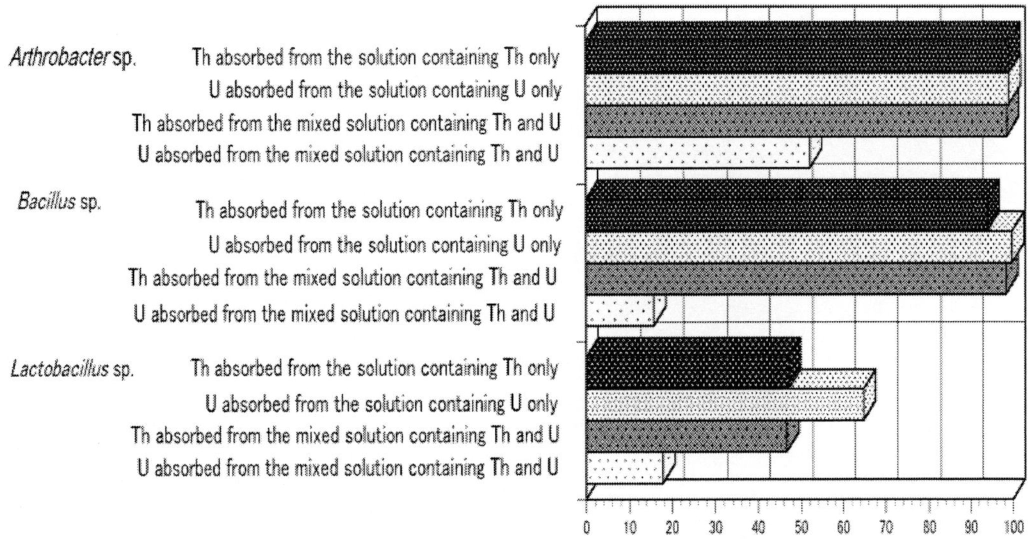

Figure 7. Absorption of thorium and/or uranium from the solution containing thorium and/or uranium.

RECOVERY OF U BY IMMOBILIZED BACTERIA

As described above, bacteria such as *Arthrobacter, Bacillus,* and *Lactobacillus* sp. can absorb remove large quantities of U from aqueous systems. However, the free cells of these bacteria are not reusable because of their mechanical instability and susceptibility to cell degradation. Furthermore, free cells are not suitable for use in column systems, because they cause plugging. To overcome these deficiencies with free cells, the cells of *Arthrobacter* sp. US-10 having high U absorbing ability were immobilized with polyacrylamide.

Five grams of precultured *Arthrobacter* cells were suspended in 4.5 mL isotonic NaCl solution and 680 mg acrylamide monomer. A total of 34 mg N, N'- methylene-bis(acrylamide), 0.3 mL 3-dimethylaminopropionitrile solution (5%) and 0.34 mL potassium persulfate solution (2.5%) were added to the suspension. After solidification, the gel was crushed into small pieces (50-100 mesh), washed thoroughly with isotonic NaCl solution followed by deionized water and then used for adsorption experiments.

To obtain basic information on the recovery of U using immobilized microbial cells, U adsorption-desorption cycle tests were carried out. It was previously shown (Sakaguchi et al., 1996) that the U retained on the adsorbent can easily be desorbed with dilute Na_2CO_3 solution, so 0.1 M Na_2CO_3 solution was used as the desorbent in this experiment.

Fifteen milliliters of a solution (pH 5.8) containing 42 μM U was adsorbed on a column (bed volume, 2 mL) of immobilized *Arthrobacter* cells at a space velocity of 20 h^{-1}. Adsorbed uranium was desorbed with 10 mL of 0.1 M Na_2CO_3 solution. The test was replicated six times.

The ability of the immobilized *Arthrobacter* cells to adsorb U did not decrease after six repetitions of adsorption-desorption cycles (figure 8). Thus, immobilized microbial cells appear to have excellent handling characteristics and can be used repeatedly in adsorption-desorption cycles.

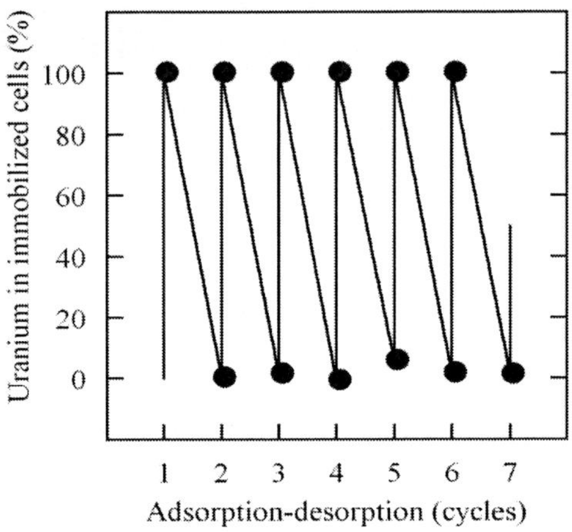

Figure 8. Test of repeated uranium adsorption-desorption using immobilized *Arthrobacter*, US-10 cells.

Absorption of U from U Refining Wastewater by Bacteria

As mentioned above, some microbial species have a high U absorbing ability, which suggests the possibility that they may be used for absorption of U from U mine tailings, U refining wastewater and other waste sources.

We attempted to absorb U from U refining wastewater sampled at the Ningyo-toge Environmental Engineering Center of the Japan Atomic Energy Agency using bacteria exhibiting a significant ability to absorb U. Resting cells (15.0 mg dry weight basis) were suspended in 100 ml of a solution (pH 6.0) of wastewater containing 21.0 mM U for 1 h at 25 ^{o}C.

Lactobacillus and *Bacillus* sp. isolated from Japanese U deposits removed 88.1 and 74.4% U, respectively (Table 4a), when solution pH was adjusted initially to 6.0. Solution pH gradually decreased, with *Bacillus* cells being more adversely affected by pH change than *Lactobacillus* cells. However, both strains quantitatively removed U when the pH was maintained at 6.0. These species can thus remove U from U refining wastewater with a high efficiency.

Attempt was also made to absorb U from U refining wastewater using immobilized microorganisms having a high ability to absorb U. Uranium refining wastewater (100 mL, pH 6.0) supplemented with 2.1 mM of U were adsorbed on a column (bed volume 2 mL) of immobilized bacterial cells, such as *Arthrobacter* sp. (US-10) and *Bacillus* sp. (US-9) at a space velocity of 10 /h at 25 ^{o}C.

Table 4a. U removal from U refining wastewater using microbial cells isolated from Japanese U mine

Strains	Removed U (%)	
	pH adjusted only started at pH 6.0	pH adjusted continuously at pH 6.0
Lactobacillus sp.	88.1	99.5
Bacillus sp.	74.4	95.5

Immobilized bacterial cells isolated from U mines in the United States can also remove U from the U refining wastewater with high efficiency (Table 4b).

Table 4b. Uranium removal from uranium refining wastewater using immobilized microorganisms isolated from uranium mines

Strains	Removed U (%)
Arthrobacter sp.	100
Bacillus sp.	100

ABSORPTION OF U FROM SEAWATER BY BACTERIA

The absorptions of U from seawater supplemented with 4.2 μM U using the bacteria isolated from U deposits were examined. The concentration of carbonate in seawater is ~2.34 x 10^{-3} M (Ogata et al. 1971). The amounts of U absorbed by *Chlorella* cells from solutions containing 1.196 x 10^{-3} M sodium hydrogen carbonate was less at pH values above 6 than at pH 5 (Nakajima et al.1979). The decrease in the amount of absorbed U from solutions containing carbonate was estimated from the amount of the UO_2CO_3 formed at pH 6 and of UO_2 $(CO_3)_3^{4-}$ formed at pH values greater than 7. (Nakajima et al.1979). Although *Lactobacillus* sp. absorbed 36.2% of U from seawater, it absorbed nearly twice as much (70.2%) when the seawater was decarbonated (figure 9). *Arthrobacter* and *Bacillus* cells, which can absorb large amounts of U from non-saline water, absorbed far less U from either seawater or decarbonated seawater than did *Lactobacillus*. Accordingly, *Lactobacillus* has great potential in applications to remove significant amounts of U from seawater.

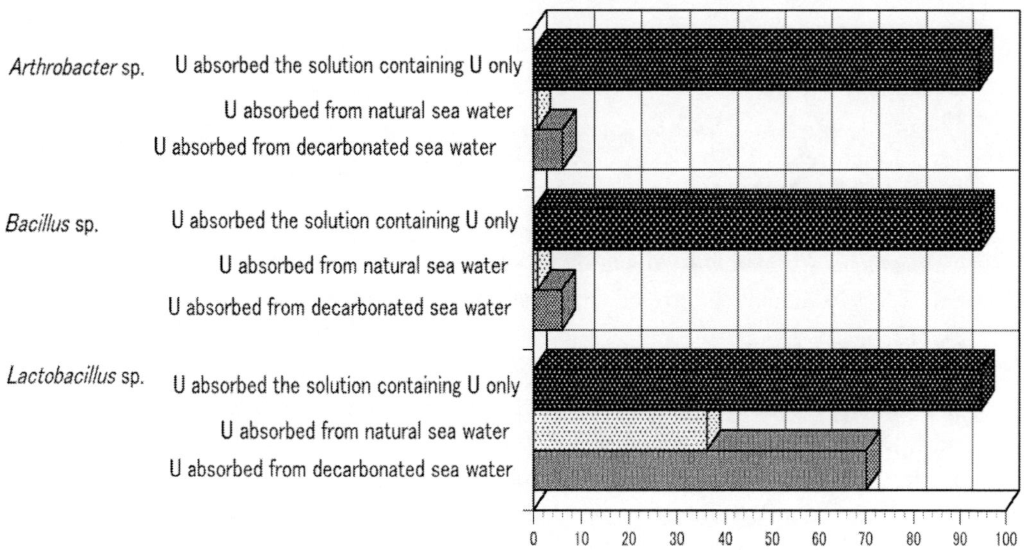

Figure 9. Removal of uranium using some microorganisms from seawater.

CONCLUSION

In U deposits located in Canada, the United States, Australia, and Japan, we isolated strains of bacteria such as *Arthrobacter, Bacillus* and *Lactobacillus* sp. having a significant ability to absorb U. These species could absorb approx. 2500 mmol U per gram dry wt. of microbial cells within one hour. These strains absorbed U selectively from solution containing six other heavy metals in solution. Cells also absorbed Th as well as U with high efficiency. These species removed U from U refining wastewater with high efficiency. *Lactobacillus* also absorbed U from seawater more effectively than other microbial cells which have high removing capacities, from non-saline U solution.

Cells immobilized with polyacrylamide gel have excellent handling characteristics and can be used repeatedly in U adsorption-desorption cycles.

These strains of *Arthrobacter, Bacillus* and *Lactobacillus* can be used as an absorbing agent for the absorption of nuclear fuel elements which may be present in nuclear fuel processing effluents, mine tailings, seawater and other environmental sources.

ACKNOWLEDGMENTS

I would like to thank Professor Tsuyoshi Hirajima, Keiko Sasaki, and Mr. Yuki Aiba of the Department of Earth Resources Engineering, Faculty of Engineering, Kyushu University, for their measurement of Zeta Potencials.

REFERENCES

Andres Y, Maccordick H J, Hubert J –C (1993) Adsorption of several actinide (Th, U) and lanthanide (La, Eu, Yb) ions by *Mycobacterium smegmatis*. *Appl Microbiol Biotechnol* 39:413-417.

Byerley J J, Scharer J M, Charles A (1987) Uranium (VI) biosorption from process solutions. *Chem Eng J* 36:B49-B59.

Friiss N, Myers-Keith P (1986) Biosorption of uranium and lead by *Streptomyces longwoodensis*. *Biotechnol Bioeng* 28:21-28.

Epstein E (1966) Dual pattern of ion absorption by plant cells and by plants. *Nature* 212:1234-1237.

Galun M, Keller P, Malki D et al (1983a) Recovery of uranium (VI) from solution using precultured *Penicillium* biomass. *Water Air Soil Pollut* 20:221-232.

Galun M, Keller P, Malki D, et al (1983b) Removal of uranium (VI) from solution by fungal biomass and fungal wall-related biopolymers. *Science* 219:285-286.

Gorab Z, Orlowwska B, Smith R W (1991) Biosorption of lead and uranium by *Streptomyces* sp. *Water Air Soil Pollut* 60:99-106.

Hu M Z -C, Norman J M, Faison B D et al (1996) Biosorption of uranium by *Pseudomonas aeruginosa* strain CSU: characterization and comparison studies. *Biotechnol Bioeng* 51:237-247.

Marques A M, Roca X, Simon-Pujol M D et al (1991) Uranium accumulation by *Pseudomonas* sp. EPS-5028. *Appl Microbiol Biotechnol* 35:406-410.

Nakajima A, Horikoshi T, Sakaguchi T (1979) Ion effects on the uptake of uranium by *Chlorella regularis*. *Agric Biol Chem* 43:625-629.

Ogata N, Inoue N, Kakihana H (1971) Collection of uranium in Sea-Water (X), *J Atomic Energy Soc Jpn* 13:560 (in Japanese).

Sakaguchi T (1998) Removal of uranium by using microorganisms isolated from Australian and American uranium deposites. *Environment and Innovation in Mining and Mineral Technology,* Sanchez M A, Vergara F, Castro S H Eds, University of Concepcion-Chile:181-191.

Sakaguchi T, Tsuruta T, Nakajima A. (1996) Removal of uranium by using microorganisms isolated from uranium mines. *Proc. Technical Solutions for Pollution Prevention in the Mining and Mineral Processing Industries:*183-191.

Shumate S E II, Strandberg G W, Parrott Jr J R (1978) Biological removal of metal ions from aqueous process streams. *Biotechnol Bioeng Symp* 8:13-20.

Strandberg G W, Shumate S E II, Parrott, Jr J R (1981) Microbial cells as biosorbents for heavy metals: accumulation of uranium by *Saccharomyces cerevisiae* and *Pseudomonas aeruginosa*. *Appl Env Microbiol* 41:237-245.

Tsuruta T (2006) Removal and recovery of uranium using microorganisms isolated from Japanese uranium deposits. *J Nucl Sci Technol* 43:896-902.

Tsezos M, Volesky B (1981) Biosorption of uranium and thorium. *Biotechnol Bioeng* 23:583-604.

White, C, Gadds G M (1990) Biosorption of radionuclides by fungal biomass. *J Chem Technol Biotechnol* 49 331-343.

In: Uranium: Characteristics, Occurrence and Human Exposure ISBN: 978-1-62081-207-5
Editors: Alik Ya. Vasiliev and Mikhail Sidorov

Chapter 12

ONE-GROUP FISSION CROSS SECTIONS FOR PLUTONIUM AND MINOR ACTINIDES INSERTED IN CALCULATED NEUTRON SPECTRA OF FAST REACTOR COOLED WITH LEAD-208 AND LEAD-BISMUTH EUTECTIC

G. L. Khorasanov* and A. I. Blokhin

State Scientific Center of the Russian Federation - Institute of Physics and Power Engineering named after A. I. Leypunsky, Obninsk, Kaluga Region, Russia

ABSTRACT

The chapter is dedicated to one-group fission cross sections of Pu and MA in LFRs spectra with the aim to increase these values by choosing a coolant which hardens neutron spectra. It is shown that replacement of standard Pb-Bi coolant over proposed Pb-208 coolant in the fast reactor RBEC-M, designed at Kurchatov Institute, Russia, leads to increasing the core mean neutron energy. As concerns Pu isotopes used as fuel, their one-group fission cross sections become slightly changed, while in the core spectrum Am-241 one-group fission cross section is changed more essentially. Another situation occurs in lateral blanket contained small quantities of minor actinides. It is shown that as a result of lateral blanket neutron spectrum hardening the one-group fission cross sections of Np-237, Am-241 and Am-243 increases up to 8-11%. This result permits to reduce the time of minor actinides exposition in FRs for their burning.

Keywords: Lead fast reactor, neutron spectra hardening, lead-208 as coolant, fission cross sections of Pu and MA

---

* E-mail: khorasan@ippe.ru.

1. INTRODUCTION

Since 1999 authors have proposed a new coolant for nuclear power installations with intermediate and fast neutron spectra [1]. This coolant is based on lead enriched with the lead isotope, lead-208, which is characterized with very low neutron capturing and slow neutron moderating. These features of lead-208 permit to have some gain in the quantity of plutonium [2] to reach criticality of fast reactor (FR) or in hardening the neutron spectra of the core and its lateral and topical blankets. Natural lead consisting from 52.3% of ^{208}Pb, 22.6% of ^{207}Pb, 23.6% of ^{206}Pb, 1.5% of ^{204}Pb and ^{208}Pb enriched up to 100% are assumed. In the present chapter the one-group fission cross sections for plutonium and minor actinides inserted in the calculated neutron spectra of the FR RBEC-M core and lateral blanket designed at Kurchatov Institute, Russia [3] are given. Calculations performed show that in this type of 900 $MW_{thermal}$ reactor core the replacement of its standard lead-bismuth coolant over ^{208}Pb proposed coolant leads to neutron spectrum hardening. As a result the MA one- group fission cross sections are growing.

2. METHODS OF SPECTRA CALCULATIONS AND RESULTS

The 900 $MW_{thermal}$ reactor core of the RBEC-M fast reactor design [3] is shown in Figure 1. Mixed uranium-plutonium nitride fuel $(U_{0.863}+Pu_{0.137})N$ is used, which is composed of power grade plutonium recovered from typical light water reactor spent fuel and depleted uranium with 0.1 wt. % of U-235. The core zones are surrounded by lateral (radial) and topical blankets; the structural material is steel; and the standard coolant material is lead-bismuth eutectic. Geometry of the calculation model of the RBEC-M reactor is shown in Figure 2. The model includes 12 physical zones, differing from each other by volume fractions and temperatures of materials.

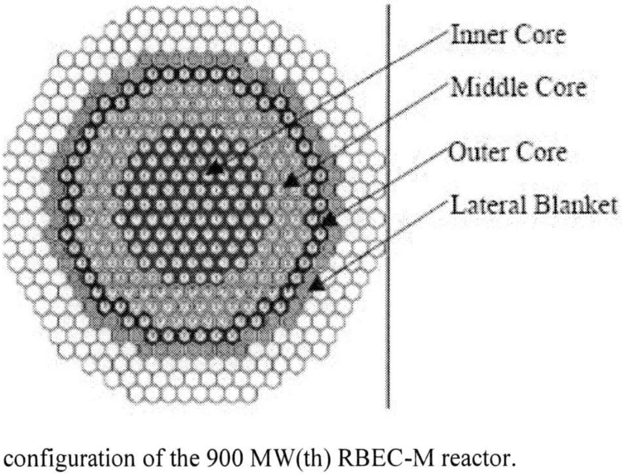

Figure 1. Actual core configuration of the 900 MW(th) RBEC-M reactor.

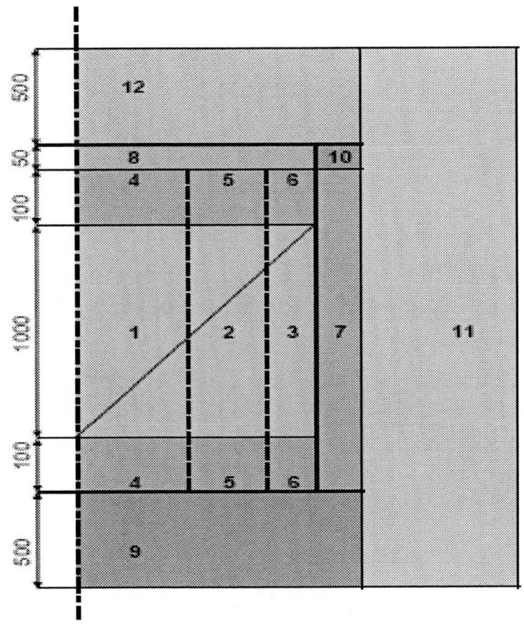

Figure 2. Geometry of calculation model of the RBEC-M reactor.

For cells 1, 2, 3 and 7 cooled with ^{208}Pb/ ^{nat}Pb-Bi the neutron spectra were calculated using MCNP5 code [4] and nuclear data library ENDF/B-VII.0. Calculations were performed in the ABBN- energy group structure system with 28 neutron energy groups [5].

On the base of calculated neutron spectra for several zones of the LFR core the mean energies of neutrons (< E_n >) for corresponding neutron spectra of the core zones were performed and presented in Table 1. Mean neutron energies are calculated for the RBEC-M core cells 1, 2, 3 and 7 cooled with ^{208}Pb and ^{nat}Pb-Bi.

Table 1. Neutron energy and one-group fission cross sections of fuel nuclides: Pu-238, -239, -240, -241, -242 and Am -241 averaged over the neutron spectrum of FR RBEC-M core (Subcore Cell 1 with small fuel enrichment) cooled with ^{208}Pb or ^{nat}Pb-Bi eutectic

Reactor and its coolant	Neutron energy averaged over core spectrum, MeV	Pu-238 one-group fission cross section, barns	Pu-239 one-group fission cross section, barns	Pu-240 one-group fission cross section, barns	Pu-241 one-group fission cross section, barns	Pu-242 one-group fission cross section, barns	Am-241 one-group fission cross section, barns
LFR cooled with ^{208}Pb	0.4246	1.1223	1.7091	0.4101	2.3079	1.6371	0.2882
LFR cooled with Pb-Bi	0.3992	1.0997	1.7030	0.3871	2.3122	1.6387	0.2629
Difference between data, %	+6.3627	+2.0551	+0.3582	+5.9416	-0.1860	-0.0010	+9.6234

As follows from Table I, the replacement of coolant from Pb-Bi over Pb-208 in RBEC-M leads to increasing the core mean neutron energy from its standard value 0.3992 MeV to 0.4246 MeV, i.e. on 6.4%. As concerns Pu isotopes, their one-group fission cross sections become slightly changed. More essentially Am-241 one-group fission cross section is changed, on 9.6% as a result of neutron spectrum hardening. It must be mentioned that 5-6 % of Am-241 is usually presented in the power grade plutonium obtained from spent fuel of PWR or VVER. Such mix of power grade plutonium and 5-6 % of Am-241 is usually proposed as a fuel for FR's core.

In Table 2 neutron energy and one-group fission cross sections of long-lived actinides in the Cell 7 – lateral blanket are given:

From Table 2 it follows that as a result of coolant changing from Pb-Bi over Pb-208 lateral blanket mean neutron energy increases from 0.2509 to 0.2662 MeV, i.e. on 6.1%. As a result of the neutron spectrum hardening the one-group fission cross sections of Np-237, Am-241 and Am-243 increases up to 8-11%. As is known these minor actinides need higher neutron energy, more than 0.1 MeV, to be fissile. In Figures 3, 4 and 5 the microscopic fission cross sections for these nuclides are given. The data are taken from ENDF/B-VII.0 library and presented in the 28 energy group ABBN system.

The Figures 3-5 presented really show that Np-237, Am-241 and Am-243 need high neutron energy, more than 0.1 MeV, to be fissile. Of course, they can be burned in using the low energy neutrons, around 1-10 eV, where their fission cross sections are also high enough, but at these neutron energies the cross sections of neutron capture are very high as, for example, is shown in Figure 6 for Am-241.

Thus, the result of increasing MA one-group cross sections in the hardened neutron spectrum is very important as it permits to reduce the time of minor actinides exposition in the FR lateral blanket for their burning and thus excluding from nuclear wastes.

Table 2. Neutron energy and one-group fission cross sections of long-lived actinides: Np-237, Am -241, -243 averaged over the neutron spectrum of FR RBEC-M lateral blanket cooled with lead-208 or lead-bismuth eutectic

Reactor and its coolant	Neutron energy averaged over blanket spectrum, MeV	NP-237 one-group fission cross section, barns	AM-241 one-group fission cross section, barns	AM-243 one-group fission cross section, barns
LFR cooled with Pb-208	0.2662	0.2308	0.1670	0.1167
LFR cooled with Pb-Bi	0.2509	0.2140	0.1521	0.1050
Difference between data, %	+6.0980	+7.8505	+9.8619	+11.1429

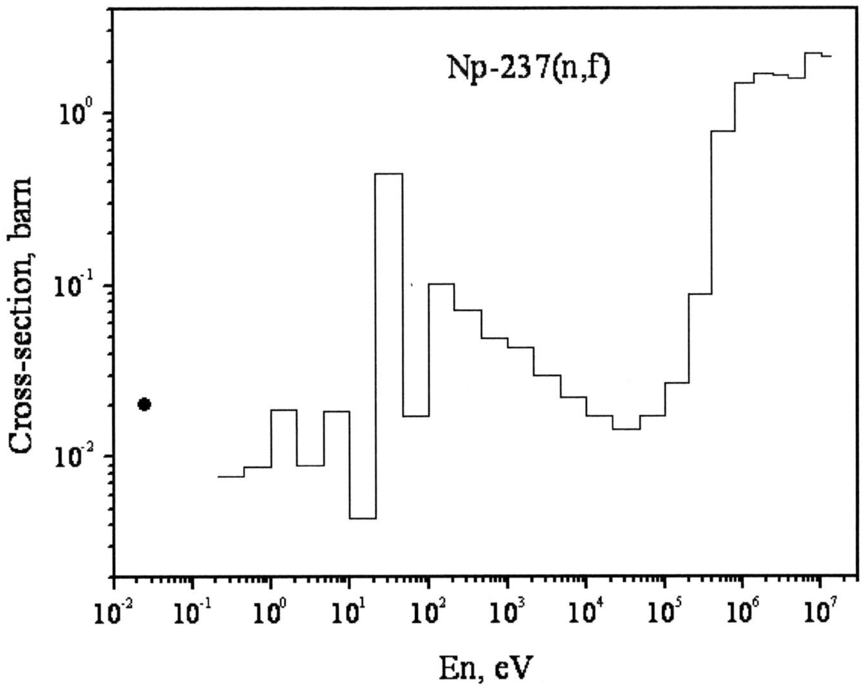

Figure 3. Microscopic fission cross section for Np-237.

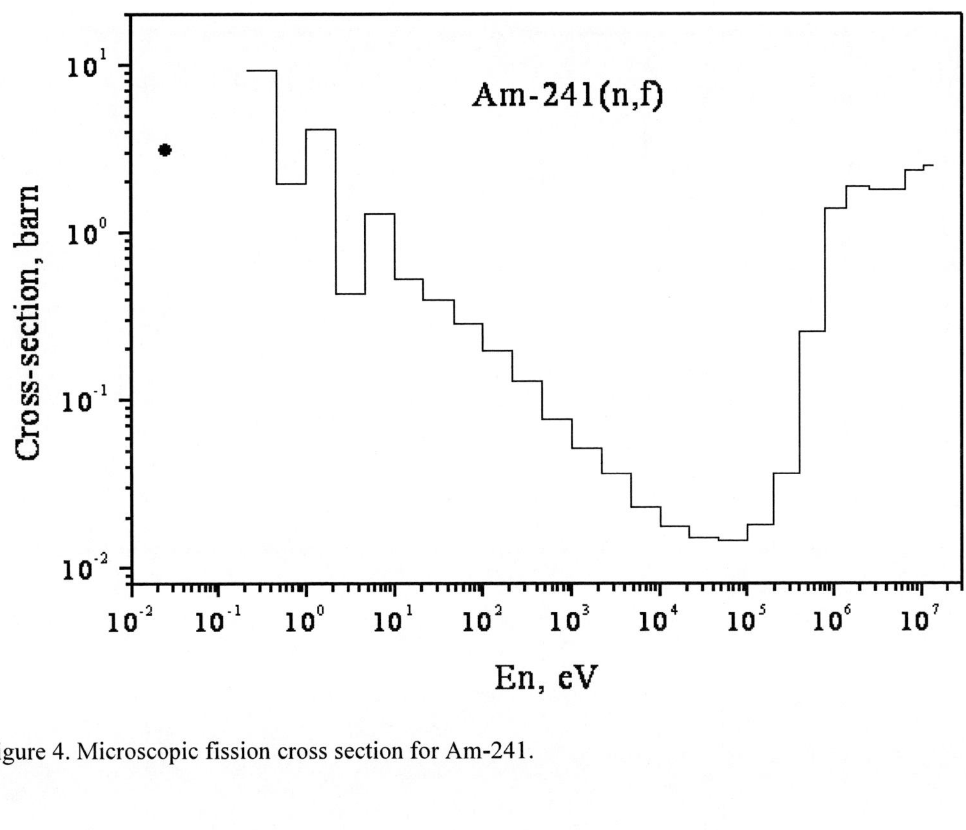

Figure 4. Microscopic fission cross section for Am-241.

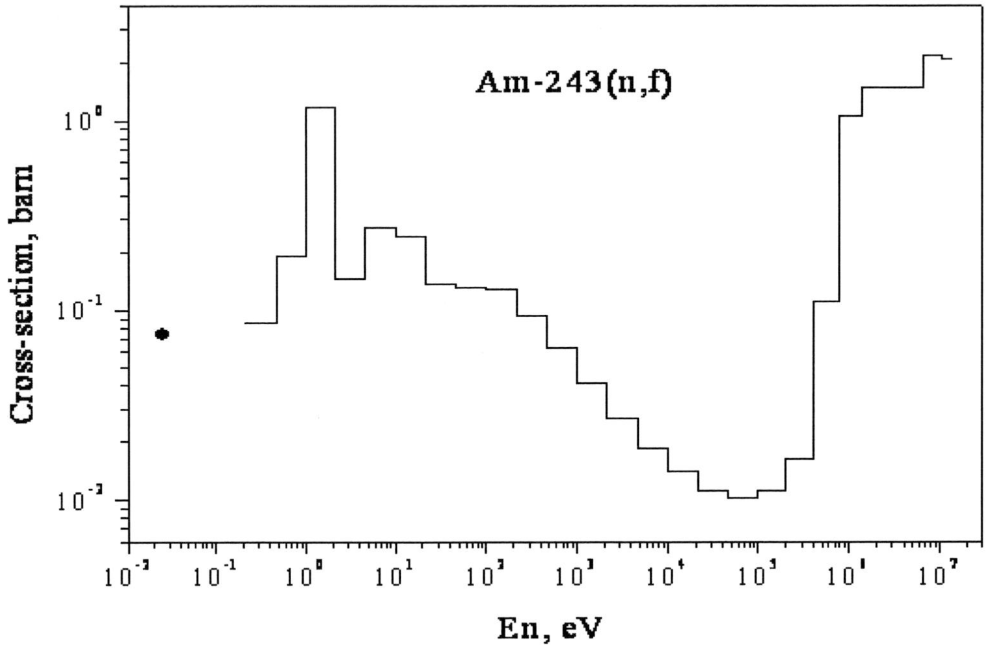

Figure 5. Microscopic fission cross section for Am-243.

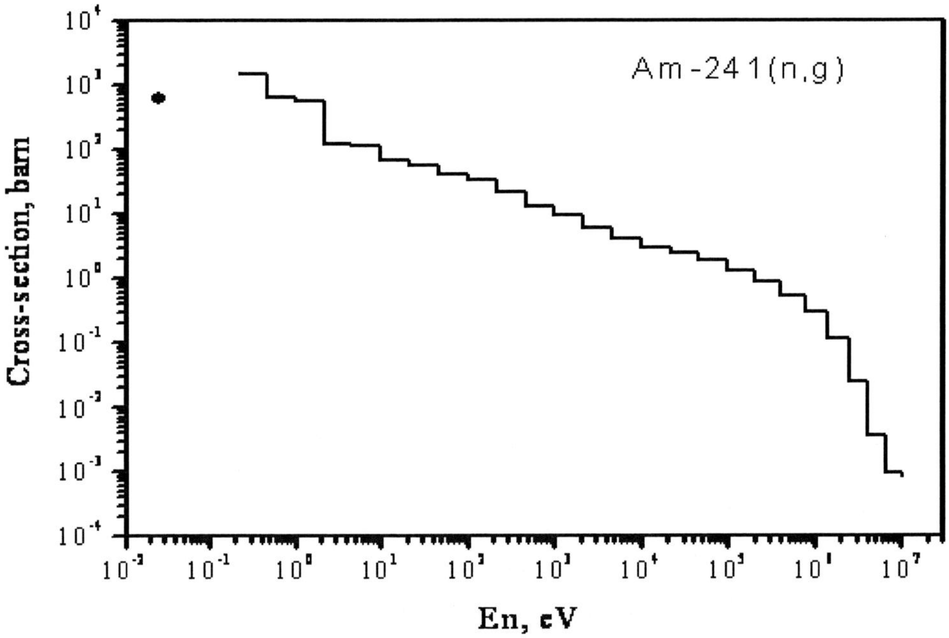

Figure 6. Microscopic cross section of radiation neutron capture by Am-241.

CONCLUSION

The chapter is dedicated to the proposal of using lead enriched with the stable lead isotope, lead-208, in FRs instead of lead natural or lead-bismuth eutectic.

It seems that unique neutron features of lead-208 make it as one of the best among the molten metal coolants now assumed for FRs and ADSs: sodium, lead-bismuth, lead natural and others [6].

The main advantage of lead-208 is its low neutron absorption ability: for neutron energies E_n<50 keV, the difference in the microscopic cross section of neutron capture values reaches 3-5 orders of magnitude as compared with lead natural or lead-bismuth. Averaged over neutron spectra of the LFR the one-group cross sections of neutron radiation capture for a coolant from lead-208 are by 5-6 times smaller than those for the coolant consisted from lead natural [7].

The second advantage of using lead-208 consists in achievement the core neutron spectra hardening on 6% due to low energy losses in it. Low neutron absorbing and slow moderating features of lead-208 permit to reach the gain in the multiplication factor K_{ef} on 2-3% for critical or subcritical core fueled with U-Pu mix. In this case to have the neutron multiplication factor K_{ef} =1.01 for the LFR or K_{eff} =0.97 for the ADS, both cooled with lead-208, the enrichment of power grade Pu in the U-Pu fuel can be reduced approximately on 0.7-0.8% [2].

In this chapter it is shown that replacement of coolant from Pb-Bi over Pb-208 in fast reactor RBEC-M, designed in Russia, leads to increasing the core mean neutron energy from its standard value 0.3992 MeV to 0.4246 MeV, i.e. on 6.4%. As concerns Pu isotopes, their one-group fission cross sections become slightly changed. More essentially, on 9.6%, Am-241 one-group fission cross section is changed as a result of neutron spectrum hardening.

Another situation occurs in the lateral blanket containing small quantities of minor actinides. It is shown that as a result of lateral blanket mean neutron energy increasing from 0.2509 to 0.2662 MeV the one-group fission cross sections of Np-237, Am-241 and Am-243 increases up to 8-11%. This result is very important as it permits to reduce the time of minor actinides exposition in FRs for their burning..

The possibility of using lead-208 as coolant in commercial fast critical or subcritical reactors requires a special considering, but relatively high content of this isotope in natural lead, 52.3%, and perspectives of using high performance photochemical technique of lead isotope separation permit to expect obtaining in future such a material in large quantities and under economically acceptable price [8,9].

REFERENCES

[1] G.L. Khorasanov, A.P. Ivanov, A.I. Blokhin, and A.L. Shimkevich. "Lead and tin targets for reducing polonium waste". *Proc. of the 3rd Int. Conf. on Accelerator Driven Transmutation Technologies and Applications*, Praha, Czech Republic, June 7-11, 1999, Published by ICARIS Ltd., p.46, Paper Tu-O-E8.

[2] G.L. Khorasanov, V.V. Korobeynikov, A.P. Ivanov, A.I. Blokhin. "Minimization of an initial fast reactor uranium-plutonium load by using enriched lead-208 as a coolant". *Nucl. Eng. and Design*, 239, No 9, pp. 1703-1707 (2009), (ISSN 0029-5493).

[3] P.N. Alekseev, K.O. Mikityuk, A.V. Vasiljev, P.A. Fomichenko, T.D. Shchepetina, S.A. Subbotin. "Optimization of Conceptual Decisions for the Lead-Bismuth Cooled Fast Reactor RBEC-M". *Atomnaya Energiya*, 97(2), pp.115-125 (2004); http://www.iaea.org/NuclearPower/ SMR/crpi25001 /html/).

[4] MCNP – A General Monte Carlo N-Particle Transport Code, Version 5. X-5 Monte Carlo Team. Los Alamos National Laboratory Report LA-UR-03-1987 (April, 2003).

[5] G.N. Manturov, M.N. Nikolaev, A.M. Tsiboulia, "Group constant system ABBN-93. Part 1: Nuclear constants for calculation of neutron and photon emission fields", *Issues of Atomic Science and Technology, Series: Nuclear Constants,* Issue 1, p. 59 (1996) (in Russian), (ISSN 0207-3668).

[6] G.L. Khorasanov and A.I. Blokhin. "A low neutron absorbing coolant for fast reactors and accelerator driven systems". In the book "Cooling Systems", Editor Aaron L. Shanley, Chapter 5, Edition: Nova Science Publishers, Inc., USA, 2011, (ISBN 978-1-61209-379-6).

[7] Georgy L. Khorasanov and Anatoly I. Blokhin. "Macroscopic cross sections of neutron radiation capture by Pb-208, U-238 and Tc-99 nuclides in the accelerator driven subcritical core cooled with molten Pb-208". *In CD-ROM Proceedings of the International Conference PHYSOR 2010 – Advances in Reactor Physics to Power the Nuclear Renaissance*, Pittsburgh, Pennsylvania, USA, May 9-14, 2010, Paper #286 at the Session 5C "Advanced Reactors Design".

[8] ISTC #2573 project: "Investigation of Processes of High - Performance Laser Separation of Lead Isotopes by Selective Photoreactions for Development of Environmentally Clean Perspective Power Reactor Facilities", Project Manager: A.M. Yudin (Saint-Petersburg, Efremov Institute, NIIEFA), Project Submanagers: G.L. Khorasanov (Obninsk, Leypunsky Institute, IPPE) and P.A. Bokhan (Novosibirsk, Institute for Semiconductor Physics, ISP), 2004-2005.

[9] Bortnyansky A.L., V.L.Demidov, S.A. Motovilov, F.P. Podtikan, Yu.I. Savchenko, V.A. Usanov, A.M. Yudin, B.P. Yatsenko, 2005. "Experimental Laser Complex for Lead Isotope Separation by means Selective Photochemical Reactions". *Proceeding of the X International Conference "Physical and Chemical Processes on Selection of Atoms and Molecules"*, Moscow, Russia, 3-7 October 2005, pp.76-82 (2005).

INDEX

#

20th century, viii, xi, 49, 88, 245, 270

A

absorption spectra, 123, 128
abuse, 79
accelerator, 328
access, xiii, 7, 33, 152, 175, 295
accessibility, xii, 110, 269, 273, 299
acetic acid, 76
acid, 4, 6, 18, 19, 44, 53, 55, 56, 60, 67, 76, 80, 85, 93, 106, 108, 110, 179, 194, 196, 202, 212, 228, 233, 247, 250, 251, 252, 258, 262, 298, 299
acidic, 20, 34, 37, 61, 67, 95, 101, 108, 236, 242, 250, 251, 255, 256, 262, 275, 276
acidity, 58, 63, 105, 202, 250, 307, 308
acquisitions, ix, 117
active site, 150
adjustment, 15, 273
adolescents, 208
adsorption, xiii, 6, 7, 8, 17, 42, 60, 67, 70, 77, 80, 82, 85, 106, 158, 179, 247, 305, 315, 316, 318
adults, 199, 252, 261
aerosols, 66, 68, 70, 71, 73, 100
Africa, v, vii, 1, 2, 3, 20, 35, 39, 42
agar, 306
age, 3, 16, 17, 35, 37, 71, 92, 93, 94, 95, 96, 98, 111, 156, 164, 166, 172, 173, 174, 175, 180, 181, 200, 215, 216, 217, 222, 223, 224, 229, 233, 241, 247
agencies, 52, 194
aggression, viii, 50, 66
aggressiveness, x, 155
agriculture, xii, 3, 246, 257, 264, 270
Alaska, 232, 233
alcohols, 19
algae, 159, 163
Algeria, vii, 2, 3, 20, 21, 22, 23, 30, 32, 35, 36, 38, 40
algorithm, 273
alkalinity, 224
alpha activity, 98
alternative energy, 205
alters, 118, 119, 146, 151
aluminium, 248, 252
alveoli, 73
amino, 71
amino acid(s), 71
ammonia, 18, 106, 298
ammonium, 18, 56, 195
amphibia, 251
amphibians, 251
amplitude, 274
anatase, 229
anxiety, 201, 205
aqueous solutions, ix, 18, 19, 20, 37, 44, 117, 119, 121, 122, 123, 125, 128, 130, 132, 136, 144, 145, 147, 151, 179
aquifers, viii, ix, 2, 3, 6, 7, 8, 10, 12, 13, 15, 16, 17, 20, 24, 28, 29, 31, 35, 39, 42, 47, 155, 167, 171, 174, 176, 177, 179, 180, 194
Armenia, 79
ARs, 28, 30, 31, 33, 35, 36
arsenic, 203
Asia, 204
assessment, xii, xiii, 70, 80, 85, 199, 245, 249, 263, 264, 270, 295, 297, 298, 300, 301, 302, 304
assets, 206, 264
asymmetry, 120
atmosphere, 7, 51, 100, 174, 201, 202, 203, 205, 299
atmospheric pressure, 108
atoms, 118, 150, 167, 168, 169, 173, 189
Au nanoparticles, 136, 141
authorities, xi, 245, 271
authority, 109, 203

B

Bacillus subtilis, 306
background radiation, 51, 52, 298, 302
backscattering, 38
bacteria, xiii, 7, 305, 306, 307, 314, 315, 316, 317, 318
bacterial cells, 312, 316, 317
bacterium, 252
Balkans, 70, 80, 85
bandgap, 146
banks, 204
Barbados, 38
barium, 90, 102, 107, 189, 190, 198
barriers, 60
base, 23, 56, 76, 99, 107, 218, 228, 323
batteries, 106
beams, 117, 118
beef, 306
Belarus, 198, 199, 200, 207, 208
Belgium, 56, 158, 187, 264
beta particles, 10, 192
bicarbonate, 7, 9, 164
bioaccumulation, 72, 249
bioavailability, 62, 63, 81, 84, 85, 267
biological processes, 51, 68
biomass, 113, 306, 318, 319
biopolymers, 318
bioremediation, 79
biosphere, 78
births, 74
bismuth, 138, 322, 324, 326, 327
blood, 74, 195
bloodstream, 73
boils, 197
bone, 71, 72
boreholes, 3, 28, 171, 172, 174, 180, 273, 275, 276, 277
Bosnia, 65, 81
brain, 72
branching, ix, 117, 121, 123, 127, 128, 135, 136, 138, 139, 144, 145, 150, 151
Brazil, 38, 44, 83, 185, 187, 206, 208, 224, 229, 232, 233, 296
breakdown, 64, 189, 190
Brittany, 91
bromine, 193
Bulgaria, 89, 204, 251
by-products, 52

C

Ca^{2+}, 54, 163, 172
cabbage, 258, 261
cables, 273
cadmium, 2, 71
calcium, 54, 57, 62, 102, 107, 164, 202, 248
calcium carbonate, 248
caliber, 64
calibration, 99
Cambrian, 31, 156, 217, 235
Cameroon, xiii, 45, 224, 229, 295, 296, 297, 298, 300, 303, 304
campaigns, 18, 98, 99
cancer, viii, 50, 71, 74, 79, 80, 82, 198, 200, 201, 208, 261
cancer death, 198, 201
carbon, 16, 166, 180, 257
catalyst, 119
cation, 19, 61, 257
cattle, 257
Caucasus, 158
CEC, 42, 257
cell surface, 308
Central Asia, 204, 302
Central Europe, xi, 211, 212, 213, 219
central nervous system, 72
ceramic, 66
cesium, 200, 201
CFR, 287
chain transfer, 261
challenges, 297
chelates, 76
chemical characteristics, 107, 199
chemical properties, 8, 34, 58, 59, 60, 63, 84, 119, 189, 251, 260
chemical reactions, 119
chemical stability, 119
chemicals, 298, 299
Chernobyl accident, 200, 207, 208
childhood, 200, 208
childhood cancer, 208
children, 73, 200, 208, 261
Chile, 233, 235, 319
China, xi, 187, 204, 211, 213, 214, 223, 224, 227, 236, 240
chlorine, 193
circulation, xii, 9, 95, 269, 270, 271, 279, 281, 284, 292
cities, 205, 260
citizens, 205
City, 200, 244
civilization, 78
cladding, 200
classification, xii, 30, 34, 35, 39, 94, 96, 99, 110, 213, 269, 303

clay minerals, 67, 199, 215, 238, 249
cleaning, 68, 249
cleanup, 74, 83
climate, 20, 23, 31, 35
closure, ix, 88, 89, 90, 111, 247, 248, 302
CO_2, 4, 7, 58, 165, 223, 241
coal, vii, viii, x, 49, 51, 52, 53, 185, 186
coatings, 218
color, 121, 162
combined effect, 7
combustion, viii, 49, 51, 52, 53, 66
commercial, 151, 194, 274, 327
communication, 190
communities, 260, 263
community, 109, 206, 260
compaction, 162
compilation, 46
complement, xii, 16, 270, 284, 292
complexity, 10, 17, 35, 36, 288
compliance, 88
composition, 8, 15, 22, 34, 41, 43, 58, 62, 64, 66, 73, 91, 99, 101, 104, 122, 126, 127, 136, 159, 162, 163, 164, 165, 166, 169, 170, 172, 176, 181, 198, 202, 207, 213, 223, 229, 231, 233, 241, 248, 251, 284
compounds, 51, 71, 76, 118, 121, 122, 190, 192, 193, 194, 196, 206, 212, 300
compression, 94, 246
computer, 38, 173, 293
concentration ratios, 46
conceptual model, 284, 285, 291, 292
Concise, 205
conductivity, 24, 28, 159, 249, 273, 275, 276, 279, 281, 283, 286, 287
conductors, 144
conference, 38
configuration, 146, 187, 322
confinement, 89, 262, 291
conflict, viii, 49, 65, 73, 80, 85, 203
Congress, 85, 114, 243
conservation, 12
consolidation, 90
constant rate, 23
constituents, 36, 55, 201
construction, 52, 75, 105, 203, 205, 262, 270, 302, 303
consumers, 266
consumption, 70, 105, 178, 179, 181, 194, 256, 260, 261, 300
consumption habits, 261
consumption patterns, 300
containers, 106
contaminant, 199, 314
contaminated sites, 70, 78
contaminated soil, 40, 46, 54, 60, 61, 73, 75, 77, 79, 83, 249, 266
contaminated soils, 46, 60, 75, 79, 83, 249, 266
contaminated water, xii, 71, 202, 246, 252, 254, 255, 256, 258, 259, 261, 269
contamination, viii, x, 49, 50, 56, 57, 68, 69, 71, 73, 78, 80, 81, 83, 84, 186, 198, 199, 202, 203, 247, 249, 251, 252, 255, 260, 262, 265, 266, 276, 287, 292
Continental, viii, 2, 3, 20, 21, 23, 24, 25, 27, 28, 29, 30, 32, 34, 35, 36, 40, 41, 47, 111
control group, 201
control measures, 203
cooling, 125, 197, 200
cooperation, 192, 209
coordination, 101
copper, 52, 102, 121, 141, 233
correlation, 63, 158, 182
correlations, 36, 301
corrosion, xi, 69, 81, 186
cortex, 69
cosmic rays, 51
cost, xi, 11, 186, 187
Council of Ministers, 264
covering, 128
critical density, 118
crop, 57, 218, 275
crops, 54, 61, 202, 217, 256, 258, 259, 261, 266
crust, 50, 51, 91, 92, 100
crystalline, 10, 11, 37, 43, 62, 102, 193, 222, 257, 274, 293
crystallization, 190, 196, 215
crystals, 128, 173, 229
CT, 20, 23
CTA, 262
cultivation, 57, 247, 256
culture, 60, 61
cycles, xi, xiii, 15, 211, 305, 315, 318
cytometry, 249
cytoplasm, 74
cytotoxicity, 249
Czech Republic, 211, 212, 213, 215, 243, 304, 327

D

damages, 204
danger, xi, 78, 177, 186, 203
data processing, 99
database, 76, 85, 103, 104, 187, 296, 297
deaths, 198, 201
decision makers, 75
decomposition, 159, 163

decontamination, viii, 50, 79
deconvolution, 274
deficiencies, 315
deficit, 12
deformation, 92, 94, 224, 232, 283
degradation, 19, 315
Denmark, 89
Department of Defense, 64, 83
Department of Energy, 204
Department of Health and Human Services, 79
depleted uranium ammunition, viii, 50, 71
deposition, xi, xii, 20, 32, 71, 118, 158, 179, 188, 198, 199, 211, 216, 224, 236, 239, 242, 260, 262, 269, 270, 299, 300
depression, 28, 33, 41, 159, 160, 164, 172, 174, 177, 178, 179, 180, 279
depth, xii, 21, 23, 24, 28, 30, 33, 34, 57, 69, 95, 156, 158, 163, 164, 179, 216, 218, 250, 252, 269, 270, 274, 277, 278, 279, 280, 281, 282, 283, 284, 289, 290, 291, 292
derivatives, 19
desorption, xiii, 77, 106, 305, 315, 316, 318
destruction, 76, 90, 156, 179
detectable, 68
detection, 18, 62, 97, 277
deviation, 27, 122, 123, 127, 128, 130, 131, 133, 135, 136, 138, 140, 144, 150
diet, 72, 260, 261, 300
dietary intake, 199
differential equations, 169
diffusion, x, 10, 11, 100, 186, 196
diffusion process, 11
dipoles, 273
disaster, 201
discharges, 158, 166
discordance, 23
discs, 19
diseases, 71
disequilibrium, 2, 8, 9, 10, 11, 12, 15, 30, 35, 36, 39, 40, 41, 42, 43, 44, 45, 46, 47, 181, 182, 296
dislocation, 54
dispersion, 17, 29, 202, 205
displacement, 10
disposition, 298
dissolved oxygen, 15
distilled water, 19
distribution, vii, viii, x, 2, 5, 12, 19, 20, 30, 35, 36, 42, 53, 55, 58, 60, 62, 69, 72, 73, 83, 108, 114, 120, 155, 156, 158, 160, 166, 179, 180, 185, 204, 235, 259, 266, 284, 296, 301, 303
distribution function, 120
diversity, ix, 87, 112, 119
DNA, 71, 82, 249
DNA damage, 71, 82, 249
DOC, 81
dogs, 194
DOI, 266
donor countries, 204
dosage, 70
drainage, 39, 107, 170, 202, 247, 251, 252, 262, 271, 284, 286, 292
drawing, 191
drinking water, 3, 43, 46, 50, 51, 52, 69, 164, 178, 179, 180, 194, 202, 252, 267
drought, 250
drying, 106
dusts, 195
dykes, 96, 214, 216, 218, 223, 231, 235, 241, 286, 289, 290
dynamism, 166

E

earthworms, 249, 265
ECM, 262
economic development, xiii, 295
economic evaluation, 195
economics, 303
ecosystem, viii, 49
ecotoxicological, 251
editors, 182
education, 88
effluent, 107, 248, 249, 255, 265
effluents, xiii, 262, 305, 306, 318
egg, 251
Egypt, 32, 39
elaboration, 186
electric conductivity, 248, 250
electric field, 118, 146, 147, 148, 149, 150
electrical conductivity, 202, 251, 273, 277, 281, 286, 287
electricity, 186, 197, 296, 303
electrodeposition, 18, 19, 20, 41, 45, 47
electrodes, 20, 273
electrolysis, 307
electrolyte, 19, 20
electromagnetic, xii, 148, 151, 196, 270, 271, 273, 277, 284
electromagnetic fields, 151
electron, 118, 150
electrons, 118, 119, 121, 123, 146
elongation, 91
e-mail, 49
emergency, 197, 198, 200, 201
emission, 8, 10, 73, 100, 118, 123, 124, 135, 136, 138, 139, 140, 141, 151, 190, 192, 195, 301, 328

emission field, 328
emitters, 20, 300, 301, 304
employees, 198, 205
employment, 88
encapsulation, 75
endangered, viii, 50, 66
energy, vii, ix, x, xiii, 18, 19, 20, 51, 52, 64, 66, 78, 87, 88, 90, 117, 118, 120, 135, 136, 137, 138, 141, 144, 146, 149, 150, 185, 186, 195, 196, 198, 271, 293, 295, 296, 301, 306, 321, 323, 324, 327
energy emission, 18
energy supply, 296
engineering, 198
England, 37, 38, 90
environment, viii, x, 4, 5, 7, 8, 10, 11, 12, 40, 45, 50, 51, 52, 53, 56, 57, 58, 60, 65, 66, 68, 74, 78, 79, 80, 82, 83, 89, 99, 100, 101, 108, 109, 111, 119, 121, 146, 183, 185, 186, 194, 199, 204, 247, 249, 250, 251, 255, 262, 264, 296, 298, 299, 302, 303
environmental aspects, 243
environmental change, xi, 245
environmental conditions, 158, 179, 192, 251
environmental contamination, viii, 45, 49, 50, 54, 203
environmental degradation, viii, 49, 203, 263
environmental effects, 88
environmental impact, xi, xii, 107, 200, 202, 203, 204, 245, 246, 257, 263, 270, 291, 299
environmental issues, 303
environmental protection, 62, 78, 194, 203, 306
Environmental Protection Agency, 52, 113, 203, 261
environmental quality, 57, 204
environments, xi, 2, 4, 5, 8, 9, 17, 44, 54, 108, 188, 211, 212, 274, 284, 292
eosinophils, 195
EPA, 52, 80, 109, 113, 203, 287
epidemiology, 260
EPS, 319
equilibrium, vii, ix, 1, 2, 11, 15, 17, 31, 43, 47, 100, 117, 122, 123, 126, 127, 128, 130, 131, 135, 136, 138, 140, 144, 145, 148, 149, 150, 151, 159, 164, 167, 168, 169, 170, 177, 181, 192, 297, 299
equipment, 20, 99, 273
erosion, 57, 80, 159, 163, 164, 172, 199, 251, 262
ESR, 182
etching, 11, 12
ethers, 19
Eurasia, 182
Europe, 74, 88, 89, 91, 111, 114, 198, 260
European Union, 257, 264
evacuation, 198, 200
evaporation, 8, 9, 23, 38, 52, 136, 251, 263
evidence, 38, 39, 41, 45, 71, 72, 190, 200, 205, 278, 300
evolution, 12, 37, 40, 45, 88, 113, 181, 223, 232, 234
EXAFS, 37, 46
excitation, 146, 149
execution, 304
experimental condition, 150
exploitation, viii, xii, 2, 51, 246, 247, 248, 269, 274, 306
exporters, 204
external influences, 126
extraction, 17, 18, 19, 56, 62, 69, 75, 84, 85, 106, 163, 174, 195, 196, 205, 298, 299
extrusion, 71, 235

F

facies, 21, 24, 101, 215, 232, 252
families, 100
farmers, 258, 259
Federal Republic of Yugoslavia, 65
fencing, 262
fertility, 70
fertilization, 83, 202, 259
fertilizers, viii, 49, 50, 53, 54, 55, 56, 57, 58, 60, 62, 79, 81, 82, 83, 84, 86, 194, 258
fiber, 85
filters, 52
filtration, 18, 106, 163, 172, 306
financial, 79, 262
financial resources, 79
financial support, 262
fingerprints, 12, 24, 27, 36
Finland, 8, 45, 89
fires, 68
fish, 199, 207
fission, x, xiii, 10, 151, 186, 188, 190, 192, 195, 197, 198, 200, 205, 207, 296, 321, 322, 323, 324, 325, 326, 327
fixation, 42, 60
flame, 64, 66
flank, 226, 233
flight, 66
flights, 65
flooding, 247
flotation, 53, 196
fluctuations, 250
fluid, 106, 215, 216, 218, 222, 223, 226, 229, 232, 236, 241, 279
fluorescence, 102, 113
fluorimeter, 99
fluorine, 202, 248
foils, 122, 127, 144, 145, 150, 151

food, viii, 50, 58, 60, 62, 70, 73, 202, 203, 205, 256, 261, 263, 300
food chain, viii, 50, 58, 60, 70, 73, 202, 203, 205, 261, 263
force, 119
formation, 3, 10, 23, 45, 58, 66, 77, 89, 94, 109, 115, 118, 121, 123, 135, 141, 158, 159, 164, 174, 175, 179, 181, 182, 190, 193, 224, 226, 229
formula, 178
fractures, 34, 37, 94, 95, 97, 107, 229, 241, 271, 281, 283, 286, 290, 291
fragments, 7, 64, 68, 162, 198, 233, 277
France, 38, 41, 45, 47, 80, 89, 91, 115, 158, 187, 212, 213, 214
freshwater, vii, x, 185, 186, 249
FTIR, 46
FTIR spectroscopy, 46
Fukushima, xiii, 200, 201, 206, 295
full capacity, 164
fungi, 86, 306
fusion, 91, 92, 111, 118, 195

G

Gabon, 8, 45
gamma radiation, 100, 109, 192, 271, 272, 288
gamma rays, 199, 200
gel, xiii, 95, 158, 305, 315, 318
genetic predisposition, 71
genus, 251
geology, 40, 99, 166, 270
geometry, 10, 98, 148, 150, 279, 289
Georgia, 79
Germany, 81, 83, 89, 90, 158, 187, 204, 205, 213, 215, 243, 259, 265, 304
global climate change, 296
glucose, 71
gold nanoparticles, 128, 129
governments, 296, 303
grades, xi, 88, 98, 107, 211, 223, 224, 231, 233, 235
grading, 237
grain size, 105, 106, 112
graphite, 198, 219, 228
grass, 199, 258, 266
grasses, 265
gravity, 235
Great Britain, 44, 158
grids, 98
growth, 249, 251
guardian, 80
guidance, 53
guidelines, 80, 203
Gulf Coast, 13
Gulf of Mexico, 42

H

half-life, 2, 10, 15, 18, 50, 73, 100, 123, 126, 135, 138, 140, 141, 143, 145, 147, 167, 192, 195, 199
halogen, 156
hardness, 248
harvesting, 58
hazards, vii, x, 85, 114, 185, 203, 204, 208, 246, 266, 298, 301, 302
health, xi, 50, 57, 58, 67, 69, 71, 72, 73, 74, 79, 82, 83, 85, 88, 109, 186, 194, 201, 257, 258, 260, 261, 263, 264, 296, 302
health condition, 73
health effects, 71, 79, 195, 260, 261, 263
health problems, 71
health risks, 71, 73, 88, 258
health status, 71, 74, 82
heavy metals, 6, 71, 76, 276, 298, 306, 314, 318, 319
height, 146, 147, 156, 179
helium, 37, 181
hemisphere, 198, 199
hemoglobin, 195
heterogeneity, 21, 57, 288
hexavalent, vii, 1, 3, 4, 10, 96, 100, 108
history, xi, 35, 39, 111, 166, 212, 224, 245
HM, 304
holding company, 270
Holocene, 156, 174, 180
homes, 109
homogeneity, 36, 291
host, ix, 15, 17, 29, 32, 33, 87, 159, 217, 218, 219, 222, 223, 224, 225, 226, 233, 235, 241
hot spots, 69
housing, 136
human, vii, viii, x, xiii, 50, 51, 52, 68, 71, 72, 74, 79, 80, 185, 186, 193, 194, 199, 203, 204, 207, 257, 259, 260, 263, 264, 266, 295, 299, 302
human activity, 52
human body, 50, 71, 72, 207
human development, 79
human exposure, vii, 52, 186, 199, 203, 299
human health, viii, 50, 68, 71, 74, 80, 195, 204, 260, 263, 264, 266, 302
humidity, 62, 250
humus, 60, 63, 108
Hungary, 204
Hunter, 80
hybrid, 78
hybridization, 75
hydrogen, xi, 108, 186, 193, 200, 317
hydrogen fluoride, xi, 186

hydrogen gas, 200
hydrogen sulfide, 108
hydrological conditions, 200
hydrolysis, 158, 164, 172, 179
hydroponics, 60
hydrosphere, 8, 181
hydrothermal activity, 225, 236
hydrothermal process, 212
hydroxide, 18, 34, 67, 107, 158, 308, 314
hydroxyapatite, 37
hydroxyl, 4, 6, 54
hydroxyl groups, 6
hypertension, 195
hypothesis, 11, 12, 17, 33, 120

I

ideal, 60
identification, xii, 35, 100, 269, 297
images, 121
immersion, 107
immobilization, 76, 77, 78, 89, 108
impact assessment, 71
imports, 53
impregnation, 90, 98
improvements, 105
impulses, 68
impurities, 84, 190
incidence, 201
India, 38, 46, 158, 187, 205
indirect measure, 140
individuals, 199, 301
industrialization, viii, 49
industries, 52, 53, 206, 277
industry, viii, xi, 49, 50, 51, 52, 54, 55, 78, 90, 112, 113, 114, 193, 198, 212, 243, 245, 296
inferences, 112
infrastructure, 299
ingest, 73
ingestion, 70, 194, 195, 260, 261, 267, 300, 301
ingredients, 215
initial state, 3, 16
initiation, 118, 128, 140, 141
institutions, viii, 49
integration, 202, 277
interface, 10, 11, 17, 31, 120, 121, 278, 279, 282, 283
International Atomic Energy Agency, x, 39, 46, 81, 113, 185, 192, 205, 208, 209, 265, 296, 302, 303, 304
International Energy Agency, 296
International Energy Agency (IEA), 296
international standards, 107
intervention, 99, 108, 109, 176, 178, 180, 262
intrusions, 91, 104, 224, 241, 242
inversion, 273, 293
investment, 76, 104, 111
iodine, 163, 182, 193, 198, 201
ion-exchange, 67
ionization, 123, 146, 150
ionizing radiation, 54, 58, 79, 85, 198, 200, 264, 303
ions, vii, 1, 4, 5, 6, 7, 8, 9, 13, 16, 34, 58, 62, 66, 67, 73, 77, 78, 95, 101, 108, 119, 125, 135, 136, 150, 151, 192, 308, 312, 318
Iraq, 65, 85, 90
Iraq War, 65
Ireland, 44, 85
iris, 265
iron, 11, 67, 76, 80, 93, 105, 108, 162, 224, 233, 257
irradiation, 71, 100, 120, 136, 189, 197, 198
irrigation, 106, 202, 252, 254, 256, 257, 258, 259, 262
isolation, 18, 108, 299, 301
isotherms, 310
isotope, xi, 2, 8, 14, 18, 20, 29, 35, 36, 39, 42, 47, 50, 63, 64, 100, 122, 123, 128, 136, 138, 140, 143, 148, 159, 166, 167, 169, 177, 180, 181, 186, 189, 190, 192, 236, 241, 296, 322, 326, 327
Israel, 42, 56
issues, 85, 182, 246, 252, 263, 279, 296
Italy, 47, 85, 89, 158

J

Japan, xiii, 57, 58, 90, 187, 305, 306, 307, 316, 318
Jordan, 41

K

K^+, 9, 163
Kazakhstan, 187, 204, 296, 302
ketones, 19
kidney, 71, 72, 73, 267
kidneys, 50
kinetics, 47, 166
Kola Peninsula, 158
Korea, 187
Kosovo, viii, 50, 65, 66, 74, 80, 85
Kuwait, 65
Kyrgyzstan, 204

L

Lactobacillus, xiii, 305, 306, 307, 308, 309, 310, 311, 312, 313, 314, 315, 316, 317, 318

lakes, 199, 207, 208
landscape, 52, 202, 250, 251, 262
lanthanide, 318
lanthanum, 190
larvae, 251
laser ablation, 118, 119, 121, 125, 136, 149, 152
laser radiation, 118, 119, 120, 121, 122, 123, 125, 127, 128, 130, 131, 135, 136, 138, 141, 143, 145, 146, 148, 149, 151
lasers, 117, 120, 128, 151
lattices, 2
leaching, vii, ix, x, 1, 8, 10, 11, 12, 13, 14, 29, 32, 33, 34, 36, 40, 42, 57, 76, 87, 90, 105, 106, 108, 110, 113, 155, 177, 180, 196, 235, 247, 250, 251, 252, 255, 275, 298, 299
lead, 2, 36, 42, 46, 47, 57, 64, 66, 70, 119, 120, 136, 138, 141, 148, 151, 194, 203, 271, 296, 297, 300, 318, 321, 322, 324, 326, 327
leaks, xi, 186
legislation, viii, 49, 79, 246, 257
lens, 141, 226, 302
leukemia, 200
lichen, 69
lifetime, xi, 3, 16, 186, 198, 199, 261
light, 117, 119, 126, 128, 135, 137, 146, 147, 148, 189, 198, 322
limestone, 8, 23, 24, 33, 37, 228
linear dependence, 55
liquid phase, 8, 10, 11
liquids, 119, 120, 121, 147, 149, 152
lithology, vii, 1, 21, 24, 31, 33, 35, 100, 108, 167, 182, 286
liver, 50
livestock, 53, 202, 246, 252, 259
Louisiana, 56
low temperatures, 4, 41, 43, 113
luminescence, 123, 212
Luo, 45
lying, 174
lymphocytes, 195
lymphoma, 72
lysis, 200

M

magnesium, 102, 105, 106, 202
magnetic field, 148, 149, 150
magnetic field XE "magnetic field" s, 150
magnitude, x, 3, 7, 8, 16, 17, 28, 31, 68, 118, 122, 145, 146, 150, 158, 185, 200, 201, 203, 257, 327
majority, 17, 260, 296
man, 51, 68, 261, 263, 293, 296
management, 51, 74, 76, 78, 89, 109, 186, 203, 208, 264, 265, 298, 299, 304
manganese, 11, 108, 247, 248, 252
mantle, 91, 241
mapping, 97, 98, 102, 279, 293
marginalization, 172
marsh, 162
Marx, 6, 42
MAS, 206
mass, 12, 29, 64, 70, 90, 119, 168, 198, 301
materials, viii, ix, 8, 13, 49, 51, 52, 53, 57, 71, 73, 75, 76, 77, 78, 80, 89, 108, 117, 181, 193, 195, 196, 197, 199, 203, 207, 247, 250, 251, 262, 284, 285, 287, 288, 291, 299, 302, 322
matrix, 10, 11, 17, 34, 198, 199, 233, 236
matter, 18, 58, 74, 88, 108, 300
measurement, 20, 74, 102, 110, 128, 178, 197, 199, 261, 273, 297, 318
measurements, 11, 17, 69, 98, 99, 100, 102, 109, 123, 126, 127, 131, 133, 134, 136, 138, 143, 147, 177, 198, 272, 288, 297, 300, 303, 304
media, 37, 74, 108, 147, 296
medical, 71, 73, 212, 298
medicine, 51
Mediterranean, 23, 24, 35
melt, 91, 119, 121
melting, 66, 91, 120, 159, 164, 172, 175, 180, 192, 197
melting temperature, 66, 120
melts, 91
mercury, 71
Mesozoic era, xi, 156, 179, 211
metal ion, 312, 319
metal nanoparticles, 118, 119
metallic nanoparticles, ix, 117, 123, 151
metals, 43, 52, 76, 84, 100, 109, 121, 146, 213, 246, 249, 250, 257, 265
metasediments, ix, 87, 88, 92, 96, 97, 98, 105, 110, 111, 215, 218, 224, 225, 227, 228, 229, 232, 246
meter, 50, 162, 273
methodology, 3, 20, 73, 74
Mexico, 181
Mg^{2+}, 163, 172
microbial cells, xiii, 305, 307, 308, 310, 313, 315, 316, 318
microcrystalline, 225
micrograms, 29, 50
microorganisms, xiii, 305, 306, 307, 316, 317, 319
microscope, 68
migration, 6, 15, 17, 41, 46, 47, 60, 67, 70, 108, 200, 216, 274, 302
militarization, viii, 49
military, 68, 69, 71, 73, 88, 186

mine soil, 306
mineral water, x, 155, 164, 174, 177, 178, 179, 181
mineralization, xi, 6, 28, 33, 81, 88, 92, 94, 95, 96, 97, 100, 102, 156, 172, 175, 177, 181, 188, 211, 212, 215, 216, 218, 219, 222, 223, 224, 225, 226, 227, 228, 229, 231, 233, 235, 236, 237, 238, 241, 242, 287
mining companies, ix, 88, 111
Ministry of Education, 79
Miocene, 24, 236
miscarriages, 74
mission, viii, 50, 69
Missouri, 31, 38
mixing, 3, 10, 12, 13, 14, 28, 32, 33, 34, 36, 190, 202, 205, 236
modelling, 12, 38, 42, 43, 45, 80
models, xii, 11, 12, 17, 35, 63, 92, 207, 270, 273, 274, 280, 284, 285, 292, 300, 304
modifications, 173, 195
moisture, 51
molecular weight, 76
molecules, 100, 196
molybdenum, 52
monolayer, 11
Montenegro, 65, 70, 85
morbidity, 82
mortality, 82
Moscow, 156, 162, 163, 172, 181, 182, 183, 204, 207, 243, 244, 328
motivation, 119
MR, 81, 266
multiplication, 327

N

Na^+, 9, 163, 172
NaCl, 216, 217, 218, 222, 223, 226, 233, 236, 241, 306, 315
Namibia, 187, 296
nanometer, 119
nanometer scale, 119
nanometers, 121
nanoparticles, ix, 117, 118, 121, 123, 140, 141, 144, 146, 148, 151, 152
nanostructures, 121, 125, 147
National Academy of Sciences, 52
national product, viii, 50
National Research Council, 80
native species, 249
NATO, viii, 49, 50, 65, 66, 68, 74, 79, 83, 85
natural compound, 101, 181
natural gas, 52
natural radioactivity, viii, xiii, 49, 52, 83, 99, 109, 115, 295
Nd, 98, 123, 124, 126, 128, 130, 131, 133, 134, 136, 140, 141, 143, 152
negative consequences, 68
negative effects, 71
neodymium, 129
Netherlands, 81
neutral, 60, 107, 108, 287
neutrons, x, 112, 149, 151, 186, 188, 189, 190, 323, 324
neutrophils, 195
New Zealand, 202, 252, 254, 263
nitrogen, 4
noble metals, 119, 147
nodules, 188
non-nuclear technologies, viii, 49
non-saline U solutions, xiii, 305
North Africa, 20, 22, 35, 54
North America, xiii, 39, 260, 305, 306
North-Western Sahara Aquifer System (NWSAS), vii, 1, 3, 20
NRC, 53
nuclear program, 303
Nuclear Regulatory Commission, 266
nuclear weapons, 51, 78, 186, 296
nuclei, 40, 123, 141, 143, 146, 147, 148, 149, 150, 151, 186, 189, 190
nucleus, x, 74, 145, 186, 189, 191, 197
nuclides, ix, x, 2, 8, 44, 117, 121, 122, 123, 126, 127, 128, 131, 133, 134, 135, 137, 138, 139, 140, 141, 143, 144, 145, 148, 150, 151, 178, 186, 194, 323, 324, 328
null, 287
nutrient, 60
nutrients, 62, 70
nutrition, 58, 59, 62

O

oceans, 9, 50, 199
OECD, 205, 304
officials, 204
OH, 54, 102, 193, 247, 308
oil, viii, 33, 49, 51, 52, 58
Oklahoma, 31
one-group fission cross section, xiii, 321, 322, 323, 324, 327
open spaces, 212, 242
operations, 53, 83, 107, 108, 178, 197, 204, 205, 249, 260, 298, 299, 302

ores, ix, 87, 88, 95, 96, 102, 105, 106, 110, 112, 115, 194, 196, 202, 203, 212, 215, 217, 224, 239, 298, 299, 301
organ, 73, 301, 303
organic compounds, 62, 67, 76, 205
organic matter, 4, 7, 58, 62, 108, 115, 158, 163, 166
organic soils, 42
organic solvents, 106
organism, 72, 73, 249
organize, 166
organs, 54, 62, 71
overlap, 162
oxalate, 190
oxidation, vii, 1, 2, 3, 4, 5, 10, 37, 39, 51, 66, 79, 96, 105, 108, 125, 156, 158, 159, 166, 171, 179, 180, 192, 193, 196, 223, 232, 252, 300
oxygen, 4, 159, 179, 200, 223

P

Pacific, 80, 82, 200, 223, 240
parallel, 92, 188, 218
parity, 150
pasture, 62, 199
pastures, 246
pathways, 67, 68, 73, 145, 199, 300
peat, 108, 158
Pentagon, 65, 82
percolation, 23
permeability, 105, 159, 276
permit, 322, 327
Peru, 233
petroleum, 52
Petroleum, 38
phase boundaries, 10
phosphate, vii, viii, x, 1, 5, 7, 45, 49, 50, 51, 52, 53, 54, 55, 56, 57, 58, 60, 61, 62, 77, 80, 81, 82, 83, 84, 86, 101, 102, 185, 186, 188, 194
phosphates, 7, 54, 66, 108, 242, 247
phosphorous, 81
phosphorus, 54, 55, 57, 58, 60, 61, 62, 63, 70, 83, 85, 86, 188, 194
photons, 117, 118, 123, 133, 136, 137, 146
physical phenomena, 151
physical properties, 2, 271, 284, 291, 292
physicochemical properties, 205
physics, 167, 190
phytoremediation, 46, 75, 76, 77, 266
pigmentation, 251
pitch, 156
plant growth, 258
plants, xi, 46, 54, 56, 57, 60, 62, 63, 68, 70, 71, 75, 76, 77, 78, 79, 80, 83, 84, 86, 106, 186, 199, 202, 203, 251, 257, 258, 259, 263, 265, 266, 298, 318
plasma proteins, 73
plasmon resonance, ix, 117, 118, 119, 123, 128, 131, 141, 146
platform, 22, 158, 227
platinum, 19
Pliocene, 23, 156, 179, 188, 233
plutonium, x, 44, 46, 185, 199, 322, 324, 327
PMMA, 144
Poland, 89
polarization, 98, 148
policy, 85
pollutants, 42, 75, 76
pollution, xi, 245, 262
polonium, 188, 260, 327
polyacrylamide, xiii, 305, 315, 318
ponds, 107, 263
pools, 62
population, xi, 51, 58, 68, 71, 74, 79, 82, 88, 112, 178, 194, 199, 245, 261, 274, 300
population density, 274
porosity, 168, 171, 172, 173, 291
porous media, 47
Portugal, v, vii, viii, xi, xii, xiii, 87, 88, 89, 90, 92, 93, 95, 97, 99, 102, 103, 105, 109, 110, 111, 112, 113, 114, 115, 202, 208, 245, 246, 247, 255, 257, 260, 261, 262, 263, 264, 265, 266, 269, 270, 274, 284, 292, 293
positive feedback, 121
potassium, 51, 62, 99, 100, 101, 217, 310, 315
potassium persulfate, 315
potato, 260
poverty, 88
power generation, x, 185
power plants, x, xiii, 53, 54, 78, 185, 186, 196, 295, 298, 303
precipitation, 7, 8, 9, 10, 11, 12, 13, 15, 32, 33, 34, 41, 62, 68, 70, 80, 102, 106, 158, 159, 166, 190, 231, 232, 248, 276, 298
predators, 199
preparation, 17, 19, 43, 119, 299
principles, 10
probability, x, 10, 20, 109, 118, 146, 149, 150, 151, 155, 167, 168, 174, 180, 197, 301
probability theory, 20
probe, 145
producers, 204, 297
project, 90, 99, 109, 110, 167, 204, 208, 261, 262, 303, 328
proliferation, 18, 296
protactinium, 189

protection, xiii, 79, 178, 203, 246, 251, 263, 264, 295, 296, 301, 302
proteins, 71
protons, 118, 149
Pseudomonas aeruginosa, 318, 319
public concern, 303
public health, xi, 57, 200, 204, 245, 262
public opinion, 246
public service, xii, 98, 269
pure water, 4, 5, 125
purification, 17, 18, 19, 298
PVC, 107
pyrite, 95, 96, 105, 215, 222, 223, 229, 233, 239, 241, 242, 247, 252

Q

quartz, 23, 43, 89, 92, 93, 94, 95, 96, 97, 102, 162, 212, 215, 216, 217, 218, 219, 222, 223, 224, 226, 229, 231, 234, 235, 239, 241, 242, 247, 277, 286
Quartz, 215
Queensland, 229, 236
quotas, 204

R

Radiation, 37, 38, 40, 44, 45, 50, 79, 80, 82, 83, 85, 109, 114, 115, 123, 177, 193, 203, 205, 206, 207, 208, 250, 303, 304
radiation damage, vii, 1, 2
radio, 38, 53, 72, 78, 97, 131, 135, 145, 148, 178, 296
radioactive contamination, viii, 49, 75, 197, 199, 204
radioactive isotopes, 50, 100, 166, 181
radioactive waste, viii, 49, 51, 57, 64, 203, 264, 298, 301, 302
radium, ix, 39, 40, 45, 46, 51, 55, 87, 89, 100, 105, 107, 109, 111, 188, 189, 190, 203, 212, 246, 247, 255, 260, 261, 264, 298, 299
radius, 54, 91, 120, 146, 149, 201
radon, 46, 47, 51, 99, 100, 109, 203, 247, 260, 262, 288, 298, 299, 300, 301
rainfall, 21, 23, 174, 180, 194, 249
raw materials, xiii, 74, 204, 295
reactant, 200
reaction time, 77
reactions, vii, x, 1, 51, 66, 70, 118, 158, 186, 189, 197
reactivity, 119
real time, 136, 138
receptors, 88
recognition, 13, 14, 79
rccombination, 123
recommendations, 177, 206
reconstruction, 174, 207
recovery, ix, xii, 56, 87, 88, 98, 105, 106, 107, 127, 179, 200, 262, 263, 269, 306, 315, 319
recovery plan, xii, 269
recreational, 262
recrystallization, 190, 232
recycling, 41, 76, 112, 296
red blood cell count, 195
red blood cells, 73
redistribution, ix, 6, 55, 73, 155, 156, 186
reflectivity, 121
refractive index, 141
registries, 200
Registry, 79, 200, 208
regression, 99, 273
regression equation, 99
regression method, 273
regulations, 53, 181, 202, 252, 254, 303
rehabilitation, xii, 80, 245, 250, 263
relaxation, 126, 216
reliability, 127, 173, 301
relief, 24, 121, 200, 201
remediation, viii, ix, xiii, 50, 74, 75, 76, 77, 78, 81, 83, 84, 88, 89, 99, 108, 115, 202, 205, 255, 264, 265, 270, 295, 296, 302, 303
repetitions, 315
requirements, 177, 178, 297, 302
researchers, 71, 166, 167, 186, 201
reserves, viii, xi, 2, 20, 21, 80, 88, 89, 90, 98, 111, 186, 211, 229, 233, 236, 242, 246
residuals, 274
residues, 202, 203, 206, 246, 259, 299
resins, 18, 19, 85, 106, 107, 247
resistance, 197
resolution, 18, 20, 68, 136, 279
resources, xiii, 35, 40, 55, 88, 98, 187, 188, 203, 225, 236, 237, 246, 249, 295, 296, 297, 306
respiratory problems, 71
response, 13, 251, 265, 303
restoration, 127, 144, 181
restrictions, 203
retardation, 3, 17, 42
Richland, 80, 82
rights, 90
risk, vii, x, xi, 38, 57, 58, 60, 63, 67, 68, 71, 72, 77, 78, 88, 109, 181, 185, 194, 197, 200, 202, 203, 204, 205, 245, 249, 258, 261, 263, 266, 271, 302
risk assessment, vii, x, 58, 60, 63, 88, 181, 185, 258, 261
risks, viii, xi, xii, 2, 70, 71, 88, 109, 186, 202, 245, 259, 262, 263, 271
rods, 195

Romania, 89
room temperature, 193
root, 62, 258, 266
root system, 258
roots, 76, 200, 202, 258, 310
Royal Society, 65, 85
RSFSR, 183
rules, 202, 303
runoff, 158, 199, 298
Russia, v, ix, xiii, 54, 155, 156, 176, 177, 179, 180, 182, 198, 204, 208, 236, 237, 321, 322, 327, 328
ruthenium, 198
rutile, 93, 225

S

safety, x, xi, 64, 88, 155, 156, 176, 177, 178, 180, 181, 186, 194, 197, 203, 245, 261, 264, 296, 303
saline water, 166, 174, 180, 317
salinity, viii, x, 2, 4, 7, 24, 28, 155, 172, 175, 202, 216, 223, 241, 256, 257
salt concentration, 130, 131, 135
salts, ix, x, 67, 117, 121, 122, 123, 130, 143, 144, 185, 190, 202, 246, 258, 265
sanctions, viii, 49, 79
sapphire, 127, 131
saturation, 19, 91, 248
scattering, 147
schema, 9
science, viii, 49
sea level, 156, 179
seasonal changes, 199
Second World, ix, 87, 90
security, 256, 262, 296
sediment, 53, 125, 141, 162, 172, 173, 188, 249
sedimentation, 125, 158, 249
sediments, ix, xii, 21, 23, 37, 38, 42, 44, 54, 108, 155, 156, 159, 160, 162, 163, 164, 166, 171, 172, 174, 179, 182, 218, 236, 245, 246, 250, 256, 275, 279, 299
selectivity, 123, 130, 147, 150
selenium, 2
semiconductor, 124
sensitivity, 72, 98, 138, 303
Serbia, v, vii, viii, 49, 50, 53, 54, 58, 63, 65, 69, 70, 73, 74, 79, 80, 81, 82, 83, 84, 85, 89
sex, 71
shape, 92, 118, 119, 217, 301
shear, 92, 94, 96, 97, 212, 213, 218, 224, 225, 226, 228, 229, 231, 232, 233, 241, 242
shelter, 252
shoots, 76, 251
showing, xii, 89, 270, 278, 279, 291
signs, xi, 186, 203, 279
silica, 91, 95, 101
silver, 52, 212
simple linear regression, 274
simulation, 38, 88
SiO_2, 6, 162, 224
skeleton, 50
skin, 73, 198, 302
Slovakia, 89
sludge, 52, 69, 249, 262
SO_4^{2-}, 108, 163, 172
social welfare, 69
society, 203, 296
sodium, 196, 212, 233, 317, 327
software, 273, 274, 297
soil erosion, 57
soil type, 59, 60, 62, 63, 67, 84, 108
solid phase, vii, 1, 8, 11, 12, 62, 174, 248, 249, 266
solid state, 31, 66
solid waste, 52, 247, 250, 298
solidification, 119, 315
solubility, x, 4, 6, 7, 8, 10, 11, 13, 35, 45, 51, 62, 73, 83, 91, 108, 155, 159, 176, 180, 192, 194, 224, 263
solvents, 106
sorption, 10, 11, 40, 58, 63, 67, 80, 84, 108, 158, 166, 177
sorption process, 67
South Africa, 31, 187, 204, 296
South America, 38, 233
Soviet Union, 198, 199, 204, 303
Spain, 56, 89, 202, 208
speciation, 3, 37, 45, 58, 69, 80, 84
species, x, 4, 5, 7, 8, 11, 54, 58, 60, 62, 68, 80, 83, 101, 108, 119, 135, 186, 189, 197, 246, 248, 251, 258, 260, 306, 307, 308, 316, 318
specific adsorption, 67
spectroscopy, 113
Spring, 47, 170
stability, 62, 125, 197, 204, 262, 303
stabilization, 75, 77, 78, 120, 208, 251, 303
stable complexes, 6
standard deviation, 20, 61
state, vii, 1, 3, 4, 5, 10, 13, 15, 17, 30, 31, 35, 39, 47, 52, 53, 54, 80, 108, 120, 123, 126, 159, 196, 215, 232, 299
states, viii, ix, 3, 37, 50, 51, 88, 150, 155, 156, 179, 192, 193, 300, 302
steel, xi, 19, 20, 37, 64, 66, 186, 201, 322
sterile, 107, 275
stimulus, 251
storage, xi, 18, 43, 110, 186, 286
stratification, 188

stress, 72, 94
stretching, 166
structure, 21, 55, 91, 101, 147, 200, 203, 215, 236, 277, 279, 283, 286, 299, 323
style, 223
substitutes, 101
substitution, 54
substrate, 121
succession, 21, 159
sulfate, 101, 105, 108, 159, 163, 164
sulfuric acid, 106, 108, 196
sulphur, 4
Sun, 10, 46
surface area, 2, 9, 35, 166
surface energy, 120
surface layer, 11, 68, 199
surface tension, 119, 120
surface treatment, 299
surfactant, 82
surveillance, 203, 263
susceptibility, 315
suspensions, 119
sustainability, 296
sustainable development, 74, 84
Sustainable Development, 84
Sweden, 17, 37, 89, 181, 207, 208
Switzerland, 85
symmetry, 150
symptoms, 201, 205, 251
syndrome, 72, 198
synergistic effect, 77, 78
synthesis, 119, 120
Syria, 41

T

Taiwan, 46, 56
Tajikistan, 204
tanks, 65, 106
target, 64, 66, 68, 70, 73, 118, 119, 120, 121, 125, 126, 127, 128, 130, 131, 133, 135, 136, 141, 142, 147, 148, 301
Task Force, 70, 80, 85
techniques, ix, 3, 17, 74, 75, 76, 87, 89, 98, 108, 113, 115, 249, 293, 298, 303
technological revolution, 78
technologically enhanced naturally occurring radioactive materials (TENORM), viii, 49
technologies, 75, 76, 205
technology, viii, 18, 49, 74, 78, 187, 197, 296, 298
TEM, 120
temperature, xi, 4, 6, 13, 23, 24, 28, 32, 33, 34, 51, 58, 62, 66, 81, 91, 92, 96, 108, 120, 125, 136, 197, 200, 211, 215, 216, 223, 229, 232, 236, 275, 276, 298
tension, 94, 95
territory, 69, 94, 158, 159, 162, 172, 199, 204
test data, 79
tetravalent U, vii, 1
texture, 51, 229, 256
thorium, xiii, 37, 39, 41, 42, 45, 46, 47, 51, 54, 55, 82, 84, 86, 99, 100, 101, 109, 181, 189, 203, 260, 296, 297, 298, 299, 300, 305, 306, 314, 315, 319
thyroid, 200
thyroid cancer, 200
time periods, 298
tin, 52, 327
tissue, 71, 72, 76, 199, 301
titanium, 52, 64, 108
tones, 203
topology, 121
tourmaline, 93
toxic effect, viii, 50, 61
toxic metals, 247
toxicity, x, 64, 70, 71, 73, 74, 82, 109, 185, 195, 205, 249, 251, 257, 260, 300
toxicology, 43, 206
trace elements, 2, 18, 242, 250
trade, 88
trajectory, 64
transcription, 212
transformation, 77, 78, 109, 232
transformations, 118, 145
transgression, 163, 172, 180
translocation, 80
transpiration, 108
transport, vii, 1, 6, 38, 39, 40, 43, 44, 45, 47, 73, 76, 80, 106, 108, 113, 172
transport processes, vii, 1
transportation, 75
treaties, ix, 87, 90
treatment, ix, x, 11, 18, 32, 51, 52, 74, 75, 88, 89, 105, 106, 110, 112, 113, 185, 202, 205, 247, 248, 262, 275, 299
tributyl phosphate, 47
tuff, 82, 241
tumors, 74
tungsten, 52
tunneling, 146, 147, 149
Turkmenistan, 204

U

U.S. Geological Survey, 181, 293
UK, 40, 46, 65, 82, 187
Ukraine, xi, 89, 187, 198, 199, 200, 208, 211, 225

UN, 85
UNESCO, 40
uniform, 11, 13, 19, 172, 179, 196, 291
United, xi, xiii, 50, 56, 84, 85, 114, 115, 186, 193, 201, 204, 205, 206, 207, 208, 266, 304, 305, 306, 317, 318
United Nations, 50, 85, 115, 193, 205, 206, 207, 304
United States, xi, xiii, 56, 84, 114, 186, 201, 204, 208, 266, 305, 306, 317, 318
uranium exploration, ix, 88, 90, 97, 98, 111, 113, 115, 293, 296, 303
uranium–enriched granitic rocks, xi, 211
urban, 46, 284
urbanization, viii, 49
urine, 74
USA, 31, 40, 42, 46, 47, 55, 65, 85, 90, 109, 158, 187, 188, 205, 206, 236, 241, 257, 296, 328
USGS, 43, 293
USSR, 166, 181
UV, 119, 121, 123, 130, 131, 135
Uzbekistan, 187, 204

V

vacuum, 118, 122, 149
valence, 5
valve, 201
vanadium, 41, 52
vapor, 119, 120, 121, 123, 124, 136, 137, 138, 141, 142, 143
variable factor, 10
variations, 3, 10, 12, 21, 23, 28, 36, 44, 62, 146, 149, 159, 252
varieties, 119
vector, 6
vegetables, 54, 70, 73, 199, 256, 258, 260, 300
vegetation, 54, 69, 200, 251, 266
vehicles, 97, 102
vein, ix, xi, 87, 88, 89, 90, 91, 92, 95, 96, 97, 98, 102, 110, 111, 112, 114, 211, 212, 213, 215, 223, 241, 242
velocity, 8, 15, 17, 169, 274, 277, 278, 279, 283, 285, 291, 315, 316
visualization, 280, 282
vulnerability, 10, 292

W

war, 65
Washington, 46, 79, 80, 114
waste, x, xii, xiii, 43, 44, 47, 52, 53, 55, 57, 88, 107, 109, 110, 113, 151, 185, 202, 203, 204, 247, 248, 250, 251, 252, 255, 259, 265, 269, 270, 271, 296, 298, 299, 301, 302, 305, 306, 314, 316, 327
waste disposal, xii, 270, 296
waste disposal sites, xii, 270
waste management, x, 44, 109, 185, 202, 296, 302
waste water, xiii, 113, 305
wastewater, 255, 306, 316, 317, 318
water chemistry, 263
water ecosystems, 203
water quality, 206, 251, 252, 254, 258, 260, 262
water supplies, 260, 263
watershed, 45, 199
waterways, 205
wavelengths, 141
wcapons, ix, x, 65, 87, 90, 185, 196
weapons of mass destruction, ix, 87
weather patterns, 200
web, 266, 297
wells, 23, 33, 164, 167, 170, 171, 174, 175, 176, 180, 248, 252, 254, 255, 256, 286
West Indies, 38
Western Australia, 37
Western Europe, 212, 213
white blood cell count, 195
WHO, 50, 74, 86, 195, 206
wires, ix, 117, 130, 150, 153
wood, 69, 94
workers, xiii, 53, 71, 75, 82, 109, 189, 195, 198, 199, 200, 201, 260, 264, 295, 296, 301, 302, 303
World Health Organization (WHO), 86, 177, 195, 206, 260, 304
worldwide, vii, viii, 2, 29, 36, 49, 55, 63, 64, 80, 187, 193, 296, 297

X

X-ray diffraction, 113

Y

yield, 18, 19, 20, 118
Yugoslavia, viii, 50, 65

Z

zirconium, 200
Zone 1, 15
Zone 2, 15
Zone 3, 15